D. J. Mitchell

Environmental Change in Drylands

British Geomorphological Research Group Symposia Series

Geomorphology in Environmental Planning
Edited by **J. M. Hooke**

Floods
Hydrological, Sedimentological and
Geomorphological Implications
Edited by **Keith Beven** and **Paul Carling**

Soil Erosion on Agricultural Land
Edited by **K. Boardman, J. A. Dearing** and **I. D. L. Foster**

Vegetation and Erosion
Processes and Environments
Edited by **J. B. Thornes**

Lowland Floodplain Rivers
Geomorphological Perspectives
Edited by **P. A. Carling** and **G. E. Petts**

Geomorphology and Sedimentology of Lakes and Reservoirs
Edited by **J. McManus** and **R. W. Duck**

Landscape Sensitivity
Edited by **D. S. G. Thomas** and **R. J. Allison**

Process Models and Theoretical Geomorphology
Edited by **M. J. Kirkby**

Environmental Change in Drylands
Biogeographical and Geomorphological Perspectives
Edited by **A. C. Millington** and **K. Pye**

Environmental Change in Drylands

Biogeographical and Geomorphological Perspectives

Edited by
Andrew C. Millington
*Department of Geography
University of Leicester, UK*

and

Ken Pye
*Post Graduate Institute for Sedimentology
University of Reading, UK*

JOHN WILEY & SONS
Chichester · New York · Brisbane · Toronto · Singapore

Copyright © 1994 by John Wiley & Sons Ltd.
Baffins Lane, Chichester,
West Sussex PO19 1UD, England
Telephone National Chichester (0243) 779777
International +44 243 779777

All rights reserved.

No part of this book may be reproduced by any means,
or transmitted, or translated into a machine language
without the written permission of the publisher.

Other Wiley Editorial Offices

John Wiley & Sons, Inc., 605 Third Avenue,
New York, NY 10158-0012, USA

Jacaranda Wiley Ltd, 33 Park Road, Milton,
Queensland 4064, Australia

John Wiley & Sons (Canada) Ltd, 22 Worcester Road,
Rexdale, Ontario M9W 1L1, Canada

John Wiley & Sons (SEA) Pte Ltd, 37 Jalan Pemimpin #05-04,
Block B, Union Industrial Building, Singapore 2057

Library of Congress Cataloging-in-Publication Data

Environmental change in drylands: biogeographical and geomorphological perspectives / edited by A. C. Millington and K. Pye.
    p. cm.—(British Geomorphological Research Group symposia series)
  Includes bibliographical references and index.
  ISBN 0-471-94267-7
  1. Arid regions—Congresses. I. Millington, A. C. II. Pye, Kenneth. III. Series.
GB611.E68 1994
551.4'15—dc20   93-27489
                                                                                                              CIP

British Cataloguing in Publication Data

ISBN 0-471-94267-7

Typeset in 10/12pt Times from author's disks by Photo-graphics, Honiton, Devon
Printed and bound in Great Britain by Bookcraft, Bath, Avon

Contents

List of Contributors		ix
Series Preface		xiii
Preface		xv
Chapter 1	Deserts in a Warmer World **A. S. Goudie**	1
Chapter 2	Timescales, Environmental Change and Dryland Valley Development **D. J. Nash, D. S. G. Thomas and P. A. Shaw**	25
Chapter 3	Mineral Magnetic Analysis of Iron Oxides in Arid Zone Soils from the Tunisian Southern Atlas **K. White and J. Walden**	43
Chapter 4	Late Pleistocene and Holocene Changes in Hillslope Sediment Supply to Alluvial Fan Systems: Zzyzx, California **A. M. Harvey and S. G. Wells**	66
Chapter 5	Natural Stabilization Mechanisms on Badland Slopes: Tabernas, Almería, Spain **R. W. Alexander, A. M. Harvey, A. Calvo, P. A. James and A. Cerda**	85
Chapter 6	Responses of Rivers and Lakes to Holocene Environmental Change in the Alcañiz Region, Teruel, North-East Spain **M. G. Macklin, D. G. Passmore, A. C. Stevenson, B. A. Davis and J. A. Benavente**	113
Chapter 7	The Palaeolimnological Record of Environmental Change: Examples from the Arid Frontier of Mesoamerica **S. E. Metcalfe, F. A. Street-Perrott, S. L. O'Hara, P. E. Hales and R. A. Perrott**	131

Contents

Chapter 8	Lacustrine Sedimentation in a High-Altitude, Semi-Arid Environment: The Palaeolimnological Record of Lake Isli, High Atlas, Morocco **H. F. Lamb, C. A. Duigan, J. H. R. Gee, K. Kelts, G. Lister, R. W. Maxted, A. Merzouk, F. Niessen, M. Tahri, R. J. Whittington and E. Zeroual**	147
Chapter 9	Abrupt Holocene Hydro-Climatic Events: Palaeolimnological Evidence from North-West Africa **N. Roberts, H. F. Lamb, N. El Hamouti and P. Barker**	163
Chapter 10	Aeolian Activity, Desertification and the 'Green Dam' in the Ziban Range, Algeria **J.-L. Ballais**	177
Chapter 11	The Ambiguous Impact of Climate Change at a Desert Fringe: Northern Negev, Israel **A. Yair**	199
Chapter 12	The Environmental Consequences and Context of Ancient Floodwater Farming in the Tripolitanian Pre-Desert **D. D. Gilbertson, C. O. Hunt, N. R. J. Fieller and G. W. W. Barker**	229
Chapter 13	Evolutionary Trends in the Wheat Group in Relation to Environment, Quaternary Climatic Change and Human Impacts **M. A. Blumler**	253
Chapter 14	Post-European Changes in Creeks of Semi-Arid Rangelands, 'Polpah Station', New South Wales **J. Pickard**	271
Chapter 15	Anthropogenic Factors in the Degradation of Semi-Arid Regions: A Prehistoric Case Study in Southern France **J. Wainwright**	285
Chapter 16	Erosion–Vegetation Competition in a Stochastic Environment Undergoing Climatic Change **J. B. Thornes and J. Brandt**	305
Chapter 17	Environmental Change, Disturbance and Regeneration in Semi-Arid Floodplain Forests **F. M. R. Hughes**	321
Chapter 18	Monitoring the Flooding Ratio of Tunisian Playas Using Advanced Very High Resolution Radiometer (AVHRR) Imagery **N. A. Drake and R. G. Bryant**	347

Chapter 19	Waterlogging and Soil Salinity in the Newly Reclaimed Areas of the Western Nile Delta of Egypt **R. Goossens, T. K. Ghabour, T. Ongena and A. Gad**	365
Chapter 20	The Implications of the Altered Water Regime for the Ecology and Sustainable Development of Wadi Allaqi, Egypt **G. Dickinson, K. Murphy and I. Springuel**	379
Chapter 21	Soil-Forming Processes on Reclaimed Desertified Land in North-Central China **D. J. Mitchell and M. A. Fullen**	393
Chapter 22	Facing Environmental Degradation in the Aravalli Hills, India **C. A. Scott**	413
Chapter 23	Biogeographical and Geomorphological Perspectives on Environmental Change in Drylands **A. C. Millington and K. Pye**	427
Index		443

List of Contributors

Dr Roy W. Alexander Department of Geography, Chester College, Cheyney Road, Chester CH1 4BJ, UK

Professor Jean-Louis Ballais Institut de Geographie, Université d'Aix-Marseille II, 29, av R. Schuman, F13621, Aix-en-Provence, France

Professor G. W. W. Barker School of Archaeological Studies, University of Leicester, Leicester LE1 7RH, UK

Dr P. Barker Department of Geography, Lancaster University, Bailrigg, Lancaster, LA1 4YB, UK

Dr J. A. Benavente Taller de Arquelogía y Prehistoria de Alcañiz, Alcañiz, Teruel, España

Dr Mark A. Blumler Department of Geography, State University of New York–Binghamton, Binghamton, NY 13902, USA

Dr Jane Brandt Department of Geography, King's College London, Strand, London WC2R 2LS, UK

Dr Robert G. Bryant Department of Environmental Science, University of Stirling, FK9 4LA, UK

Dr Antonio Calvo Departamento de Geografia, Universidad de Valencia, Apartado 22060, 46080 Valencia, España

Dr A. Cerda Departamento de Geografia, Universidad de Valencia, Apartado 22060, 46080 Valencia, España

Dr B. A. Davis Department of Geography, University of Newcastle upon Tyne, Newcastle-upon-Tyne NE1 7RU, UK

Dr Gordon Dickinson Department of Geography and Topographic Science, University of Glasgow, Glasgow G12 8QQ, UK

Dr Nicholas A. Drake Department of Geography, King's College London, Strand, London WC2R 2LS, UK

Dr C. A. Duigan Countryside Council for Wales, Plas Penrhos, Fford Penrhos, Bangor, Gwynedd LL57 2LQ, UK

Dr N. El Hamouti Department of Geology, Faculty of Sciences, Université Mohammed I, Oujda, Morocco

Dr N. R. J. Fieller Department of Probability and Statistics, University of Sheffield, Sheffield S10 2TN, UK

List of Contributors

Dr Michael Fullen School of Applied Sciences, University of Wolverhampton, Dudley Campus, Castle View, Dudley, West Midlands DY1 3HR, UK

Dr Abdalla Gad Remote Sensing Center, Academy of Scientific Research and Technology, 101 Kasr El Eini Street, Cairo, Egypt

Dr J. H. R. Gee Department of Biological Sciences, University of Wales, Aberystwyth SY23 3DB, UK

Dr Tharwat K. Ghabour Remote Sensing Center, Academy of Scientific Research and Technology, 101 Kasr El Eini Street, Cairo, Egypt

Professor David D. Gilbertson Department of Archaeology and Prehistory, University of Sheffield, Sheffield S10 2TN, UK

Dr Rudi Goossens Laboratory for Regional Geography and Landscape Science, University of Gent, Krijgslaan 281 (S8-A1), B-9000 Gent, Belgium

Professor Andrew S. Goudie School of Geography, University of Oxford, Mansfield Road, Oxford OX1 3TB, UK

Dr P. E. Hales School of Geography, University of Oxford, Mansfield Road, Oxford OX1 3TB, UK

Dr Adrian M. Harvey Department of Geography, University of Liverpool, PO Box 147, Liverpool L69 3BX, UK

Dr Francine M. R. Hughes Department of Geography, University of Cambridge, Downing Place, Cambridge CB2 3EN, UK

Dr C. O. Hunt Department of Geographical and Environmental Sciences, University of Huddersfield, UK

Dr P. A. James Department of Geography, University of Liverpool, PO Box 147, Liverpool L69 4BX, UK

Professor K. Kelts Limnological Research Center, University of Minnesota, Minneapolis, MN 55455, USA

Dr Henry F. Lamb Institute of Earth Studies, University of Wales, Aberystwyth SY23 3DB, UK

Dr G. Lister Geologisches Institut, ETH Zentrum, Swiss Federal Institute of Technology, CH-8092, Zurich, Switzerland

Dr Mark G. Macklin Department of Geography, University of Leeds, Leeds LS2 9JT, UK

Dr R. W. Maxted National Rivers Authority, Bromholme Lane, Brampton, Huntington PE18 8NE, UK

Dr A. Merzouk Département des Sciences du Sol, Institute Agronomique et Vétérinaire Hassan II, Rabat, Morocco

Dr Sarah E. Metcalfe School of Geography and Earth Resources, University of Hull, Hull HU6 7RX, UK

List of Contributors

Professor Andrew C. Millington Geography Department, University of Leicester, University Road, Leicester LE1 7RH, UK

Dr David Mitchell School of Applied Sciences, University of Wolverhampton, Dudley Campus, Castle View, Dudley, West Midlands DY1 3HR, UK

Dr Kevin Murphy Department of Botany, University of Glasgow, Glasgow G12 8QQ, UK

Dr David J. Nash Department of Building (Geography Division), University of Brighton, Mithras House, Lewes Road, Brighton, BN2 4AT, UK

Dr F. Niessen Geologisches Institute, ETH Zentrum, Swiss Federal Institute of Technology, CH-8092 Zurich, Switzerland

Dr Sarah L. O'Hara Department of Geography, University of Sheffield, Sheffield S10 2TN, UK

Dr Trees Ongena Laboratory for Regional Geography and Landscape Science, University of Gent, Krijgslaan 281 (S8-A1), B-9000 Gent, Belgium

Dr D. G. Passmore Department of Geography, University of Newcastle-upon-Tyne, Newcastle-upon-Tyne NE1 7RU, UK

Dr R. A. Perrott Environmental Change Unit, University of Oxford, Mansfield Road, Oxford OX1 3TB, UK

Dr John Pickard Graduate School of the Environment, Macquarie University, NSW 2109, Australia

Dr Ken Pye Postgraduate Research Institute for Sedimentology, University of Reading, Whiteknights, Reading RG6 2AB, UK

Dr Neil Roberts Department of Geography, Loughborough University of Technology, Loughborough, Leicestershire LE1 7RH, UK

Christopher Anand Scott Catholic Relief Services, Apartado 257, Tegucigalpa, Honduras

Dr Paul A. Shaw Department of Geological and Environmental Science, Luton College of Higher Education, Park Square, Luton LU1 3JU, UK

Professor Irina Springuel Department of Botany, Faculty of Science at Aswan, Assiut University, Aswan, Egypt

Dr A. C. Stevenson Department of Geography, University of Newcastle-upon-Tyne, Newcastle-upon-Tyne NE1 7RU, UK

Dr F. Alayne Street-Perrott Environmental Change Unit, University of Oxford, Mansfield Road, Oxford OX1 3TB, UK

Dr M. Tahri Département des Sciences du Sol, Institute Agronomique et Vétérinaire Hassan II, Rabat, Marocco

Dr David S. G. Thomas Department of Geography, University of Sheffield, Sheffield S10 2TN, UK

Professor John B. Thornes Department of Geography, King's College London, Strand, London WC2R 2LS, UK

Dr John Wainwright Department of Geography, King's College London, Strand, London WC2R 2LS, UK

Dr John Walden School of Geography, University of Oxford, Mansfield Road, Oxford OX1 3TB, UK

Dr Stephen G. Wells Department of Earth Sciences, University of California, Riverside, CA 92521, USA

Dr Kevin White Department of Geography, University of Reading, Whiteknights, Reading RG6 2AB, UK

Dr R. J. Whittington Institute of Earth Studies, University of Wales, Aberystwyth SY23 3DB, UK

Professor Aaron Yair Institute of Earth Sciences, The Hebrew University of Jerusalem, Givat Ram, Jerusalem, Israel 91904

Dr E. Zeroual Institut de Géologie, Université de Neuchatel, 11, rue Emile-Argand, CG-2007, Switzerland

Series Preface

The British Geomorphological Research Group (BGRG) is a national multidisciplinary Society whose object is 'the advancement of research and education in geomorphology'. Today, the BGRG enjoys an international reputation and has a strong membership from both Britain and overseas. Indeed, the Group has been actively involved in stimulating the development of geomorphology and geomorphological societies in several countries. The BGRG was constituted in 1961 but its beginnings lie in a meeting held in Sheffield under the chairmanship of Professor D. L. Linton in 1958. Throughout its development the Group has sustained important links with both the Institute of British Geographers and the Geographical Society of London.

Over the past three decades the BGRG has been highly successful and productive. This is reflected not least by BGRG publications. Following its launch in 1976, the Group's journal, *Earth Surface Processes* (since 1981 *Earth Surface Processes and Landforms*) has become acclaimed internationally as a leader in its field, and to a large extent the journal has been responsible for advancing the reputation of the BGRG. In addition to an impressive list of other publications on technical and educational issues, including 30 *Technical Bulletins* and the influential *Geomorphological Techniques*, edited by A. Goudie, BGRG symposia have led to the production of a number of important works. These have included *Nearshore Sediment Dynamics and Sedimentation*, edited by J. R. Hails and A. P. Carr; *Geomorphology and Climate*, edited by E. Derbyshire; *Geomorphology, Present Problems and Future Prospects*, edited by C. Embleton, D. Brunsden and D. K. C. Jones; *Megageomorphology*, edited by R. Gardner and H. Scoging; *River Channel Changes*, edited by K. J. Gregory; and *Timescales in Geomorphology*, edited by R. Cullingford, D. Davidson and J. Lewin. This sequence of books culminated in 1987 with a publication, in two volumes, of the *Proceedings of the First International Geomorphology Conference*, edited by Vince Gardiner. This international meeting, arguably the most important in the history of geomorphology, provided the foundation for the development of geomorphology into the next century.

This current BGRG Symposia Series has been founded and is now being fostered to help maintain the research momentum generated during the past three decades, as well as to further the widening of knowledge in component fields of geomorphological endeavour. The series consists of authoritative volumes based on the themes of BGRG meetings, incorporating, where appropriate, invited contributions to complement chapters selected from presentations at these meetings under the guidance and editorship of one or more suitable specialists. Although maintaining a strong emphasis on pure geomorphological research, BGRG meetings are diversifying, in a very positive way, to consider links between geomorphology *per se* and other disciplines such as ecology, agriculture, engineering and planning.

The first volume in the series was published in 1988. *Geomorphology in Environmental Planning*, edited by Janet Hooke, reflected the trend towards applied studies. The second volume, edited by Keith Beven and Paul Carling, *Floods—Hydrological, Sedimentological and Geomorphological Implications*, focused on a traditional research theme. *Soil Erosion on Agricultural Land* reflected the international importance of the topic for researchers during the 1980s. This volume, edited by John Boardman, Ian Foster and John Dearing, formed the third in the series. The role of vegetation in geomorphology is a traditional research theme, recently revitalized with the move towards interdisciplinary studies. The fourth in the series, *Vegetation and Erosion—Processes and Environments*, edited by John Thornes, reflected this development in geomorphological endeavour, and raised several research issues for the next decade. The fifth volume, *Lowland Floodplain Rivers—Geomorphological Perspectives*, edited by Paul Carling and Geoff Petts, reflects recent research into river channel adjustments, especially those consequent to engineering works and land use change. The sixth volume, *Landscape Sensitivity*, edited by David Thomas and Robert Allison, addresses a vital geomorphological topic. This concerns the way in which landscape and landforms respond to external changes—important concepts for understanding landform development and crucial to an appreciation of human-induced response in the landscape. The seventh in the series, *Geomorphology and Sedimentology of Lakes and Reservoirs*, edited by John McManus and Robert Duck, provided a stimulating mixture of pure and applied research which appealed to a wide audience, and the eighth in the series, *Process Models and Theoretical Geomorphology*, edited by Mike Kirkby, reported some important new numerical and field results from a variety of environments.

The present volume (the ninth in the series), *Environmental Change in Drylands*, edited by Andrew Millington and Ken Pye, continues the current trend of analysing recent palaeo-environmental change in order to understand and to control present-day systems. Examples of dryland evolution are presented by an international group of authors and include the Kalahari, Tunisia, California, north-west Africa, Australia and Europe. The whole is well synthesised in an introductory chapter by Andrew Goudie and in a concluding chapter by the editors. Desertification is one of the most public faces of environmental change and this timely research book, drawing on the recent past, will become an important primary source that will, in turn, inform the policy debate in the future.

Jack Hardisty
BGRG Publications

Preface

At a time when environmental change is at the forefront of much biogeographical and geomorphological research it is timely to focus on drylands for two reasons. Firstly, under many scenarios of global warming it is predicted that large parts of the Earth's drylands, and areas marginal to them, will become drier. Secondly, human-induced environmental changes are inevitable in the Earth's drylands because they house over 10% of the world's population, of which at least 54% live in rural communities and are directly affected by, as well as effect, the natural environment. Despite the fact that we recognise these two important triggers of change, and the direct and indirect impacts of change on the ever-expanding population, many of the effects of environmental change in drylands are poorly understood: ambiguities and misinterpretations abound.

This book comprises a selection of peer-reviewed papers based on presentations made at a symposium on the *Effects of Environmental Change in Drylands*, jointly organised by the Biogeography Research Group and the British Geomorphological Research Group, which was held at the Institute of British Geographers Annual Conference in Swansea during January 1992. The papers presented at the symposium were wide ranging, covering geomorphology and biogeography, as well as research that has sought to integrate the two disciplines. The latter research area is particularly important in drylands because of the links between vegetation and erosion, although the actual nature, extent and importance of vegetation-geomorphology relationships are far from clear.

Environmental change, and its effects, can be interpreted in many ways. In this book, the main division of papers has been made in terms of the timescales of the studies, rather than between biogeography and geomorphology, because we wish to stress the importance of linking these two areas. The division falls rather uneasily between long-term and medium-term studies, in which the majority of the evidence presented relates to environments before the last thousand years, and short term studies, which focus on the last thousand years. The book begins with an introductory chapter—'Deserts in a Warmer World' by Andrew Goudie—before moving on to the first main section which consists of ten papers on long-term and medium-term change (Chapters 2 to 11). The second main section, short-term change, comprises eleven chapters (Chapters 12 to 22). The final chapter develops the main theme of the preceding chapters in the context of a wider field of scientific literature.

Many people helped during the organisation of the symposium and the preparation of this book. In particular, we wish to thank the committees of the Biogeography Research Group and British Geomorphological Research Group for encouraging us to organise this symposium, David Jones and Haydn Williams for acting as chairmen at the symposium. We would also like to thank the Institute of British Geographers for providing funds for some contributors to attend the symposium, and to Ron Cooke who fortuitously was President of the Institute of British Geographers in 1992! In terms of the book's preparation we would like to thank the following: the contributors for meeting our deadlines; the many reviewers scattered around the world (and one in particular who reviewed four manuscripts);

the many cartographers at different institutions but, in particular, Heather Browning and Judith Fox in Reading, who redrew many figures; Erika Meller for producing most of the photographic plates; Donna Edwards for completing the final versions of the manuscript. Finally, we would like to thank Helen Bailey at John Wiley for accepting the idea of this book, and also to Abi Hudlass, Louise Metz and Claire Morrison for the advice and logistical help they provided.

Andrew C. Millington
Ken Pye

1 Deserts in a Warmer World

A. S. GOUDIE
School of Geography, University of Oxford, UK

ABSTRACT

Arid zone landforms, rates and processes seem to be prone to a rapid response to apparently modest stimuli. Examples include arroyo incision, fan head trenching, terrace deposition, colluviation, lake basin change, dust storm activity, dune accumulation and drainage density. It is therefore likely that deserts will respond rapidly and significantly as a result of global warming and the hydrological, vegetation and sea-level changes that will ensue. The pattern of future precipitation changes is complex and uncertain: some desert areas will become moister but others drier. The positions of the ITCZ, westerly wind tracks, upwelling and tropical cyclones may all change and affect desert areas. Arid zone sabkhas and deltas may be afflicted by the effects of sea-level rise. However, changes caused by global warming will compound the effects of other anthropogenic changes which are already having a major impact on arid land geomorphology.

INSTABILITIES AND THRESHOLDS

Arid environments *sensu lato* often appear to be prone to rapid geomorphological changes in response to apparently modest stimuli. They seem to display characteristics that render them subject to instability so that when a particular threshold is reached they switch speedily from one state to another. Some examples will serve to illustrate this contention:

1. Desert valley bottoms appear to have been subject to dramatic alternations of cut-and-fill during the course of the Holocene and a large literature has developed on the arroyos of the American south-west (see, for instance, Cooke & Reeves, 1976; Balling & Wells, 1990). Schumm has proposed that valley bottom trenching may occur when a critical valley slope gradient is crossed for a drainage area of a particular areal extent (Schumm, 1977) (Figure 1.1), but other examples result from changes in rainfall intensity and grazing pressure.
2. The heads of alluvial fans often display fan head trenching. There has been considerable debate in the literature about the trigger for such trenching, be it tectonic, extreme flood events, climatic change or an inherent consequence of fluctuating sediment-discharge relationships during the course of a depositional cycle (e.g. Harvey, 1989).
3. Fluvial systems, as, for example, in the drier lands around the Mediterranean Basin, display a suite of terraces which demonstrates a complex record of cut-and-fill (Figure 1.2) during the late Holocene (Van Andel *et al.*, 1990). There has been prolonged debate as to whether the driving force has been climatic change (Vita-Finzi, 1967) or anthropogenic activities (Butzer, 1974).

Environmental Change in Drylands: Biogeographical and Geomorphological Perspectives.
Edited by A. C. Millington and K. Pye. © 1994 John Wiley & Sons Ltd

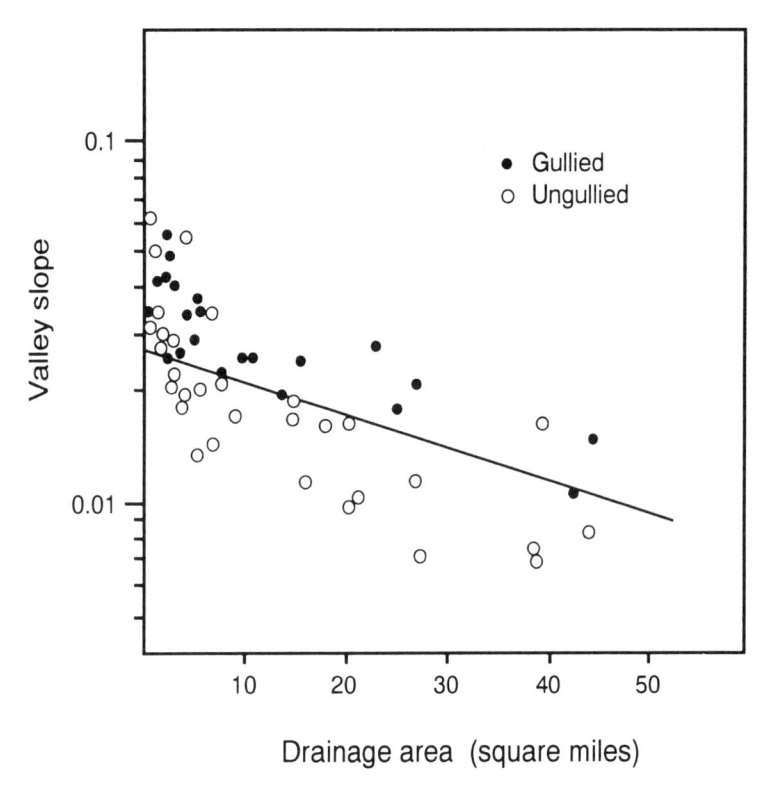

Figure 1.1 The relationship between valley slope, drainage basin area and the development of valley bottom gully systems for the Piceance Creek area of Colorado, USA (after Schumm, 1977)

4. Colluvial sections in sub-humid landscapes (including southern Africa) show complex sequences of deposition, stability and incision (Watson et al., 1984). Ongoing luminescence dating is beginning to give an indication of the complexity of chronology (Wintle, personal communication, 1991).
5. Terminal lake basins (pans, playas, etc.) can respond dramatically (in terms of both their extent and the rapidity of change that can occur) to episodic rainfall events in their catchments. The history of Lake Eyre in Australia in the present century bears witness to this fact.
6. West African dust storm activity has shown very marked shifts in the past few decades in response to runs of dry years and increasing land-use pressures (Goudie & Middleton, 1991). The data for Nouakchott (Mauritania) are especially instructive, revealing a sudden acceleration in dust storm events since the 1960s (Figure 1.3). A broadly similar picture could be presented for the High Plains of the United States during the 'dust bowl years' of the 1930s.
7. Some dune fields have also proved to be prone to repeated fluctuations in deposition and stabilisation in the Holocene, and as more ^{14}C and luminescence dates become avail-

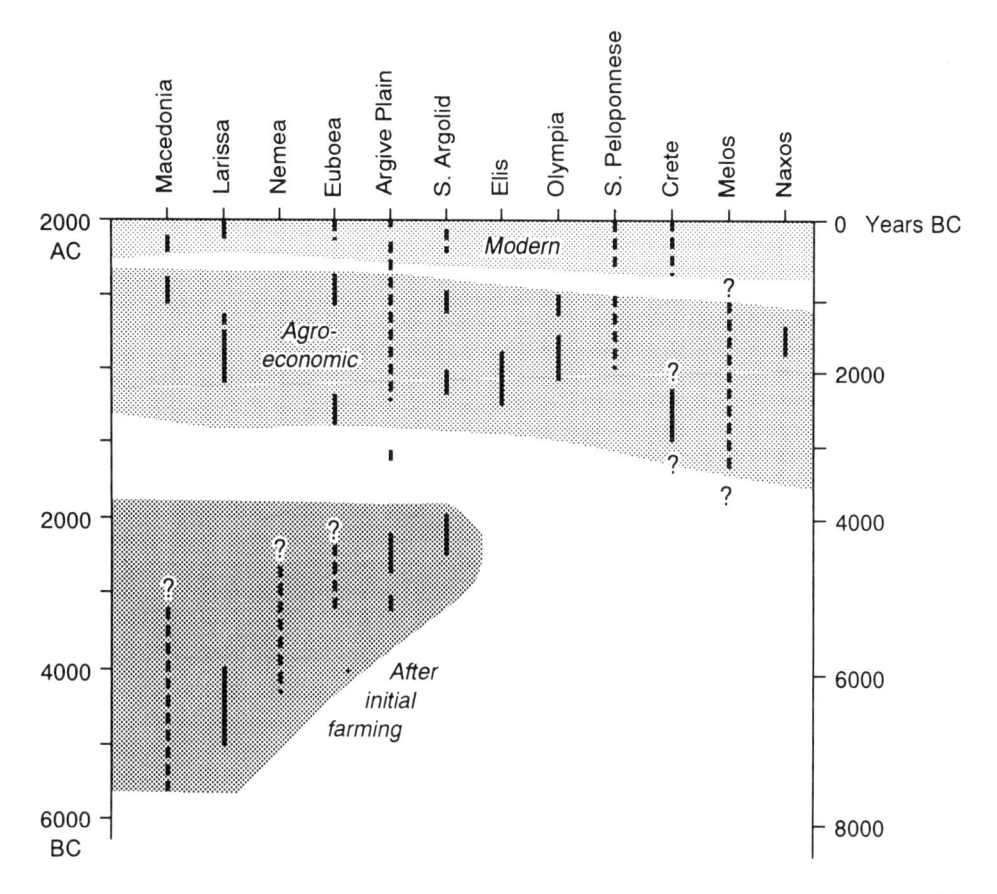

Figure 1.2 Chronology of the complex alluviation events of the Holocene in Greece and the Aegean. Broken bars are dated uncertainly or represent intermittent deposition (from various sources in Van Andel et al., 1990, Fig. 10)

able, the situation is likely to prove to be even more complex than hitherto believed. A good illustration of this is provided by Gaylord's (1990) work in the Clear Creek area of south-central Wyoming (Figure 1.4), where over the last 7500 years at least four episodes of enhanced aeolian activity and aridity are recorded.
8. Various studies have shown how rates of denudation (e.g. Langbein & Schumm, 1958) (Figure 1.5) and drainage density (Figure 1.6) can change very rapidly in semi-arid areas either side of a critical rainfall or precipitation/evapotranspiration (P/E) value related to vegetation cover that constitutes a particular threshold between equilibrium states.

This apparent instability and threshold dependence of a range of arid zone landforms, rates and processes leads one to believe that such areas may be especially susceptible to the effects of global warming, should this occur in the coming decades.

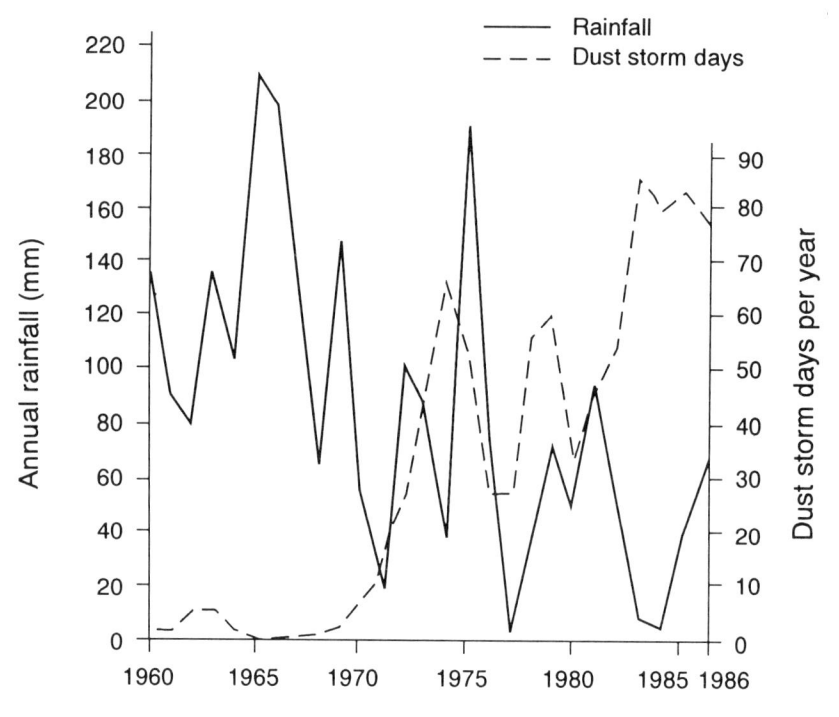

Figure 1.3 Annual frequency of dust storm days and annual rainfall from Nouakchott, Mauritania, 1960–86 (reproduced by permission of Kluwer Academic Publishers from Goudie & Middleton, 1991, Fig. 5)

THE GEOMORPHOLOGICAL IMPACT OF GLOBAL WARMING: INTRODUCTION

Although there exist great problems of prediction, some scientific consensus has arisen in the last decade, that global climates are likely to become warmer as concentrations of greenhouse gases increase in the atmosphere. A review is provided by Houghton *et al.* (1990) of the findings of the Intergovernmental Panel on Climate Change. It is not the purpose of this paper to review the arguments for and against the likelihood of global warming, for a firm statement of the dissenting viewpoint is put by Idso (1989). Nor is it possible to discuss in detail the uncertainties about the degree of change that will take place (see Goudie, 1989, for an extended analysis). Instead, in this paper the working assumption will be made that a warming of several degrees is likely as a result of the doubling of the effective concentrations of greenhouse gases over pre-industrial levels, and that such warming could become unmistakenly apparent in the middle of the next century.

Such global warming will have important implications for many geomorphological processes and phenomena as a result of the direct effects of warming, of other related climatic change (e.g. hydrological changes resulting from precipitation and evapotranspiration modifications) and of temperature and precipitation related changes in major geomorphologically significant variables (e.g. vegetation cover).

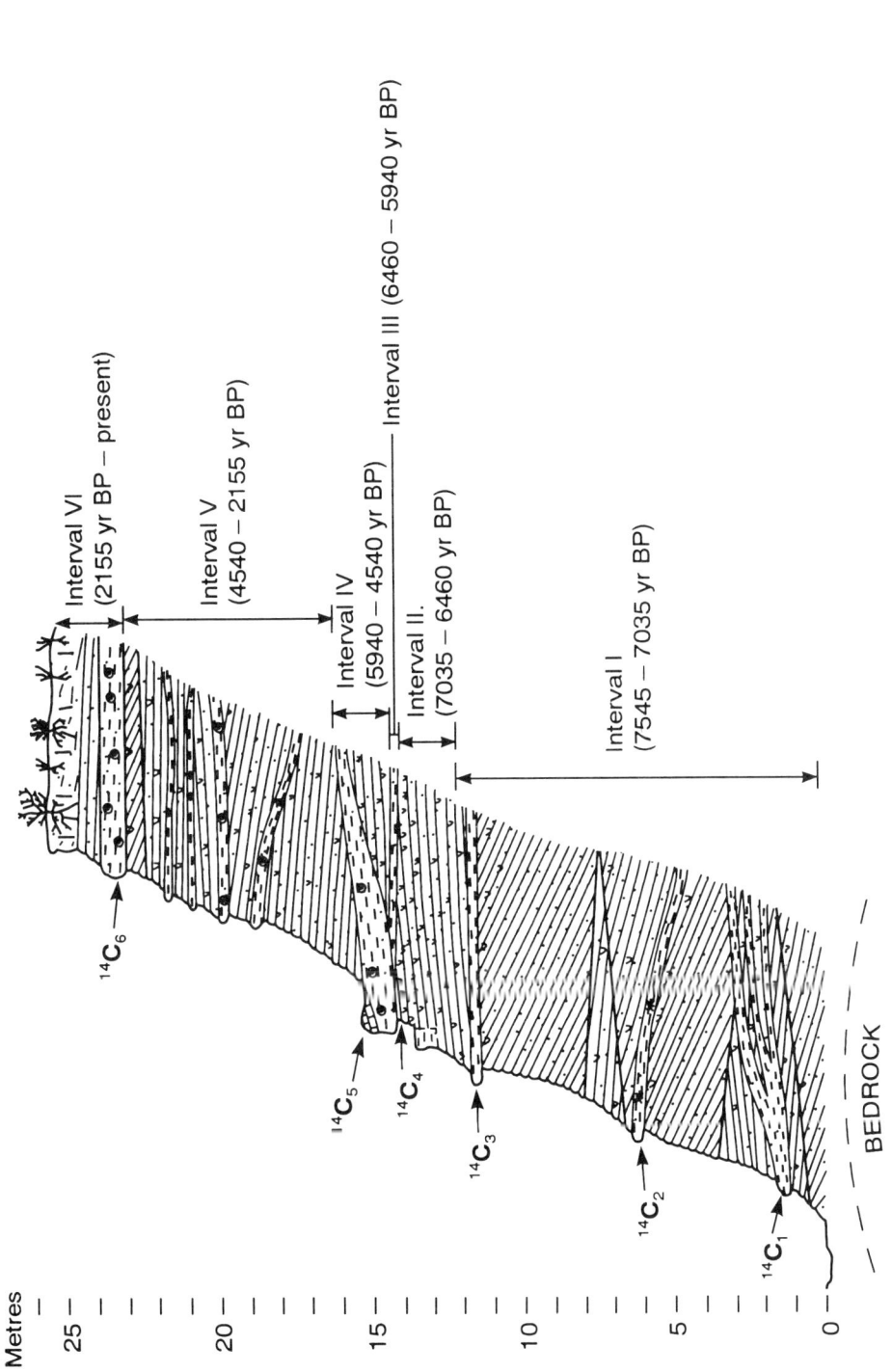

Figure 1.4 Fluctuations in dune accumulation and stability in the Clear Creek area of Wyoming in the Holocene. (Modified after Gaylord, 1990. Reproduced by permission of Academic Press Ltd, London)

6 Environmental Change in Drylands

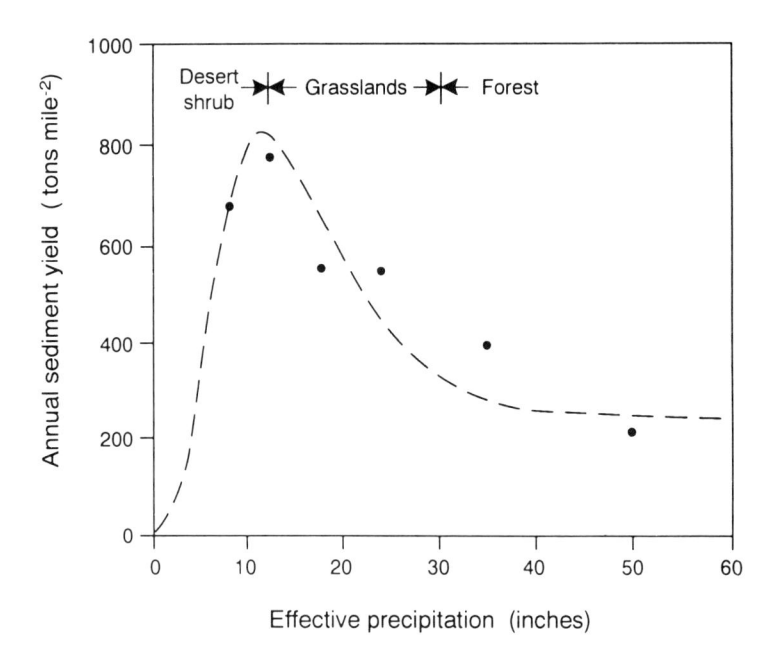

Figure 1.5 Variation of sediment yield with climate as based on data from small watersheds in the United States (after Langbein and Schumm, 1958)

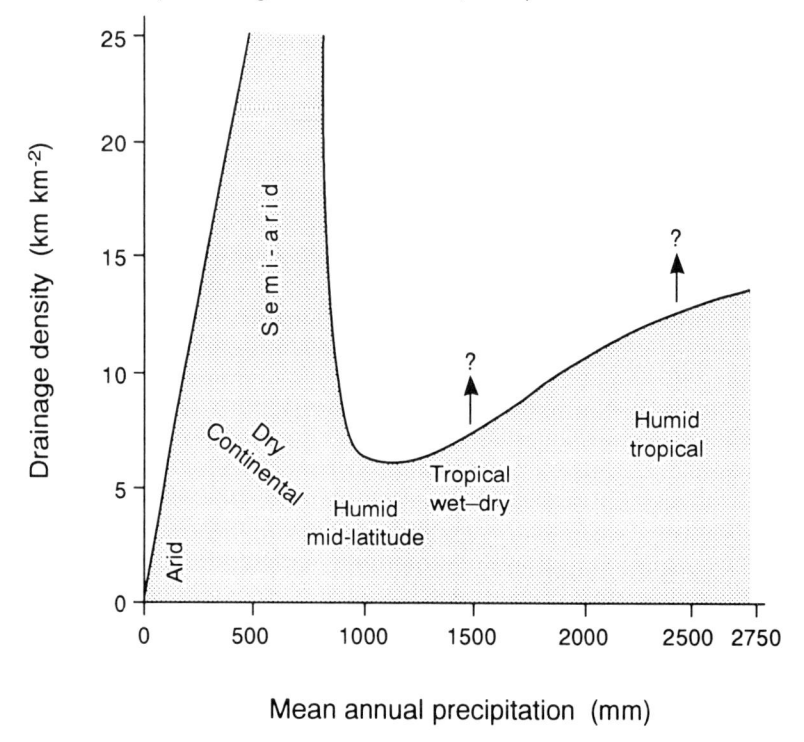

Figure 1.6 Relation between drainage density and mean annual precipitation. (After Gregory, 1976. Copyright © 1976. Reprinted by permission of John Wiley & Sons, Ltd)

Changes in Precipitation and Runoff

One of the great uncertainties associated with climate prediction for a warmer world is the nature of changes in precipitation. General circulation models (GCMs) show complex patterns of change but also demonstrate rather poor levels of agreement. There may be some tendency for dry regions to become drier (Mitchell *et al.*, 1987) and the strength of monsoons and conventional activity might tend to increase in tropical areas, giving increased precipitation. Predictions of precipitation change based on warmer analogue years of the twentieth century (Wigley *et al.*, 1980) and from the reconstruction of Holocene altithermal conditions, while showing less complex patterns of change than the GCM predictions, do indicate substantial precipitation decreases in areas like the High Plains of the United States (Kellogg, 1982). In some arid areas aridity might be intensified as a result of accelerated upwelling of cold west coast waters caused by intensification of the alongshore wind stress on the ocean surface (Bakun, 1990).

One of the few reasonably comprehensive attempts to predict future precipitation changes by combining the evidence from palaeoclimates, modern analogue data and GCMs is provided by Budyko & Izrael (1991). They argue on the basis of palaeoclimatic data that under conditions of warmth comparable to those that existed in the Pliocene (Figure 1.7) precipitation would increase (by up to 30 cm yr^{-1}) over most of Eurasia and the Sahara. On the other hand, under conditions comparable to those of the Holocene Atlantic Optimum, with temperatures up to 1°C warmer than the present, there is a large zone (between 50 and 30°N) where precipitation levels would decline (by up to 20 cm yr^{-1} in central North America). There would, however, be improved moisture conditions in the sub-tropical regions (between 10 and 20°N) and in higher latitudes (more than 60°N). This pattern of changes is attributed to a northward shift both of the ITCZ and of westerly cyclone tracks. In sum, Budyko and Izrael believe that under conditions of marked warming both the mid-latitude and lower-latitude arid zones of the northern hemisphere will be wetter, whereas under conditions of less marked warming (i.e. by around 1°C) areas like the Sahara and Thar will become moister, but areas like the High Plains of the United States or the steppes of the former USSR will become drier (Figure 1.8).

That not all arid areas will necessarily respond in the same way to global warming in terms of changes in precipitation is brought out by a consideration of GCM predictions for two different parts of the United States: the central valley of California and the Great Plains (Figure 1.9). For both areas three different GCMs are employed. In general, though, there are differences between the predictions of the GCMs: the Central Valley will show some tendency towards an increase in precipitation (particularly in winter), while in the Great Plains there is some tendency towards a decrease in precipitation. The GFDL model, when applied to Nebraska and Kansas (Smith & Tirpak, 1990), shows that in a world with twice the amount of CO$_2$ temperatures will be higher and the precipitation decrease in the summer months will be of a similar order to those experienced during the severe dust bowl years of the 1930s (Figure 1.10). Given that higher temperatures will cause greater evapotranspiration loss, this scenario would indicate that dust storm activity plus ground-water depletion would, other things being equal, be much the same as or rather worse than in the 1930s. Such a view is supported by the work of Wheaton (1990) in Canada.

Changes in temperatures, and in the quantities, timing and form of precipitation, would have important implications for runoff (Gleick, 1986). In high-latitude tundra areas, which are currently very dry, warmer winters might cause more snow to fall, thereby creating

8   Environmental Change in Drylands

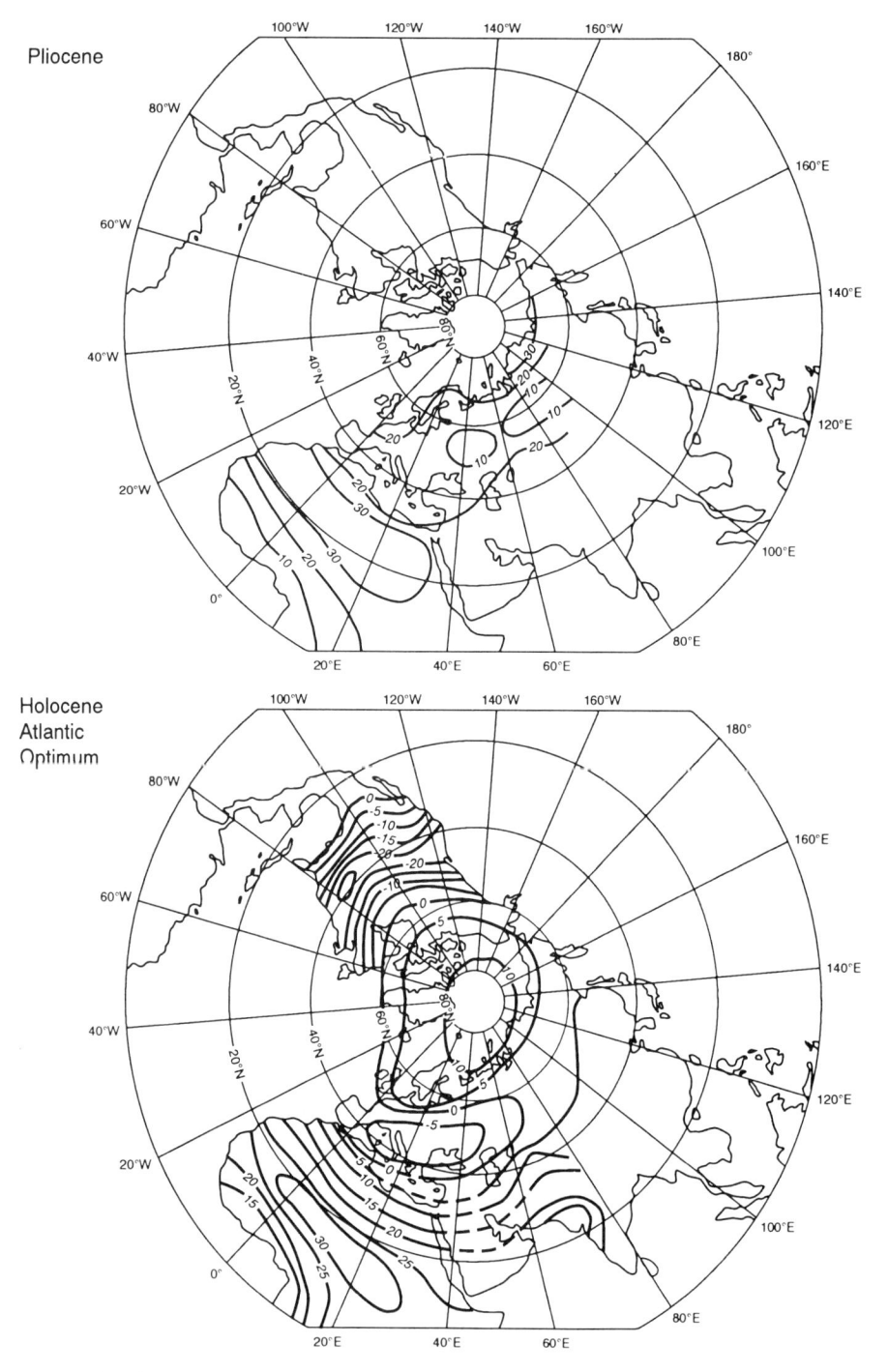

Figure 1.7 Deviations in annual precipitation means (cm) for two past warm phases (reproduced by permission of the University of Arizona Press from Budyko & Izrael, 1991)

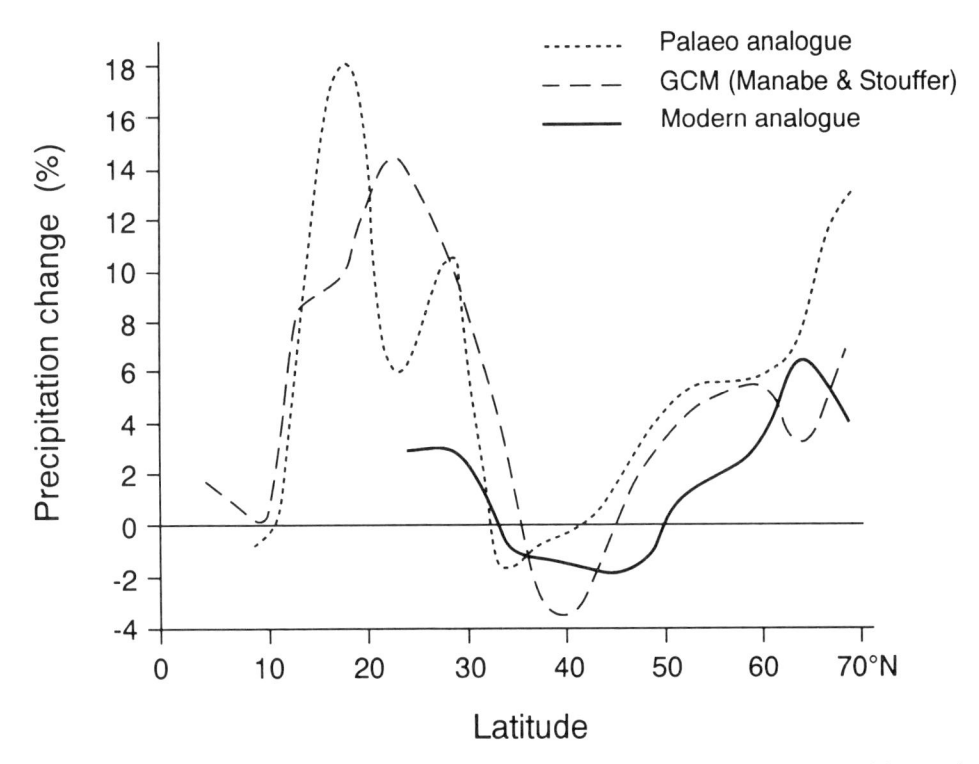

Figure 1.8 Relative changes in mean latitudinal precipitation on the continents of the northern hemisphere with 1°C higher mean surface air temperature (after Budyko & Izrael, 1991, Figs. 1–5)

increased summer runoff (Barry, 1985). Budyko (1982, p. 242) predicted that in the tundra zone of the former USSR annual precipitation could increase by 500–600 mm, causing runoff in latitude 58–60°N to increase by a factor of two or three. In somewhat warmer environments, where there are currently substantial winter snowfalls, there might in a warmer world be a tendency for a marked decrease in the proportion of winter precipitation that falls as snow. There would thus be greater winter rainfall and winter runoff, and less overall precipitation entering snowpacks to be held over until spring snowmelt. There would also be an earlier and shorter spring snowmelt, with adverse consequences both for summer runoff and for spring and summer soil moisture levels (Gleick, 1986).

Highly significant runoff changes may also be anticipated for semi-arid environments, such as the south-west United States. The models of Revelle & Waggoner (1983) suggest that the effects of increased evapotranspiration losses as a result of a 2°C rise in temperature would be particularly serious in those regions where the mean annual precipitation is less than about 400 mm (Table 1.1). Projected summer dryness in such areas may be accentuated by a positive feedback process involving decreases in cloud cover and associated increases in radiation absorption on the ground consequent upon a reduction in soil moisture levels (Manabe & Wetherald, 1986). Our modelling capability in this area is still imperfect and different types of model indicate differing degrees of sensitivity to climatic change (Nash & Gleick, 1991).

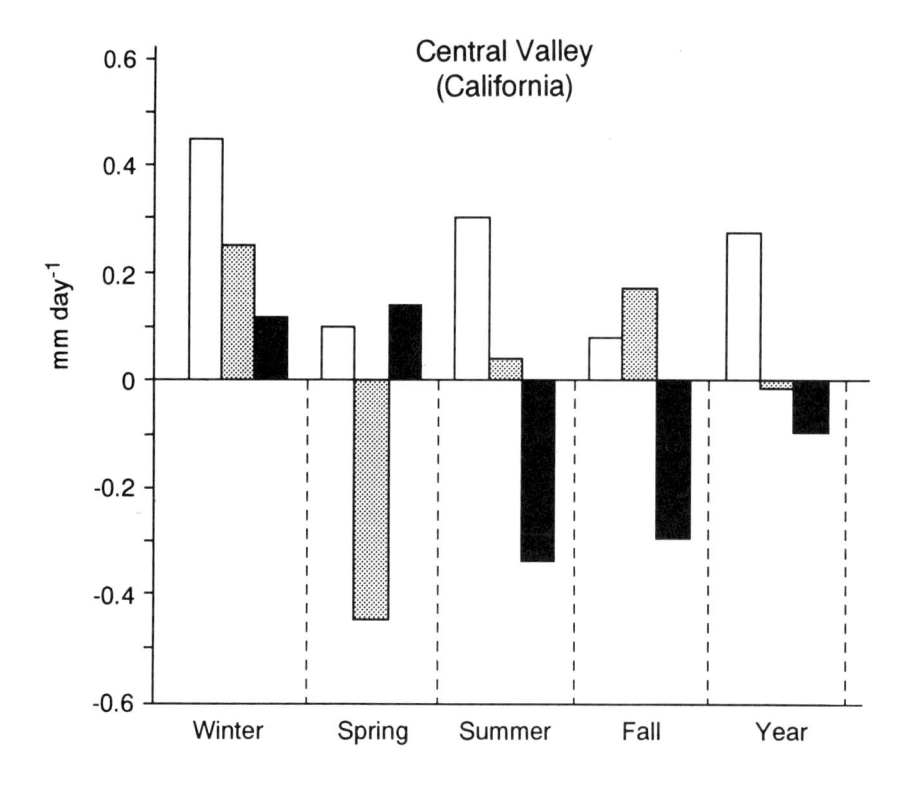

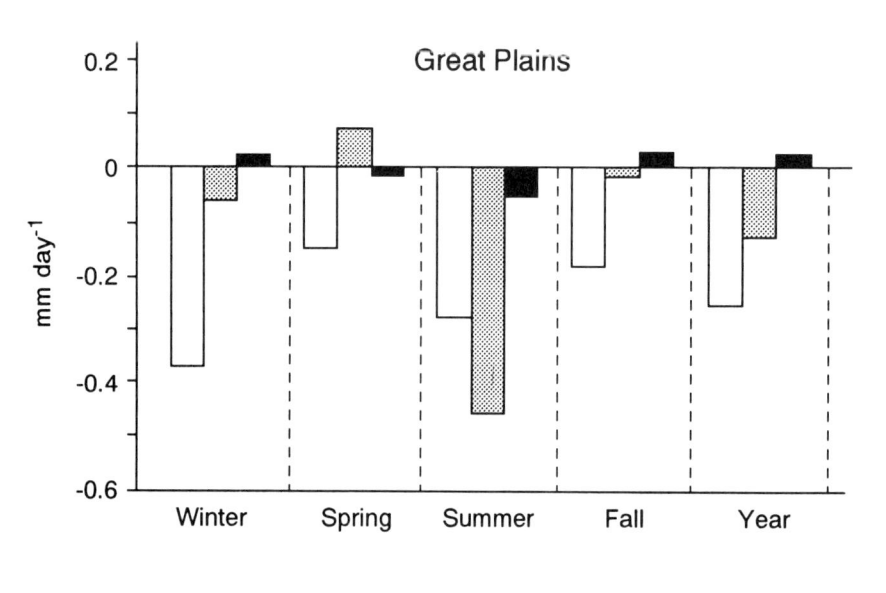

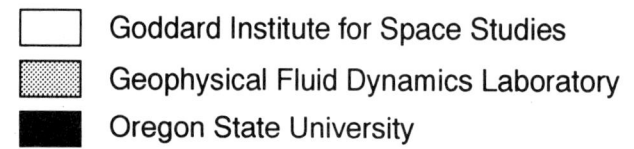

Environmental Change in Drylands

Central Valley (California)

Great Plains

- Goddard Institute for Space Studies
- Geophysical Fluid Dynamics Laboratory
- Oregon State University

Deserts in a Warmer World 11

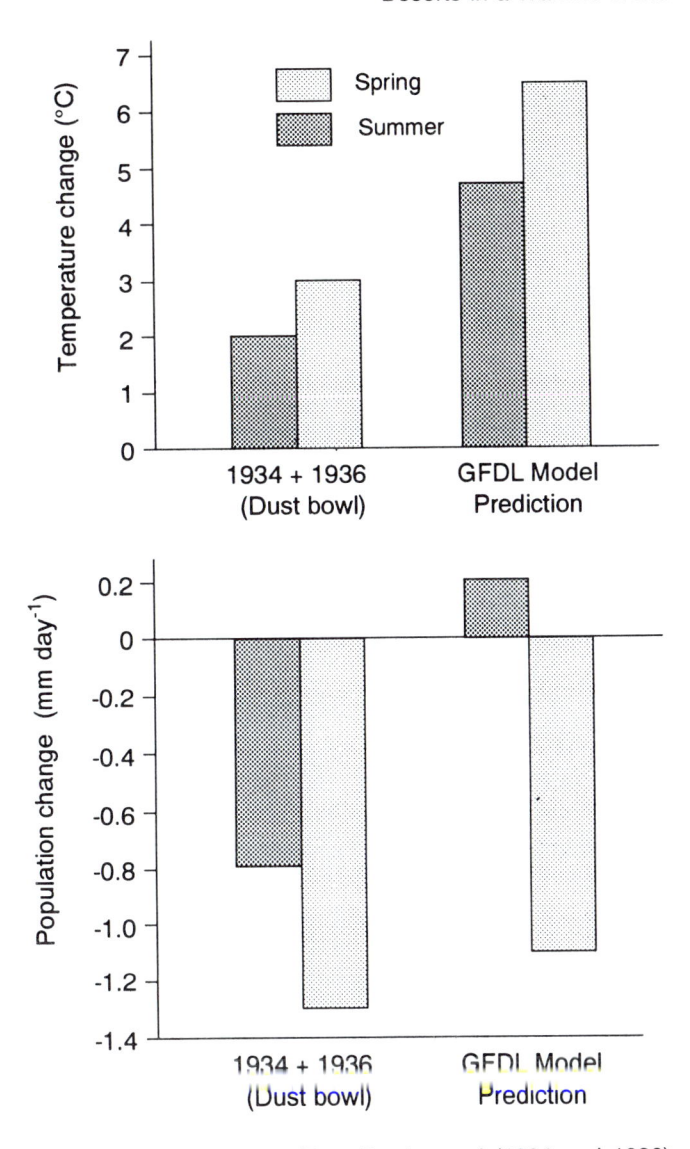

Figure 1.10 Comparison between two 'Dust Bowl years' (1934 and 1936) and the GFDL model prediction for a ×2 CO_2 situation for the Great Plains (Kansas and Nebraska) (modified from Smith & Tirpak, 1990, Fig. 7.3)

Figure 1.9 Predicted changes in precipitation (mm day^{-1}) for three different GCMs for a ×2 CO_2 situation compared to the present. a = Goddard Institute for Space Studies, b = Geophysical Fluid Dynamics Laboratory, c = Oregon State University (modified after Smith & Tirpak, 1990, Figs. 4.5 and 7.2)

Table 1.1 Approximate percentage decrease in runoff for a 2°C increase in temperature (from data in Revelle & Waggoner, 1983)

Initial temperature (°C)	Precipitation (mm yr^{-1})					
	200	300	400	500	600	700
−2	26	20	19	17	17	14
0	30	23	23	19	17	16
2	39	30	24	19	17	16
4	47	35	25	20	17	16
6	100	35	30	21	17	16
8		53	31	22	20	16
10		100	34	22	22	16
12			47	32	22	19
14			100	38	23	19

Tropical Cyclones

Tropical cyclones are important agents of geomorphological change. They scour out river channels, deposit debris fans, cause slope failures, build up or break down coastal barriers and islands, and change the turbidity and salinity of lagoons. It is, therefore, important to assess whether or not their frequency, extent and intensity would change in a warmer world to the extent that they might start to have an impact on areas that are currently not prone to them.

The situation is far from clear. Intuitively one would expect cyclone activity to become more frequent, intense (Figure 1.11(d) and (e)) and extensive if sea-surface temperatures were to rise, because sea-surface temperature is a clear control on where they develop (there is a threshold of around 26.5–27°C) and increasingly deep low-pressure centres can be maintained as temperatures rise. Some researchers have claimed to find evidence for an increasing trend of cyclone activity in the present century (Figure 1.11(c)), and Spencer and Douglas (1985) suggest that in the cold years of the Little Ice Age neoglacial their frequency was less. Furthermore, Emanuel (1987) has employed a GCM which predicts that with a doubling of present atmosphere concentrations of CO_2 there will be an increase of 40–50% in the destructive potential of tropical cyclones.

On the other hand, the Intergovernmental Panel on Climate Change (Houghton et al., 1990, p. xxv) was somewhat equivocal on the extent to which warming would stimulate cyclone activity:

> Although the area of sea having temperatures over this critical value [26.5°C] will increase as the globe warms, the critical temperature itself may increase in a warmer world. Although the theoretical maximum intensity is expected to increase with temperature, climate models give no consistent indication whether tropical storms will increase or decrease in frequency or intensity as climate changes; neither is there any evidence that this has occurred over the past few decades.

Nonetheless, various Australian workers (e.g. Holland et al., 1988) have attempted to model the potential changes in tropical cyclone activity. Figure 1.11(a) and (b) indicates one scen-

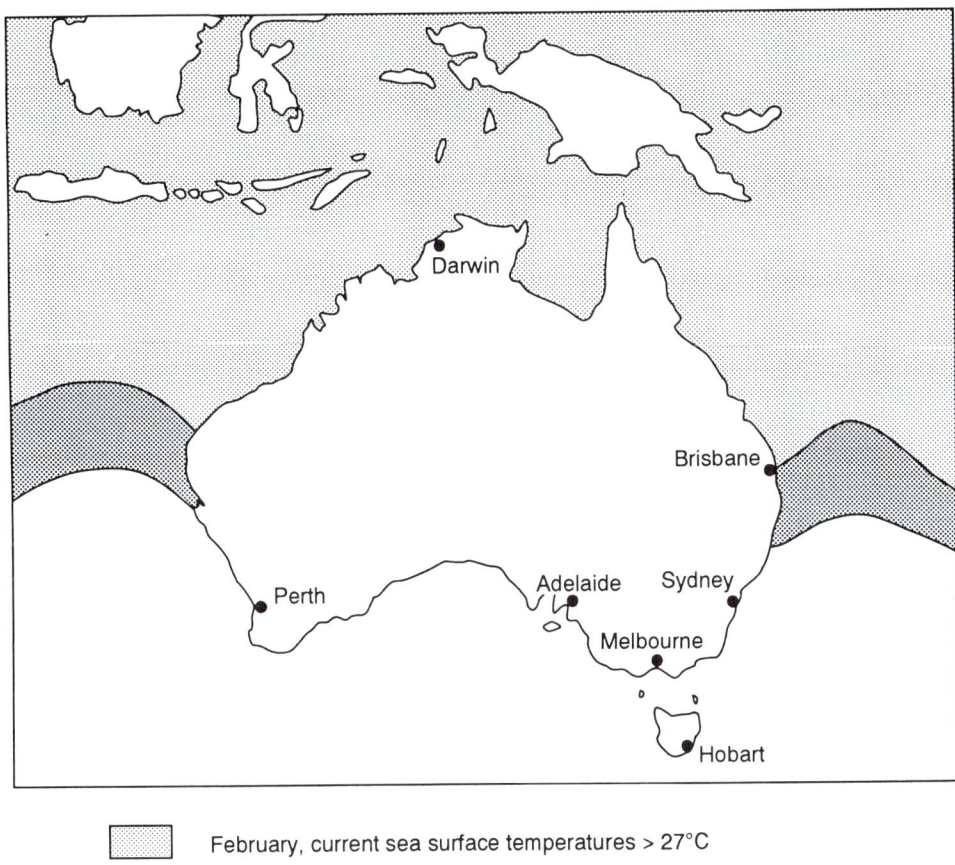

Deserts in a Warmer World 13

(a)

- February, current sea surface temperatures > 27°C
- Additional area with February sea surface temperatures > 27°C under greenhouse conditions and temperatures 2°C higher

Figure 1.11 Tropical cyclones in a warmer world. (a) Areas where February sea surface temperatures around Australia are currently greater than 27°C (stippled) and the additional area with such temperatures under greenhouse conditions with temperatures 2°C higher. (b) The frequency of tropical cyclones crossing 500 km along sections of the Australian coast per decade at present and an estimate of the frequency under greenhouse conditions (after Henderson-Sellers & Blong, 1989. Reproduced by permission of UNSW Press). (c) Cyclone frequencies, Bay of Bengal, North Atlantic, SW Pacific, Australian and SW Indian Ocean regions, 1880–1980 (from Spencer & Douglas, 1985, Fig. 2.3. Reproduced by permission of Routledge). (d) The derived relationship between sea surface temperature and potential intensity of tropical cyclones. (After Miller, 1958, in Holland et al., 1988, Fig. 6. Reproduced by permission of CSIRO Editorial Services). (e) Scatter diagram of monthly mean sea surface temperature and best-track maximum wind speeds (after removing storm motion) for a sample of North Atlantic cyclones. The line indicates the 99th percentile and provides an empirical upper bound on intensity as a function of ocean temperature. (After Merrill, 1987, in Holland et al., 1988, Fig. 5. Reproduced by permission of CSIRO Editorial Services)

14   Environmental Change in Drylands

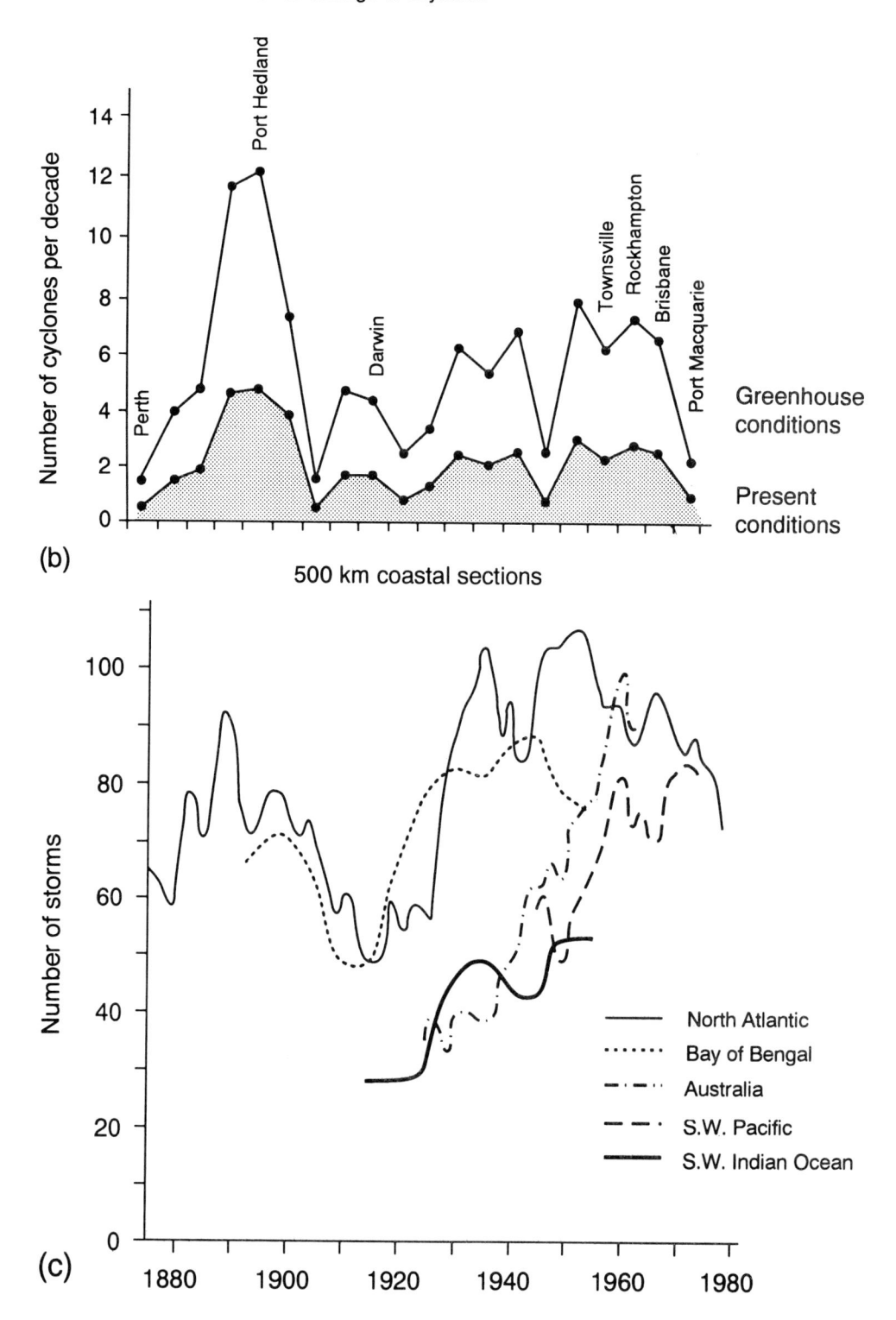

(b) 500 km coastal sections

(c)

(d)

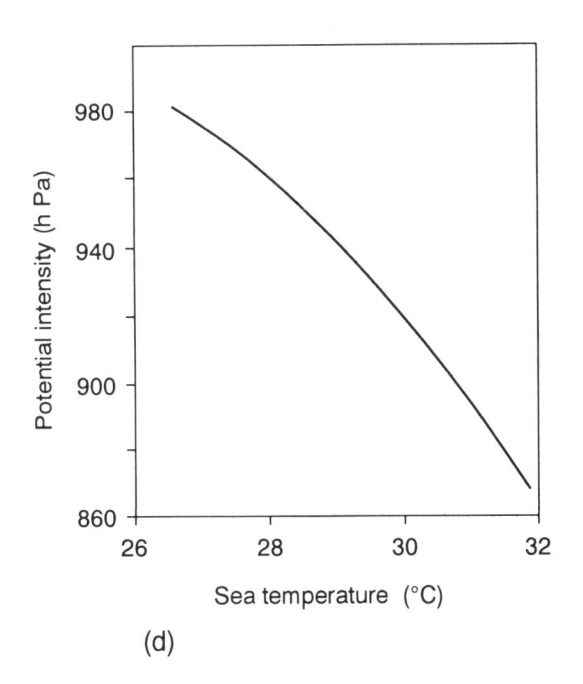

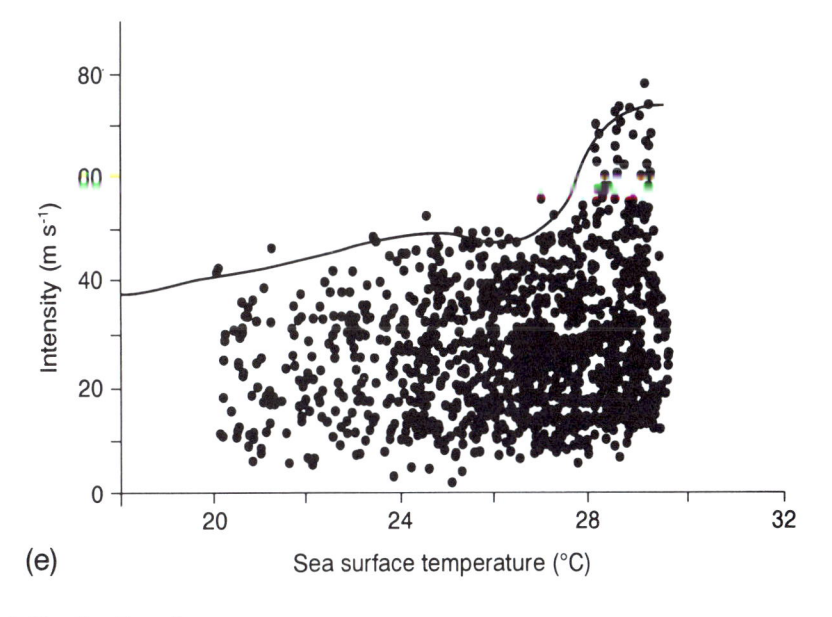

(e)

Figure 1.11 Continued

ario of the likely latitudinal change in the extent of warm, cyclone generating sea water in the Australian region, using as a working threshold for cyclone genesis a summer (February) sea surface temperature of 27°C. Although cyclones do occur to the south of this line, they are considerably more frequent to the north of it. Under greenhouse conditions it is probable that on the margins of the Great Sandy Desert near Port Hedland the number of cyclones crossing the coast will approximately triple from around four per decade to 12 per decade (Henderson-Sellers & Blong, 1989).

THE EXTENT OF SEA-LEVEL RISE AND THE SABKHA AND DELTA RESPONSE

Among the most important geomorphological consequences of global warming would be a worldwide rise in sealevel resulting from two distinct processes: the thermal expansion of the upper layers of the oceans and the melting of land ice (US Department of Energy, 1985; Titus, 1986).

The anticipated rise of sea level over the next century is the subject of contention, largely because of uncertainties about the behaviour of Antarctic ice. In the 1980s there were expectations that sea level could rise by over 3.5 m by 2100. Now, however, there is a tendency to view such values as excessive and the Intergovernmental Panel on Climate Change (Houghton et al., 1990, p. 279) concluded 'that a rise of more than 1 metre over the next century is unlikely'. Nonetheless, this is a rate three to six times faster than that experienced over the last 100 years (1–2 mm yr^{-1}). The Panel also recognised that there were still large uncertainties associated with the future contributions of the Antarctica and Greenland ice caps to sea-level rise.

The degree of sea-level rise could be moderated by reservoir construction, not least in arid areas. Newman & Fairbridge (1986) have calculated that between 1907 and 1982 human intervention stored as much as 0.75 mm yr^{-1} of sea-level rise potential in reservoirs and irrigation projects.

Salt marshes, including the sabkhas of arid coasts, are potentially highly vulnerable in the face of sea-level rise, particularly in those circumstances where sea defences and other barriers prevent the landward migration of marshes as the sea level rises. However, salt marshes are highly dynamic features and in some situations may well be able to cope, even with quite rapid rises of sea level.

There are three possible responses of a salt marsh to a rising sea level (Orson et al., 1985) (Figure 1.12). If the rate of submergence exceeds its ability to accrete vertically the marsh system will drown and sink (Figure 1.12(a)). Alternatively, if the rate of vertical growth due to the inputs of sediments equals the rate of submergence, the extent of marsh may remain stable, but it will erode and be submerged at its seaward margin, while if inland topography permits, it will encroach over terrestrial vegetation at its inland margin (Figure 1.12(b)). If sedimentation rate and plant production are relatively high, salt marshes may expand both laterally and vertically during relative sea-level rise (Figure 1.12(c)).

In arid areas, such as the Persian Gulf, extensive areas of coastline are bounded by low-lying sabkhas (Figure 1.13). These are salt marsh types that result from the interaction of various depositional and erosional (aeolian and storm surge) processes. They are subject to periodic inundation, lie close to sealevel and thus might be vulnerable to sea-level rise and to any increase in storm-surge events.

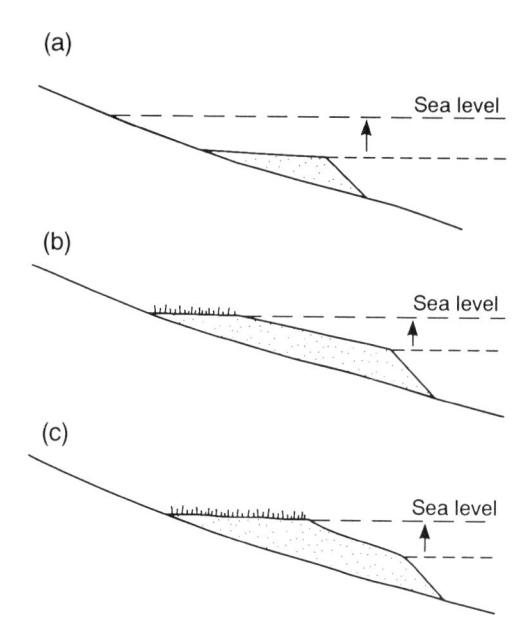

Figure 1.12 Possible salt marsh response to sea-level rise (after Orson *et al.*, 1985, Fig. 3)

However, it is likely that many sabkhas will be able to cope with modestly rising sea levels, for a range of processes contribute to their accretion. These include algal stromatolite growth, faecal pellet deposition, aeolian inputs and evaporite precipitation. Some of these can cause markedly rapid accretion (Schreiber, 1986), even in the absence of a very well-developed plant cover. Moreover, as sea level (and groundwater) rises, surface lowering by deflational processes will be reduced. An example of a sabkha that may well maintain itself, or even continue to aggrade in spite of sea-level rise, is provided by the Umm Said Sabkha of Qatar, where aeolian dune inputs from inland cause the sabkha to build out into the Persian Gulf.

Among arid zone coastal environments that may be particularly susceptible to sea-level rise are deltaic areas subject to subsidence and sediment starvation (e.g. the Nile) and areas where ground subsidence is occurring as a result of fluid abstraction (e.g. California). Whereas the IPCC prediction of sea-level rise is 30–100 cm per century, rates of deltaic subsidence in the Nile Valley are 35–50 cm per century, and in other parts of the world rates of land subsidence produced by oil, gas or groundwater abstraction can be up to 500 cm per century.

ANTHROPOGENIC IMPACTS

It is, as we have seen, extraordinarily difficult to make very firm statements about the response of desert landforms and land-forming processes to global warming. What is more likely, however, is that the effects of global warming will often be insignificant in comparison with the effects of other human activities. Many desert areas and geomorphic systems have already shown themselves to be highly susceptible to the effects of human intervention

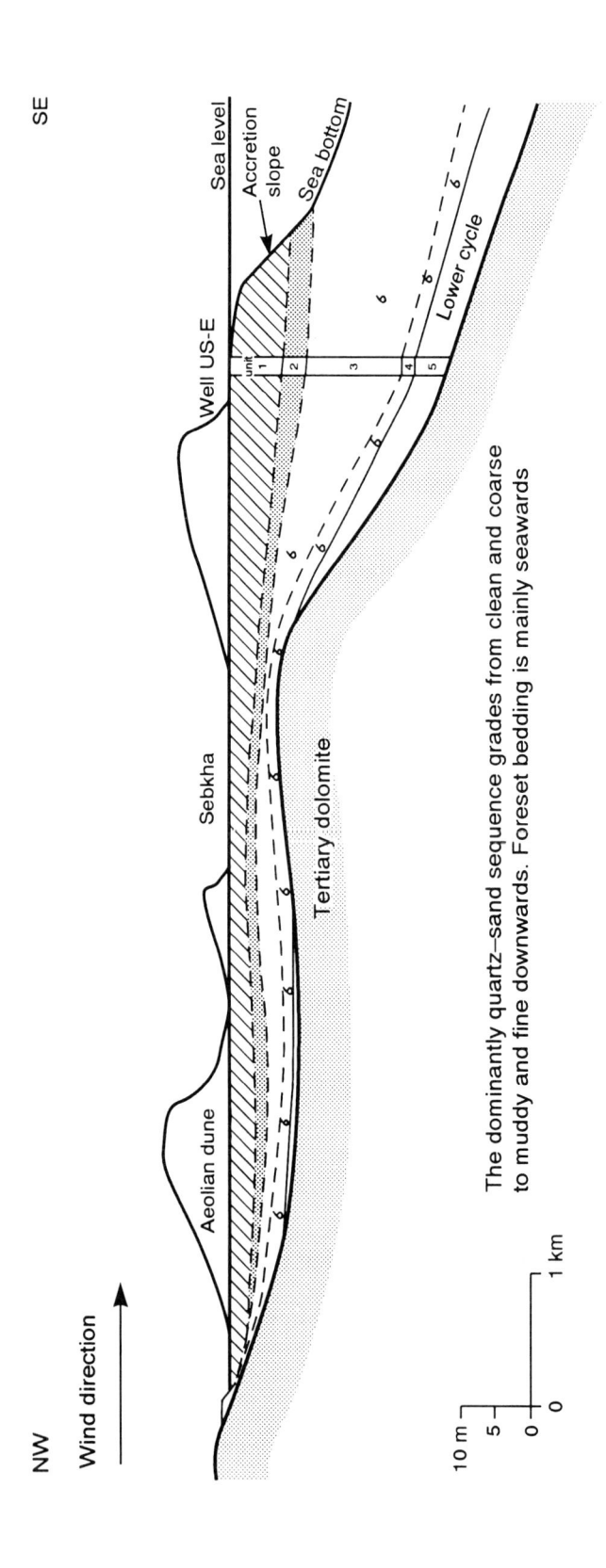

Figure 1.13 Profile across Umm Said Sabkha showing the lateral distribution of the principal sedimentary units (reproduced by permission of Springer-Verlag from Shinn, 1973)

and will so increasingly as human population levels explode. Desertification is, after all, seen by many as essentially a condition caused by people. Some specific examples will demonstrate what has already been achieved:

1. The Colorado River (USA). Perhaps the most dramatic evidence of anthropogenic change to a river's regime and sediment load is provided by the Colorado River of the southwest United States. The estimated annual virgin flow was 18.5 km³, and it used to carry 125–160 million tons of suspended sediment per year to its delta in the Gulf of California. It now discharges neither sediment nor water to the sea (Schwarz *et al.*, 1991) (Figure 1.14).
2. The Aral Sea (Uzbekistan and Turkmenistan). Perhaps the most severe change to a major inland water body is that taking place in the Aral Sea (Figure 1.15). Since 1960 the Aral Sea has lost more than 40% of its area, about 60% of its volume, and its level has fallen by more than 14 m (Kotlyakov, 1991). This has lowered the artesian water table over a band 80–170 km in width, has exposed 24 000 km² of former lake bed to desiccation and has created salty surfaces from which salts are deflated to be transported in dust storms to the detriment of soil quality downwind. The mineral content of what remains has almost increased threefold over the same period. It is probably the most dire ecologi-

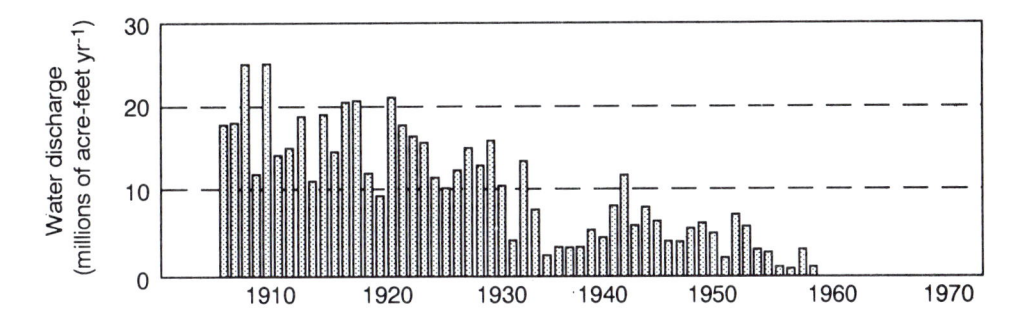

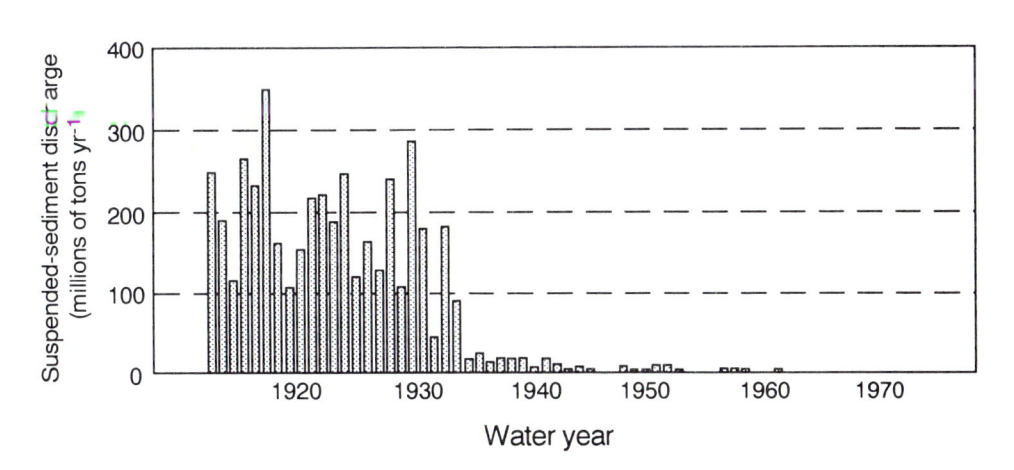

Figure 1.14 Historical sediment and water discharge of the Colorado River. (After the US Geological Survey, in Schwarz *et al.*, 1991)

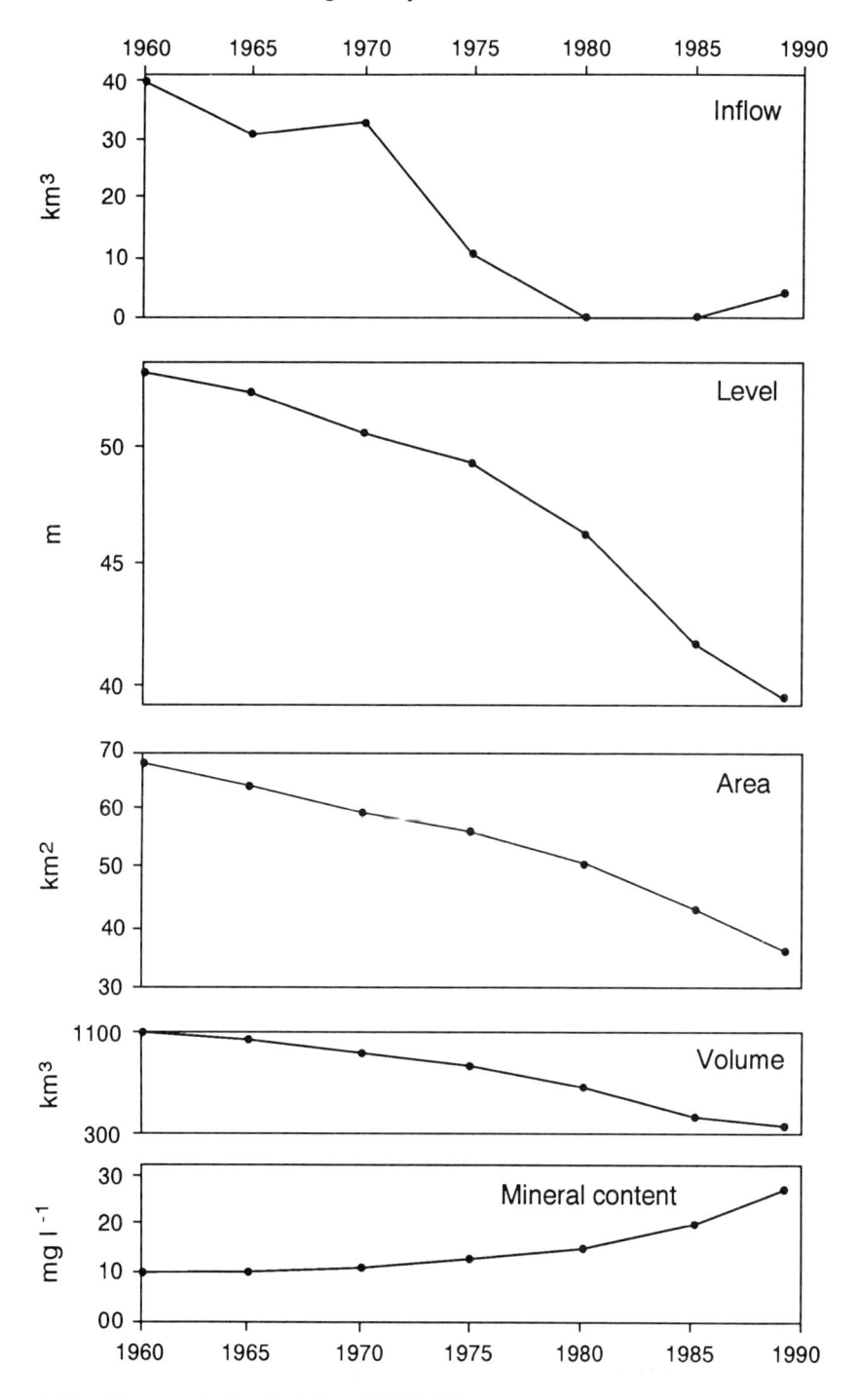

Figure 1.15 Changes in the Aral Sea (1960–89)

cal tragedy to have afflicted the former Soviet Union, and much of the blame rests with excessive use of water which would otherwise replenish the sea. Figure 1.16 shows the major reductions in mean annual runoff that have taken place in the southern Russian plain under the influence of an expansion in the use of water for irrigation.

3. The High Plains Aquifer. The mining of groundwater has transformed the High Plains (Ogallala) Aquifer in the United States (Figure 1.17). Over large areas its level has fallen by over 30 m, as for example in the Panhandle of Texas. In some areas the process is now being reversed because of rising power costs and policy changes.

4. Sediment starvation by dam construction. The construction of more and more ever-larger dams has transformed the sediment load of many rivers, creating changes in downstream channel aggradation and erosion, and reducing sediment nourishment of shorelines. The Nile provides an excellent illustration of this, as it now only transports 8% of its natural load below the Aswan High Dam.

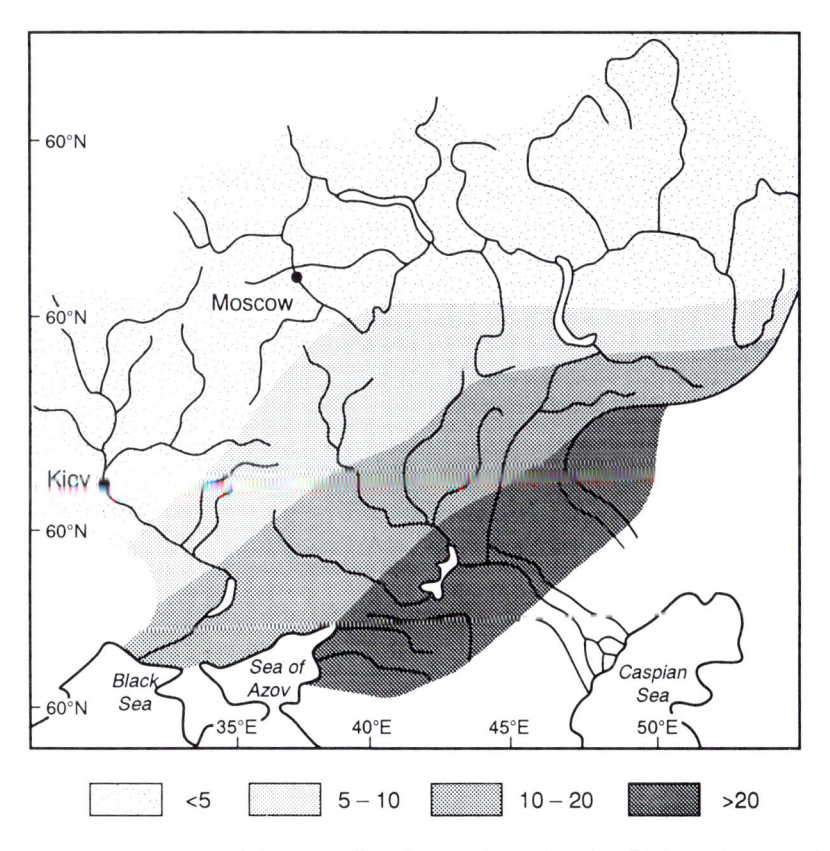

Figure 1.16 The decrease of river runoff in the southern Russian Plain under the influence of irrigated agriculture (as a percentage of the period up to 1950). (After Alayev et al., 1991. Reproduced by permission of Cambridge University Press)

22   Environmental Change in Drylands

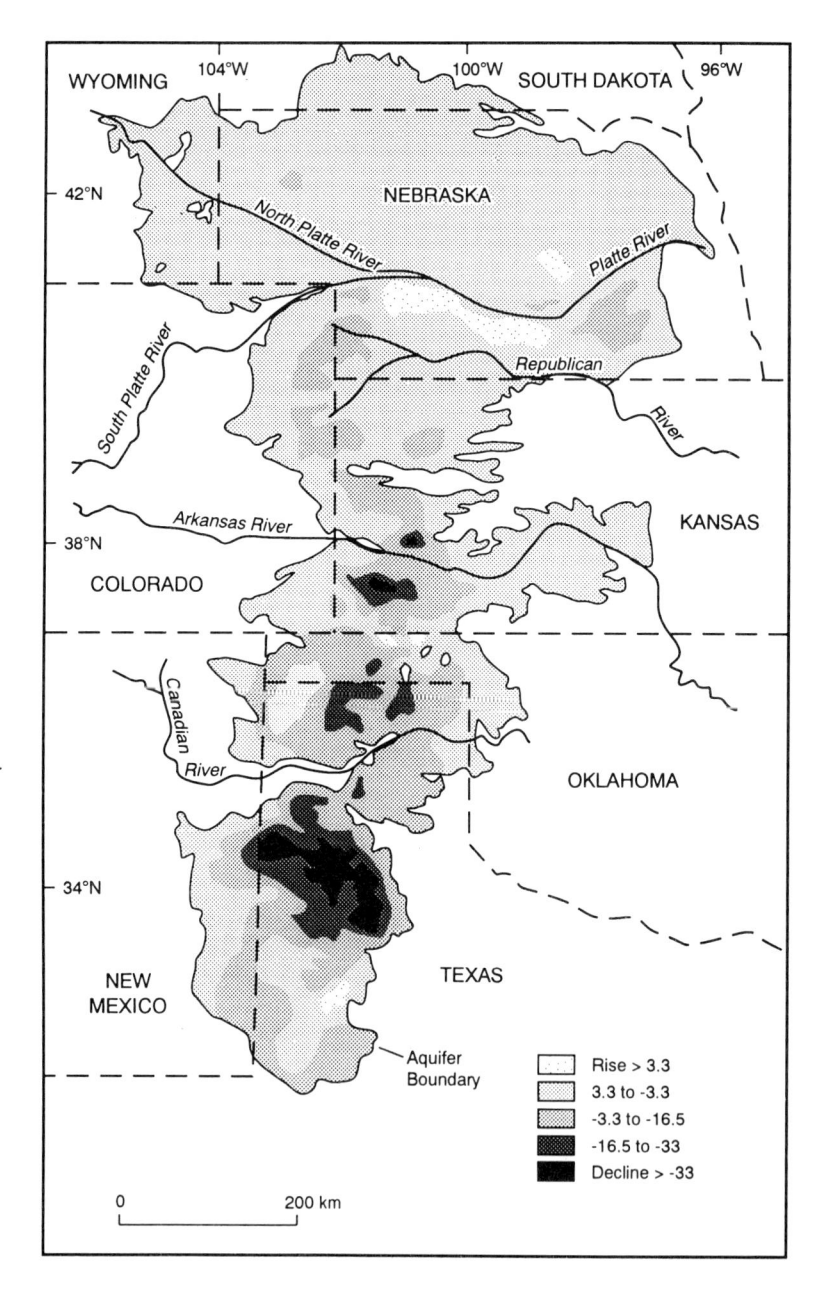

Figure 1.17 Pre-development to 1980 water-level changes in the High Plains Aquifer. (After the US Geological Survey, in Schwarz *et al.*, 1991)

However, perhaps the most important point that one can make by way of conclusion is that the effects of global warming may, in critical areas, compound the most serious consequence of current human activities. For example, reduced runoff levels resulting from increased evapotranspiration under conditions of global warming will be especially significant in areas like the south-west United States or the Texas High Plains where river discharges and groundwater levels have already been very substantially modified by water abstraction. Likewise the threat of coastal inundation and accelerated erosion caused by the rising sea levels promoted by global warming will be greater in locations like the Nile Delta, where the coast is starved of replenishing fluvial sediments.

REFERENCES

Alayev, E. B., Badenkov, Y. P. and Katavaeva, N. A. (1991). The Russian Plain. In Turner, B. L. (ed.), *The Earth as Transformed by Human Action*, Cambridge University Press, Cambridge, pp. 203–14.

Bakun, A. (1990). A global climate change and intensification of coastal upwelling. *Science*, **247**, 198–200.

Balling, R. C. and Wells, S. G. (1990). Historical rainfall patterns and arroyo activity within the Zuni River drainage basin, New Mexico. *Annals of the Association of American Geographers*, **80**, 603–17.

Barry, R. G. (1985). The cryosphere and climate change. In MacCracken, M. C. and Luther, F. M. (eds.), *Detecting the Climatic Effects of Increasing Carbon Dioxide*, US Department of Energy, Washington D.C, pp. 111–48.

Budyko, M. I. (1982). *The Earth's Climate: Past and Future*, Academic Press, New York.

Budyko, M. I. and Izrael, Yu. A. (1991). *Anthropogenic Climatic Change*, University of Arizona Press, Tucson.

Butzer, K. W. (1974). Accelerated soil erosion: a problem of man-land relationships. In Manners, I. R. and Mikesell, M. W. (eds.), *Perspectives on Environments*, Association of American Geographers, Washington D.C.

Cooke, R. U. and Reeves, R. W. (1976). *Arroyos and Environmental Change in the American South-West*, Clarendon Press, Oxford.

Emanuel, K. A. (1987). The dependence of hurricane intensity on climate. *Nature*, **326**, 483–5.

Gaylord, D. R. (1990). Holocene palaeoclimate fluctuations revealed from dune and interdune strata in Wyoming. *Journal of Arid Environments*, **18**, 123–38.

Gleick, P. H. (1986). Regional water resources and global climatic change. In Titus, J. G. (ed.), *Effects of Changes in Stratospheric Ozone and Global Climate, Volume 3, Climatic change* UNEP/USEPA, Washington D.C., pp. 217–49.

Goudie, A. S. (1989). The global geomorphological future. *Zeitschrift für Geomorphologie Supplementband*, **79**, 51–62.

Goudie, A. S. and Middleton, N. J. (1991). The changing frequency of dust storms through time. *Climatic Change*, **18**, 197–225.

Gregory, K. J. (1976). Drainage networks and climate. In Derbyshire, E. (ed.), *Geomorphology and Climate*, Wiley, Chichester, pp. 289–315.

Harvey, A. M. (1989). The occurrence and role of arid zone alluvial fans. In Thomas, D. S. G. (ed.), *Arid Zone Geomorphology*, Belhaven Press, London, pp. 136–58.

Henderson-Sellers, A. and Blong, R. (1989). *The Greenhouse Effect: Living in a Warmer Australia*, New South Wales University Press, Kensington, NSW.

Holland, G. J., McBride, J. L. and Nicholls, N. (1988). Australian region tropical cyclones and the greenhouse effect. In Pearman, G. I. (ed.), *Greenhouse, Planning for Climate Change*, Brill, Leiden, pp. 438–55.

Houghton, J. T., Jenkins, G. J. and Ephraums, J. J. (1990). *Climatic Change: The IPCC Scientific Assessment*, Cambridge University Press, Cambridge.

Idso, S. B. (1989). *Carbon Dioxide and Global Change: Earth in Transition*. IBR Press, Tempe.

Kellogg, W. W. (1982). Precipitation trends on a warmer earth. In Peck, R. A. and Hummel, J. R. (eds.), *Interpretation of Climate and Photochemical Models, Ozone and Temperature Measurements*, American Institute of Physics, New York, pp. 35–46.

Kotlyakov, V. M. (1991). The Aral Sea Basin, a critical environmental zone. *Moscow Environment*, **33**, 4–9, 36–38.

Langbein, W. B. and Schumm, S. A. (1958). Yield of sediment in relation to mean annual precipitation. *Transactions American Geophysical Union*, **39**, 1076–84.

Manabe, S. and Stouffer, R. J. (1980). Sensitivity of a global climate model to an increase of CO_2 concentration in the atmosphere. *Journal of Atmospheric Science*, **37**, 99–118.

Manabe, S. and Wetherald, R. T. (1986). Reduction in summer soil wetness by an increase in atmospheric carbon dioxide. *Science*, **232**, 626–8.

Mitchell, J. F. B., Wilson, C. A. and Cunningham, W. M. (1987). On CO_2 climate sensitivity and model dependence of results. *Quarterly Journal of the Royal Meteorological Society*, **113**, 293–322.

Nash, L. L. and Gleick, P. H. (1991). Sensitivity of streamflow in the Colorado Basin to climatic changes. *Journal of Hydrology*, **125**, 221–41.

Newman, W. S. and Fairbridge, R. W. (1986). The management of sea-level rise. *Nature*, **320**, 319–21.

Orson, R., Panageotor, W. and Leatherman, S. P. (1985). Response of tidal salt marshes to rising sea levels along the US Atlantic and Gulf Coasts. *Journal of Coastal Research*, **1**, 29–37.

Revelle, R. R. and Waggoner, P. E. (1983). Effect of a carbon dioxide-induced climatic change on water supplies in the western United States. In *Carbon Dioxide Assessment Committee, Changing Climate*, National Academy Press, Washington D.C., pp. 419–32.

Schreiber, B. C. (1986). Arid shorelines and evaporites. In Reading, H. G. (ed.), *Sedimentary Environments and Facias*, Blackwell Scientific, Oxford, pp. 189–228.

Schumm, S. A. (1977). *The Fluvial System*, Wiley, New York.

Schwarz, H. E., Emel, J., Dickens, W. J., Rogers, P. and Thompson, J. (1991). Water quality and flows. In Turner, B. L. (ed.), *The Earth as Transformed by Human Action*, Cambridge University Press, Cambridge, pp. 253–70.

Shinn, E. A. (1973). Sedimentary accretion along the leeward, SE coast of Qatar Peninsula, Persian Gulf. In Purser, B. H. (ed.), *The Persian Gulf*, Springer Verlag, New York, pp. 199–209.

Smith, J. B. and Tirpak, D. A. (eds.) (1990). *The Potential Effects of Global Climate Change on the United States*, Hemisphere Publishing Corporation, New York.

Spencer, T. and Douglas, I. (1985). The significance of environmental change: diversity, disturbance, and tropical ecosystems. In Douglas, I. and Spencer T. (eds.), *Environmental Change and Tropical Geomorphology*, Allen and Unwin, London, pp. 13–33.

Titus, J. G. (ed.) (1986). *Effects of Changes in Stratospheric Ozone and Global Climate*, Volume 4, Sea-Level Rise, UNEP/USEPA, Washington D.C.

US Department of Energy (1985). *Glaciers, ice sheets, and sea-level: effect of a CO_2-induced climatic change*, US Department of Energy, Washington D.C.

Van Andel, T. H., Zangger, E. and Demitrack, A. (1990). Land use and soil erosion in prehistoric and historical Greece. *Journal of Field Archaeology*, **17**, 379–96.

Vita-Finzi, C. (1967). *The Mediterranean Valleys*, Cambridge University Press, Cambridge.

Watson, A., Price-Williams, D. and Goudie, A. S. (1984). The palaeoenvironmental interpretation of colluvial sediments and palaeosols of the Late Pleistocene hypothermal in southern Africa. *Palaeogeography, Palaeoclimatology, Palaeoecology*, **5**, 225–49.

Wheaton, E. E. (1990). Frequency and severity of drought and dust storms. *Canadian Journal of Agricultural Economics*, **38**, 695–700.

Wigley, T. M. L., Jones, P. D. and Kelly, P. M. (1980). Scenario for a warm high-CO_2 world. *Nature*, **283**, 17–21.

2 Timescales, Environmental Change and Dryland Valley Development

D. J. NASH*, D. S. G. THOMAS
Department of Geography, University of Sheffield, UK

and

P. A. SHAW
School of Geology, University of Luton, UK (formerly Department of Environmental Science, University of Botswana)

ABSTRACT

The traditional approach when considering the origins of dryland valley systems is to view development purely in terms of erosional and depositional sequences resulting from channels acting within the valley network. These sequences are usually considered as resulting from major climatic and environmental changes. Recent studies in southern Africa have indicated that climatic shifts during the Quaternary may not have been as large as often implied. This suggests that another factor other than fluvial activity under humid conditions may be needed to explain valley origins and development in the semi-arid interior. In this paper, using the example of valley systems from the Kalahari, the role of groundwater sapping, deep-weathering and structural control in valley development is considered. Evidence for both fluvial and groundwater origins is included, considering elements of valley form at the valley and intra-valley scale. It is concluded that groundwater and ancient structures were significant in the location and early development of valley form, with subsequent fluvial activity shaping intra-valley features. It is also recognised that groundwater and fluvial activity would have operated over a range of timescales, and that additional factors such as tectonic changes and drainage capture need to be considered.

INTRODUCTION

Fluvial geomorphological studies generally concentrate upon river channel form and processes and rarely address the development of valleys themselves. Exceptions to this generalisation do, however, occur, such as the work of Small (1964) and Sparks & Lewis (1957-8). There is also a comparative lack of data for river channels within drylands as opposed to temperate regions (Reid & Frostick, 1989). As a result it is not surprising that the origin and mode of development of dryland valley systems is poorly understood.

Research on dryland valleys, especially those that are considered ephemeral, fossil or contain misfit channels, generally infers formation due to erosion during periods of excess

---

Present address: Department of Building (Geography Division), University of Brighton, UK.

Environmental Change in Drylands: Biogeographical and Geomorphological Perspectives.
Edited by A. C. Millington and K. Pye. © 1994 John Wiley & Sons Ltd

available moisture with sufficient surface water to maintain permanent rivers (e.g. Heine, 1982; Moore, 1988). By inference, such valleys are traditionally considered to be indicative of large-scale changes in climate and on this basis are accorded palaeoclimatic significance.

However, evidence from other landforms is increasingly suggesting that past fluctuations in climate may not have been as great as previously thought. In the Kalahari Desert of central southern Africa, recent studies of linear dunefields (e.g. Thomas & Tsoar, 1990), palaeolakes (e.g. Shaw & Cooke, 1986) and cave deposits (Klein *et al.*, 1991) suggest that climates during the Quaternary period have fluctuated around a semi-arid mean. This is in marked contrast to the large changes in environmental conditions envisaged by earlier research (e.g. Grove, 1969). If the major shifts in climate traditionally thought necessary for valley development did not occur, or their occurrence is less clear-cut, then other causes need to be considered.

This paper discusses possible modes of development of dryland valley systems, using the example of the extensive valley networks of the Kalahari, and considers the implications such valleys have for the interpretation of past environmental changes.

KALAHARI DRY VALLEY NETWORKS

Dry or 'fossil' valleys have been recognised in the Kalahari since the records of Andersson (1856). Many terms meaning 'valley' occur in the different languages of the region, e.g. *mekgacha* in SeTswana, *laagte* in Afrikaans and *omiramba* or *dum* in the various languages of north-east Namibia and north-west Botswana. A number of systems exist (Figure 2.1), which can be subdivided according to whether they are directed endoreically or exoreically, i.e. whether they trend towards the Makgadikgadi Depression or Okavango Delta, or ultimately connect with the Orange River via the Molopo River. With the exception of the headwaters of parts of the Serorome, Mmone/Quoxo and Moselebe systems, valleys rise at the fringe of the Kalahari Desert in areas where bedrock outcrops or is covered by a thin sand veneer. Sediment thicknesses increase towards the centre of the Kalahari depositional basin, with thicknesses exceeding 500 m in parts (Thomas, 1988). As a result of this, endoreic valley courses are situated mainly upon Jurassic to recent Kalahari Group sediments (Thomas & Shaw, 1991).

In general, seasonal flow is restricted to short valley sections in headwater regions and other areas of bedrock outcrop or subcrop. Floods have been recorded within the exoreic valleys (the Molopo, Kuruman, Auob and Nossop systems), the most recent being in the Kuruman in 1988–9. Within endoreic systems flows have only been recorded during exceptional storm events, such as the Letlhakane Valley in 1969 (Mazor *et al.*, 1977).

The form of the valleys is highly variable, as might be expected in systems covering such large areas; the Okwa and Mmone systems have a combined catchment of over 90 000 km^2 (Thomas & Shaw, 1991). This is despite the marked homogeneity of the Kalahari Basin sediments which might be expected to produce homogeneous valley forms. Valley widths vary considerably, ranging from small channels in headwater areas to broad forms between 1 and 2 km across. However, some generalisations can be made. With the exception of the Rooibrak and Deception valleys, which have negligible relief along their entire courses, valleys can be considered as exhibiting three main forms (Boocock & Van Straten, 1962; Thomas & Shaw, 1991):

Timescales, Environmental Change and Dryland Valley Development

Mekgacha networks

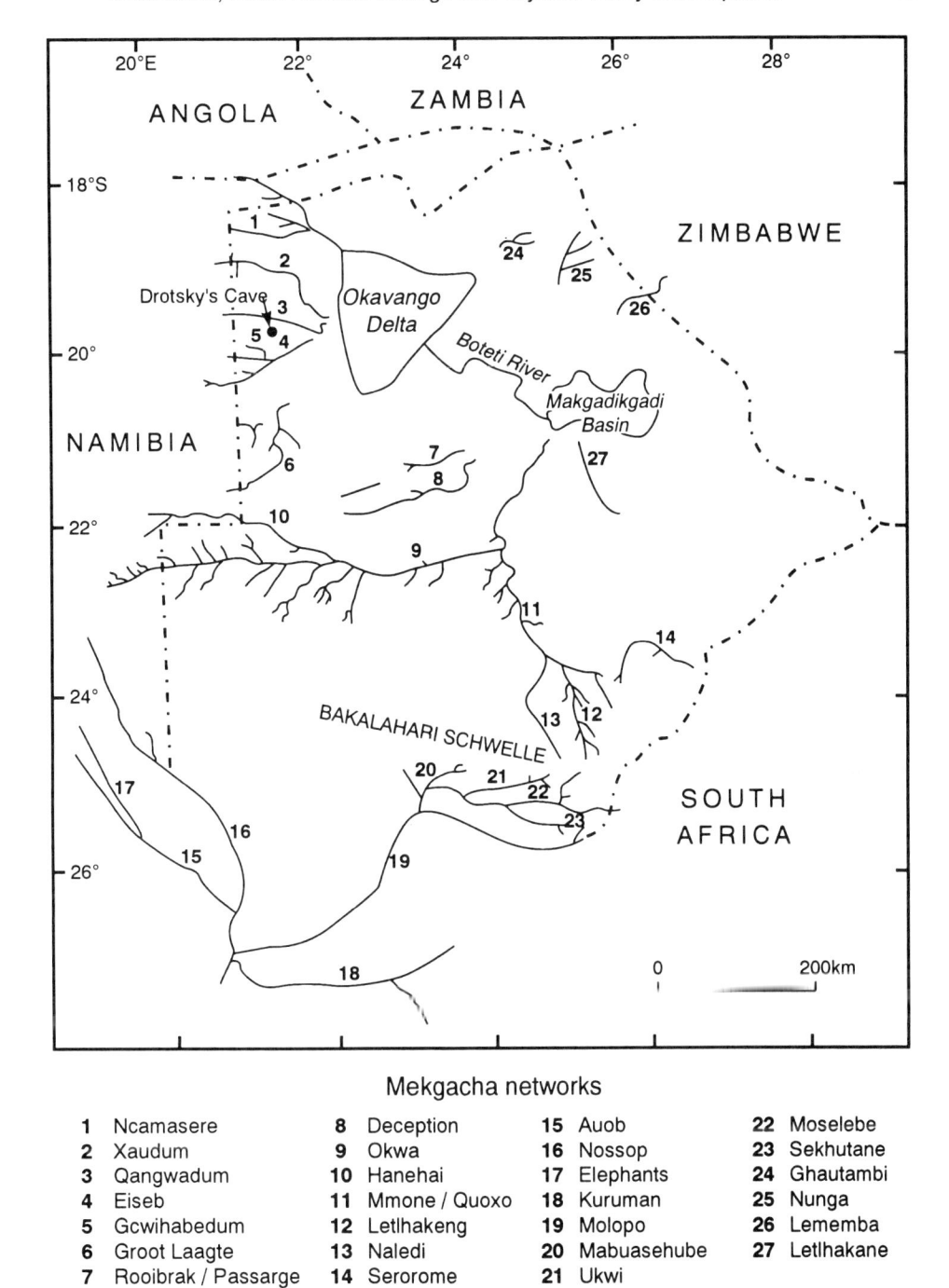

1 Ncamasere	8 Deception	15 Auob	22 Moselebe
2 Xaudum	9 Okwa	16 Nossop	23 Sekhutane
3 Qangwadum	10 Hanehai	17 Elephants	24 Ghautambi
4 Eiseb	11 Mmone / Quoxo	18 Kuruman	25 Nunga
5 Gcwihabedum	12 Letlhakeng	19 Molopo	26 Lememba
6 Groot Laagte	13 Naledi	20 Mabuasehube	27 Letlhakane
7 Rooibrak / Passarge	14 Serorome	21 Ukwi	

Figure 2.1 Kalahari dry valley (*mekgacha*) networks

1. In headwater regions, unless the valley is situated upon bedrock, the form is characteristically broad and flat with limited relative relief.
2. The form changes abruptly in the middle sections of many valleys (Figure 2.2), with a flat valley floor and steep sides. In the case of valleys to the south of Letlhakeng in Botswana, the valley floor is incised up to 45 m below the level of the surrounding terrain with a maximum valley width in excess of 2 km.
3. This gorge-like section gradually transforms into a further broad and flat stage, with endoreic valleys such as the Okwa and Mmone becoming sand-choked and almost imperceptible towards the central Kalahari. These variations in form are best exhibited by the endoreic systems, with other valleys such as the Auob and Nossop showing less systematic variation (Leistner, 1967).

EVIDENCE FOR ENVIRONMENTAL CHANGE IN THE KALAHARI

Studies of environmental change in the Kalahari contrast markedly with those in surrounding areas because the nature of the available evidence has tended to make them dependent on geomorphological lines of evidence rather than archaeological and palynological sources (Thomas, 1987; Deacon & Lancaster, 1988). Since the seminal paper on the Kalahari by Grove (1969), which brought to attention suites of landforms that have since been viewed in a palaeoenvironmental context, investigations have attempted, not always successfully, to place evidence of environmental changes into a broader framework and chronology. Recent detailed reviews have been provided by Deacon & Lancaster (1988) and Thomas & Shaw (1991), while critiques of problems and misinterpretations are also available (Shaw & Cooke, 1986; Shaw et al., 1988; Shaw & Thomas, 1993).

Due to the lack of dates derived directly from sediments and controversy surrounding what vegetated desert sand dunes mean in terms of processes and environmental conditions (Thomas & Tsoar, 1990; Thomas & Shaw, 1991; Livingstone & Thomas, 1993), the dune systems of the Kalahari cannot currently be said to make a specific and chronologically controlled contribution to our understanding of Kalahari environmental change. More rigorous data, however, come from lacustrine and cave features in the Middle Kalahari.

Interpretation of the extensive suite of sediments and features associated with the Makgadikgadi Basin and its former extensions is now controlled by nearly 50 ^{14}C dates (e.g. Shaw & Cooke, 1986; Shaw et al., 1988). Additionally, Drotsky's Cave in north-western Botswana has yielded U/Th dates back to 300 000 yr BP (Brook et al., 1990) and 26 ^{14}C dates from speleothems and flowstones (Cooke, 1984). While Drotsky's Cave yields unequivocal evidence of more humid past climates, the Makgadikgadi data cannot be viewed solely as evidence of rainfall-induced high lake levels but requires interpretation in the light of subtle tectonic changes and complex hydrological links with perennial rivers such as the Zambezi (Cooke, 1984; Shaw & Thomas, 1988, 1993).

The movement away from explaining Kalahari palaeoenvironmental changes in terms of simple humid to arid climatic shifts has been accompanied by a wider awareness of the range of factors that can invoke changes. Of particular note in this regard is an increased realisation of the role that groundwater can play in the development of sediments and landforms in dryland environments, especially those such as the Kalahari where relief is low.

According to De Vries (1984), groundwater in the Kalahari is only undergoing significant recharge where local conditions are favourable, with the general aquifer depth and hydraulic gradient concomitant with major recharge ceasing at least 12 500 yr BP. Groundwater movements within the Kalahari are probably focused on sub-surface lineaments, which have been noted to be sites of preferential sub-surface weathering and duricrust development in association with pan depressions by Butterworth (1982), Farr *et al.* (1982) and Arad (1984). It is now becoming clearer that the widespread pan depressions of the southern Kalahari, previously seen as important sites for yielding simple palaeoclimatic information about precipitation changes (e.g. Lancaster, 1978), need to be viewed in a more complex hydrological sense (Lancaster, 1986); this has already occurred in the interpretation of similar features in Australia (Bowler, 1986; Torgersen *et al.*, 1986). Overall, the picture of changing environments that is emerging from the Kalahari is one of shifts in climatic parameters that were more subtle than once evoked (e.g. Klein *et al.*, 1991; Shaw & Thomas, 1993). As such, the development of so-called relict landforms is a consequence of a range of environmental parameters rather than being explainable just in terms of simple shifts in rainfall amounts.

ORIGINS OF KALAHARI VALLEY NETWORKS

If environmental changes in the Kalahari have not been as great as previously thought, then mechanisms in addition to simple fluvial erosion and wetter climatic regimes need to be considered as processes of valley development. From the above discussion, two possible origins for valleys are available, although it should be noted that they are not mutually exclusive:

1. Fluvial activity leading to gradual erosion and valley incision. This could be in the form of perennial rivers or high-magnitude/low-frequency flood events, the latter not necessarily invoking a major period of wetter climate.
2. Groundwater erosion processes leading to valley incision, initially by deep weathering along preferential flowpaths and, subsequently, by sapping processes (Shaw & De Vries, 1988).

The potential role of groundwater as an agent in valley formation is less well understood than incision by fluvial activity; Baker (1990) and Higgins (1984) provide useful summaries.

One of the earliest reports of groundwater sapping was by Peel (1941) from studies in the Gilf Kebir region of Libya. He described wadis with steep walls and flat floors which terminated abruptly with a cliff at their headward end and appeared to have been 'cut out from below' rather than 'let down from above'. Other terrestrial valley systems attributed to formation by groundwater sapping processes have been identified in the Colorado Plateau of the United States (e.g. Howard *et al.*, 1988; Laity & Malin, 1985; Laity, 1983; Pieri *et al.*, 1980), Hawaii (e.g. Baker, 1980; Kochel & Piper, 1986) and New Zealand (Schumm & Phillips, 1986). Additionally, through the use of terrestrial analogues, sapping has been suggested as the process responsible for many channels identified on Mars (e.g. Baker, 1980, 1982; Higgins, 1982; Sharp, 1973; Sharp & Malin, 1985). Groundwater erosion by deep-weathering along fractures has been implicated in the development of African dambos (see Boast, 1990, for a review).

From these various studies, a number of morphological features typical of valleys developed by groundwater sapping can be identified (Howard *et al.*, 1988; Baker, 1990). These include abrupt 'amphitheatre' valley heads, alcoves and springs in headward areas, steep valley flanks and a flat floor, an elongate basin shape with a low drainage density,

30

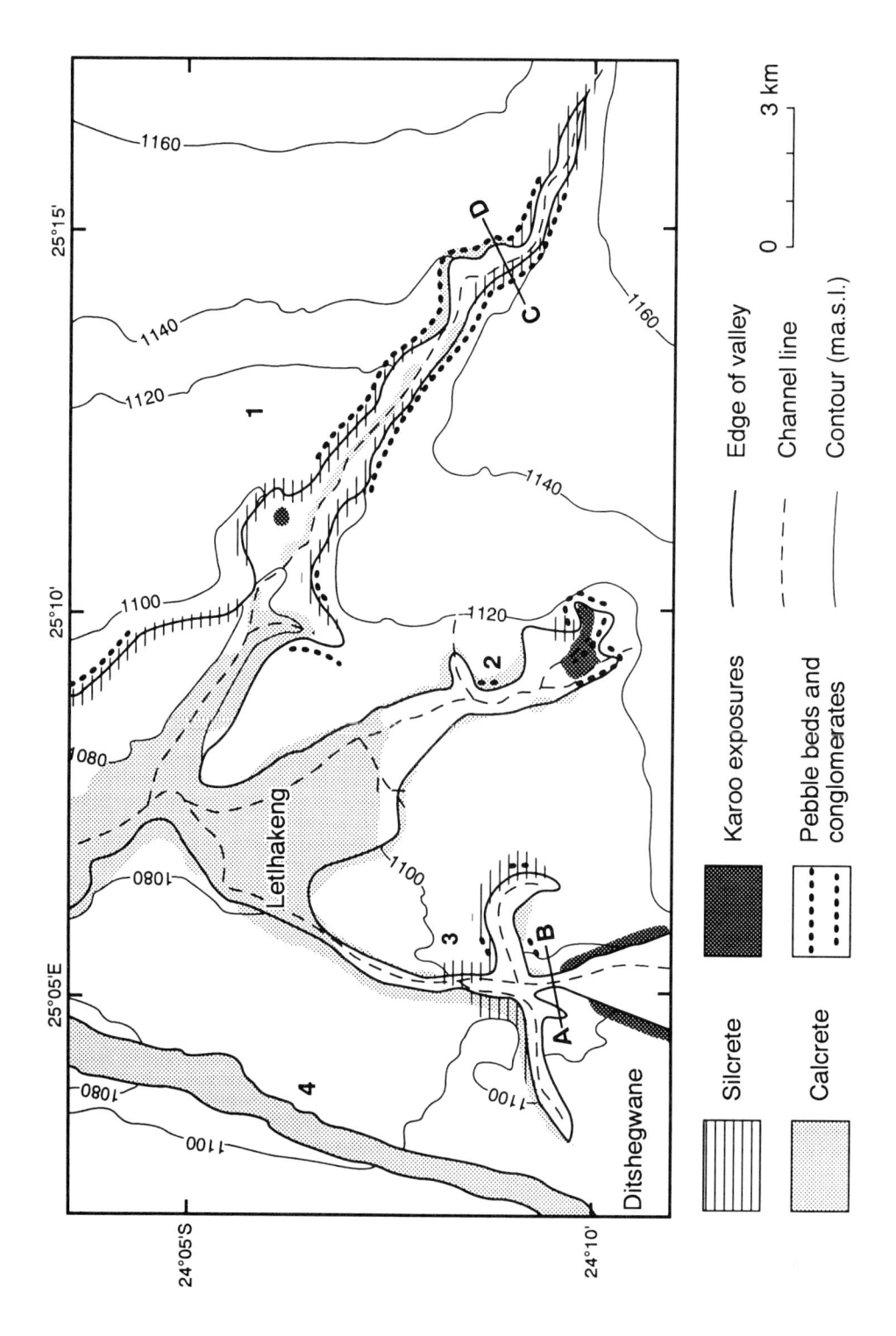

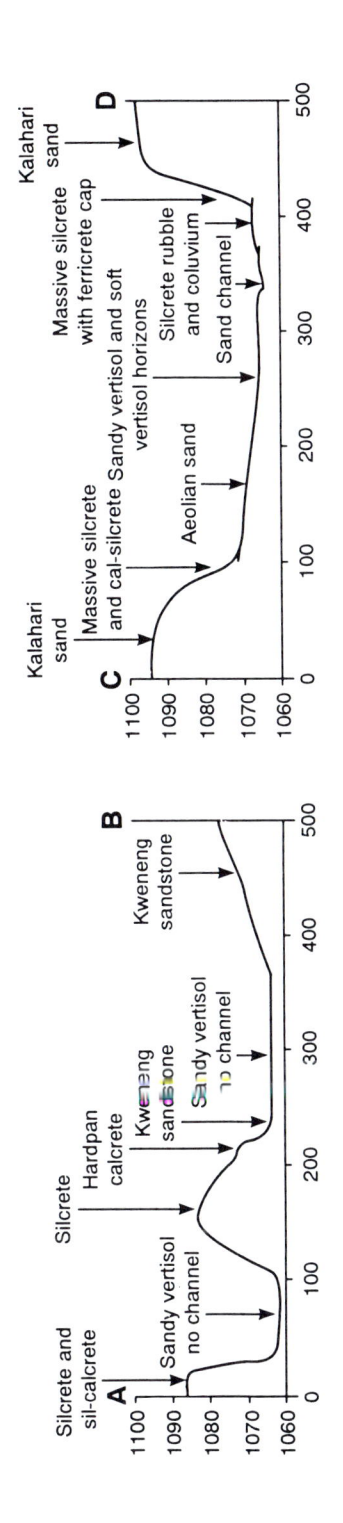

Figure 2.2 Valleys in the vicinity of Lethakeng, Kweneng District, Botswana, showing valley morphology and surface materials. (Modified from Shaw and De Vries, 1988. Reproduced by permission of Academic Press (London) Ltd)

short first-order tributaries entering a long main valley, tributary asymmetry and possible hanging valleys.

The environmental significance of groundwater sapping and deep-weathering processes is that they do not necessarily require increased moisture availability in a particular area in order to operate. Indeed, in the case of sapping processes, Howard *et al.* (1988) suggest that greater rainfall and groundwater outflow may not lead to an increased rate of erosion. Enhanced groundwater outflow would hinder the accumulation of minerals and salts which, through the operation of heave processes, act as important mechanical weathering agents. McFarlane (1989) further proposes that deep-weathering processes (chemical and biochemical corrosion) operate in the development of dambos during periods of little or no fluvial activity within the valleys. In general, sub-surface water movement along preferential flowpaths and groundwater emergence are more closely linked to bedrock permeability and water table height fluctuations (Higgins, 1984); these can be tectonically as well as climatically controlled. In the case of the Kalahari Basin, inflows of groundwater from areas around the Kalahari rim are equally important (Farr *et al.*, 1981). It is therefore possible that enhanced groundwater emergence or flow could arise as a result of increased rainfall from areas beyond the Kalahari.

EVIDENCE FOR VALLEY ORIGINS

Kalahari valley systems exhibit a variety of features indicative of their mode of origin, which can be separated into 'valley' and 'intra-valley' forms, the latter including all features derived from sedimentary deposition and erosion. This subdivision enables different elements of valley morphology to be viewed independently, intra-valley forms being considered separately from the overall form of the valley.

Evidence for Fluvial Action

Intra-valley forms are most indicative of the role of fluvial activity in valley development. They include features such as valley terraces (e.g. Crockett & Jennings, 1965), evidence of former channels and lag deposits. Terraces are the most frequently described elements of valley morphology in studies of Kalahari valleys, primarily because terrace sediments often include material such as shell suitable for absolute dating. As a result sites containing datable material have often been studied at the expense of considering the entire valley form.

Whilst fairly widespread in exoreic valleys such as the Molopo, Nossop and Auob, sedimentary evidence for former flow is lacking for most endoreic valleys. An exception occurs in the Okwa Valley near to the Tswaane borehole where two areas of lag gravel point bar deposits, containing clasts up to 40 cm in diameter, are found on the inside of bends in the valley. Whether these can be considered indicative of perennial flow or high-magnitude flooding is uncertain.

Other features indicating former flow include relict meandering channels only identifiable using remotely sensed imagery. One example is shown in Figure 2.3 for a section of the Nossop Valley, where a number of abandoned channels are evident from aerial photographs. Similar examples of abandoned channels occur within the Xaudum and Ncamasere valleys near the Botswana–Namibia border. In these three cases, the abandoned channels were located between flanks of an indistinct wider valley, approximately 4 km wide in the case

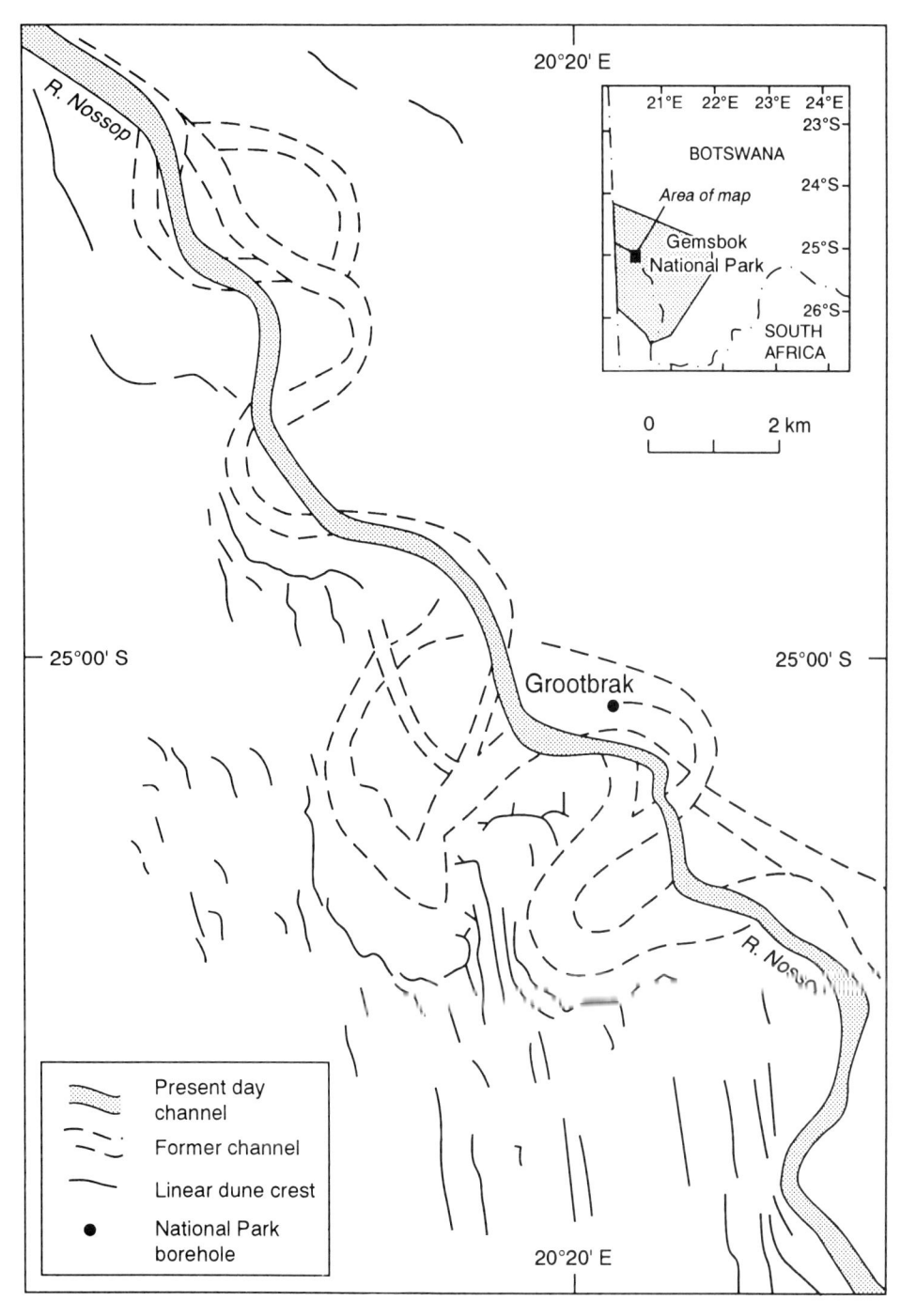

Figure 2.3 Abandoned meander channels in the Nossop Valley, Kalahari Gemsbok National Park

of the Nossop example, and 1.5 and 1 km for the Xaudum and Ncamasere respectively. In parts of the Okwa and Auob valleys, incised meanders occur as opposed to the intra-valley meandering described above.

Evidence for the Role of Groundwater and Structural Control

The examples outlined above are clear evidence for fluvial activity in Kalahari valley development. However, there are a number of features of Kalahari valleys that cannot be easily explained solely by the presence of rivers. These features are mainly concerned with the overall valley form and not the intra-valley features already discussed.

Considering networks as a whole, a close alignment between valley orientations and geological structures such as fractures and faults has been noted in previous studies (Coates *et al.*, 1979; Mallick *et al.*, 1981). In areas where bedrock outcrops or is close to the surface this could be a direct result of geological control of channel location. However, as noted above, most Kalahari dry valleys are located within the area covered by Kalahari Group sediments, with over 150 m of sediment overlying bedrock in places. Fractures within pre-Kalahari bedrock are known to be important aquifers and act as preferential flowpaths allowing groundwater transmission (Buckley & Zeil, 1984). In many cases, even where these faults have not been propagated into the Kalahari Group by recent movements along old faultlines, they are still visible on remotely sensed imagery. This has been attributed to the effects of buried faults upon groundwater circulation in overlying sediments which encourages vegetational growth above the fault zone (Mallick *et al.*, 1981). Such sub-surface flow is also an important agent in subterranean deep weathering, as noted from studies of lithological borehole logs (e.g. Von Hoyer *et al.*, 1985).

In order to assess the relationship between valley orientation and the alignment of geological structures a technique of network orientation analysis was used, based upon a method by Abdel-Rahman & Hay (1981). The results of this analysis (data and methodology discussed elsewhere; Nash, 1992) show a strong correlation between valley and structural orientation for the Okwa Valley and its tributaries, and also for parts of the Mmone/Quoxo system. For example, the Dikgonnyane Valley, part of the Mmone/Quoxo system (Figure 2.4), shows particularly close alignment between lineaments and valley orientation, with the valley following a faultline for 35 km despite the fault being covered by 15–25 m of Kalahari Group sediments. Other systems showed a lack of parallelism, notably the valleys of the Moselebe system and those in north-western Botswana, which apparently relates to variations in the thickness of Kalahari Group sediments. In the case of the Okwa and Mmone valleys, sediment thicknesses beneath valleys were in the range 20–30 m whilst exceeding 40 m for the Moselebe Valley area.

At the network scale, the pattern of certain valley systems shows many of the features indicative of groundwater sapping outlined above. Most notable of these are the elongate nature of networks such as the Okwa (Figure 2.1), where widely spaced sub-parallel first-order tributaries enter the main valley, with the majority of tributaries entering from the south.

Another feature of Kalahari valleys that cannot be easily explained by fluvial erosion is the variation in form previously mentioned, particularly the gorge-like section present in many valleys. The characteristically broad and flat headwater form gives way, in many cases abruptly, to an incised, steep-sided flat-floored valley. Such changes are not coincident with changes in geological formation and valley floors show no evidence of gradient

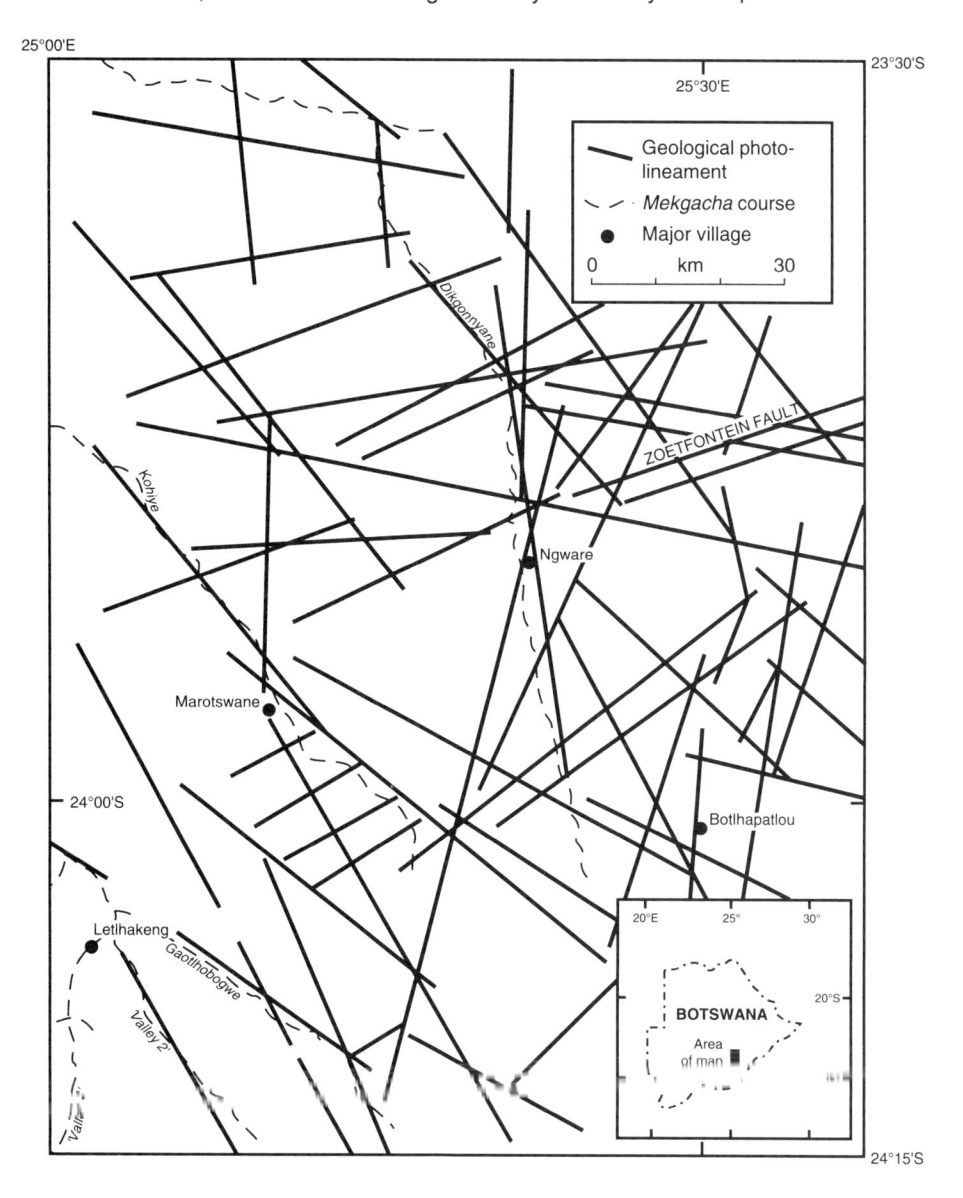

Figure 2.4 Relationships between dry valleys and geological photo-lineaments in the south-east Kalahari, Botswana

changes. The change in form is best seen in the vicinity of Letlhakeng village in Botswana (Figures 2.2 and 2.4) where the form of the Gaotlhobogwe Valley (Letlhakeng Valley 1) changes abruptly with an amphitheatre-type valley head at the transition point (Shaw & De Vries, 1988). All of these features are typical of valleys developed by sapping in other semi-arid environments (Baker, 1990; Howard et al., 1988). Relict spring lines and associated silcrete deposits are found in the valley head area, with similar silcrete deposits also occurring in probable spring sites in the Serorome Valley (Thomas & Shaw, 1991). Inspection of these silcretes in thin section, combined with geochemical analyses (Nash, 1992), suggests that they have developed as a result of the long-term focusing of silica-rich groundwater upon the valley head area and that their formation is intrinsically linked to the presence of the valley.

TIMESCALES AND VALLEY DEVELOPMENT

Fluvial activity and groundwater erosion in valley development operate at distinctly different timescales and as such have different environmental significance. From the above discussion it appears that groundwater sapping and deep-weathering processes may have been largely responsible for the location of valleys and for the shape of many valley networks. Once a zone of weakness has developed due to deep-weathering along faultlines, then this may act as an initial focus for valley development. In the case of valleys such as those near Letlhakeng, sapping has also controlled valley form (Shaw & De Vries, 1988). Whilst groundwater processes may be responsible for initiating valley development, the evidence of meanders and lag deposits would suggest a later dominance by fluvial activity in shaping valley form in many systems.

Valley development over potentially long periods must be viewed in context with other environmental changes, both climatic and in terms of tectonic setting. Geological structures acting as preferential flowpaths for groundwater in many cases pre-date the onset of deposition of the Kalahari Group sediments and are at least Mesozoic in age (Thomas & Shaw, 1991). Superimposed upon this ancient fault pattern in the northern Kalahari are north-east to south-west trending faults associated with neotectonic activity and rifting in the Okavango Delta region (Reeves, 1972).

Uplift along the Kalahari–Zimbabwe Axis since the mid-Tertiary (Du Toit, 1933) has also caused changes in land elevation along the eastern Kalahari margin and has been proposed as a mechanism for initiating groundwater sapping in valleys in the south-east Kalahari (Shaw & De Vries, 1988). There are also many examples of the influence of uplift upon drainage in southern Africa, with drainage patterns greatly disrupted and river capture evident in the evolution of the Zambezi system (Thomas & Shaw, 1988). A former link between the Okavango River and the Limpopo via the Makgadikgadi Depression was suggested by Du Toit (1933), with downwarping and uplift causing separation of the two systems. Capture of the headwaters of endoreic rivers by more aggressive exoreic ephemeral channels may also be a mechanism that deprived central Kalahari valleys of flow (Thomas & Shaw, 1991).

Tectonic activity has also influenced former lake levels within the Makgadikgadi Depression and the shape of the Okavango Delta (Cooke, 1980; Mallick *et al.*, 1981). For example, in Makgadikgadi, the position of deltaic sediments associated with the perennial Boteti River indicate a gradual downwarping to the south (Cooke & Verstappen, 1984). As the endoreic Kalahari dry valley systems are directed towards these areas, it is likely that their form at least partly reflects such activity. In addition to the longer-term effects of tectonic activity upon base levels, the terrace levels within the Okwa Valley reflect a number of other factors. In particular, climatic changes are of significance, both in influencing lake levels in Makgadikgadi and in causing shifts in the spatial pattern of rainfall distribution (Cooke & Verstappen, 1984).

In broad terms the influences upon valley development can be viewed at three scales. Firstly, factors influencing intra-valley form are comparatively recent climatic and neotectonic changes, including contemporary floods operating at scales between 10^1 and 10^5 years. At an intermediate scale, neotectonic changes and drainage capture act upon valley location, gradient and possible supply of water, whilst in the long term (10^7–10^8 years) geological structural lineaments are likely to have influenced valley location by deep-weathering processes.

An example of a valley system with evidence for both possible groundwater erosion and fluvial activity is the Kuruman River (Figure 2.5). The Kuruman is intermediate between a perennial channel and the dry valleys of the Kalahari, being fed by springs in the Kuruman Hills and thus containing flow along at least part of its length on an annual basis (Thomas & Shaw, 1991). At the location shown in Figure 2.5, the Kuruman Valley is 90 m wide and 11 m deep on average, the upper valley flanks consisting of highly indurated cal-silcretes

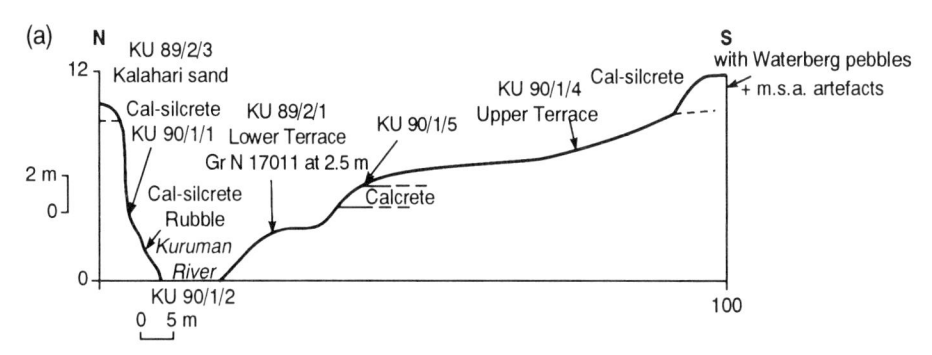

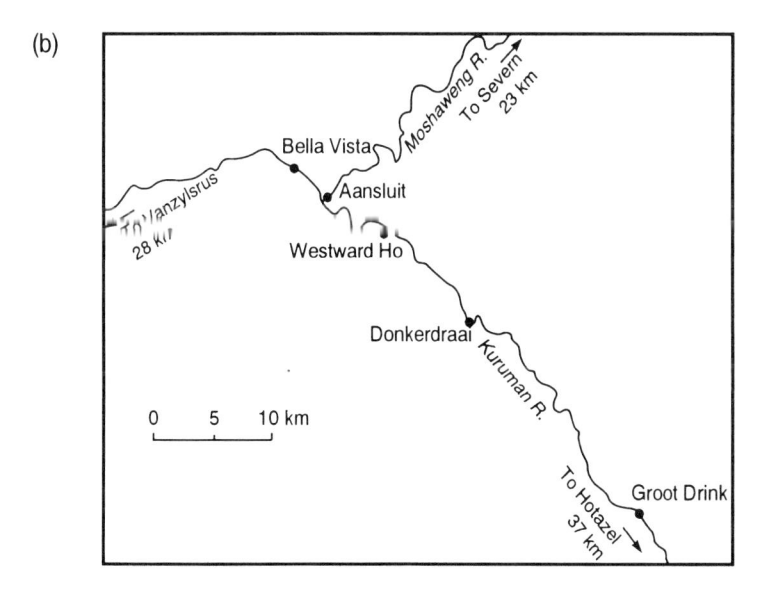

Figure 2.5 (a) Cross-section of the Kuruman Valley at Groot Drink, Northern Cape Province, RSA, indicating terrace formations and surface materials; (b) locations along the Kuruman Valley

with partly calcified intra-valley sediments (Shaw *et al.*, 1992). Two terrace levels are identifiable, the lower terrace containing the freshwater gastropod *Lymnaea* sp. yielding a radiocarbon date of 320 ± 150 yr BP (sample GrN-17011). Other terrace levels along the Kuruman and Moshaweng Rivers indicate formation related to Late Holocene floods (Shaw *et al.*, 1992). In contrast, Middle Stone Age artefacts rest upon the cal-silcretes on the southern flank of the valley, indicating a significantly older age for the valley flanks than the terrace formations.

The Kuruman exemplifies the arguments outlined in this paper, with the overall valley form most likely the product of a combination of deep-weathering and fluvial processes operating over timescales of millions of years. Within this larger valley, terrace sediments have been deposited in at least two separate periods of perennial fluvial activity or flash-flooding. However, unlike other valley systems, flood events in the spring-fed Kuruman occur to the present day, the latest major flood occurring in 1988–9.

CONCLUSIONS

The traditional view that the development of large Kalahari valley systems occurred entirely during former wetter periods (e.g. Heine, 1982) may be an oversimplification of a complex series of processes. These processes act at a variety of timescales, potentially ranging from the Late Tertiary to the present day. If groundwater sapping was (or indeed *is*) a significant factor in valley development, then less recourse to major climatic change is required to explain the origin of valleys. Sapping processes would operate over longer timescales and, if connected with regional uplift raising levels of watertables, would not require great inputs of water. This appears to be the case for valleys in eastern Botswana near the Kalahari–Zimbabwe Axis of uplift. Evidence of terraces and meandering channels in valley systems indicates past fluvial activity, probably connected with more recent climatic changes but also influenced by regional uplift. Increasing aridity since valley formation may be indicated by the sand-choked channels of the central Kalahari, although loss of water due to headwater capture by exoreic rivers is another possibility. However, the notion that major 'fossil' valley systems in semi-arid environments simply indicate recent environmental changes may need to be reassessed; the role of other, less easily determined factors may be equally important.

ACKNOWLEDGEMENTS

The authors are grateful to the British Geomorphological Research Group and the University of Botswana for providing funding for ^{14}C dating. Dating was carried out by Dr W. G. Mook at the University of Groningen. Fieldwork was funded by the University of Sheffield Linley Scholarship, Palmers College and the Explorers Club (D.J.N.), the Royal Society (D.J.N., D.S.G.T.), and the University of Botswana (P.A.S.). Fieldwork was also undertaken in conjunction with the 1989 and 1990 'Sheffield University Botswana Expeditions', funded by the Royal Geographical Society, Gilchrist Educational Trust and the Exploration Fund of the Explorer's Club, New York. Original figures were drawn by Paul Coles and Graham Allsopp, Department of Geography, University of Sheffield.

REFERENCES

Abdel-Rahman, M. A. and Hay, A. M. (1981). Statistical analysis of multi-modal orientation data. *Proceedings of the 3rd International Conference on Basement Tectonics, Durango, Colorado*, Volume 3, Basement Tectonics Committee Publications, pp. 73–86.

Andersson, C. J. (1856). *Lake Ngami; or Explorations and Discoveries during Four Years Wanderings in the Wilds of South Western Africa*, Hurst and Blackett, London; Reprinted 1967, C. Struik, Cape Town.

Arad, A. (1984). Relationship of salinity of groundwater to recharge in the southern Kalahari Desert. *Journal of Hydrology*, **71**, 225–38.

Baker, V. R. (1980). Some terrestrial analogs to dry valley systems on Mars. NASA Technical Memo TM-81776, pp. 286–8.

Baker, V. R. (1982). *The Channels of Mars*, University of Texas Press, Austin.

Baker, V. R. (1990). Spring-sapping and valley network development. In Higgins, C. G. and Coates, D. R. (eds.), *Groundwater Geomorphology; The Role of Subsurface Water in Earth-Surface Processes and Landforms*, Geological Society of America Special Paper 252, Boulder, pp. 235–65.

Boast, R. (1990). Dambos: a review. *Progress in Physical Geography*, **14**, 153–77.

Boocock, C. and Van Straten, O. J. (1962). Notes on the geology and hydrology of the central Kalahari region, Bechuanaland Protectorate. *Transactions of the Geological Society of South Africa*, **65**, 125–71.

Bowler, J. M. (1986). Spatial variability and hydrologic evolution of Australian lake basins: analogue for Pleistocene hydrologic change and evaporite formation. *Palaeogeography, Palaeoclimatology, Palaeoecology*, **54**, 21–41.

Brook, G. A., Burney, D. A. and Cowart, J. B. (1990). Desert palaeoenvironmental data from cave speleotherms with examples from the Chihuahuan, Somali-Chalbi and Kalahari deserts. *Palaeogeography, Palaeoclimatology, Palaeoecology*, **76**, 311–29.

Buckley, D. K. and Zeil, P. (1984). The character of fractured rock aquifers in eastern Botswana. In *Challenges in African Hydrology and Water Resources (Proceedings of the Harare Symposium July 1984)*, International Association of Hydrological Sciences Publication 144, Wallingford, pp. 25–36.

Butterworth, J. S. (1982). The chemistry of Mogatse Pan, Kgalagadi District. Botswana Geological Survey unpublished report JSB/14/82, Lobatse.

Coates, J. N. M., Davies, J., Gould, D., *et al.* (1979). The Kalatraverse, Report 1'. *Botswana Geological Survey Bulletin*, **21**, Gaborone.

Cooke, H. J. (1980). Landform evolution in the context of climatic change and neotectonism in the middle Kalahari of north-central Botswana. *Transactions of the Institute of British Geographers (NS)*, **5**, 80–99.

Cooke, H. J. (1984). The evidence from northern Botswana of climatic change. In Vogel, J. (ed.), *Late Cainozoic Paleoclimates of the Southern Hemisphere*, A. A. Balkema, Rotterdam, pp. 265–78.

Cooke, H. J. and Verstappen, H. Th. (1984). Landforms of the western Makgadikgadi basin in northern Botswana, with a consideration of the chronology of the evolution of lake Palaeo-Makgadikgadi. *Zeitschrift für Geomorphologie*, **28**, 80–99.

Crockett, R. N. and Jennings, C. M. H. (1965). Geology of part of the Okwa Valley, western Bechuanaland. In *Records of the Geological Survey of Bechuanaland 1961–62*, Bechuanaland Geological Survey, Gaborone, pp. 101–13.

Deacon, J. and Lancaster, N. (1988). *Late Quaternary Palaeoenvironments of Southern Africa*, Oxford University Press, Oxford.

De Vries, J. J. (1984). Holocene depletion and active recharge of the Kalahari groundwaters — a review and an indicative model. *Journal of Hydrology*, **70**, 221–32.

Du Toit, A. L. (1933). Crustal movement as a factor in the geographical evolution of South Africa. *South African Geographical Journal*, **16**, 3–20.

Farr, J. L., Cheney, C. S., Baron, J. H. and Peart, R. J. (1981). Evaluation of underground water resources. GS10 Project final report, Botswana Geological Survey, Gaborone.

Farr, J. L., Peart, R. J., Nelisse, C. and Butterworth, J. (1982). Two Kalahari Pans: a study of their morphometry and evolution. Botswana Geological Survey unpublished report GS10/10, Lobatse.

Grove, A. T. (1969). Landforms and climatic change in the Kalahari and Ngamiland. *Geographical Journal*, **135**, 191–212.
Heine, K. (1982). The main stages of the Late Quaternary evolution of the Kalahari region, southern Africa. *Palaeoecology of Africa*, **15**, 53–76.
Higgins, C. G. (1982). Drainage systems developed by sapping on Earth and Mars. *Geology*, **10**, 147–52.
Higgins, C. G. (1984). Piping and sapping: development of landforms by groundwater outflow. In la Fleur, R. G. (ed.), *Groundwater as a Geomorphic Agent*, Allen and Unwin, London, pp. 18–58.
Howard, A. D., Kochel, R. C. and Holt, H. E. (1988). Sapping features of the Colorado Plateau — a comparative planetary geology fieldguide. NASA Publication SP-491.
Klein, R. G., Cruz-Uribe, K. and Beaumont, P. B. (1991). Environmental, ecological and paleoanthropological implications of the late Pleistocene mammalian fauna from Equus Cave, Northern Cape Province, South Africa. *Quaternary Research*, **36**, 94–119.
Kochel, R. C. and Piper, J. F. (1986). Morphology of large valleys on Hawaii: evidence for groundwater sapping and comparisons with Martian valleys. *Journal of Geophysical Research*, **91**(b13), e175–92.
Laity, J. E. (1983). Diagenetic controls on groundwater sapping and valley formation, Colorado Plateau, as revealed by optical and electron microscopy. *Physical Geography*, **4**, 103–25.
Laity, J. E. and Malin, M. C. (1985). Sapping processes and the development of theater-headed valley networks in the Colorado Plateau. *Geological Society of America Bulletin*, **96**, 203–17.
Lancaster, I. N. (1978). The pans of the southern Kalahari, Botswana. *Geographical Journal*, **144**, 80–98.
Lancaster, N. (1986). Pans in the southwestern Kalahari: a preliminary report. *Palaeoecology of Africa*, **17**, 59–67.
Leistner, O. A. (1967). The plant ecology of the southern Kalahari. Botanical Research Institute Botanical Memoir 38, Republic of South Africa Department of Agricultural Technical Services, Pretoria.
Livingstone, I. and Thomas, D. S. G. (1993). Models of linear dune activity and their palaeoenvironmental significance: an evaluation with reference to southern African examples. In Pye, K. (ed.), *The Dynamics and Environmental Context of Aeolian Sedimentary Systems*, Geological Society Special Publication, London, **72**, 91–101.
McFarlane, M. J. (1989). Dambos — their characteristics and geomorphological evolution in parts of Malawi and Zimbabwe, with particular reference to their role in the hydrogeological regime of surviving areas of African surface. *Proceedings of the Groundwater Exploration and Development in Crystalline Basement Aquifers Workshop, Harare, Zimbabwe*, Volume 1 (Session 3), Commonwealth Science Council, pp. 254–308.
Mallick, D. I. J., Habgood, F. and Skinner, A. C. (1981). A geological interpretation of Landsat imagery and air photography of Botswana. In *Institute of Geological Sciences, Overseas Geology and Mineral Resources*, Volume 56, HMSO, London.
Mazor, E., Verhagen, B. Th., Sellschop, J. P. F., et al. (1977). Northern Kalahari groundwaters: hydrologic, isotopic and chemical studies at Orapa, Botswana. *Journal of Hydrology*, **34**, 203–34.
Moore, A. E. (1988). Plant distribution and the evolution of the major river systems in southern Africa. *South African Journal of Geology*, **91**, 346–9.
Nash, D. J. (1992). The development and environmental significance of the dry valley systems (*mekgacha*) in the Kalahari, central southern Africa. Unpublished PhD thesis, University of Sheffield.
Peel, R. F. (1941). Denudational landforms of the central Libyan desert. *Journal of Geomorphology*, **4**, 3–23.
Pieri, D. C., Malin, M.C. and Laity, J. E. (1980). Sapping: network structure in terrestial and Martian valleys. NASA Technical Memo TM-81979.
Reeves, C. V. (1972). Evidence of rifting in the Kalahari. *Nature*, **237**, 96.
Reid, I. and Frostick, L. E. (1989). Channel form, flows and sediments in deserts. In Thomas, D. S. G. (ed.), *Arid Zone Geomorphology*, Belhaven/Halstead, London, pp. 117–35.
Schumm, S. A. and Phillips, L. (1986). Composite channels of the Canterbury Plains, New Zealand: a Martian analogue. *Geology*, **14**, 326–30.
Sharp, R. P. (1973). Mars: fretted and chaotic terrains. *Journal of Geophysical Research*, **78**, 4063–72.
Sharp, R. P. and Malin, M. C. (1985). Channels on Mars. *Geological Society of America Bulletin*,

86, 593–609.
Shaw, P. A. and Cooke, H. J. (1986). Geomorphic evidence for the late Quaternary palaeoclimates of the middle Kalahari of northern Botswana. *Catena*, **13**, 349–59.
Shaw, P. A. and De Vries, J. J. (1988). Duricrust, groundwater and valley development in the Kalahari of south-east Botswana. *Journal of Arid Environments*, **14**, 245–54.
Shaw, P. A. and Thomas, D. S. G. (1988). Lake Caprivi: a late Quaternary link between the Zambezi and Middle Kalahari drainage systems. *Zeitschrift für Geomorphologie*, **NF32**, 329–37.
Shaw, P. A. and Thomas, D. S. G. (1993). Geomorphological processes, environmental change and landscape sensitivity in the Kalahari region of southern Africa. In Thomas, D. S. G. and Allison, R. J. (eds), *Landscape Sensitivity*, John Wiley, Chichester, 83–95.
Shaw, P. A., Cooke, H. J. and Thomas, D. S. G (1988). Recent advances in the study of Quaternary landforms in Botswana. *Palaeoecology of Africa*, **19**, 15–26.
Shaw, P. A., Thomas, D. S. G. and Nash, D. J. (1992). Late Quaternary fluvial activity in the dry valleys (*mekgacha*) of the Middle and Southern Kalahari, southern Africa. *Journal of Quaternary Science*, **7**, 273–281.
Small, R. J. (1964). The escarpment dry valleys of the Wiltshire Chalk. *Transactions of the Institute of British Geographers*, **34**, 33–52.
Sparks, B. W. and Lewis W. V. (1957-8). Escarpment dry valleys near Pegsdon, Hertfordshire. *Proceedings of the Geological Association*, **68**, 26–38.
Thomas, D. S. G. (1987). Research strategies and methods for Quaternary science: the case of southern Africa. School of Geography Research Paper, **39**, University of Oxford.
Thomas, D. S. G. (1988). The nature and depositional setting of arid and semi-arid Kalahari sediments, southern Africa. *Journal of Arid Environments*, **14**, 17–26.
Thomas, D. S. G. and Shaw, P. A. (1988). Late Cainozoic drainage evolution in the Zambezi Basin: geomorphological evidence from the Kalahari rim. *Journal of African Earth Sciences*, **7**, 611–18.
Thomas, D. S. G. and Shaw, P. A. (1991). *The Kalahari Environment*, Cambridge University Press, Cambridge.
Thomas, D. S. G. and Tsoar, H. (1990). The geomorphological role of vegetation in desert dune systems, In Thomas, J. B. (ed.), *Vegetation and Erosion*, Wiley, Chichester, pp. 471–89.
Torgersen, T., De Deckker, P., Chivas, A. R. and Bowler, A. M. (1986). Salt lakes: a discussion of processes influencing palaeoenvironmental interpretations and recommendations for future study. *Palaeogeography, Palaeoclimatology, Palaeoecology*, **54**, 7–19.
Von Hoyer, M., Keller, S. and Rehder, S. (1985). Core borehole Letlhakeng 1. Botswana Geological Survey unpublished report MVH/4/85, Lobatse.

3 Mineral Magnetic Analysis of Iron Oxides in Arid Zone Soils from the Tunisian Southern Atlas

K. WHITE
Department of Geography, University of Reading, UK

and

J. WALDEN
School of Geography, University of Oxford, UK

ABSTRACT

Mineral magnetic analysis was used to study the concentration and type of iron oxides in soils developed on geomorphological surfaces of two telescopically segmented alluvial fans in the Tunisian Southern Atlas. The mineral magnetic data, supported by other analyses, indicates a buildup of iron oxides in soils over time, with older fan segments having greater concentrations of magnetic minerals than younger segments. It had previously been thought that haematite was the main iron oxide mineral in arid zone soils, but these data suggest that the situation in Tunisia is more complex. The mineral magnetic approach has potential for establishing relative chronosequences of alluvial surfaces in arid areas and outperforms traditional XRD and chemical techniques.

INTRODUCTION

During the last 20 years, a rapid growth has occurred in the application of mineral magnetic measurements to study environmental processes (Thompson & Oldfield, 1986). Although the approach cannot be seen as fully quantitative in the same way as more traditional petrological or geochemical techniques, mineral magnetics can offer a number of advantages, not only in terms of its speed, cost and non-destructive nature, but also in terms of its great sensitivity (Maher, 1986; Oldfield, 1991a). This paper assesses the use of mineral magnetic analysis for studying the iron oxide mineralogy in arid zone soils. The purposes of this work are twofold:

(i) to see if mineral magnetic analysis can be used to establish an ordinal chronology of alluvial fan surfaces in the Tunisian Southern Atlas; and
(ii) to see if mineral magnetic measurements are more sensitive than traditional techniques for analysing pedogenic iron oxides.

Environmental Change in Drylands: Biogeographical and Geomorphological Perspectives.
Edited by A. C. Millington and K. Pye. © 1994 John Wiley & Sons Ltd

Problems of dating geomorphological surfaces in arid areas arise mainly from the lack of organic matter from which to derive ^{14}C dates, and even where such material is found, it is rarely in stratigraphically significant positions (Crook, 1986). In such situations it is often more appropriate to establish the chronological order of formation of geomorphological surfaces with certainty, rather than attempt to derive uncertain age estimates (Coleman, 1986). A variety of techniques have been employed to this end and are reviewed elsewhere (Lowe & Walker, 1984, pp. 267–75).

Several techniques have been developed based on the degree of pedogenesis on different geomorphological surfaces. For example, thickness of soils has been used to provide estimates of relative ages of alluvial fan surfaces in humid tropical areas. These estimates were then calibrated with ^{14}C dates from buried peat deposits (Kesel, 1985). However, the pedological profiles on fan surfaces in the Tunisian Southern Atlas are characterised by poorly developed aridisols. Furthermore, soil development can vary between cover types, such as stone pavements and gypsum crusts, and is dependent on particle size and gypsum content.

Quantitative and precise measurements of some parameter that changes monotonically with time are clearly of greater potential. One approach has used clast sound velocity (Crook & Gillespie, 1986), which is thought to be applicable for up to 10^6 years BP. However, the equipment is bulky and the method is complicated. Measurements of dynamic Young's modulus of elasticity can be used to assess the degree of clast weathering and has been shown to be more accurate than the Schmitt hammer (Allison, 1988, 1990). Rock varnish has also been employed in some arid areas (Dorn, 1983; Dorn *et al.*, 1986, 1987, 1989).

Pedogenic iron oxide mineralogy has been used to establish an ordinal chronology of river terraces in Spain (Diaz & Torrent, 1989; Torrent *et al.*, 1980). Under certain circumstances, iron oxide concentration and mineralogy can provide information about the period of weathering that a soil on a river terrace has undergone, assuming the original alluvial material had similar amounts of primary iron-bearing minerals. The reliability of such an ordinal chronology is affected by:

(i) the climate, with soils in more arid areas seeming to provide more reliable results;
(ii) the soil composition, implying that this needs to be held constant for comparison of different drainage basins; and
(iii) the soil pedoclimate, related to the above and to the particle size distribution.

Again, these factors need to be held constant for comparative work.

STUDY AREA

The study area for this work is the Tunisian Southern Atlas, bounded to the north by the Atlas Mountains and to the south by the Zone of Chotts, large lowland depositional basins occupied by salinas or salt playas (Figure 3.1). The area lies in the morphostructural province of the Pre-Sahara (Coque & Jauzein, 1967) and consists of a series of east–west trending anticlines and synclines resulting from movements of the late Alpine orogeny (Castany, 1952). The drainage basins are underlain by Cretaceous and Tertiary carbonates, sandstones, argillites and evaporites. The lowland depositional basins contain Quaternary alluvial fans, nested pediments (known locally as *glacis d'erosion*), dunes, alluvial plains, crusts and

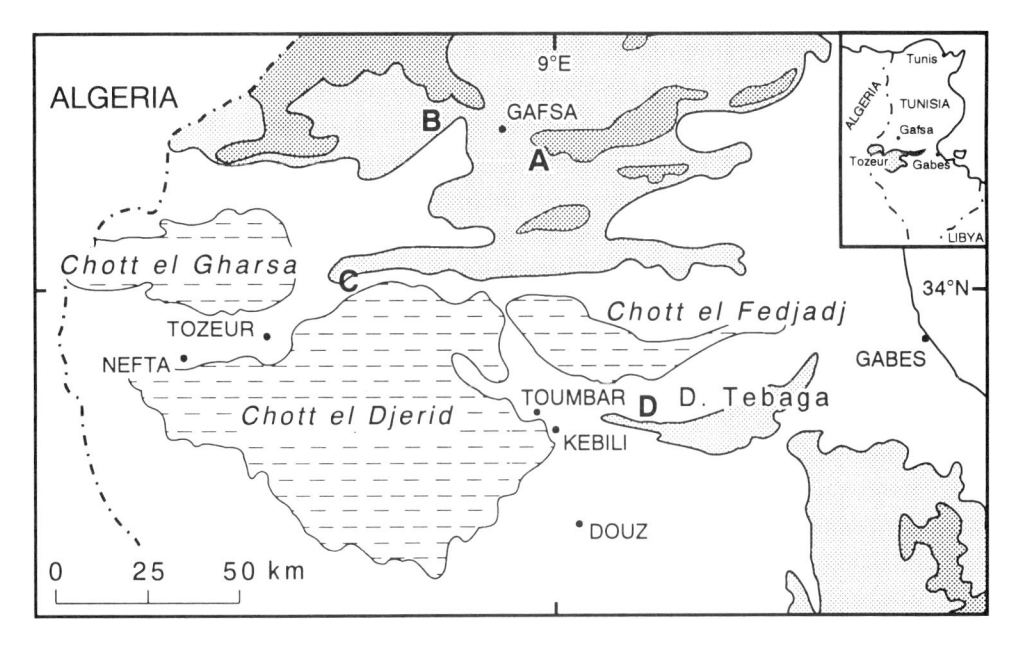

Figure 3.1 Location map, showing the positions of the Oued es Seffaia alluvial fan (A), the Oued et Tfal alluvial fan (B), the Sidi bou Helal pediment (C) and the Djebel Tebaga pediment (D)

salina/playa deposits (Burollet, 1967). The area experiences approximately 100–150 mm of rainfall per annum, with peaks in the spring and autumn (Rognon, 1987).

Two alluvial fans (the Oued es Seffaia and Oued et Tfal fans) were selected to test the mineral magnetic approach, as they each have a series of telescopic fan segments interpreted as being of different ages (Coque, 1962). Telescopic segmentation occurs when the locus of deposition on a fan moves away from the mountain front, and results in a series of nested slope segments, successively lower segments having a lower gradient than upper segments, so that the surfaces converge distally (Blissenbach, 1954). Both of these fans have similar drainage basin lithologies, being derived from Cretaceous dolomitic limestones and marls. The telescopic dissection of these fans is thought to be due to late Quaternary climatic fluctuations (White, 1991), but such models assume that lower fan segments are younger than the upper segments. However, workers looking at fans in south-west United States have shown that, in certain tectonic situations, aggradation at the fanhead can result in the upper segments being younger than the lower segments (Hooke, 1967). Thus it is important to establish the relative chronologies of the fan segments in order to interpret the geomorphological evidence for climatic change.

The Oued es Seffaia alluvial fan lies 13 km east of Gafsa, on the southern flank of Djebel Orbata (White, 1987), and covers an area of 644.3 ha. It has three distinct geomorphological surfaces (Figure 3.2): an upper fan segment, an intermediate fan segment and a depositional wash inset within this. Figure 3.3 shows the break of slope between the upper and intermediate segments. These segments, both characterised by stone pavement cover, are thought to date from the Mousterian (70–30 Ka BP) and Capsian (6–4 Ka BP) respectively, with the depositional wash being composed of contemporary alluvium (Coque, 1962, pp. 422–3). However, there is little independent evidence to support these dates.

46   Environmental Change in Drylands

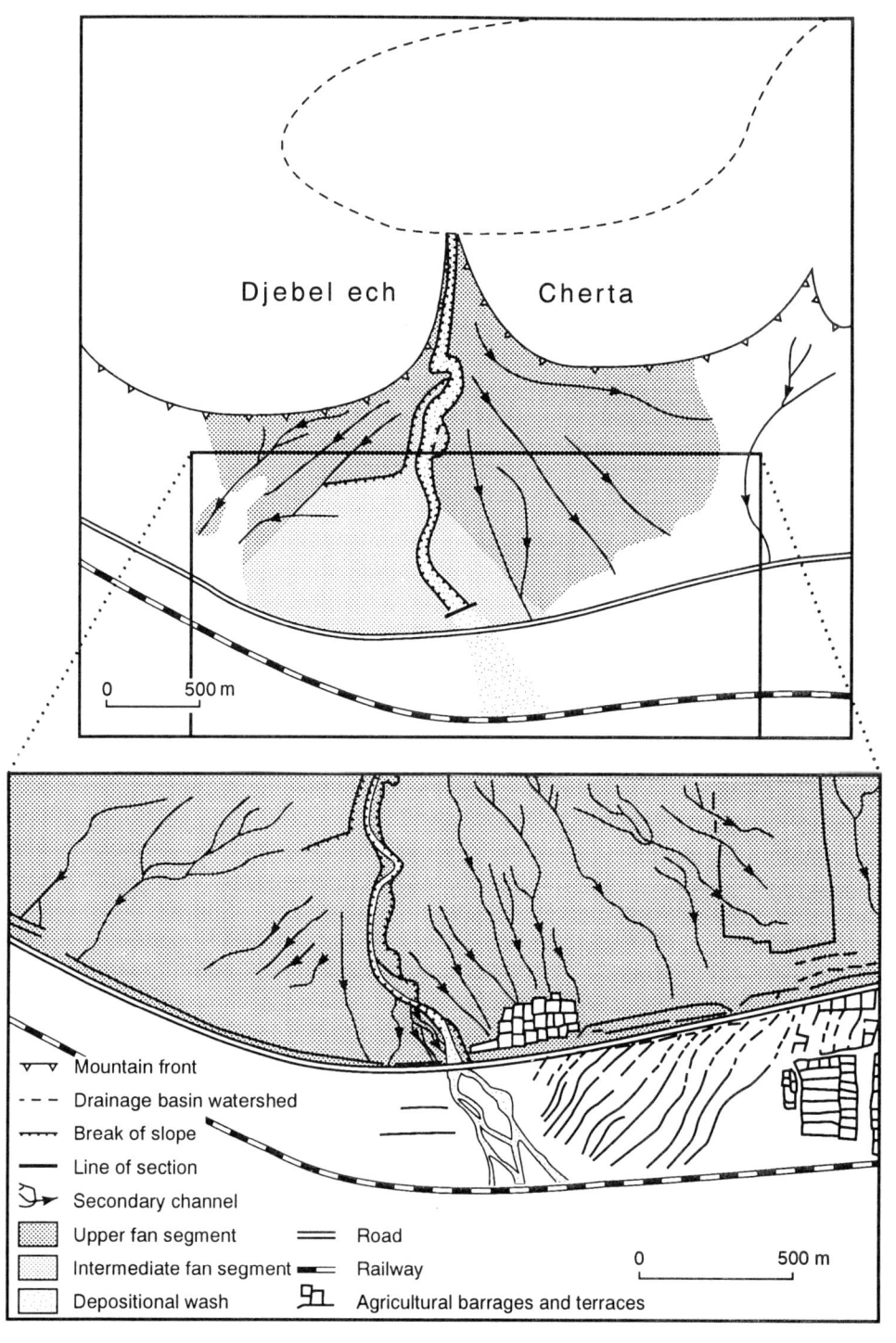

Figure 3.2 A geomorphological map of the Oued es Seffaia alluvial fan, showing the distribution of fan segments

Figure 3.3 Field photograph of the break of slope between the upper fan segment (background) and intermediate fan segment (foreground) for the Oued es Seffaia alluvial fan

The Oued et Tfal alluvial fan lies 14 km west of Gafsa on the south-eastern flank of Djebel es Stah (White, 1991) and covers an area of 1337 ha. The fan is morphologically similar to that of the Oued es Seffaia, with an upper fan segment, an intermediate telescopic segment and an incised depositional wash (Figures 3.4 and 3.5). In this case, the upper segment is covered by a stone pavement, but the intermediate segment is composed of a sandy loam surface, upon which varnished clasts are only present as a few isolated clumps. Again Mousterian and Capsian dates have been suggested for the upper and intermediate segments respectively, with the depositional wash being composed of contemporary alluvium (Coque, 1962, pp. 422–423). Figure 3.6 shows the relationship between the telescopic segments on the two fans in longitudinal profile.

The assemblage of magnetic minerals within the soils developed on these fan surfaces could be derived from a combination of four sources: firstly, primary minerals from the lithologies within each drainage basin that contributed to the fan systems; secondly, weathering products derived from *in situ* pedogenic processes operating on the fan surfaces; thirdly, both primary and secondary minerals from external sources via aeolian input: and finally, reworking within the fan systems themselves where younger fans receive material from the erosion of older fan sediments.

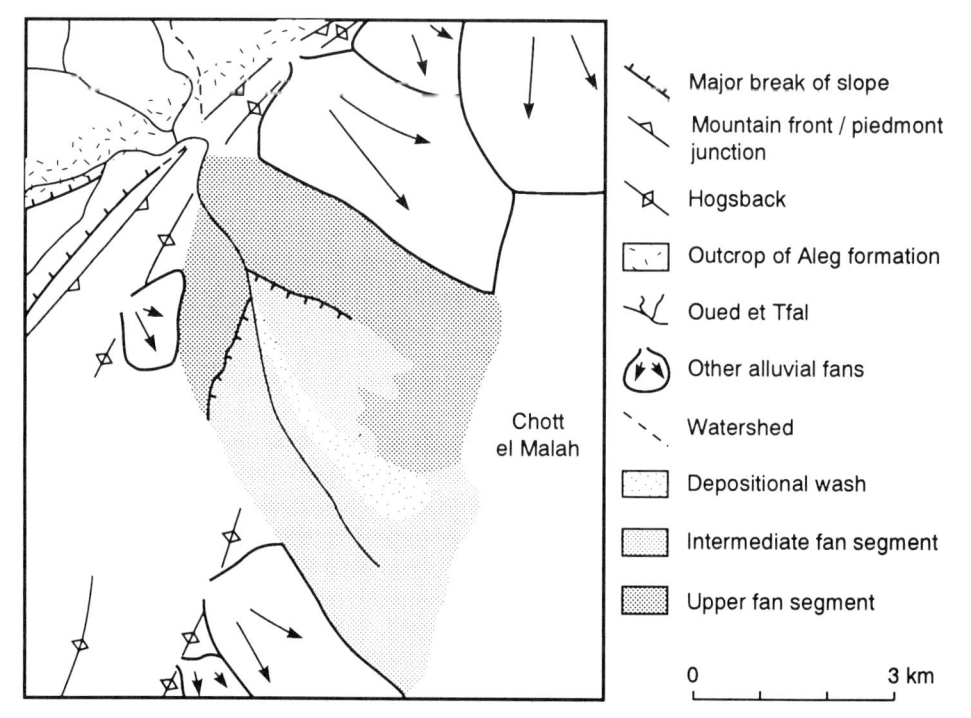

Figure 3.4 A geomorphological map of the Oued et Tfal alluvial fan, showing the distribution of fan segments

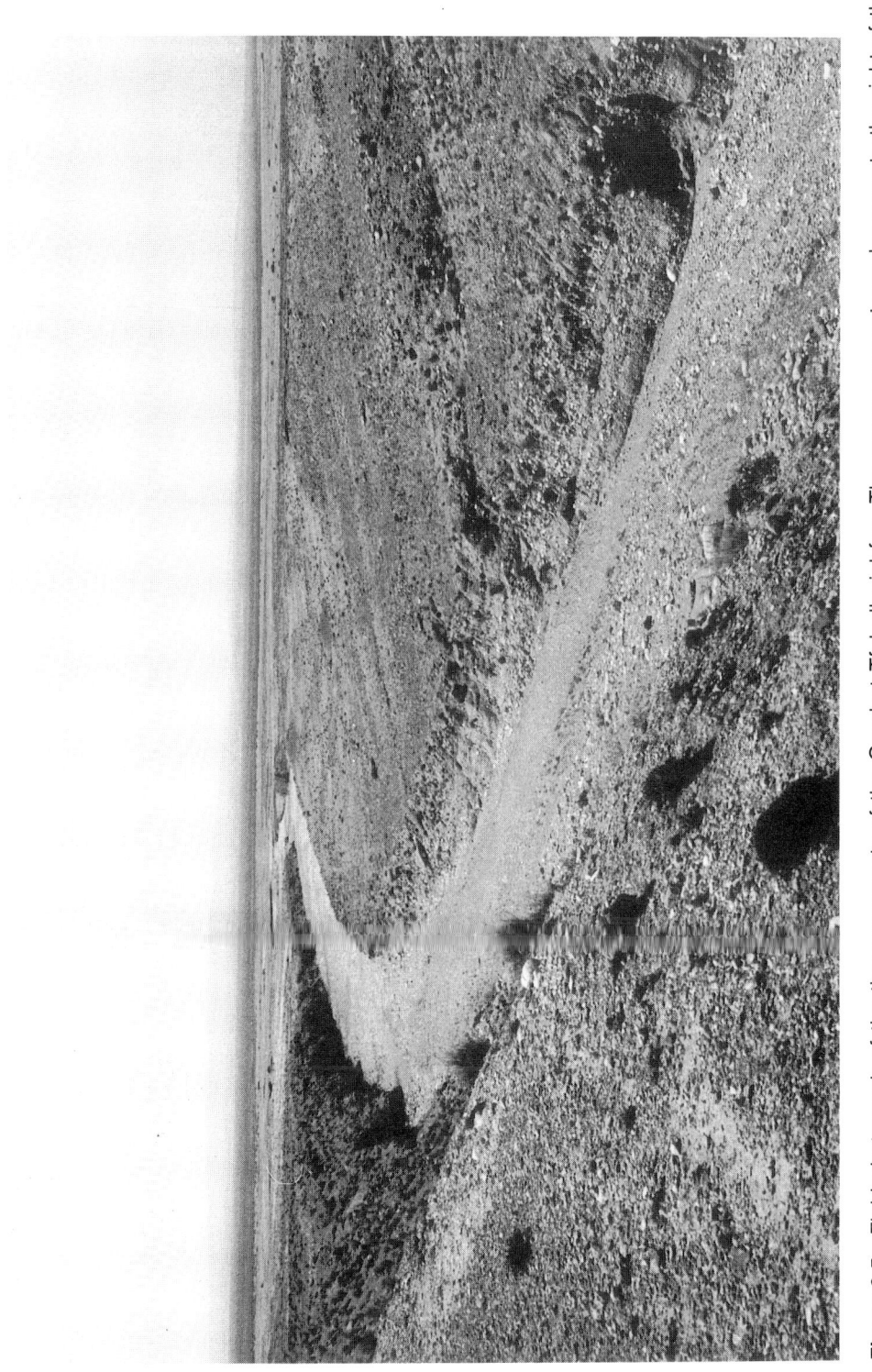

Figure 3.5 Field photograph of the three segments of the Oued et Tfal alluvial fan. The upper segment can be seen to the right of the photograph, the intermediate segment is in the centre of the photograph and incised depositional wash can be seen running from bottom right towards the top left of the photograph

50   Environmental Change in Drylands

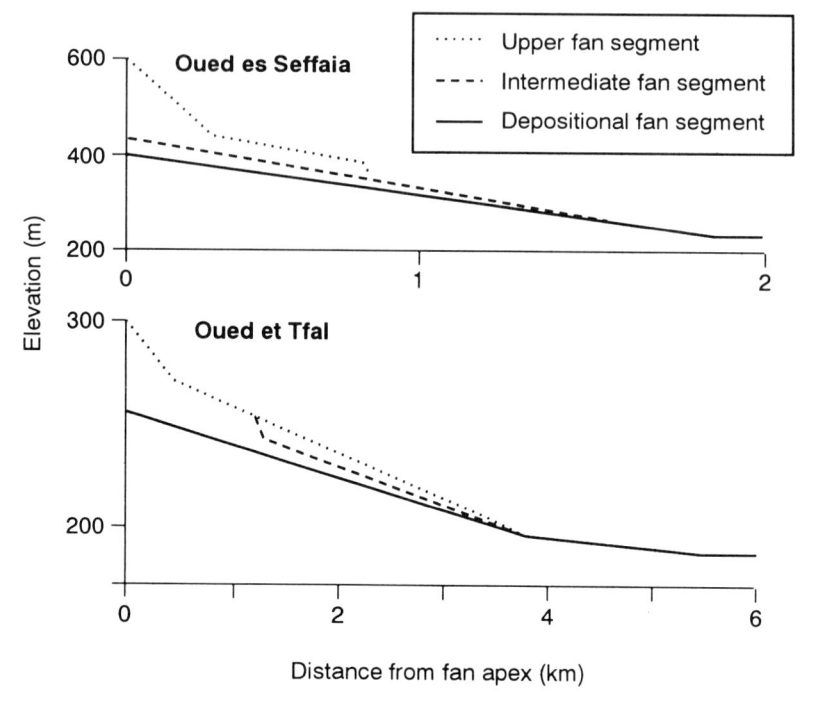

Figure 3.6 Cross-sections of the Oued es Seffaia and Oued et Tfal alluvial fans, showing the relationships of the three fan segments

METHODS

Sample Collection

A total of 94 soil samples were collected from a number of geomorphological surfaces throughout the Tunisian Southern Atlas. From the Oued es Seffaia and the Oued et Tfal alluvial fans, five sampling points were randomly selected on each of the identified fan segments. For consistency, and to reduce the effect of aeolian input, each sample was collected at a depth of 1 cm from the surface. Comparison of the sample magnetic properties with those of an aeolian dust sample, discussed below, indicate that aeolian input in the soil samples is insignificant. Approximately 1 kg of soil was collected at each sampling point.

Mineral Magnetic Analysis

Samples were dried at room temperature in a desiccation cabinet and then dry-sieved to extract the fraction finer than $\emptyset 2$ (250 µm). A pilot study revealed that this fraction contained the highest concentrations of iron oxide minerals and should therefore give the best discrimination between surfaces. In the case of the Oued et Tfal depositional wash, there was only enough material finer than $\emptyset 2$ to provide two samples for analysis. All other surfaces were characterised by five samples. Samples were weighed and then packed in 10 ml plastic pots and immobilised by packing with clean plastic foam prior to analysis.

All samples were subjected to the same measurement sequence. Initial low field specific susceptibility (χ) was measured using a Bartington MS2 susceptibility meter. By using an MS2B dual frequency sensor, both low- and high-frequency susceptibility were measured, allowing the proportion of frequency-dependent susceptibility to be calculated (χ_{fd}). Anhysteretic remanent magnetisation was induced in the samples using a Molspin a.f. demagnetiser with a small steady biasing field of 0.04 mT. The resultant remanence was measured using a Molspin 1 A magnetometer (as were all subsequent remanences), allowing χ_{ARM} to be calculated.

The samples were then placed in successively larger artificial magnetic fields (20, 500, 600 and 800 mT respectively) generated by two Molspin pulse magnetisers (0–300 and 0–800 mT). After each field, the isothermal remanent magnetisation (IRM) was measured. A 'saturation' field of 800 mT was used as this was the largest field that could be generated with the available equipment, although this is unlikely to have fully saturated some antiferromagnetic mineral species. The saturation isothermal remanent magnetisation (SIRM = $IRM_{800\,mT}$) was then destroyed by applying a series of 'reverse' fields (20, 100 and 300 mT). Approximately 75 samples could be processed in an average working day. A full list of the parameters used and their basic interpretation is given in Table 3.1 (after Dearing *et al.*, 1985; Thompson & Oldfield, 1986; Maher, 1988; Oldfield, 1991a).

Deferration of Samples

To provide a comparison between the mineral magnetic data and that of more traditional mineralogical and chemical techniques, six samples were randomly selected for further analysis. A further two samples from pediment surfaces on the south flank of Djebel Sidi bou Helal and the north flank of Djebel Tebaga (Figure 3.1) were included to cover the full range of magnetic properties found in all 94 soil samples.

These eight samples were split and one half was deferrated using the sodium bicarbonate buffered dithionite-citrate method of Mehra & Jackson (1960). This technique is thought to extract practically all secondary (free) iron oxides (Schwertmann & Taylor, 1989: 421; Colombo & Torrent, 1991; Torrent *et al.*, 1987). This enabled the magnetic and XRD (X-ray diffraction) characteristics of the deferrated (treated) samples to be compared with their undeferrated (untreated) equivalents, to see if the observed patterns could be ascribed to either primary (i.e. inherited from the source lithologies) or secondary (i.e. those formed by pedogenesis) iron oxide minerals. The amount of iron extracted in the supernatant during the deferration process was measured using a Perkin-Elmer model 3030 atomic absorption spectrophotometer (AAS). This served as an independent check on the results of the XRD and mineral magnetic analyses.

Mineral magnetic analysis was also performed on the eight treated samples. Munsell soil colours of treated and untreated samples were determined and used to calculate redness ratings, according to the method of Torrent *et al.* (1980).

Sub-samples of both untreated and treated soils were then ground to a uniform powder and XRD patterns obtained using a Philips PW1710 X-ray diffraction system using APD1700 software. A copper tube and nickel filter were used on a setting of 40 kV and 40 mA. Self-supporting powder mounts were prepared by back-filling the sample material into aluminium frames. Samples were then scanned from 18 to 36 degrees 2 theta (°2Θ) and the scan time was approximately 40 minutes. Some twelve samples could therefore be analysed in an average working day. The traces obtained were examined both manually

Table 3.1 Mineral magnetic parameters referred to in the text and their basic interpretation (after Dearing et al., 1985; Thompson & Oldfield, 1986; Maher, 1988; Oldfield, 1991a)

Parameter	Interpretation
χ (10^{-8} m^3 kg^{-1})	Initial low field mass specific magnetic susceptibility. This is measured within a small magnetic field and is reversible (no remanence induced). Its value is roughly proportional to the concentration of ferrimagnetic minerals within the sample, although in materials with little or no ferrimagnetic component and a relatively large antiferromagnetic component, the latter may dominate the signal.
χ_{fd} (10^{-8} m^3 kg^{-1})	Frequency-dependent susceptibility. This parameter measures the variation of magnetic susceptibility with the frequency of the applied alternating magnetic field. Its value is proportional to the amount of magnetic grains whose size means they lie at the stable single domain/superparamagnetic boundary. These grains show a delayed response to the magnetising field. The parameter is expressed on a mass specific basis as the difference between low- and high-frequency susceptibility.
χ_{ARM} (10^{-8} m^3 kg^{-1})	Anhysteretic remanent magnetisation (ARM) is roughly proportional to the concentration of ferrimagnetic grains in the 0.02–0.4 μm (in the stable single domain size) range. For this work ARM was induced in the samples by combining a peak AF field of 100 mT with a d.c. biasing field of 0.04 mT and the final result expressed as mass specific ARM per unit of steady field χ_{ARM} (Maher, 1988).
SIRM (10^{-5} A m^2 kg^{-1})	Saturation isothermal remanent magnetisation is the highest amount of magnetic remanence that can be produced in a sample by applying a large magnetic field. It is measured on a mass specific basis. In this study a 'saturating' field of 0.8 T has been used and this will produce saturation in most mineral types. However, some antiferromagnetic minerals may not be saturated at this field (e.g. geothite) and therefore this parameter is often called IRM$_{800\,mT}$. The values of SIRM are related to concentrations of all remanence-carrying minerals in the sample but is also dependent upon the assemblage of mineral types and their magnetic grain size.
Low-field acquisition of IRM (10^{-5} A m^2 kg^{-1})	The amount of remanence acquired in an initially demagnetised acquisition sample after it has experienced a field of 20 mT. At such low fields, the magnetically 'hard' canted antiferromagnetic minerals such as haematite and goethite are unlikely to contribute to the IRM, even at fine grain sizes. The values are therefore approximately proportional to the concentration of the magnetically 'softer' ferrimagnetic minerals (e.g. magnetite) within the sample, although also being grain size dependent.
High-field acquisition of IRM (10^{-5} A m^2 kg^{-1})	The amount of remanence acquired in a sample between fields of acquisition of 500 mT and 600 mT. At fields of 500 mT, all magnetically 'soft' ferrimagnetic minerals will have already saturated and any subsequent growth of IRM will be due to magnetically 'harder' canted antiferromagnetic components within the sample. The value is therefore approximately proportional to the concentration of canted antiferromagnetic minerals (e.g. haematite and goethite) within the sample.
Backfield ratios	Various demagnetisation parameters can be obtained by applying one or more reversed magnetic fields to a previously saturated sample. The loss of magnetisation at each backfield can be expressed as a ratio of IRM$_{backfield}$/SIRM, and therefore gives a result between +1 and −1, normalised for concentration. Such ratios can be used to discriminate between ferrimagnetic and canted antiferromagnetic mineral types. For example, using the 100 mT backfield ratio, minerals that are relatively easy to demagnetise (e.g. magnetite) have relatively low values (referred to as 'soft' magnetic behaviour). Minerals which show a stronger resistance to demagnetisation (e.g. haematite) show relatively high 100 mT backfield ratios (referred to as 'hard' magnetic behaviour).

and using an automated software based search-match facility (which interrogated the JCPDS data base) to identify the main mineral species present. Some workers (e.g. Torrent et al., 1980) have used considerably longer XRD scan times in order to identify the relatively small iron oxide component in soils. The less rigorous conditions used here were thought to be a realistic compromise between analysis time (and therefore cost) and sensitivity.

RESULTS

Mineral Magnetic Data of Undeferrated Samples

Table 3.2 lists values of six mineral magnetic parameters (χ, χ_{ARM}, SIRM, 100 mT backfield ratio, low-field IRM acquisition and high-field IRM acquisition) for each of the 27 samples from the Oued es Seffaia and Oued et Tfal alluvial fans. Means of these parameters are shown for the three fan segments. Table 3.2 also shows results of the two pediment samples, along with an aeolian dust sample, fortuitously collected during a dust storm. Values of χ_{fd} are not given as the relatively low χ values made its calculation inappropriate.

For both alluvial fan systems, the trends in the concentration-dependent magnetic parameters between different surfaces are the same; the upper segments contain greater concentrations of magnetic minerals than the intermediate segments, which in turn contain greater concentrations than the depositional wash. F-tests showed that differences between surfaces within an individual fan system are significant for χ, χ_{ARM}, SIRM, low-field IRM acquisition and high-field IRM acquisition at the 95% probability level.

The variations in the high-field acquisition values between different surfaces suggest corresponding differences in the concentrations of the magnetically 'hard' canted antiferromagnetic minerals such as haematite and goethite. The presence of such mineral phases in highly oxidised arid zone soils would be expected (Schwertmann & Taylor, 1989). The equivalent changes in the magnetically 'soft' component as indicated by the low-field IRM acquisition data adds support to Maher's (1986) assertion that the presence of magnetite/maghaemite within soils is a widespread phenomenon.

Some quantitative (albeit tentative) estimates of the concentration of ferrimagnetic and antiferromagnetic minerals can be made from values of low- and high-field acquisition of IRM (Oldfield, 1991a). Although such estimates depend upon a number of simplifications, they suggest that the concentration of ferrimagnetic material in the soil samples from Seffaia and Tfal is in the order of 0.01–0.02% whereas the concentration of antiferromagnets ranges from 0.03 to 0.2%. Schulze (1981) states that XRD can identify minerals such as haematite and goethite at levels of 1–2%. Even if the above magnetically based estimates only represent the correct orders of magnitude, they suggest that detection of these mineral species by XRD in these samples may be problematic.

Given the low-field IRM acquisition data, it would seem safe to conclude that the variation shown by χ values are also a response to variations in ferrimagnetic concentrations (rather than relatively larger variations in the concentrations of canted antiferromagnetic minerals). The χ_{ARM} values suggest increasing trends in the quantities of fine-grained ferrimagnets (0.02–0.4 mm) from the depositional wash soils to the upper segment soils.

Although the trends seen between each segment are the same within the two fans, the absolute values are somewhat different. The Tfal samples show lower concentrations of ferrimagnetic minerals χ, χ_{ARM}, low-field acquisition and SIRM) but higher concentrations

Table 3.2 Mineral magnetic data for the undeferrated samples. All units as in Table 3.1

Sample number	Sample area/type	χ	χ_{ARM}	SIRM	100 mT backfield	Acquisition of IRM Low field	High field
	Seffaia fan						
1	Depositional wash	1.76	0.221	130.5	−0.59	32.5	0.1
2	Depositional wash	0.46	0.065	38.1	−0.47	8.3	0.0
3	Depositional wash	1.27	0.180	109.9	−0.57	23.7	2.3
4	Depositional wash	1.78	0.259	116.5	−0.61	30.4	1.9
5	Depositional wash	0.57	0.093	43.1	−0.53	9.8	0.0
6	Intermediate segment	1.95	0.289	185.8	−0.59	40.5	1.1
7	Intermediate segment	1.44	0.253	144.0	−0.62	29.7	4.3
8	Intermediate segment	1.45	0.238	135.0	−0.61	30.8	3.0
9	Intermediate segment	1.50	0.249	141.7	−0.60	31.8	2.8
10	Intermediate segment	1.67	0.253	172.2	−0.58	34.5	6.9
11	Upper segment	2.34	0.461	225.0	−0.60	48.6	5.0
12	Upper segment	2.33	0.417	225.3	−0.60	48.6	4.8
13	Upper segment	2.30	0.398	227.7	−0.61	49.1	3.6
14	Upper segment	2.29	0.372	231.7	−0.61	47.5	4.3
15	Upper segment	2.35	0.347	268.0	−0.59	51.0	8.8
Mean values							
	Depositional wash	1.17	0.164	87.6	−0.55	20.9	0.8
	Intermediate segment	1.60	0.257	155.7	−0.60	33.4	3.6
	Upper segment	2.32	0.399	235.5	−0.60	48.9	5.2
	Tfal Fan						
16	Depositional wash	0.89	0.087	71.0	−0.62	16.6	2.9
17	Depositional wash	0.67	0.057	61.7	−0.64	16.7	3.4
18	Intermediate segment	1.34	0.136	112.5	−0.62	24.4	5.9
19	Intermediate segment	1.41	0.145	114.6	−0.62	25.1	4.4
20	Intermediate segment	1.39	0.145	104.1	−0.63	24.0	3.7
21	Intermediate segment	1.51	0.158	122.5	−0.63	27.8	4.0
22	Intermediate segment	1.21	0.119	95.5	−0.64	22.1	3.2
23	Upper segment	1.59	0.166	169.7	−0.58	29.9	5.4
24	Upper segment	1.56	0.190	140.9	−0.57	28.1	5.2
25	Upper segment	1.34	0.143	140.9	−0.56	25.9	5.7
26	Upper segment	1.77	0.187	195.7	−0.56	35.1	7.5
27	Upper segment	1.21	0.147	112.1	−0.58	22.8	4.9
Means	Depositional wash	0.78	0.072	66.3	−0.63	16.6	3.1
	Intermediate segment	1.37	0.141	109.8	−0.63	24.6	4.2
	Upper segment	1.49	0.167	151.8	−0.57	28.3	5.7
28	Helal glacis	0.98	0.093	105.7	−0.57	17.0	8.5
29	Tebaga glacis	0.19	0.024	26.9	−0.11	2.4	2.1
30	Aeolian dust	0.23	0.042	30.2	−0.45	5.3	2.5

of canted antiferromagnetic minerals (high-field acquisition) relative to the Seffaia samples. Student 't' tests comparing the magnetic properties of each particular surface between fans showed that the majority of differences are significant at the 95% probability level.

If the mineral magnetic composition of each surface reflected solely primary minerals derived from the basin source rocks, it would imply that, within the separate fan systems, similar shifts had occurred with time in the rock types contributing sediment to each stage of fan development. Although not impossible, such a coincidence may be unlikely. Given the geomorphological context discussed above, the trends are more likely to represent an increased contribution from secondary weathering products in older surfaces, similar to that seen in some Spanish drainage basins (Torrent et al., 1980; Diaz & Torrent, 1989). The differences between the absolute magnitude of values of equivalent surfaces between the two alluvial fans could therefore be due to differences in weathering rates, differences in the initial levels of primary mineral input (although the input to each surface within each alluvial fan system was the same) or, assuming both constant weathering rates and similar initial primary input, differences in the age of what appear on geomorphological grounds to be equivalent fan surfaces. However, Diaz & Torrent (1989) show how intrinsic soil characteristics can affect pedogenic iron oxide behaviour. Therefore, further work would be needed to reach a firm conclusion about the absolute differences between the two fan systems.

If the magnetic properties of the single aeolian dust sample are representative, it would also seem possible to exclude this source as a means of producing higher concentrations of magnetic minerals in the older fan surfaces. Although showing relatively high concentrations of antiferromagnetic minerals, its overall content of magnetic minerals, and ferrimagnets in particular, is low. Its addition in greater quantities to older surfaces is therefore likely to have diluted their magnetic component rather than increased it.

Comparison of the Deferrated and Undeferrated Samples

Assuming that the dithionite treatment removes only the secondary iron mineral assemblage, comparison of the data from the untreated and treated samples suggests that the majority of the mineral magnetic signal is secondary in origin (Table 3.3). For the Seffaia and Tfal samples over 95% of the χ_{ARM} signal was removed by the dithionite treatment and this suggests that the *in situ* formation of fine-grained ferrimagnetic mineral species does take place in arid soils systems. Even so, the magnetically 'softer' 100 mT backfield ratios shown by all the treated samples indicates that a greater proportion of the iron oxides removed were antiferromagnetic in nature.

Figure 3.7 shows scatter plots for the concentration-dependent magnetic parameters comparing the amount of signal removed by the dithionite treatment (undeferrated–deferrated) against the amount remaining (deferrated), which, assuming that the dithionite treatment is removing only weathering products, represents secondary versus primary magnetic mineralogy. Little pattern can be seen and, with the exception of χ, correlation coefficients are not significant at the 95% probability level. This suggests that there is little relationship between the primary magnetic mineralogy of the fan materials when deposited and their subsequent development of secondary iron oxides.

Table 3.3 includes the Munsell soil colour and redness rating. The deferration process reduces the redness rating to zero and the Munsell soil colour to a standard 5Y 8/1.5 (white). It would appear that the mineral magnetic technique is more sensitive than the Munsell soil

Table 3.3 Mineral magnetic, Munsell soil colour, redness rating, dithionite extractable Fe (Fe$_d$) and mineral species detected by XRD

Sample number	Sample		χ	χ$_{ARM}$	SIRM	100 mT backfield	Acquisition of IRM Low field	Acquisition of IRM High field
2	Seffaia DWU	U	0.46	0.065	38.1	−0.47	8.3	0.0
		D	−0.04	0.001	13.0	−0.72	2.3	0.0
		R	0.50	0.064	25.0		6.0	0.0
3	Seffaia DW	U	1.27	0.180	109.9	−0.57	23.7	2.3
		D	0.34	0.005	34.7	−0.69	7.2	1.0
		R	0.93	0.175	75.3		16.6	1.3
7	Seffaia IS	U	1.44	0.253	144.0	−0.62	29.7	4.3
		D	0.28	0.006	43.5	−0.66	8.3	1.6
		R	1.16	0.247	100.6		21.5	2.7
14	Seffaia US	U	2.29	0.372	231.7	−0.61	47.5	4.3
		D	0.42	0.005	55.8	−0.67	10.0	1.6
		R	1.87	0.367	175.9		37.4	2.8
19	Tfal IS	U	1.41	0.145	114.6	−0.62	25.1	4.4
		D	0.34	0.010	87.9	−0.81	10.9	1.1
		R	1.07	0.135	26.7		14.2	3.3
23	Tfal US	U	1.59	0.166	169.7	−0.58	29.9	5.4
		D	0.40	0.016	106.2	−0.78	16.1	0.9
		R	1.19	0.150	63.5		13.7	4.5
28	Helal glacis	U	0.98	0.093	105.7	−0.57	17.0	8.5
		D	0.20	0.003	39.5	−0.64	7.3	2.2
		R	0.78	0.090	66.2		9.8	6.3
29	Tebaga glacis	U	0.19	0.024	26.9	−0.11	2.4	2.1
		D	−0.01	0.006	21.7	−0.37	2.8	2.9
		R	0.20	0.018	5.2		−0.4	−0.8

Munsell soil colour	Redness rating	Fe$_d$	Mineral species detected by XRD
10YR 7/3 very pale brown	0.0		Q, C, D(?), G(?), H(??)
5Y 8/1.5 white	0.0		
		0.92	
10YR 6/3 pale brown	0.0		Q, C, D
5Y 8/1.5 white	0.0		
		1.33	
7.5YR 6/5 light brown	2.1		Q, C, D, K-F, G, H(?)
5Y 8/1.5 white	0.0		
		1.20	
7.5YR 6/5 light brown	2.1		Q, C, D, K-F, G(?), H(?)
5Y 8/1.5 white	0.0		
		1.38	
7.5YR 7/4 pink	1.7		Q, C, D, K-F, H, (R), G
5Y 8/1.5 white	0.0		
		1.29	

Table 3.3 Continued

Munsell soil colour	Redness rating	Fe$_d$	Mineral species detected by XRD
7.5YR 6/5 light brown	1.7		Q, C, D, K-F, G, H
5Y 8/1.5 white	0.0		
		1.35	
7.5YR 7/4 pink	1.4		Q, C, D, G(?)
5Y 8/1.5 white	0.0		
		1.11	
10YR 8/3 very pale brown	0.0		Q, C, Gy (R), K-F, S(?), H(?)
5Y 8/1.5 white	0.0		
		0.27	

U, Undeferrated sample (primary + secondary magnetic minerals)
D, Deferrated sample (primary magnetic minerals)
R, Removed = U-D (secondary magnetic minerals)
Q = quartz, C = calcite, G = goethite, H = haematite, K-F = potassium feldspar, S = siderite, Gy = gypsum, (?) = probably identified, (??) = possibly identified, (R) = some removal by dithionite treatment visible in XRD traces.

colour technique for estimating concentration of iron oxides. Neither Munsell soil colour, nor redness ratings derived from it, were able to differentiate between samples from the intermediate and upper fan segments from Oued es Seffaia, but this was brought out very clearly in the mineral magnetic results. Munsell soil colour and redness ratings were able to identify the greater amount of iron oxides in the upper segment of the Oued et Tfal fan, compared with the intermediate segment, but still the mineral magnetic data differentiates the samples more objectively.

Table 3.3 also summarises the results of the AAS and XRD analyses. Statistically significant (non-linear) relationships exist between the values of Fe$_d$ and the amount of signal lost from each of the concentration-dependent mineral magnetic parameters (Figure 3.8). This provides a useful independent check, suggesting that the magnetic data are responding in a manner consistent with more traditional forms of analysis.

In terms of the iron oxide mineralogy, the XRD data are less informative. Although allowing the identification of the dominant mineral species within each sample (generally quartz, calcite, dolomite and potassium-feldspars—Figure 3.9), the analytical procedure used here only provided positive identification of iron mineral species in three of the eight samples. Haematite and goethite were the only iron oxides identified; no ferrimagnetic component was detected in any sample. Tentative identifications were possible in the remaining samples but, given the low estimates of the ferrimagnetic and antiferromagnetic concentrations (based upon the mineral magnetic data of the untreated samples), this is unsurprising.

Comparison of the X-ray diffractograms obtained from the untreated and treated samples revealed only two instances where XRD had detected mineral removal. In sample 19, from the intermediate segment of Tfal, loss of a peak at 35.77°2Θ suggested the removal of haematite. Comparison of the untreated and treated diffractograms of sample 29, from the Tebaga pediment, showed that gypsum, due to its relatively high solubility, was also removed by the dithionite treatment. The loss of this major component, which is diamagnetic and would therefore act as a dilutant in the magnetic measurements, may explain why, for

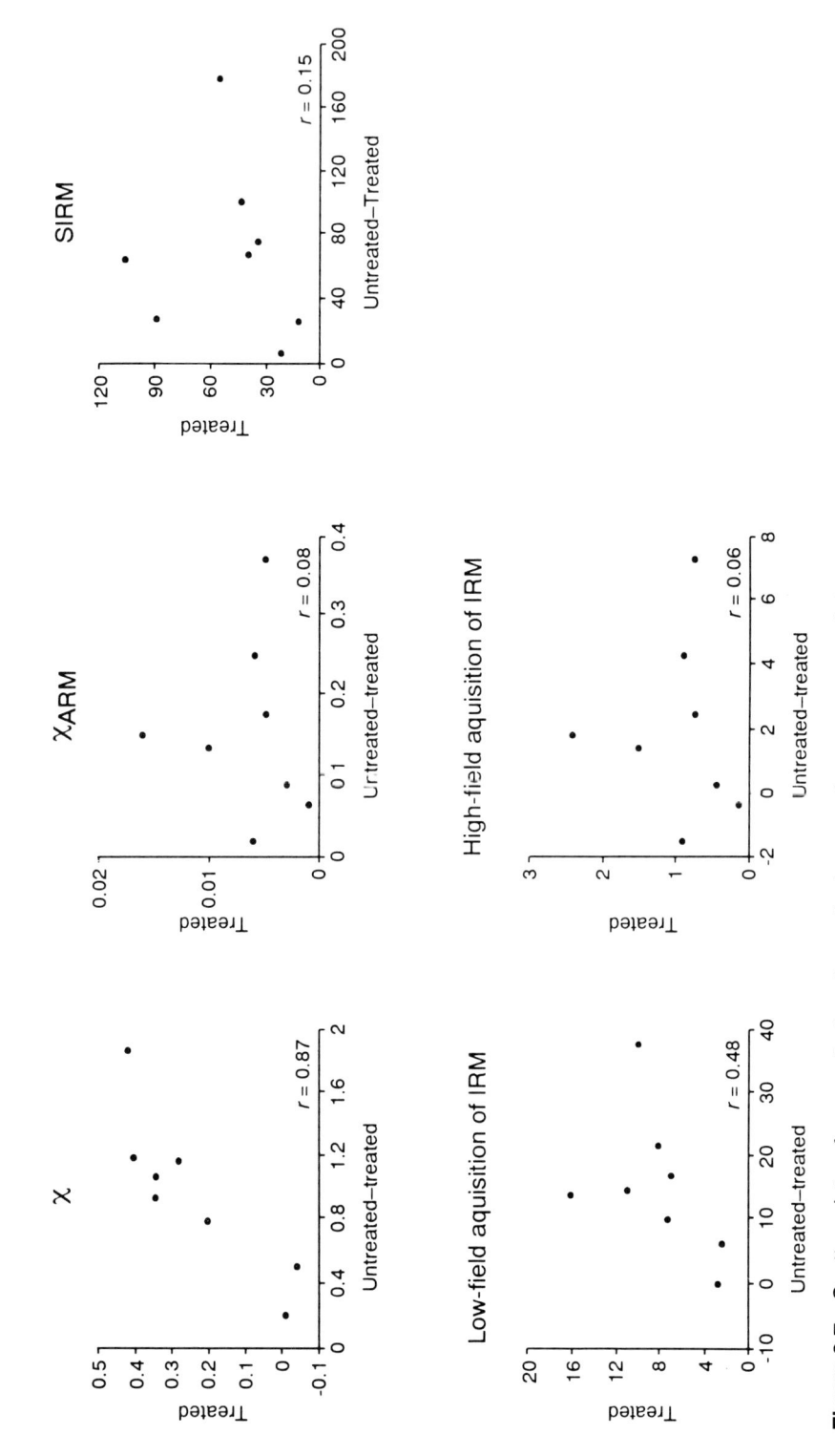

Figure 3.7 Scatter plots of concentration-dependent magnetic parameters of the amount of signal removed by the dithionite treatment (undeferrated–deferrated) against the amount remaining after the dithionite treatment (deferrated). Correlation coefficients (r^2) greater than 0.71 are significant at the 0.05 significance level. See text for discussion

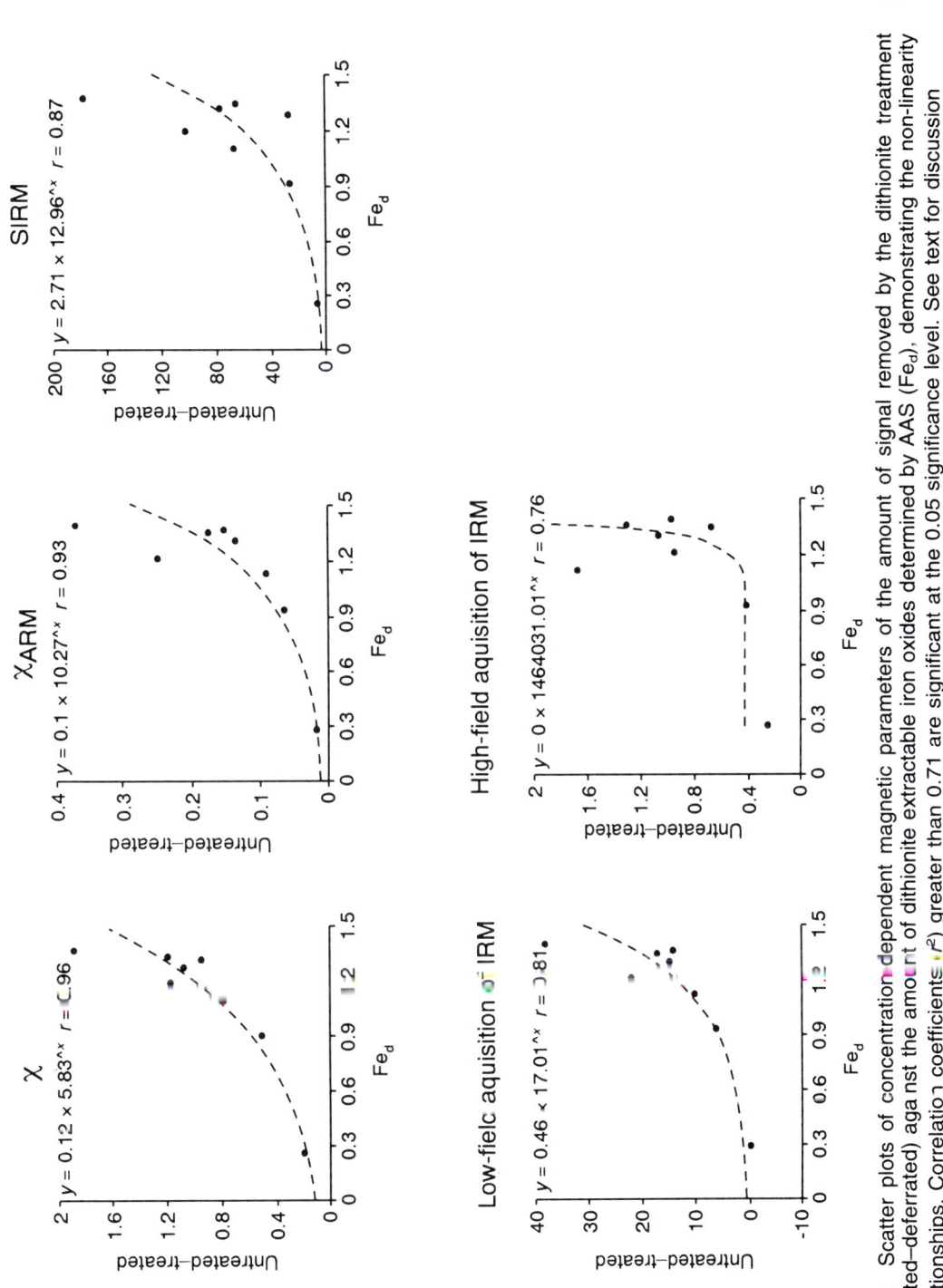

Figure 3.8 Scatter plots of concentration-dependent magnetic parameters of the amount of signal removed by the dithionite treatment (undeferrated−deferrated) against the amount of dithionite extractable iron oxides determined by AAS (Fe$_d$), demonstrating the non-linearity of the relationships. Correlation coefficients (r^2) greater than 0.71 are significant at the 0.05 significance level. See text for discussion

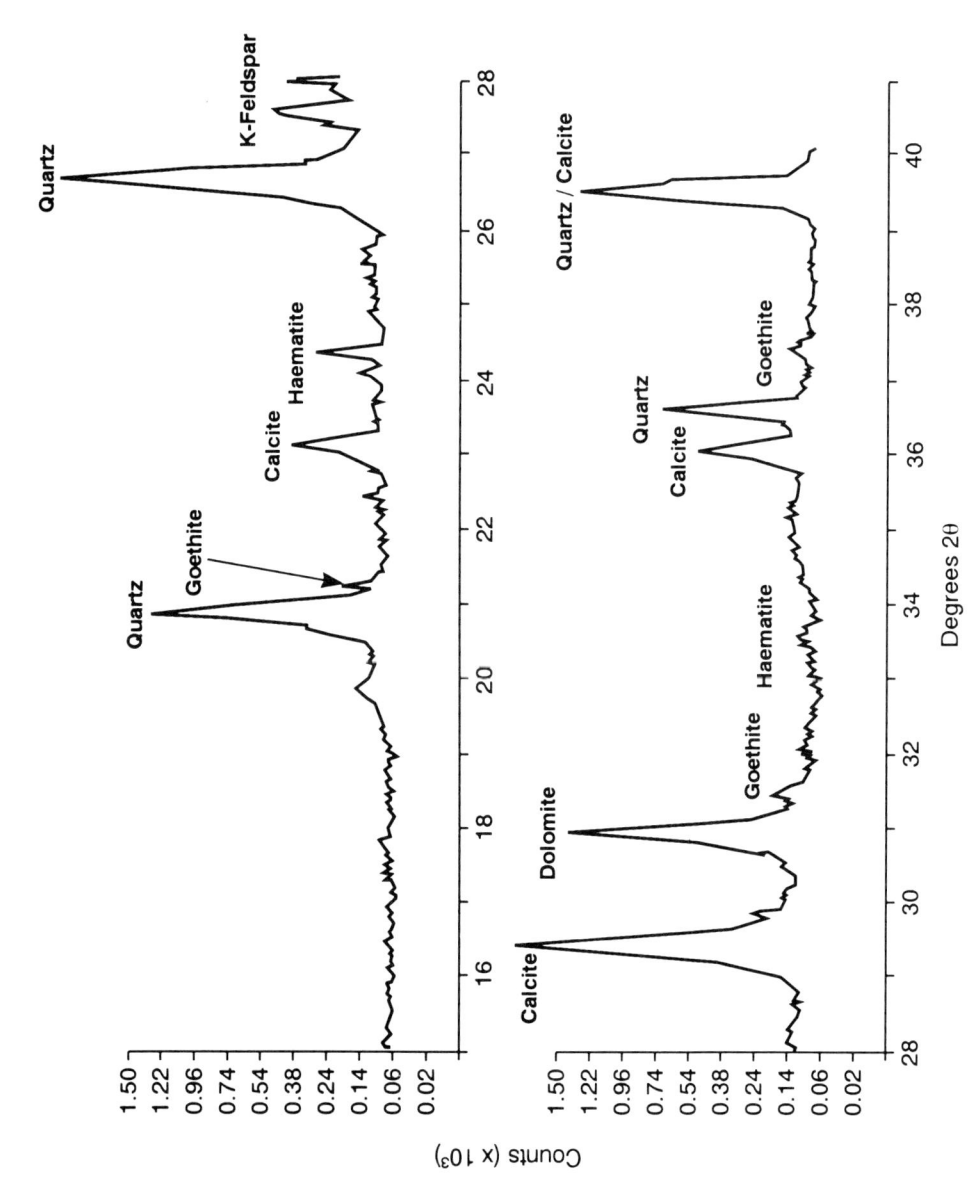

Figure 3.9 X-ray diffractogram of sample 19, showing peaks due to main mineral components, along with goethite and haematite

this sample, some mineral magnetic parameters show greater concentrations after deferration. The presence of large quantities of gypsum may also explain this sample's relatively low iron oxide content, as the formation of an indurated gypsum crust (Watson, 1985) is likely to inhibit the pedogenic processes, which concentrate iron oxides in the soil by forming an anoxic environment and preventing the percolation of water and also by raising the soil pH. XRD did not reveal gypsum as a significant component in any of the other samples and therefore its removal by the dithionite treatment is unlikely to have had an influence on their magnetic response.

DISCUSSION

The Iron Mineralogy of the Soils

The results indicate that mineral magnetic analysis allows an ordinal chronology of alluvial fan surfaces to be established for both the Oued et Tfal and the Oued es Seffaia alluvial fans. Given the comparison of the magnetic behaviour of the treated and untreated samples, it would appear that secondary iron oxide minerals formed by pedogenic weathering are responsible for mineral magnetic differences between soils on the different geomorphological surfaces. Mineral magnetic evidence suggests that external aeolian material is not a major source.

The iron-bearing minerals in soils (and, therefore, possible iron sources for rubification) are essentially Fe-bearing silicates, iron oxihydroxides and iron(II)-containing calcites and dolomites, and unlikely to show strong magnetic properties. Workers studying possible sources of iron oxides in red soils developed on haematite-free calcarenites in southern Spain identified weathering of iron-bearing clay minerals (smectites) as the main source of secondary iron oxides (Torrent & Cabedo, 1986).

However, in soils from calcareous parent materials, such as those in the Tunisian Southern Atlas, iron liberated from carbonates can also form haematite (Gebhardt *et al.*, 1969; Meyer & Kruse, 1970). Workers studying rock varnishes have noted the considerable enrichment of iron in these coatings in the Atlas Mountains of Morocco (Smith & Whalley, 1988) and Tunisia (White, 1990). Some workers prefer a physicochemical process for this enrichment (Whalley, 1983). However, there is a large amount of evidence to suggest microbial involvement (White, 1990). Given that this iron-enrichment process is taking place on clasts lying directly on the soils analysed in the work reported here, it is reasonable to suggest that the pedogenic and clast-surface processes may not be totally independent, and microbes may be intimately involved in the pedogenic enrichment in iron oxides. The role of microbial processes in controlling the magnetic characteristics of soils and sediments is an area of debate in current literature (Oldfield, 1991b, 1992; Machel & Burton, 1992). Ongoing research is seeking to determine whether a biotic or abiotic explanation is more appropriate for the Tunisian soil samples.

The Techniques

The inconclusive nature of the XRD results suggests that the more rigorous sample preparation and analysis procedure described by Torrent *et al.* (1980) or Schulze (1981) is needed if XRD is to provide useful information on the relatively small iron oxide component in these soils. The greater sensitivity of the mineral magnetic technique, its semi-quantitative nature and its relative speed would all suggest that it could provide a useful adjunct to more traditional forms of analysis.

In addition, some mineral magnetic parameters (χ_{fd}, although not applicable here, and χ_{ARM}) respond specifically to the finer-grain sizes of magnetic minerals most commonly formed by weathering processes (Thompson & Oldfield, 1986; Maher, 1986; Oldfield, 1991a) and may therefore allow a relative estimate of secondary iron oxide production without dithionite treatment.

The sensitivity of the magnetic approach has also allowed the detection of a small secondary ferrimagnetic component whose presence was not detected by XRD. Indeed, few workers have discussed the presence and role played by magnetite or maghaemite in arid zone soils and this may simply reflect an inability to identify its presence in relatively small amounts (<0.5%) using traditional techniques.

Nevertheless, the simple XRD procedure used here has allowed positive identification of haematite and goethite. Their presence can only be inferred from the mineral magnetic measurements and they cannot be distinguished. However, goethite displays 'harder' magnetic behaviour than haematite and therefore SIRM is achieved only at higher field strengths. D. France (University of Liverpool, personal communication, 1991) suggests that a somewhat more sophisticated magnetic analysis than that performed here can discriminate between haematite and goethite components of SIRM. This would, of course, add to the analysis time.

CONCLUSIONS

In both the Oued es Seffaia and Oued et Tfal alluvial fans, the trends in the concentration-dependent magnetic parameters between different surfaces are the same; the upper segments contain greater concentrations of magnetic minerals than the intermediate segments, which in turn contain greater concentrations than the depositional washes. Although the trends seen between each fan surface are the same within the two alluvial fan systems, the absolute values are somewhat different. The Tfal samples show lower concentrations of ferrimagnetic minerals but higher concentrations of canted antiferromagnetic minerals relative to the Seffaia samples, probably reflecting differences in location-specific conditions.

Assuming that the dithionite treatment removes only secondary iron minerals, comparison of magnetic characteristics of deferrated and undeferrated samples suggests that the majority of the mineral magnetic signal in the soil is secondary in origin. There is little relationship between the primary magnetic mineralogy of the fan materials and the subsequent development of secondary iron oxides.

The overall content of magnetic minerals is low in the sample of aeolian dust. The addition of greater quantities of dust to older surfaces is, therefore, unlikely to be responsible for the accumulation of iron oxides in soils with time.

Given the above, it can be concluded that the upper fan segments of both alluvial fans are the oldest surfaces. The intermediate segments are younger, but are older than the contemporary depositional wash surfaces. It is therefore possible to establish an ordinal chronology of geomorphological surfaces within individual alluvial fan systems in the Tunisian Southern Atlas, using mineral magnetic analysis of soils, provided that climate, pedoclimate and primary mineralogical composition are constant.

The magnetic parameters measured for the soil samples indicate the presence of both magnetically 'hard' canted antiferromagnetic minerals such as haematite and goethite and, more surprisingly, the presence of magnetically 'soft' minerals, probably magnetite/maghaemite. Few workers have identified the presence of these minerals by non-magnetic means in arid zone soils.

The source and process of formation of the pedogenic iron oxides in soils on alluvial fan surfaces in the Tunisian Southern Atlas is uncertain. They may be formed by physico-chemical weathering of primary lithological materials. Microbial processes, such as those proposed for iron and manganese-rich coatings on rock surfaces in arid regions, may also be involved.

The greater sensitivity of the mineral magnetic approach compared to XRD, AAS and Munsell soil colour techniques, its semi-quantitative nature and its relative speed would all suggest that it could provide a useful adjunct to more traditional forms of analysis of pedogenic iron oxides.

ACKNOWLEDGEMENTS

The authors would like to thank Dr John Smith, Dr Roger Dackombe, Brian Bucknall and Peter Jenkins of the Wolverhampton Polytechnic for arranging access to some of the instrumentation used in the work. Fieldwork was supported by a travel grant from the Faculty of Anthropology and Geography, University of Oxford.

REFERENCES

Allison, R. J. (1988). A non-destructive method of determining rock strength. *Earth Surface Processes and Landforms*, **13**, 729–36.

Allison, R. J. (1990). Developments in a non-destructive method of determining rock strength. *Earth Surface Processes and Landforms*, **15**, 571–7.

Blissenbach, E. (1954). Geology of alluvial fans in semi-arid regions. *Geological Society of America Bulletin*. **65**, 175–90.

Burollet, P. F. (1967). General geology of Tunisia. In Martin, L. (ed.), *Guidebook to the Geology and History of Tunisia*, Petroleum Exploration Society of Libya, Libya.

Castany, G. (1952). *Paléogéographie, Tectonique et Orogenese de la Tunisie*, Congres Geologique International, Tunis.

Coleman, S. M. (1986). Levels of time information in weathering measurements, with examples from weathering rinds on volcanic clasts in the western United States. In Coleman, S. M. and Dethier, D. P. (eds.), *Rates of Chemical Weathering of Rocks and Minerals*, Academic Press, New York, pp. 379–93.

Colombo, C. and Torrent, J. (1991). Relationships between aggregation and iron oxides in Terra Rossa soils from southern Italy. *Catena*, **18**, 51–9.

Coque, R. (1962). *La Tunisie Presaharienne. Etude Geomorphologique*, Armand Colin, Paris.

Coque, R. and Jauzein, A. (1967). The geomorphology and Quaternary geology of Tunisia. In Martin, L. (ed.), *Guidebook to the Geology and History of Tunisia*, Petroleum Exploration Society of Libya, Tripoli, pp. 227–57.

Crook, R. Jr (1986). Relative dating of Quaternary deposits based on P-wave velocities in weathered granite clasts. *Quaternary Research*, **25**, 281–92.

Crook, R. Jr and Gillespie, A. R. (1986). Weathering rates in granite boulders measured by P-wave speeds. In Coleman, S. M. and Dethier, D. P. (eds.), *Rates of Chemical Weathering of Rocks and Minerals*, Academic Press, New York, pp. 395–417.

Dearing, J. A., Maher, B. A. and Oldfield, F. (1985). Geomorphological linkages between soils and sediments: the role of magnetic measurements. In Richards, K. S., Arnett, R. R. and Ellis, S. (eds.), *Geomorphology and Soils*, George Allen and Unwin, London, pp. 245–66.

Diaz, M. C. and Torrent, J. (1989). Mineralogy of iron oxides in two soil chronosequences of central Spain. *Catena*, **16**, 291–9.

Dorn, R. I. (1983). Cation-ratio dating; a new rock varnish age determination technique. *Quaternary Research*, **20**, 49–73.

Dorn, R. I., Bamforth, D. B., Cahill, T. A., *et al.* (1986). Cation-ratio and accelerator radiocarbon dating of rock varnish on Mojave artifacts and landforms. *Science*, **231**, 830–3.

Dorn, R. I., Tanner, D., Turrin, B. D. and Dohrenwend, J. C. (1987). Cation-ratio dating of Quaternary materials in the eastern Mojave Desert. *Physical Geography*, **8**, 72–81.

Dorn, R. I., Jull, A. J. T., Donahue, D. J., Linick, T. W. and Toolin, L. J. (1989). Accelerator mass spectrometry radiocarbon dating of rock varnish. *Geological Society of America Bulletin*, **101**, 1363–72.

Gebhardt, H., King, M.T. and Meyer, B. (1969). Mineralogisch-chemische Untersuchungen zum Prozess der Rubefizierung in Kalkstein-Rotlehm und fossilem Laterit in Nord-hessen. *Gottinger Bodenkd. Ber.*, **9**, 65–124.

Hooke, R. Le B. (1967). Processes on arid-region alluvial fans. *Journal of Geology*, **75**, 438–60.

Kesel, R. H. (1985). Tropical fluvial geomorphology. In Pittty, A. F. (ed.), *Themes in Geomorphology*, Croon Helm, Beckenham, pp. 102–21.

Lowe, J. J. and Walker, M. J. C. (1984). *Reconstructing Quaternary Environments*, Longmans, London.

Machel, H. G. and Burton, E. A. (1992). Comment on 'Sediment magnetism. Soil erosion, bushfires or bacteria?' by Oldfield, F. *Geology*, **20**, 670–1.

Maher, B. A. (1986). Characterisation of soils by mineral magnetic measurements. *Physics of the Earth and Planetary Interiors*, **42**, 76–92.

Maher, B. A. (1988). Magnetic properties of some synthetic sub-micron magnetites. *Journal of Geophysical Research*, **94**, 83–96.

Mehra, O. P. and Jackson, M. L. (1960). Iron oxide removal from soils and clays by a dithionite-citrate system buffered with sodium bicarbonate. In Swineford, A. (ed.), *Proceedings of 7th National Conference on Clays and Clay Minerals*, Pergamon Press Elmsford, New York.

Meyer, B. and Kruse, W. (1970). Untersuchungen zum Prozess der Rubefizierung (Entkalkungsrotung) mediterraner Boden am Beispiel kalkhaltiger marokkanischer Kusten-Dunen. *Gottinger Bodenkd. Ber.*, **13**, 77–140.

Oldfield, F. (1991a). Environmental magnetism—a personal perspective. *Quaternary Science Reviews*, **10**, 73–85.

Oldfield, F. (1991b). Sediment magnetism. Soil erosion, bushfires or bacteria? *Geology*, **19**, 1155–6.

Oldfield, F. (1992). Reply to comment on 'Sediment magnetism. Soil erosion, bushfires or bacteria?' by Machel, H. G. and Burton, E. A. *Geology*, **20**, 671.

Rognon, P. (1987). Late Quaternary climatic reconstruction for the Maghreb (North Africa). *Palaeogeography, Palaeoclimatology, Palaeoecology*, **58**, 11–54.

Schulze, D. G. (1981). Identification of soil iron oxide minerals by differential X-ray diffraction. *Soil Science Society of America Journal*, **45**, 437–40.

Schwertmann, U. and Taylor, R. M. (1989). Iron oxides. In Dixon, J. B. and Weed, S. B. (eds.), *Minerals in Soil Environments*, 2nd edition, Soil Science Society of America, Madison, pp. 379–438.

Smith, B. J. and Whalley, W. B. (1988). A note on the characteristics and possible origins of desert varnishes from southeast Morocco. *Earth Surface Processes and Landforms*, **13**, 251–58.

Thompson, R. and Oldfield, F. (1986). *Environmental Magnetism*, George Allen and Unwin, London.

Torrent, J. and Cabedo, A. (1986). Sources of iron oxides in reddish-brown soil profiles from calcarenites in southern Spain. *Geoderma*, **37**, 57–66.

Torrent, J., Schwertmann, U. and Barron, V. (1987). The reductive dissolution of synthetic goethite and hematite in dithionite. *Clay Minerals*, **22**, 329–37.

Torrent, J., Schwertmann, U. and Schulze, D. G. (1980). Iron oxide mineralogy of some soils of two river terrace sequences in Spain. *Geoderma*, **23**, 191–208.

Watson, A. (1985). Structure, chemistry and origins of gypsum crusts in southern Tunisia and the central Namib desert. *Sedimentology*, **32**, 855–75.

Whalley, W. B. (1983). Desert varnish. In Goudie, A. S. and Pye, K. (eds.), *Chemical Sediments and Geomorphology*, Academic Press, London, pp. 197–226.

White, K. (1987). Piedmont surface mapping in the Tunisian Southern Atlas: Use of Landsat Thematic Mapper data. In *Advances in Digital Image Processing*, Proceedings of 13th Annual Conference of the Remote Sensing Society, Nottingham, pp. 631–49.

White, K. (1990). Spectral reflectance characteristics of rock varnish in arid areas. School of Geography Research Papers 46, University of Oxford.

White, K. (1991). Geomorphological analysis of piedmont landforms in the Tunisian Southern Atlas using ground data and satellite imagery. *The Geographical Journal*, **157**, 279–94.

4 Late Pleistocene and Holocene Changes in Hillslope Sediment Supply to Alluvial Fan Systems: Zzyzx, California

A. M. HARVEY
Department of Geography, University of Liverpool, UK

and

S. G. WELLS
Department of Earth Sciences, University of California, Riverside, USA

ABSTRACT

Catchment hillslopes supplying alluvial fan surfaces at Zzyzx, in the Mojave Desert, California, have been mapped. Late Pleistocene to Holocene ages have been assigned to their surface features by correlation with alluvial fan surfaces on the basis of the field characteristics of rock varnish and soil development. Pleistocene to Holocene ages have been assigned to the alluvial fan surfaces on the basis of stratigraphic relationships to dated lake shorelines of Pleistocene Lake Mojave. Late Pleistocene fan aggradation was dominated by debris flow deposition, fed from active hillslopes. The late Pleistocene to Holocene transition saw a switch away from widespread hillslope mass movement processes towards hillslope stabilisation and localised dissection. This resulted in a reduction of sediment supply to the fans and a switch to fluvial processes involving episodic fanhead dissection and distal progradation. The causes appear to be climatically induced changes in weathering and hillslope processes in the catchment areas.

INTRODUCTION

The fluvial system in desert mountain regions is often characterised by two strongly contrasting but related zones, a dominantly erosional mountain catchment zone, supplying sediment to a mountain-front sediment storage zone, dominated by alluvial fans (Bull, 1977, 1991; Harvey, 1989). The behaviour of the whole system depends on the processes and rate of sediment generation within the mountain catchment, the processes and rate of sediment transport from the hillslopes to the channel and fan system (i.e. the strength of coupling within the system) and on the processes within the fan environment itself. Evidence for Pleistocene geomorphic processes may be expressed by the morphology of the catchment hillslopes (Oberlander, 1989), but above all by the characteristics of the alluvial fans. Fans are depositional forms whose morphology, especially surface slope, may be a direct response to depositional style and may be near the threshold slope between erosional and depositional regimes (Bull, 1979). Should the rate of sediment supply from the mountain catchment or the style of sediment transport through the alluvial fan environment change, then the regime on the fan may switch towards greater aggradation or dissection (Gerson *et al.*, 1978). Thus the alluvial fan may be an important and sensitive indicator of sediment supply and transport relationships, not only in the context of current behavioural style but in that the sedimentary sequences within the fan may preserve the most complete signal available of past environmental change within the mountain catchment (Bull, 1991).

THE MOJAVE SEQUENCE

Much of the previous work on arid-zone mountain geomorphic sequences has been primarily concerned with alluvial fans. Many such studies demonstrate complex histories of Quaternary fan aggradation and dissection (Harvey, 1984a, 1990). The alluvial fans of the Great Basin–Mojave Desert region of the American south-west are classic examples (e.g. Blissenbach, 1954; Hooke, 1967), especially those in Death Valley (Denny, 1965; Hunt & Mabey, 1966; Hunt, 1975). There, Quaternary sequences of alternating aggradation and dissection have resulted in multi-faceted, multiple age surfaces on large alluvial fans that reflect climatically controlled variations in water and sediment supply from the mountain catchments (Dohrenwend *et al.*, 1991). In Fish Lake Valley, north-west of Death Valley, detailed chronological studies involving ^{14}C and tephra dates have been used to link alluvial fan deposition primarily to climatic transitions from relatively cold to relatively warm conditions (Slate, 1991).

However, in most cases, the detailed chronology is not well understood. Recent work by Dorn and colleagues in Death Valley, on dating and palaeoenvironmental reconstructions from desert rock varnish (Dorn & Oberlander, 1981; Dorn *et al.*, 1987, 1989; Dorn, 1983, 1984), proposes relationships between Quaternary climatic fluctuations and alluvial fan aggradation and dissection sequences (Dorn, 1988), with at least three major partially diachronous aggradation cycles corresponding to the more humid major Pleistocene 'glacials' and shorter dissection-progradation phases with the more arid 'interglacials'. However, as pointed out by Wells & McFadden (1987), these interpretations are based on experimental dates only, and do not have supporting data linking varnish microtopography to fan depositional processes.

In two areas to the south of Death Valley, Wells *et al.* (1984, 1987, 1990a) and McFadden *et al.* (1989) have established a chronology for alluvial fan deposition. For fans issuing from the Soda Mountains towards Silver Lake (Figure 4.1), a modern playa, but a remnant of pluvial Lake Mojave, stratigraphic relations between alluvial fan sediments and ^{14}C dated lacustrine sediments and lake shorelines, as well as dated packrat middens in dissected fan channels, have been used to establish a chronology of piedmont deposition. Two shorelines of pluvial Lake Mojave have been dated to *ca.* 14000 BP and *ca.* 8500 BP, a time interval effectively spanning the Pleistocene–Holocene transition. For fans issuing from the Cima volcanic field to the east of Baker (Figure 4.1), the chronology has been established using stratigraphic relations between fan sediments and K–Ar dated volcanic flows. In both areas, field properties of the relative development of soil profiles, desert pavements and rock varnish coatings for each dated alluvial fan surface are used to characterise fans of differing age and correlate fans across desert piedmonts. These studies correlate fan aggradation

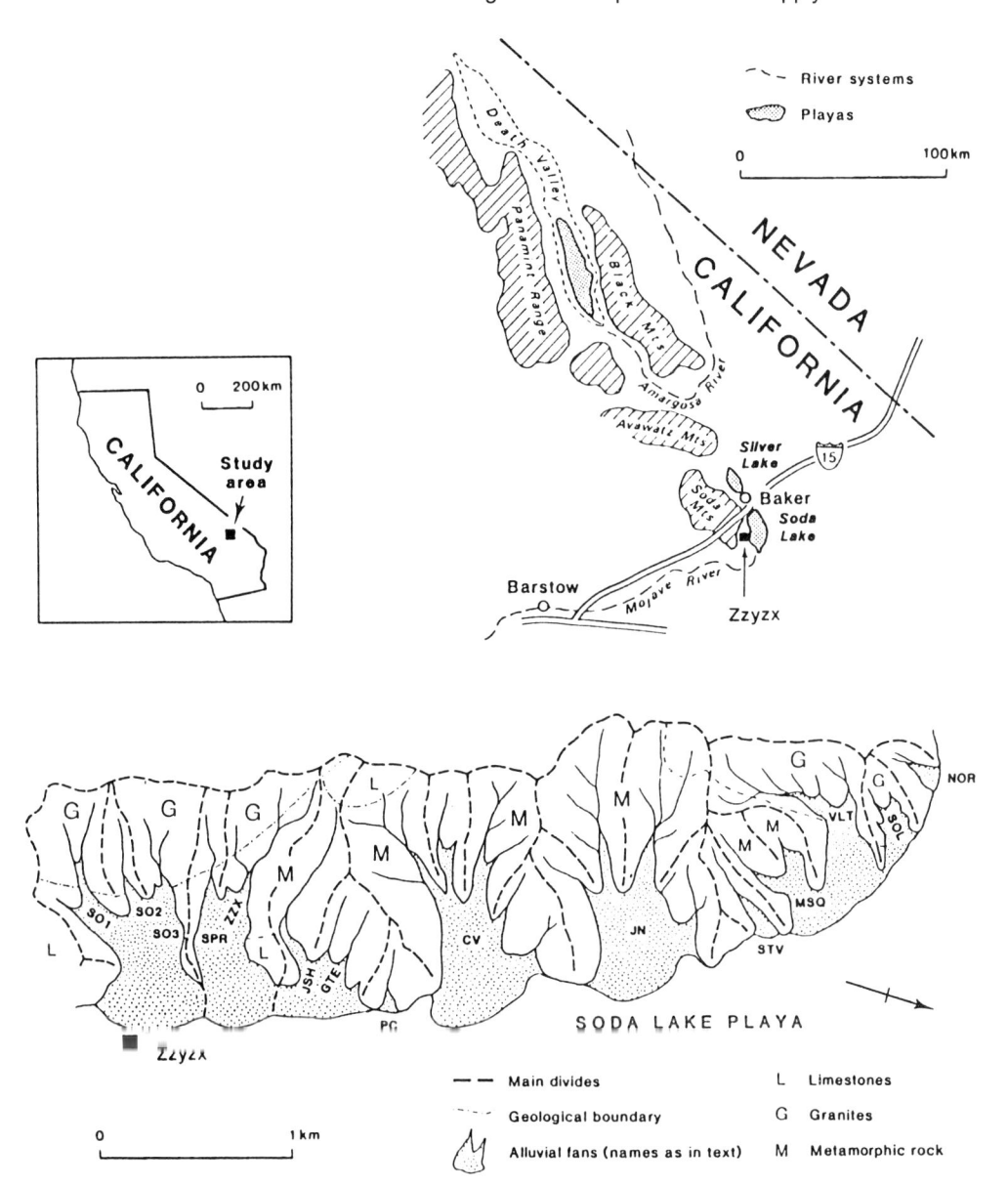

Figure 4.1 Location maps, Zzyzx fans and catchments. Letters on main map identify individual fans, groups of fans and catchment areas; SO1,2,3, Southern Group; SPR, Springer fan; ZZX, Zzyzx fan; JSH, Josh fan; GTE, Gate fan; PC, Palm cone; CV, Camino Viejo fan; JN, Johnny fan; STV, Steve fan; MSQ, Mesquite fan; VLT, Vulture fan; SOL, Solitary fan; NOR, Northern fan

resulting from destabilization of hillslopes and fan dissection resulting from large-scale flooding, with the time transgressive climatic changes during the Pleistocene–Holocene transition. The sequence of fanhead dissection and distal deposition-progradation during the Holocene is inferred to be a delayed and complex response to the widespread fan deposition occurring during the Pleistocene–Holocene transition, and to decreased permeability of the fan deposits due to pedogenesis, resulting in increased runoff and the reworking of the older alluvial fan sediments (Wells *et al.*, 1987; Ritter, 1991).

Although both in Death Valley and at Silver Lake the sequence of fan dynamics, and at Silver Lake the chronology, is fairly well understood, there has been little work linking detailed stratigraphic and morphologic evidence for past hillslope processes with the associated down-piedmont alluvial fan sequences. This paper is an attempt to rectify that. It is based on field observations made in the southern Soda Mountains, to the south of Silver Lake, in the catchment areas of the Zzyzx fans (Figure 4.1). These fans feed into the margins of Soda Lake, a modern playa but also formerly part of Pleistocene Lake Mojave.

THE ZZYZX FANS

The Field Evidence

The Zzyzx fans issue from the eastern flank of the Soda Mountains (Figure 4.2), draining catchments developed on Mesozoic granites and metamorphic rocks, including metamorphosed limestones (Figure 4.1). The fans range from small steep debris cones to larger fluvially dominated fans, with sedimentary and geomorphic sequences indicating a progressive trend from debris flow towards fluvial processes (Wells *et al.*, 1990b; Harvey, 1992).

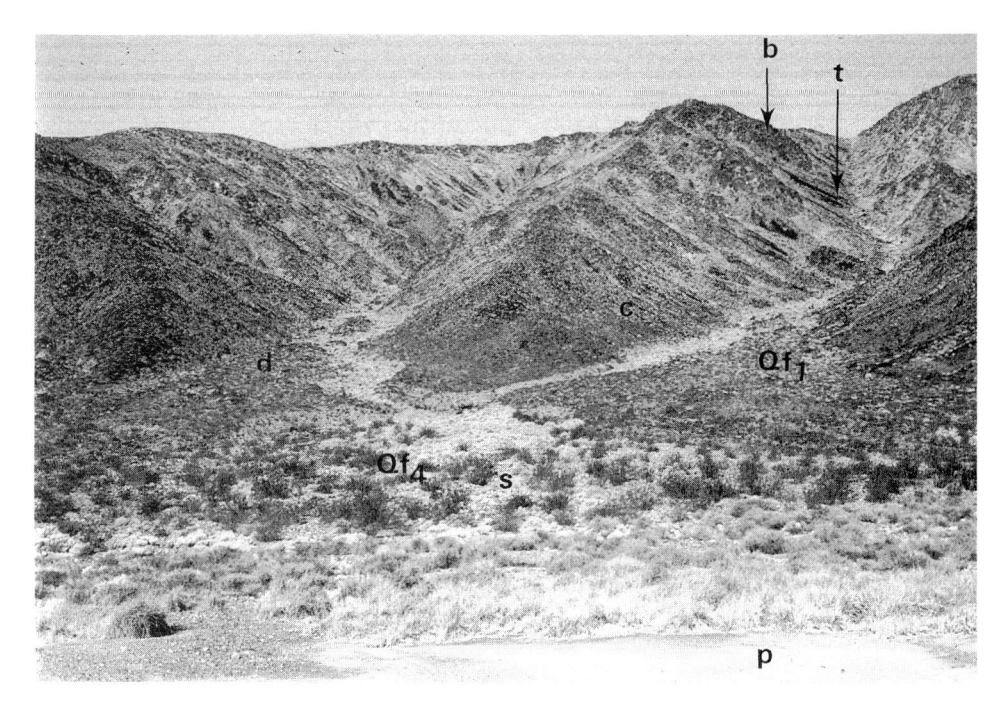

Figure 4.2 Zzyzx mountain-front fans: Johnny fan. Note: b, varnished bedrock outcrops; t, varnished talus slope; c, colluvial slope; Qf_1, Qf_4, fan surface age class, see text; d, fan deposits : debris flows; s, fan deposits : fluvial deposits; p, modern playa sediments

Both the fan surfaces and the hillslopes have been mapped on the basis of evidence for geomorphic processes and age. For the fan surfaces, field-based age-related criteria were established on the basis of the degree of rock varnish development, desert pavement and soil development (after McFadden *et al.*, 1987, 1989), and stage of $CaCO_3$ accumulation (after Gile *et al.*, 1966). A nomenclature based on that used by Wells *et al.* (1987) has been used to describe six age categories from Qf_0 for the oldest fan sediments (probably mid-Pleistocene) to Qf_5 for the youngest, recently and currently active, fan sediments.

Locally, ages of the fan segments have been assigned on the basis of the stratigraphic relationships between fan surfaces and the dated Lake Mojave shorelines (Wells *et al.*, 1987, 1989). Segments pre-dating the older shoreline (Qf_1) relate to the late Pleistocene, those dating from the period between the two shorelines (Qf_2), relate to the Pleistocene–Holocene transition and those post-dating the younger shoreline (Qf_{3-5}) are Holocene in age. The facies have been described using the criteria outlined by Wells and Harvey (1987) for modern alluvial fan sediments, into a range of debris flow, transitional and fluvial facies in alluvial fan sediments.

The Zzyzx Fan Sequences

The detailed sequence has been described elsewhere (Wells *et al.*, 1990b) and only a summary is given here (Figure 4.3). Sediments of Qf_0 age are only seen buried by younger sediments. The fan surfaces, especially in proximal areas, are dominantly of Qf_1 (late Pleistocene) age. Most of these deposits are debris flows and only on the larger fans do they grade distally into fluvial deposits. Clearly, the late Pleistocene here was a period of fan aggradation, dominated by debris flow deposition. By the end of the Pleistocene some limited fanhead trenching had begun; Qf_2 deposits form small inset terraces within the fanhead trenches of the larger fans as well as small progradation zones in the distal areas.

During the Holocene there was a dramatic change in the behaviour of the alluvial fans. On some of the smaller fans sedimentation ceased and the fan surfaces stabilised; on a few others there was minor incision of the fan surfaces but very little other geomorphic change (Harvey, 1992). On all of the larger fans and on most of the others, fanhead trenching of the proximal areas and extensive progradation of the distal areas by fluvial processes took place (Figure 4.3). This process has been episodic with major phases occurring in Qf_3 (mid Holocene?) times, again in Qf_4 (late Holocene?) and finally in Qf_5 (very recent) times. From Pleistocene to Holocene times there was a switch from proximal and mid-fan aggradation, dominantly by debris flow processes, to fanhead trenching and distal progradation, dominantly by fluvial processes. These are clearly responses to late Pleistocene to Holocene climatically induced changes in water and sediment supply from the catchment areas of alluvial fans.

THE ZZYZX HILLSLOPES

The Field Evidence

For the hillslopes a similar but simpler range of field-based age-related criteria was established primarily on the basis of the properties of rock varnish coatings, and only locally and where appropriate on the basis of other surface properties, such as soil or desert pave-

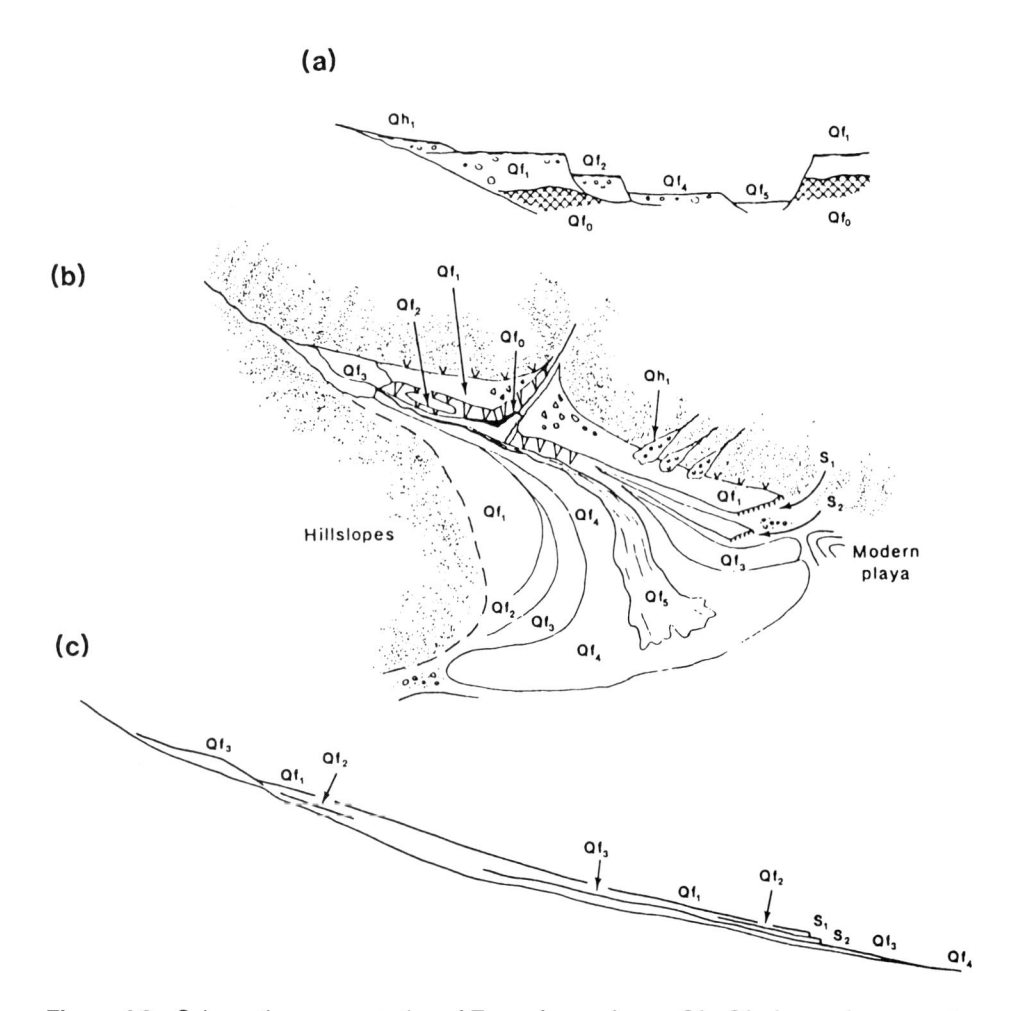

Figure 4.3 Schematic representation of Zzyzx fan surfaces. Qf$_0$–Qf$_5$, fan surface age class, see text; Qh, late Pleistocene hillslope debris flows; S$_1$, S$_2$, late Pleistocene–Holocene lake shorelines. (a) Cross-sectional relationships, near fan apex. (b) General relationships: illustrates Qf$_3$ deposits near apex (local); hillslope debris flows (Qh$_1$); relationship of fan sediments to lake shorelines (S$_1$, S$_2$). (c) Long profile relationships

ment properties. Because of the larger area involved and the wider range of surface type present, the age classification was limited to four groups, mid Pleistocene (=Q$_0$), late Pleistocene, including the Pleistocene–Holocene transition (=Q$_{1\text{ and }2}$), Holocene (=Q$_{3\text{ and }4}$), and active (=Q$_5$).

The oldest hillslope surfaces (Q$_0$), restricted to flatter areas near the divides and on lower spur tops, are characterised by a dominance of *in situ* weathering-related features. They exhibit well-developed desert pavements of small, closely interlocking clasts derived from the underlying bedrock, together with, on the upper but not all the lower surfaces, abundant caliche (calcrete) rubble (Lattman, 1977) derived from the weathering of a stage III to IV pedogenic calcrete horizon (Gile *et al.*, 1966). The bedrock clasts carry a complete cover of a dark rock varnish (reaching, for example, 2.5YR 2.5/2 on exposed surfaces and a paler 5YR 4/6 on their undersides). Below the pavement surfaces is a regolith within which shallow, weak to moderately developed Av and B horizons have developed (colours: 10YR 6/4 to 7.5YR 7/4). These overlie strongly developed stage III to IV petrocalcic horizons.

The majority of the hillslope surfaces appear to be late Pleistocene in age, in that they are mostly cut into the older geomorphic surfaces, and the exposed rock surfaces carry a rock varnish similar to that on rocks on alluvial fan surfaces $Qf_{1 \text{ and } 2}$, a dark varnish of 2.5 to 5YR 2.5/2 covering most of the rock surface (>75%), and partly or wholly obscuring the details of the lithology of the underlying rock.

Locally, two younger categories of hillslope feature can be recognised. Holocene forms lie within or dissect late Pleistocene forms. They carry a rock varnish ranging from very slight coloration of exposed surfaces to a partial cover reaching *ca.* 5YR 5/6, which can be correlated to that on alluvial fan surfaces Qf_{3-4}. The youngest surfaces of all are active and erosional, and dissect older surfaces. They are equivalent to fan stage Qf_5.

Hillslope Forms

Several different types of hillslope were mapped (Table 4.1, Figure 4.4), each relating to a different style of hillslope process, and age-assigned on the criteria outlined above. Old stable surfaces (Q_0) occur only in divide and spur-top locations. Exposed bedrock occurs on most of the steeper slopes and presumably at one time provided a major source of sediment to the hillslopes and alluvial fans below. Most outcrops are covered by a dark and complete rock varnish (correlating with a $Q_{1 \text{ to } 2}$ age), suggesting that there has been little sediment supplied from these sites since the end of the Pleistocene. Rock varnish does not develop on rock surfaces of the metamorphosed limestone, but most of these outcrops are above apparently stable talus or colluvial slopes and therefore also appear to have been stable for much of the Holocene. Only where bedrock is actively incised or undercut by a modern stream channel does the direct supply of sediment from bedrock into the modern sediment system appear to take place.

Planar slopes, mantled by talus, often occur downslope from bedrock outcrops (Figure 4.5(a)). These are mostly varnished by dark and complete rock varnish (=$Q_{1 \text{ or } 2}$), indicating stability since the end of the Pleistocene. Only in a very few localities are they not varnished, indicating some Holocene or recent activity. In some places the varnished scree slopes are dissected by recently active gullies and scars.

Many of the slopes, especially in the lower parts of the drainage basins, planar to concavo-convex in form, are blanketed by a stony regolith, within which the stones show a dark and complete ($Q_{1 \text{ to } 2}$-type) rock varnish, soils are present and there is some vegetation cover. These slopes may have been active as colluvial slopes during the late Pleistocene but have been largely stable throughout the Holocene.

Some of the slopes, especially the colluvial slopes, have been affected by hillslope debris flow activity. Debris flow trails, levees, breached lobes and lobe fronts (Wells & Harvey, 1987) are evident (Figure 4.5(b)). Most show a varnish of $Q_{1 \text{ to } 2}$-type, again indicating stability since the end of the Pleistocene. A few of the debris flows show evidence of minor reworking later, with small younger lobes carrying a lesser degree of rock varnish, equivalent to that on Qf_3 alluvial fan surfaces, suggesting a limited phase of mid Holocene(?) hillslope debris flow activity. Many of the larger hillslope debris flows (all with $Q_{1 \text{ to } 2}$-type

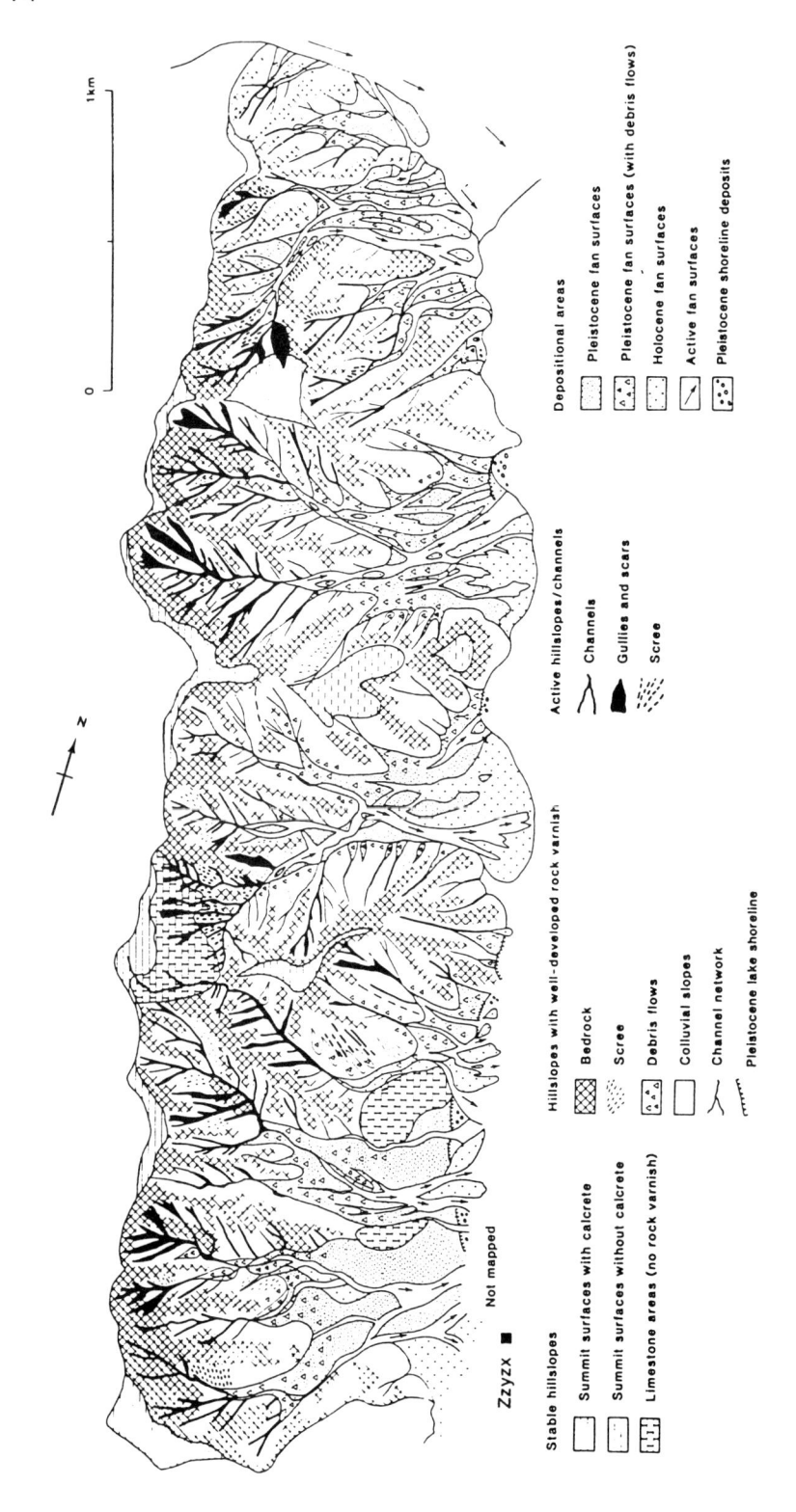

Figure 4.4 Summary map of hillslope morphology and age characteristics. Stable hillslopes either of Q_0 age or of indeterminate age (limestone areas); hillslopes with well-developed rock varnish are of late Pleistocene ($Q_{1\ to\ 2}$) age; hillslopes, now stable but active at some stage during the Holocene ($Q_{3\ to\ 4}$) are too small to be represented on this map; for the depositional areas Pleistocene fan surfaces refer to surfaces of $Qf_{1\ and\ 2}$ ages, Holocene fan surfaces to surfaces of $Qf_{3\ and\ 4}$ ages, active surfaces to Qf_5.

Table 4.1 Types of hillslope surface

Type	Age[a] (estimate)	Characteristics
Old stable surfaces	Q_0	Two surfaces, upper (older?) with calcrete rubble, both with well developed soils
Exposed bedrock	Q_{1-2}	Large areas on both granite and metamorphics, well varnished. Limestone areas: no varnish
	Q_5	Very small active areas
Talus	Q_{1-2}	Large areas of varnished scree now stabilised
	Q_5	Small active screes
Colluvial slopes	Q_{1-2}	Extensive areas, varnished boulders, some soil development, now stabilised
Hillslope debris flows	Q_{1-2}	Major wave of hillslope debris flow. Debris flow activity at end of Q_{1-2} time
	Q_{3-4}	Very localised (Q_3) debris flow activity
Dissected slopes	Q_5	Gullying and incision of drainage lines, active today

[a] By correlation of rock varnish with that on the alluvial fans. Q_0 = mid Pleistocene, Q_{1-2} = late Pleistocene, Q_{3-4} = Holocene, Q_5 = currently active.

of rock varnish) toe out on proximal or lateral alluvial fan surfaces (Figure 4.6), locally capping fan deposits of Qf_1 age and suggesting a major phase of hillslope debris flow activity right at the end of the Pleistocene, during the Pleistocene–Holocene transition.

The final type of hillslope form is the dissected hillslope, showing evidence of erosion by gullying or surface stripping. These dissected areas are directly related to the modern stream network, are currently active and appear to be wholly late Holocene in age.

Interpretation

A summary map, based on the field maps, shows the spatial distributions of these hillslope types, differentiating between Pleistocene and Holocene zones of activity (Figure 4.4). A summary of the spatial variations is given in Table 4.2. There is clear evidence of a major switch in hillslope process during the transition from the late Pleistocene into the Holocene. During the late Pleistocene most hillslopes appear to have been active, supplying abundant sediment to the fluvial system. Exposed bedrock, especially in the upper parts of the drainage basins, appears to have been weathered and to have supplied coarse clasts into the system via talus slopes. Such processes were particularly important in the basin headwaters on steep slopes, primarily but not exclusively on granite bedrock (Table 4.2). Lower down-valley, especially on the metamorphic rocks (Table 4.2), colluvial slopes fed sediment downslope.

During the Pleistocene–Holocene transition (Qf_2 times) there was a major phase of renewed hillslope debris flow activity, especially on the colluvial slopes on metamorphic rocks (Table 4.2). These debris flows supplied sediment towards the alluvial fans, but usually toed out at the base of the hillslopes rather than feeding much sediment into the alluvial fan systems themselves.

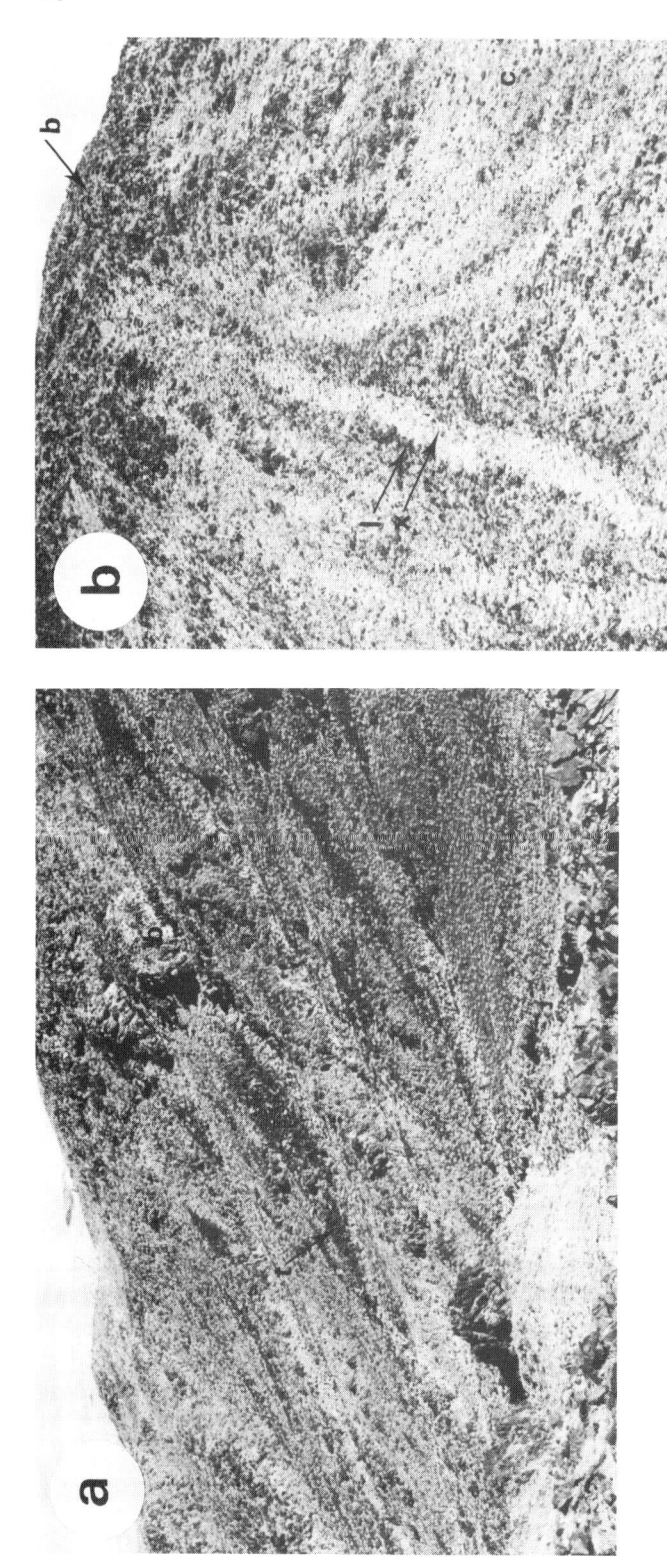

Figure 4.5 Quaternary hillslope morphology. (a) Part dissected, varnished talus slopes (t) below varnished bedrock outcrop (b). (b) Colluvial slope (c) below varnished bedrock outcrop (b), traversed by Q_{1-2} age hillslope debris flow. Note: debris flow track (X), levees (l) and minor Q_{3-4} reactivation (Y) within Q_{1-2} lobes (Z)

Figure 4.6 (a) Qh$_1$ hillslope debris flows (h) toeing out on Qf$_1$ alluvial fan surfaces (f). Sediment in foreground is Qf$_4$ sediment within the fanhead trench. (b) Details of Qh$_1$ hillslope debris flow sediments overlying Qf$_1$ fluviatile fan sediments

Table 4.2 Relative abundance of hillslope surface types

Drainage basin[a]	Rock type[b]	RC[c]	Hillslopes[d] Stable OS	RO	Sc	CS	Df	Active Sc	Ds	Fan behaviour[e] P	H
SO1	GM	2	2	3	1	3	0	1	1	DF	TF
SO2	GM	2	0	3	2	2	2	0	2	D	D
SO3	GML	2	1	3	3	2	2	0	3	DF	I
SPR	GML	2	2	2	1	3	1	1	1	D	I
ZZX	GML	2	2	3	2	2	1	0	3	DF	TF
JSH	ML	2	2	2	1	3	3	0	2	F	TF
CT/PC	M	1	1	1	0	3	1	0	1	D	P
CV-s	ML	2	2	2	1	2	3	1	2	F	TF
CV-n	M	2	2	1	0	3	3	0	1	DF	I
JN-s	M	2	2	2	1	2	3	0	3	DF	TF
JN-n	MG	2	2	3	1	2	2	0	3	DF	I
STV	M	1	0	1	0	3	0	0	0	D	(D)
MSQ	M	1	0	2	0	2	2	1	1	D	DF
VLT	G	2	1	3	2	1	2	2	3	DF	TF
SOL	G	1	1	2	0	2	1	0	1	DF	I
NOR	G	1	1	2	0	2	1	0	1	DF	I

[a] For locations, see Figure 4.1.
[b] G, granite; M, metamorphics; L, limestone (in order of abundance).
[c] RC, relief class (within drainage basin), 1, <250 m; 2, >250 m.
[d] OS, old surfaces; RO, rocky outcrop; Sc, scree; CS, colluvial slopes; Df, debris flows; Ds, dissection; 0, not present; 1, present; 2, common; 3, abundant.
[e] P, pleistocene; H, Holocene; D, debris flow; F, fluvial sedimentation; DF, proximal debris flows, distal fluvial sedimentation; TF, fanhead trenching, distal fluvial sedimentation; I, dissection throughout; P, passive (brackets indicate limited importance).

The fan response to high rates of sediment supply during the late Pleistocene (Qf_1) was aggradation, primarily by debris flow processes, burying previous fan surfaces. Only on the larger fans did debris flows dilute downfan to allow fluvial deposition. Only at the very end of the Pleistocene–Holocene transition (Qf_2 times) does fanhead trenching begin.

During the early Holocene the hillslopes stabilised, allowing rock varnish to develop on exposed surfaces of hitherto active free faces, talus slopes, bouldery colluvial slopes and hillslope debris flows. The fan environments became starved of sediment and responded by fanhead trenching and distal progradation, in a switch from a sediment-rich debris flow regime to a sediment-poor fluvial regime. Fan dissection was accentuated by the switch from debris flow deposition on relatively steep fan surfaces to fluvial transport over the same gradients, involving excess stream power and resulting in dissection of the proximal fan surfaces (Harvey, 1984b, 1989).

During the Holocene, on the steepest slopes in the headwaters of the larger watersheds, especially but not exclusively on granite rocks (Table 4.2), hillslope dissection by gullying occurred. This caused some sediment supply to the fan environment, but on a much more restricted basis than hitherto. Episodically, on several occasions (during $Qf_{3 \text{ and } 4}$ times), there was a minor renewal of sediment supply from the hillslopes, causing localised reactivation of hillslope debris flows and supplying debris flow deposits to fan apex areas, Qf_3 or Qf_4

deposits locally burying Qf$_1$ sediments (Figure 4.3). However, whether these edisodes could be related to Holocene climatic fluctuations or simply represent a delayed and complex response to the Pleistocene to Holocene changes is not certain. As a whole the Holocene is overwhelmingly a period of low coarse sediment supply from the hillslopes, dissection of the fan proximal regions and sedimentation, by fluvial processes, causing fan progradation in distal areas.

DISCUSSION

Climatic Implications

There are clear contrasts between late Pleistocene and Holocene geomorphic systems (Figure 4.7), which suggest contrasts between Pleistocene and Holocene climates. The conventional interpretation would be of a transition from a mild(?), moist 'pluvial' late Pleistocene to a hot, arid Holocene, with minor fluctuations since, to account for variations in sediment production during the Holocene. There has been some discussion (Brackenridge, 1978; Galloway, 1983) on the amount by which late Pleistocene climates would need to have been cooler or wetter than those of today, in order to account for either vegetation distributions (Van Devender & Spaulding, 1983; Spaulding, 1990) or high lake levels (Smith & Street-Perrott, 1983). The transition period itself is interpreted to have had an increased monsoonal airflow and associated rains (Enzel et al., 1990), apparently sufficient to result in the widespread growth of succulents and grasses in the northern Mojave Desert, increased discharges in the Colorado and Mojave rivers, and high lake levels in pluvial Lake Mojave (Bull, 1991; Wells et al., 1989). By 8000–9000 yr BP, the disappearence of California juniper and replacement of resident species by desert scrub documents the establishment of the typical hot, dry Holocene climatic regime (Spaulding & Graumlich, 1986).

The geomorphological evidence from Zzyzx very strongly suggests major differences in sediment production and transport mechanisms between the late Pleistocene and the Holocene brought about by changes during the Pleistocene–Holocene transition. The contrasts imply differences in *both* temperature *and* precipitation between the late Pleistocene (Qf$_1$) and the Holocene (Qf$_3$ onwards). High rates of sediment generation characterised many of today's arid areas during the late Pleistocene, both in the American south-west (Meyer et al., 1984) and in other arid regions, notably Israel (Gerson, 1982). These were followed during the Holocene, under conditions of increased aridity, by periods of hillslope stripping and enhanced fluvial activity (Bull & Schick, 1979; Bull, 1991).

During the Pleistocene (Qf$_1$) the Zzyzx hillslopes saw greatly enhanced weathering rates over those that exist today, indicating a much greater moisture availability and probably lower temperatures. It is easy to envisage frost action as an important weathering process in the Panamint Mountains, supplying the Death Valley fans, as they reach altitudes of over 3000 m a.s.l, but the Soda Mountains at Zzyzx barely reach an altitude of 600 m. It could be argued that considerably wetter and colder climates in the late Pleistocene would have been necessary for such weathering processes to have occurred there. Similarly, for the widespread mass movement processes to transport the sediment downslope, especially by debris flow processes, as observed preserved on the Zzyzx hillslopes, high soil moisture contents and high rainfall intensities or durations would be required (Caine, 1980) during the late Pleistocene and the Pleistocene–Holocene transition.

80 Environmental Change in Drylands

LATE PLEISTOCENE

Old stable surfaces

Exposed bedrock
- Limestone
- Other hillslopes (with Q_{1-2} rock varnish)

Depositional slopes (with Q_{1-2} rock varnish)
- Talus slopes
- Colluvial slopes
- Debris flows

Late Pleistocene Alluvial fans (Qf_{1-2})

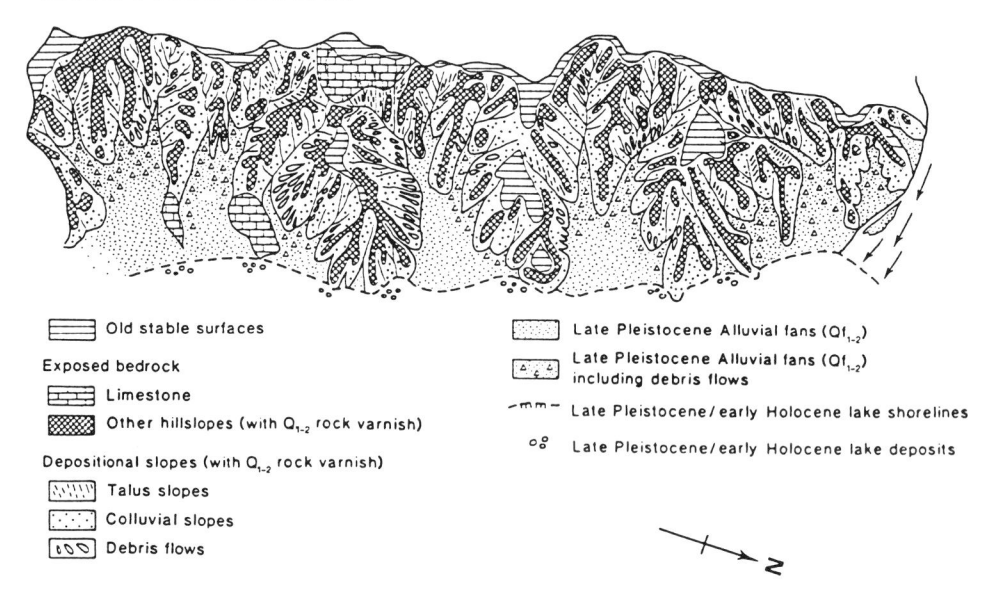

Late Pleistocene Alluvial fans (Qf_{1-2}) including debris flows

Late Pleistocene/early Holocene lake shorelines

Late Pleistocene/early Holocene lake deposits

HOLOCENE

ACTIVE FORMS
Erosional
- Active screes
- Active dissection
- Fanhead trench

Depositional
- Fan surfaces

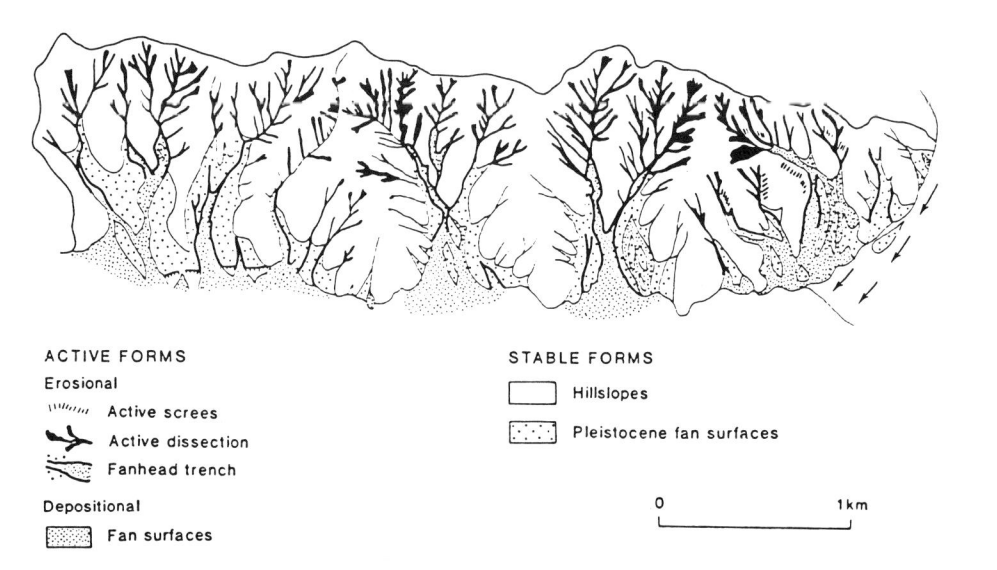

STABLE FORMS
- Hillslopes
- Pleistocene fan surfaces

0   1 km

Figure 4.7 Summary maps showing contrasting zones of late Pleistocene and Holocene geomorphic activity. Definitions as for Figure 4.4. With the exception of the old stable surfaces (Q_0), the upper map shows features erosionally or depositionally active during the late Pleistocene (Q_1) and/or Pleistocene–Holocene transition (Q_2); the lower map shows features active since

Within the eastern Mojave Desert, climatic inferences from plant macrofossils in packrat middens suggest as much as a 50% decrease in precipitation and a mean warming of *ca.* 3°C during the transition from late Pleistocene to Holocene climatic regimes (Van Devender, 1973). Estimates of temperature differences between the late Pleistocene and the Holocene, derived from analyses of stable hydrogen isotopes in cellulose from fossil packrat middens from the Great Basin (Long *et al.*, 1990), suggest that growing season temperatures 3–4°C cooler than present may have persisted between *ca.* 30 000 and 18 000 BP.

Wells *et al.* (1989) have used a simple hydrological model to reconstruct the end Pleistocene Mojave River discharge necessary to result in overflow of pluvial Lake Mojave. Their study suggests that neither increased precipitation nor reduced temperature alone is sufficient to account for the late Pleistocene lake. Temperatures 10°C lower than those of today, and precipitation 50% higher, in the Transverse Ranges of southern California, which feed the Mojave River, would result in the lake levels similar to those of late Pleistocene Lake Mojave.

In addition to climatic changes, the presence of hygroscopic salts such as aerosols, related to episodic drying of pluvial Lake Mojave during the late Pleistocene (Wells *et al.*, 1989; Brown *et al.*, 1990), may have been significant for the mechanical weathering of bedrock on hillslopes neighbouring the lake. Such processes may have been more important than frost action, because of the low elevation of the Soda Mountains. However, perhaps during the late Pleistocene, an increased frequency of deep southward penetrations of cold polar air masses into the Zzyzx region, as happened in January 1990, could have enhanced freeze–thaw activity at these elevations. Evidence for this type of activity is given by Friedman & Smith (1972), who suggest that the cold air masses and jet-stream positions of the 1968–9 winter may have been similar to those of late Pleistocene climates.

Under the present climatic conditions, as in most parts of the south-west, hillslope processes are dominated by overland flow, and sediment transport by fluvial processes during flood conditions (Baker, 1977). There is no evidence for recent hillslope debris flows in this part of the Soda Mountains, suggesting that under the present climatic regime they would occur rarely, if at all. On 20 August 1988 a major storm occurred over Zzyzx, when over 60 mm of rain fell in *ca.* 45 minutes. This was a rare event, and although we do not have a precise estimate, its return period must be well in excess of 20 years. There was considerable flood damage to the Interstate Highway where it crosses the Soda Mountains west of Zzyzx, and the Zzyzx access road was totally destroyed. Despite major geomorphic change on the Zzyzx fans by fluvial processes resulting from this event, no evidence of any geomorphic change on the hillslopes by mass movement processes could be detected after the storm.

SUMMARY

It is clear that during the late Pleistocene the climate of the eastern Mojave Desert would have been both colder and wetter than that of today. This would account for the high rates of weathering and sediment production on the hillslopes and sediment transport down the hillslopes by mass movement processes. These conditions led to a state of excess sediment supply to the alluvial fans, and hence of fan aggradation, dominantly by debris flow processes. Conditions began to change during the Pleistocene–Holocene transition. There was a last main wave of hillslope debris flow activity, but much of the sediment moved did not reach the alluvial fans. Limited fanhead trenching and distal progradation began during this period. Following desiccation (and warming) to an arid climate in the Holocene, the hillslopes stabilised, but during storm and flash flood conditions the high rates of runoff generated excess stream power, causing fanhead trenching and distal fan progradation. The minor variations in fan behaviour during the Holocene may reflect smaller scale climatic fluctuations or simply represent a delayed complex response to the major Pleistocene to Holocene changes.

ACKNOWLEDGEMENTS

The fieldwork on the Zzyzx fans was initiated by the late J.-L. Moissec. The authors are grateful to the staff of California State University Desert Studies Center, Zzyzx for general support and assistance in the field and to the drawing office and photographic sections of the Department of Geography, University of Liverpool, for producing the illustrations. A.M.H. wishes to acknowledge a grant from the Royal Society towards the costs of the fieldwork.

REFERENCES

Baker, V. R. (1977). Stream channel response to floods, with examples from central Texas. *Geological Society of America, Bulletin*, **88**, 1057–71.

Brown, W. C., Wells, S. G., Enzel, Y., Anderson, R. Y. and McFadden, L. D. (1990). The late Quaternary history of pluvial Lake Mojave-Silver Lake and Soda Lake basins, California. In Reynolds, J. (ed.), *At the End of the Mojave: Quaternary Studies in the Eastern Mojave Desert*, San Bernadino County Museum Association and 1990 Mojave Desert Quaternary Research Center Symposium, pp. 55–72.

Blissenbach, E. (1954). Geology of alluvial fans in semi arid regions. *Geological Society of America, Bulletin*, **55**, 175–90.

Brackenridge, G. R. (1978). Evidence for a cold, dry full-glacial climate in the American southwest. *Quaternary Research*, **9**, 22–40.

Bull, W. B. (1977). The alluvial fan environment. *Progress in Physical Geography*, **1**, 222–70.

Bull, W. B. (1979). Threshold of critical power in streams. *Geological Society of America, Bulletin*, **90**, 453–64.

Bull, W. B. (1991). *Geomorphic Responses to Climatic Change*, Oxford University Press, Oxford.

Bull, W. B. and Schick, A. P. (1979). Impact of climatic change on an arid watershed: Natal Yael, Southern Israel. *Quaternary Research*, **11**, 153–71.

Caine, N. (1980). The rainfall intensity-duration control of shallow landslides and debris flows. *Geografiska Annaler*, **62A**, 23–7.

Denny, C. S. (1965). Alluvial fans in Death Valley region, California and Nevada. United States Geological Survey, Professional Paper 466.

Dohrenwend, J. C., Bull, W. B., McFadden, L. D., Smith, G. I., Smith, R. S. U. and Wells, S. G. (1991). Quaternary geology of the Basin and Range Province in California. In Morrison, R. B. (ed.), *Quaternary Nonglacial Geology: Conterminous U. S., The Geology of North America*, Volume K-2, pp. 321–52.

Dorn, R. I. (1983). Cation-ratio dating: a new rock varnish age-determination technique. *Quaternary Research*, **20**, 49–73.

Dorn, R. I. (1984). Cause and implications of rock varnish microchemical laminations. *Nature*, **310**, 767–70.

Dorn, R. I. (1988). A rock varnish interpretation of alluvial-fan development in Death Valley, California. *National Geographic Research*, **4**, 56–73.

Dorn, R. I. and Oberlander, T. M. (1981). Rock varnish origin, characteristics and usage. *Zeitschrift für Geomorphologie*, **NF25**, 420–36.

Dorn, R. I., Jull, A. J. T, Donahue, D. J., Linick, T. W. and Toolin, L. T. (1989). Accelerator mass spectrometry radiocarbon dating of rock varnish. *Geological Society of America, Bulletin,* **101**, 1363-72.

Dorn, R. I., De Niro, M. J. and Ajie, H. O. (1987). Isotopic evidence for climatic influence on alluvial fan development in Death Valley, California. *Geology,* **15**, 108–10.

Enzel, Y., Anderson, R. Y., Brown, W. J., Cayan, D. R. and Wells, S. G. (1990). Tropical and subtropical moisture and southerly displaced North Pacific storm track: factors in the growth of late Quaternary lakes in the Mojave Desert. In Betancourt, J. L. and Mackay, A. M. (eds.), *Proceedings of the 6th Annual Pacific Climate (PACLIM) Workshop,* 5–8 March 1989, California Department of Water Resources, Interagency Ecological Studies Program, Technical Report 23, pp. 135–9.

Friedman, I. and Smith, G. I. (1972). Deuterium content in snow as an index to winter climate in the Sierra Nevada area. *Science,* **176**, 790–3.

Galloway, R. W. (1983). Full-glacial southwestern United States: mild and wet or cold and dry? *Quaternary Research* **19**, 236–48.

Gerson, R. (1982). Talus relics in deserts: a key to major climatic fluctuation. *Israel Journal of Earth Sciences,* **31**, 123–32.

Gerson, R., Bull, W. B., Fleischhauer, H. L., *et al.* (1978). Origin and distribution of gravel in stream systems of arid regions. US Air Force Office of Scientific Research, Contract Report F49–620–77-C-0115.

Gile, L. H., Peterson, F. F. and Grossman, R. B. (1966). Morphological and genetic sequence of carbonate accumulation in desert soils. *Soil Science,* **101**, 347–60.

Harvey, A. M. (1984a) Aggradation and dissection sequences in Spanish alluvial fans: influence on morphological development. *Catena,* **11**, 289–304.

Harvey, A. M. (1984b) Debris flows and fluvial deposits in Spanish Quaternary alluvial fans: implications for fan morphology. In Koster, E. H., and Steel, R. (eds.), *Sedimentology of Gravels and Conglomerates,* Canadian Society of Petroleum Geologists, Memoir 10, pp. 123–32.

Harvey, A. M. (1989). The occurrence and role of arid zone alluvial fans. In Thomas, D. S. G. (ed.), *Arid Zone Geomorphology,* Belhaven Press, London, pp. 136–58.

Harvey, A. M. (1990). Factors influencing Quaternary alluvial fan development in southeast Spain. In Rackocki, A. H. and Church, M. J. (eds.), *Alluvial Fans: A Field Approach,* Wiley, Chichester, pp. 247–69.

Harvey, A. M. (1992). Controls on sedimentary style on alluvial fans. In Billi, P., Hey, R. D., Thorne, C. R. and Tacconi, P. (eds.), *Dynamics of Gravel-Bed Rivers,* Wiley, Chichester, pp. 519–535.

Hooke, R. le B. (1967). Processes on arid region alluvial fans. *Journal of Geology,* **75**, 438–60.

Hunt, C. B. (1975). *Death Valley: Geology, Archaeology, Ecology,* University of California Press, Berkeley.

Hunt, C. B. and Mabey, D. R. (1966). Stratigraphy and structure, Death Valley, California. United States Geological Survey, Professional Paper 494A.

Lattman, L. H. (1977). Weathering of caliche in southern Nevada. In Doehring, D. E. (ed.), *Geomorphology in Arid Regions,* Allen & Unwin, London, pp. 221–31.

Long, A., Warnecke, L., Betancourt, J. L. and Thompson, R. S. (1990). Deuterium variations in plant cellulose from fossil packrat middens. In Betancourt, J. L., Van Devender, T. R. and Martin, P. S. (eds.), *Packrat Middens, The Last 40,000 Years of Biotic Change,* University of Arizona Press, Tucson.

McFadden, L. D., Wells, S. G. and Jercinovich, M. J. (1987). Influences of eolian and pedogenic processes on the origin and evolution of desert pavements. *Geology,* **15**, 504–8.

McFadden, L. D., Ritter, J. B. and Wells, S. G. (1989). Use of multiparameter relative-age methods for age estimation and correlation of alluvial fan surfaces on a desert piedmont, eastern Mojave Desert, California. *Quaternary Research,* **32**, 276–90.

Meyer, L., Gerson, R. and Bull, W. B. (1984). Alluvial gravel production and deposition: a useful indicator of Quaternary climatic changes in deserts (a case study in southwestern Arizona). In Shick, A. P. (ed.), *Channel Processes: Water, Sediment, Catchment Controls,* Catena Supplement, **5**, pp. 137–51.

Oberlander, T. M. (1989). Slope and pediment systems. In Thomas, D. S. G. (ed.), *Arid Zone Geomorphology,* Belhaven Press, London, pp. 56–84.

Ritter, J. B. (1991). The response of alluvial fan systems to late Quaternary climatic change and local base-level change, eastern Mojave Desert, California. In Reynolds, J. (ed.) *At the End of the Mojave: Quaternary Studies in the Eastern Mojave Desert*, San Bernardino County Museum Association and 1990 Mojave Desert Quaternary Research Centre Symposium, pp. 117–18.

Slate, J. L. (1991). Quaternary stratigraphy, geomorphology, and ages of alluvial fans in Fish Lake Valley. In *Guidebook for Field Trip to Fish Lake Valley, California-Nevada*, Pacific Cell, Friends of the Pleistocene, pp. 94–113.

Smith, G. I. and Street-Perrott, F. A. (1983). Pluvial lakes of the western United States. In Porter, S. C. (ed.), *Late Quaternary Environments in the United States. Volume I, The Late Pleistocene*, University of Minnesota Press, Minneapolis, pp. 190–214.

Spaulding, W. G. (1990). Vegetation dynamics during the last deglaciation, southeastern Great Basin, USA. *Quaternary Research,* **33**, 188–203.

Spaulding, W. G. and Graumlich, L. J. (1986). The last pluvial climatic episodes in the deserts of southwestern North America. *Nature,* **320**, 441–4.

Van Devender, T. R. (1973). Late Pleistocene plants and animals of the Sonoran Desert: a survey of ancient packrat middens in southwestern Arizona. Unpublished PhD thesis, University of Arizona, Tucson.

Van Devender, T. R. and Spaulding, W. G. (1983). Development of vegetation and climate in the south western United States. In Wells, S. G. and Haragan, D. R. (eds.), *Origin and Evolution of Forests*, University of New Mexico Press, Albuquerque, pp. 131–56.

Wells, S. G. and Harvey, A. M. (1987). Sedimentologic and geomorphic variations in storm generated alluvial fans, Howgill Fells, northwest England. *Geological Society of America, Bulletin,* **98**, 182–98.

Wells, S. G. and McFadden, L. D. (1987). Comment and reply on 'Isotopic evidence for climatic influence on alluvial fan development in Death Valley, California'. *Geology,* **15**, 1178–9.

Wells, S. G., McFadden, L. D., Dohrenwend, J. C., *et al.* (1984). Geomorphic history of Silver Lake, eastern Mojave Desert, California: an example of the influence of climatic change on desert piedmonts. In Dohrenwend, J. C. (ed.), *Surficial Geology of the Eastern Mojave Desert*, Field trip 14 Guidebook, Geological Society of America, pp. 69–87.

Wells, S. G., McFadden, L. D. and Dohrenwend, J. C. (1987). Influence of late Quaternary climatic change on geomorphic and pedogenic processes on a desert piedmont, eastern Mojave Desert, California. *Quaternary Research,* **27**, 130–46.

Wells, S. G., Anderson, R. Y., McFadden, L. D., Brown, W. J., Enzel, Y. and Moissec, J.-L. (1989). Late Quaternary paleohydrology of the Eastern Mojave River drainage basin, southern California: quantitative assessment of the Late Quaternary hydrologic cycle in large arid watersheds. New Mexico Water Resources Research Institute, Technical Completion Report, Project 14-08-0001-G1312, 253 pp.

Wells, S. G., McFadden, L. D. and Harden, J. (1990a). Preliminary results of age estimations and regional correlations of Quaternary alluvial fans within the Mojave Desert of southern California. In Reynolds, J. (ed.), *At the End of the Mojave: Quaternary Studies in the Eastern Mojave Desert*, San Bernardino County Museum Association and 1990 Mojave Desert Quaternary Research Center Symposium, pp. 45–53.

Wells, S. G., Moissec, J-L. and Harvey, A. M. (1990b). Sedimentary processes during Holocene and Pleistocene alluvial fan deposition along the southern Soda Mountains, California: a summary. In Reynolds, J. (ed.), *At the End of the Mojave: Quaternary Studies in the Eastern Mojave Desert*, San Bernardino County Museum Association and 1990 Mojave Desert Quaternary Research Center Symposium, pp. 39–44.

5 Natural Stabilisation Mechanisms on Badland Slopes: Tabernas, Almeria, Spain

R. W. ALEXANDER
Department of Geography, Chester College, UK

A. M. HARVEY
Department of Geography, University of Liverpool, UK

A. CALVO
Departamento de Geografia, Universidad de Valencia, Spain

P. A. JAMES
Department of Geography, University of Liverpool, UK

and

A. CERDA
Departamento de Geografia, Universidad de Valencia, Spain

ABSTRACT

Complex multiple-age badlands are cut in Upper Miocene mudstones of the uplifted and dissected Tabernas basin, Almeria, in semi-arid south-east Spain. Six phases of episodic badland development and subsequent stabilisation can be recognised, with ages ranging from the late Pleistocene to the present day. The three earlier phases are based by extensive pediment surfaces, the three younger phases by basally incising streams and limited pediment development.

The morphology suggests that during each phase, once incision ceases, basal stabilisation takes place by the development of a protective stone cover on the surface, lichen colonisation and ultimately by vegetation colonisation by higher plants and the development of a soil. Analyses of vegetation and soils show developmental sequences related to surface age. The sequences show a reduction in the effectiveness of erosion through the development of biotic crusts, increased rainfall interception and enhanced infiltration capacity.

Each of the stabilisation processes reduces runoff and sediment generation, thereby reinforcing stabilisation. Rainfall simulation experiments carried out on the badland surfaces show reduced erosion rates with increasing surface age.

INTRODUCTION

Much of the literature on badland geomorphology tends to emphasise erosion primarily by overland flow processes and to focus on the influence of different material properties or different surface cover on erosion rates. There are fewer studies that deal with multiple

---

Environmental Change in Drylands: Biogeographical and Geomorphological Perspectives.
Edited by A. C. Millington and K. Pye. © 1994 John Wiley & Sons Ltd

processes, the relationships between process and form or with progressive changes in rates of erosion during badland evolution (Harvey & Calvo, 1991; Calvo *et al.*, 1991a; Harvey, 1992).

Recent work has identified the importance of multiple processes for badland evolution and how process interactions may change during badland development. Erosion rates may change as process interactions change, and ultimately may result in the cessation of badland processes if erosion rates are reduced sufficiently to allow the stabilisation of the surface. Possible natural stabilisation mechanisms include reductions in erosion rates resulting from: progressive slope angle reduction during slope erosion (Campbell, 1989), the development of a stone cover (Poesen, 1986; Poesen *et al.*, 1990), lichen colonisation of the surface (Alexander & Calvo, 1990), development of a vegetation cover through the colonisation of higher plants (Thornes, 1990) and ultimately soil development allowing greater infiltration rates.

The chief factors of soil development which may reduce erosion rates in this environment are organic matter incorporation and associated soil structural development, the reduction of sodium concentration in the topsoil, the greater depth of soil and the reduction in mean particle size. Most of these pedogenic changes would tend to increase infiltration capacity and resistance to detachment by rainsplash and surface flow. This paper examines possible mechanisms of badland stabilisation on complex multiple-age badlands at Tabernas, Almeria, south-east Spain. Evidence for previous erosion and stabilisation sequences is examined in relation to slope, vegetation (vascular plant and lichen cover) and soil development. The influence of these variables on contemporary surface processes is investigated by means of rainfall simulation experiments.

THE TABERNAS BADLANDS

The Tabernas badlands form one of the most extensive areas of badlands in semi-arid south-east Spain. They are cut in an uplifted sequence of Upper Miocene (Tortonian) marls, shales and turbidites, with occasional interbedded sandstone bands. These rocks form the lower part of the basin fill in the Neogene Tabernas basin within the Betic cordillera (Harvey, 1987; Weijermars *et al.*, 1985). The basin was subjected to post-Tortonian deformation and sustained uplift, continuing through into the Pliocene and Quaternary. The juxtaposition of the maximum uplift at the western end of the Tabernas basin and the downfaulted Rioja corridor, resulted in coincidence of the zone of maximum tectonically induced relief with the outcrop of the thick sequence of weak Tortonian sedimentary rocks. During the Quaternary, under dry climates, this has led to the development of a landscape dominated by deep dissection, characterised by canyons along main *ramblas*, and by poorly vegetated, erosive badland slopes away from local *rambla* base levels.

Uplift and dissection during the Quaternary appears to have been episodic, resulting in a stepped landscape of multiple-age badlands separated by dissected pediment surfaces, some grading to river terraces of the Rambla de Tabernas. There are two extensive river terraces, one at a high level above the modern *rambla*, comprising gravels capped by calcrete. This terrace probably dates from the Middle Pleistocene or earlier (Harvey, 1984, 1987). The lower terrace is very extensive, comprising gravels without a calcrete cap, and probably dates from the late Pleistocene. Since then, dissection has taken place to modern river level and badlands have developed through several stages of hillslope evolution during the Holocene. The possible causes for this episodicity in landform development include episodic uplift and staggered erosional response, Quaternary climatic fluctuations, but also intrinsic badland erosion and stabilisation cycles. This paper examines the mechanisms behind the last of these.

The area is semi-arid, in the driest part of Europe, with a mean annual precipitation of ca. 170 mm (Geiger, 1970). There are hot, dry summers and mild winters, with most precipitation falling as intense storm rainfall, especially in late September and October. Pleistocene climates in this part of the Mediterranean were also dry (Rohdenburg & Sabelberg, 1973, 1980; Sabelberg, 1977).

Modern geomorphic processes on the hillslopes are dominated by overland flow processes, producing badland surfaces, especially on almost bare south-facing slopes. With only limited swelling and cracking material properties, overland flow and rilling processes are dominant (Harvey, 1982, 1987; Calvo & Harvey, 1992). There is little evidence of piping or of major mass movement. North-facing slopes show a thicker regolith development, often with lichen and some higher plant cover (Alexander & Calvo, 1990). There is less evidence of high rates of surface processes. Swelling and cracking may be better developed, but rilling is less well defined, and locally there is evidence of small-scale mass movement.

MORPHOLOGICAL DEVELOPMENT

The study area is south of Bar Alfaro, 5 km south-west of Tabernas (Figure 5.1), where badlands are cut into a sequence of grey marls, punctuated by occasional, very thin (< 5 cm), marly limestone bands. At the top of the marl sequence, outcropping above the main area of badlands, is a thick sandstone band which forms a caprock escarpment (Figures 5.1 and 5.2). Within a multiple-age badland zone, south of the Rambla de Tabernas, at least six stages can be identified in the morphological evolution of the badlands (Calvo et al., 1991a). The earlier stages, A, B and C, represent successive badland and pediment stages, which truncate the structures in the marl. They form successive stages in the long-term hillslope evolution through scarp retreat. Stages B and C are probably equivalent to the extensive (late Pleistocene) lower river terrace of the Rambla de Tabernas. Following stage C, erosion switched to linear incision along the main drainage lines rather than by extensive pediment development. Stages D and E are small basal pediments within the incised stream valleys. Stage F is the youngest, modern surface, adjacent to the modern incised channels.

Surface morphology has been mapped in the field, and the surfaces classified A–F on the basis of height differences and morphological continuity (Figures 5.1, 5.2 and 5.3). The slope profile characteristics change during badland evolution. Slopes are initiated by undercutting by an incising or laterally eroding mainstream, the initial slope failure producing slope angles from near vertical to ca. 60°. Slope reduction then takes place to ca. 40–50° for stream-based slopes. Further slope reduction to ca. 30–35° occurs after the stream ceases to incise, and a mini pediment develops at the slope base. These trends in slope development are clear both over the long-term evolution of slopes through stages A to F, and over the shorter term, for both stream-based and pediment-based slopes, through stages D to F (Figure 5.4).

On the badland surfaces there is evidence for auto-stabilisation processes during badland evolution. There is clear evidence for a reduction in slope angle (see above), for stone accumulation on the lower slopes, for lichen colonisation and for colonisation by higher

88 Environmental Change in Drylands

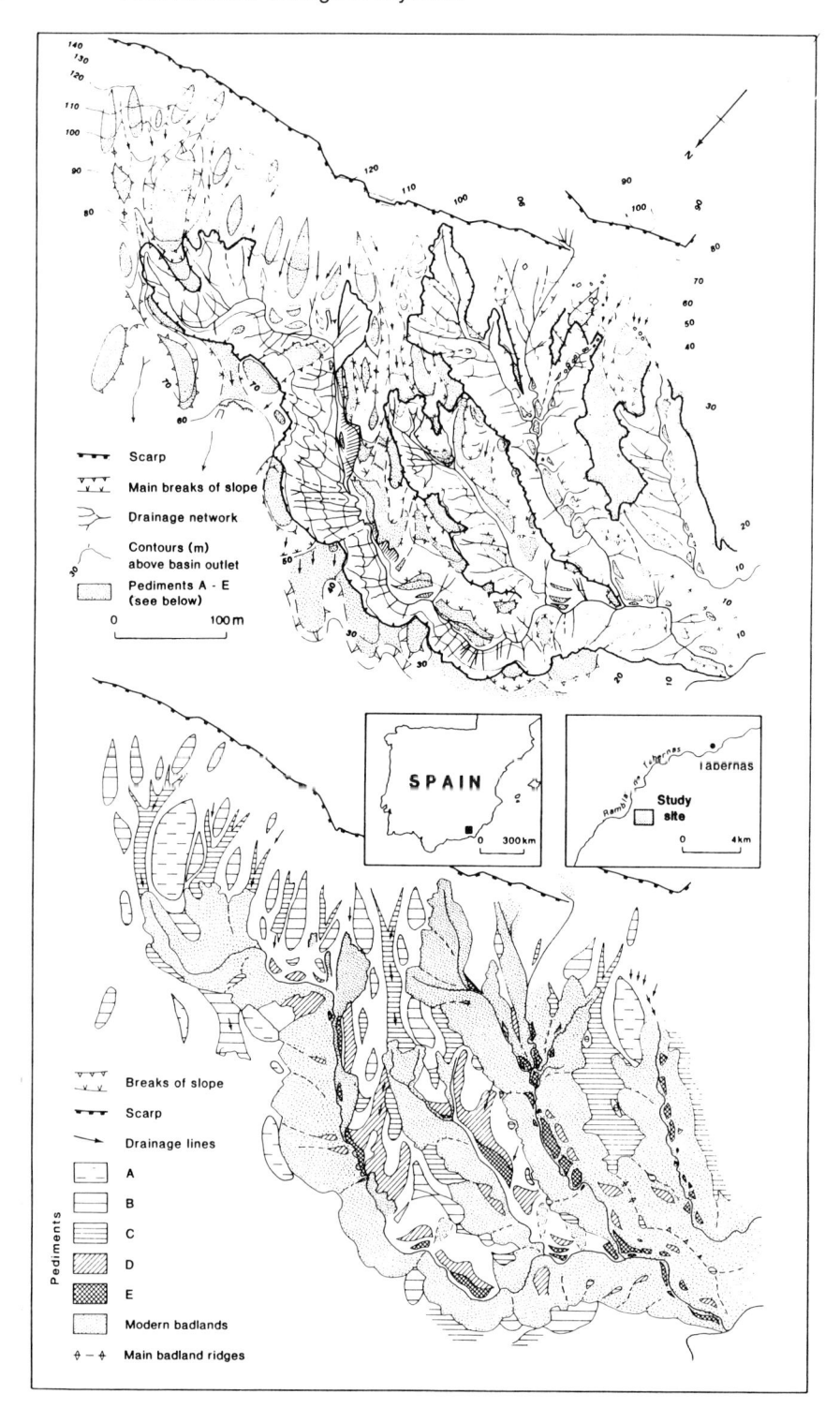

Figure 5.1 Location, morphology and relative surface ages of the Tabernas study area

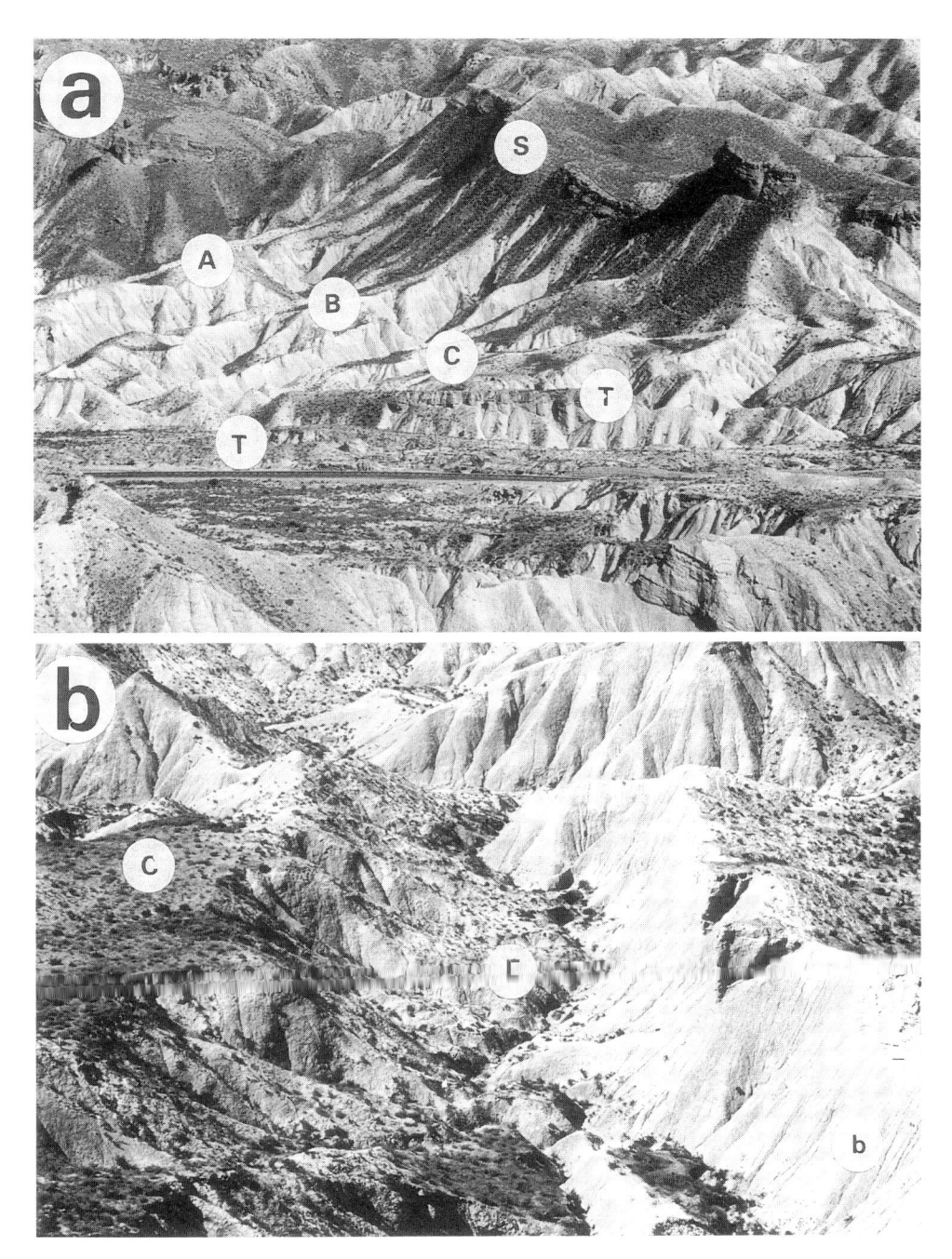

Figure 5.2 The study site in the Tabernas badlands. (a) General view of the study site from the west. Note: sandstone scarp (S), with badlands cut in the underlying marls; successive pediment surfaces (A, B, C) truncate geological structure; pediment B grades to river terrace (T) of the Rambla de Tabernas, middle distance. (b) Detailed view of part of the study site, view north from below the sandstone scarp. Note: pediment surfaces (C, E), modern badlands (b) cut in marls

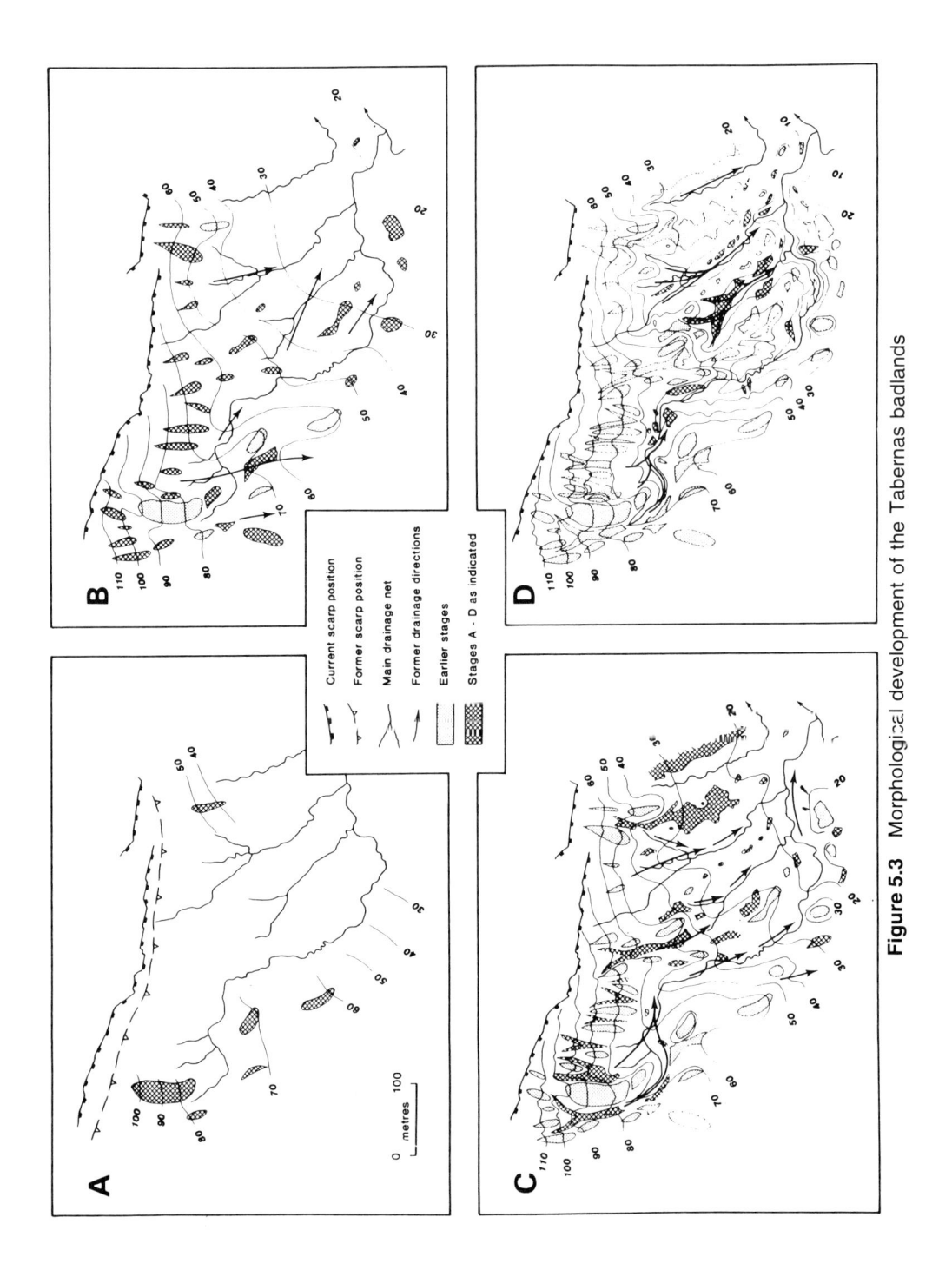

Figure 5.3 Morphological development of the Tabernas badlands

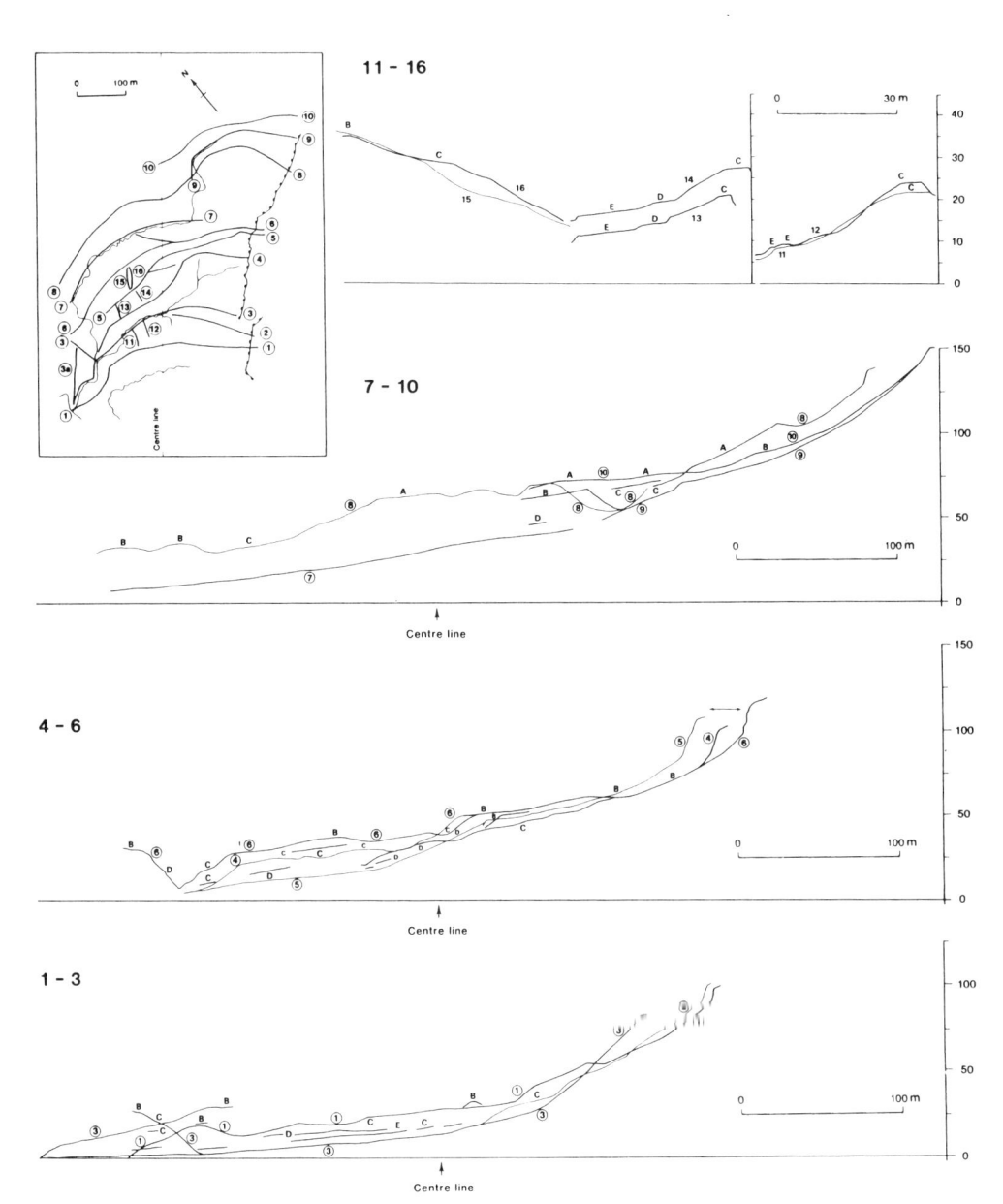

Figure 5.4 Slope profiles in the study area. (The lower parts of profiles 3 and 5, and the whole of profile 7 are gully floor profiles, stage F)

plants on pediment-based slopes (especially on north-facing slopes), and on the older pediment surfaces there is field evidence for the initiation of pedogenic processes (see below).

Investigation of vegetation and soil patterns in relation to auto-stabilisation processes raises two important questions. Firstly, do the vegetation and soil patterns reflect progressive stabilisation tendencies with surface age or the effect of local morphological controls? Put more simply, what are the geomorphological influences on vegetation and soil patterns? Secondly, if an age/stabilisation sequence can be detected, then are differences in vegetation and soils reflected in contemporary rates of geomorphological processes including erosion? Or, more simply, do vegetation and soils influence geomorphology? These questions were addressed through sampling and analysis of the vegetation and soils in the study area and by the use of rainfall simulation experiments on badland surfaces of differing age and cover type.

VEGETATION AND SOIL PATTERNS

Using a morphological base map, a preliminary reconnaisance survey of the study area was carried out in order to establish the extent and distribution of the major vegetation types and to select sites for more detailed sampling of vegetation and soils.

Vegetation

The vegetation of the site shows much variation in both structure and degree of cover. Aspects of its structure, floristics and ecology have been reported by Kunkel & Kunkel (1987). The steep scarp slopes above the badlands carry a moderately dense cover of tall (*ca.* 1 m) shrubs and tussock grasses. The flatter pediment tops support a moderate to high density of grasses together with scattered shrubs, whilst lichens occur in the spaces between the taller plants. There is abundant evidence of grazing by goats and rabbits. On the steeper remnants of the older pediments and on the slopes below pediments there is a very open shrub community with lichens achieving high cover at ground level. On the lowermost pediments close to contemporary channels an open community of halophytic shrubs with patches of crustose lichens and large areas of bare ground occurs.

The vegetation was sampled using 84 3 × 3 m quadrats distributed in a partial-random fashion. In order to ensure replicate sampling within each of the major community types identified from the preliminary survey and to provide sample coverage of the full range of morphological surface types, at least two areas of each community type were selected for detailed sampling. Within each selected area one or more quadrats (depending on the extent of the community) were randomly located by throwing a stick over the shoulder, the point at which the stick landed forming the north-east corner of the quadrat. Within each quadrat, data were recorded for vegetation structure in terms of the life form categories: shrubs, grasses and annuals, lichens (and mosses) and also for bare ground, using a six-point rank cover/abundance scale. The same cover scale was then employed to collect floristic data on individual shrub species, dominant herbs and lichens (at least to genus level). The data set was analysed using DECORANA (detrended correspondence analysis: Hill, 1979a) and TWINSPAN (two-way indicator species analysis: Hill, 1979b) within the VESPAN II computer package (Malloch, 1988). The major trends in the vegetation can be examined on the first two ordination (DECORANA) axes (Figure 5.5) and interpreted using the structural (life form) data (which were not included in the analysis), slope and aspect measurements and surface age (recorded from the morphological map of Figure 5.1). The correlation coefficients (Spearman Rank) between the ordination axis scores and the structural, site and surface age scores for the quadrats are shown on the diagram (significant values only). The major vegetation trend, represented on the first ordination axis, has significant negative

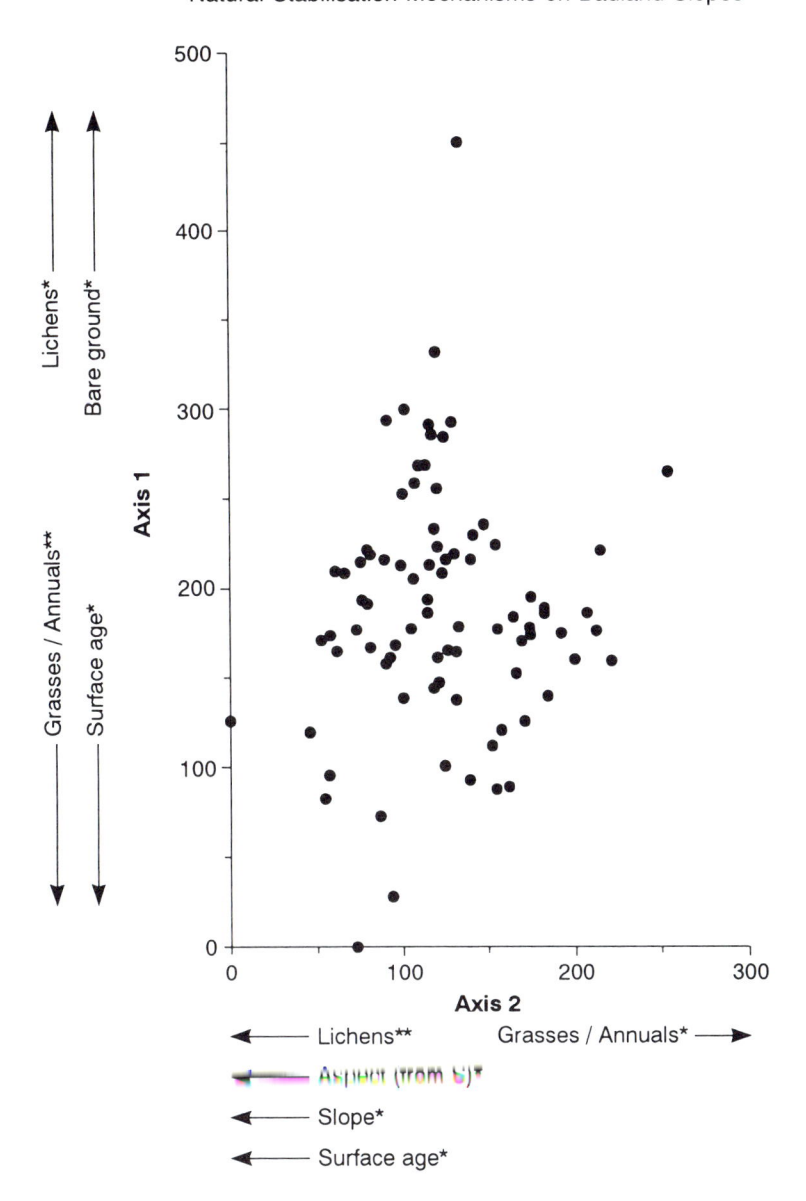

Figure 5.5 Distribution of quadrats on DECORANA axes 1 and 2 with significantly correlated (Spearman rank) vegetation structure (life-form type) and site variables. (* = $p < 0.05$; ** = $p < 0.01$)

correlations with grasses and annuals, and surface age, and significant positive correlations with lichen cover and bare ground. Thus, although the relationship is not simple, axis 1 reflects a trend from 'young' sites with high lichen cover and much bare ground (high values) to 'older' sites with a more complete cover of higher plants (low values). There is a cluster of quadrats around the middle of axis 1 and these are separated, to some extent, by axis 2. Here surface age remains as a significantly negatively correlated variable along with lichen cover. Slope and aspect (measured as degrees deviation from south) are also significantly correlated in the negative direction, whilst the correlation with grasses and annuals is also significant but positive. Thus axis 2 gives some separation between 'older', steeper, lichen-rich sites with more northerly apects (low values) and 'younger', more gently sloping, sites with high cover of grasses and annuals. Surface age remains a significant variable on axis 3 and only falls below significance on axis 4, which carries the least information. Thus, general age-related trends emerge with slope and aspect being of subsidiary importance in determining community patterns.

The classification analysis (TWINSPAN) (Figure 5.6) employed semi-quantitative data using 'pseudo-species' (Hill *et al.*, 1975) based on taxa cover values. The concept of the 'pseudo-species' '... is such that for each species there are a given number of pseudospecies each representing a part of the quantitative range for that species' (Malloch, 1988). The classification was run to four division steps with a minimum group size for division criterion of five members. This resulted in the production of twelve end-groups, of which three groups (4, 11 and 12) contained only one quadrat. At this level of subdivision, site morphology emerges as the overriding control on vegetation pattern with pediments and gentle slopes being separated from steeper slopes. These features, together with position on slope, reflect the influence of soil moisture noted by others working on semi-arid vegetation (Kirkby *et al.*, 1990; Dargie, 1984). Within the broad morphological groupings, however, surface age remains as an important separator.

The first division step separated those quadrats containing a high proportion of halophytic species. These were mainly located on the morphologically younger sites. Subsequent steps divided the remaining quadrats according to dominant life form types with those sites containing a high proportion of shrubs and grasses and annuals moving to the left of the division tree and those with more lichens moving to the right. The halophyte group was further subdivided in terms of degree of lichen cover. The distribution of the nine end-groups containing more than one quadrat plus the single-member group, 4, is shown in Figure 5.7, where the quadrat locations have been used in conjunction with the field mapping data to produce a map of vegetation types. Group 4 is retained, although it is a single-member group, as it is representative of small areas of D-age pediment. Figure 5.8 shows the occurrence of the most frequent taxa in the nine multimember TWINSPAN groups. *Helianthemum almeriense* and an unidentified acrocarpous moss achieve at least 50% presence in all nine groups, but the distribution of the remaining species is more restricted.

Group 1 (two quadrats) represents a community with a very dense cover of shrubs, fine grasses and annuals which occurs on a small number of abandoned agricultural terraces constructed in a channel of C and D age. *Helianthemum almeriense* and *Hammada tamariscifolia* dominate the shrub layer together with *Teucrium polium, Sideritis* spp. and occasional tussocks of *Stipa tenacissima*. There is also a dense cover of finer grasses (*Bromus* and *Hordeum* species) and annual herbs. The presence of lichens is precluded by the dense growth of the vascular plants. Group 2 consists of the four quadrats taken on the scarp slope above the badlands. It represents a mixed community with a moderate cover of taller shrubs (*Thymus vulgaris, Anthyllis terniflora, Euzomodendron bourgeanum* and *Salsola genistoides*) and tussocks of *Stipa tenacissima*. *Asphodelus microcarpus* is an occasional but characteristic member of the community and the crustose lichens, *Fulgensia fulgens* and *Buellia* spp., and an indeterminate brown crust (probably consisting of cyanobacteria) are frequent on the soil between the taller plants. Group 3 (four quadrats) contains a similar range of shrubs to group 1, though with the inclusion of *Artemisia barrelieri*. There is also a high diversity and cover of fine grasses and annual herbs reflecting the location

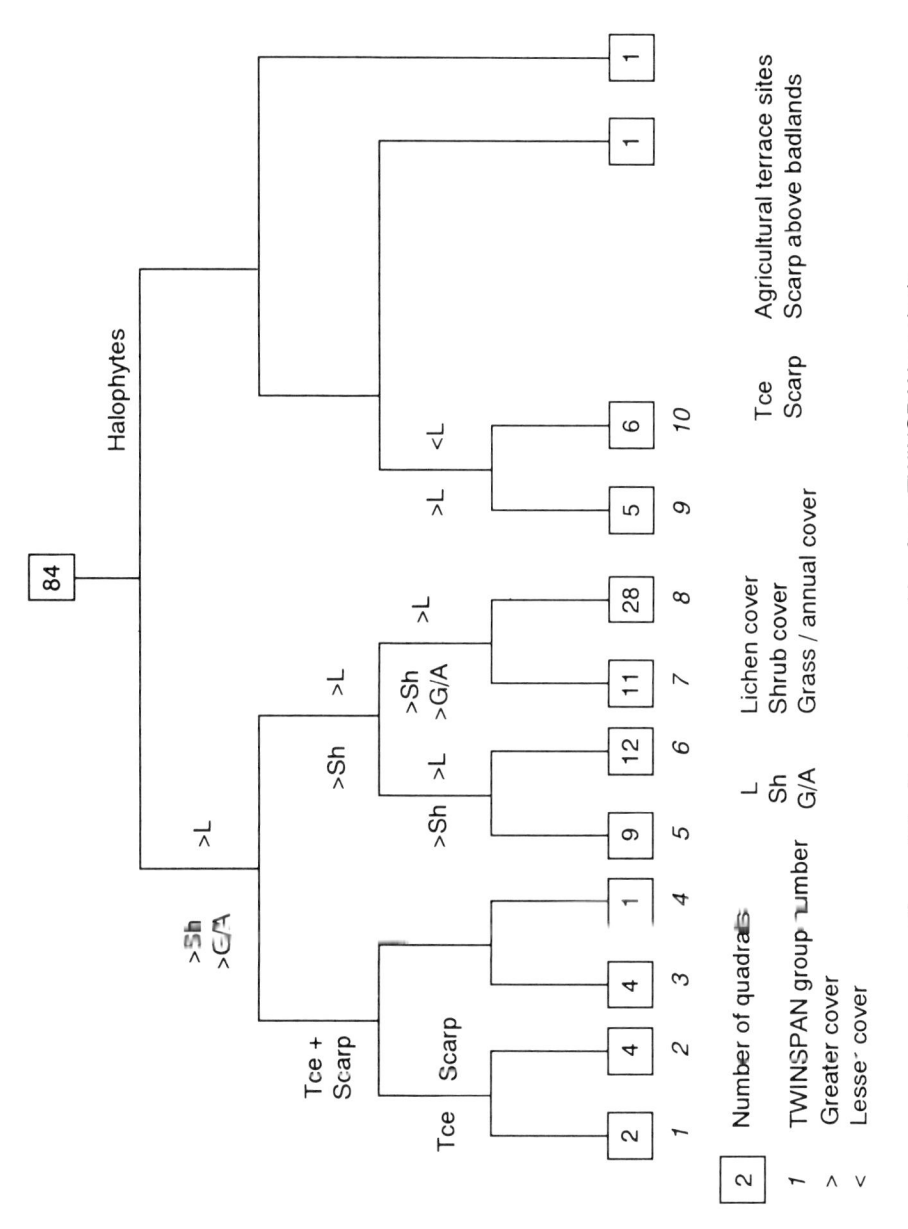

Figure 5.6 Dendrogram resulting from TWINSPAN analysis

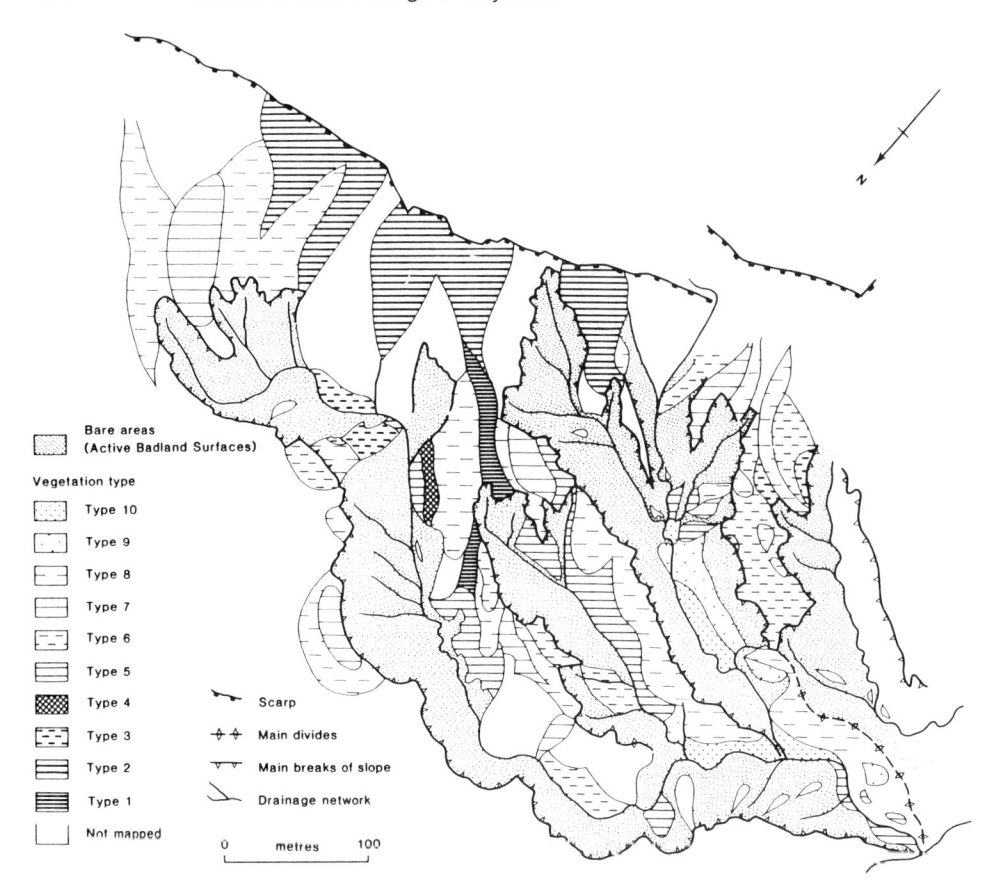

Figure 5.7 Distribution of vegetation types represented by TWINSPAN groups 1–10

on gently sloping, often north-facing, pediment surfaces (of C and D age) at the base of slopes and thus in water-receiving situations. *Stipa tenacissima* is a constant member, as are the crustose lichens, *Squamarina* spp. and *Psora decipiens*. Group 5 (nine quadrats) also occurs on gently sloping pediments of C and D age but has a different shrub community, of which *Salsola genistoides* and *Fagonia cretica* are important members. It has a high diversity of annual herbs and a moderately high cover of crustose lichens, notably including *Diploschistes dicapsis*. Group 6 occurs in similar situations to group 5 but has a less diverse shrub component, relatively high cover of fine grasses and annuals, and a more diverse range of lichens (*Psora decipiens, Toninia* spp., *Fulgensia fulgens, Squamarina* spp., *Diploschistes dicapsis, Buellia* spp. and constant brown crust). It is also characterised by the frequent presence of the crucifer, *Diplotaxis crassifolia*, growing here as a perennial.

Group 7 (eleven quadrats) occurs on the more steeply sloping pediments of A and B age and on some north-facing slopes of C and D age. The pediment sites are more isolated than those of C and D age and thus receive less lateral water flow. Shrub cover and diversity are moderate and *Stipa* tussocks are a notable feature of many sites. There is a low diversity of annual herbs but crustose lichens are diverse and achieve high cover. Group 8 (28 quadrats) is the most widespread type occurring on slopes of varying gradient, aspect and

Natural Stabilisation Mechanisms on Badland Slopes

Taxa	\multicolumn{9}{c}{TWINSPAN groups}
	1	2	3	5	6	7	8	9	10
Helianthemum almeriense	6–10%	iso	1–5%	1–5%	1–5%	6–10%	1–5%	iso	iso
Moss	iso	iso	1–5%	1–5%	1–5%	1–5%	1–5%	1–5%	iso
Hammada tamariscifolia	1–5%		1–5%	6–10%	6–10%	6–10%			
Teucrium polium	iso		iso			1–5%	iso		
Sideritis spp.	6–10%								
Phagnalon spp.	1–5%								
Thymus vulgaris		1–5%							
Anthyllis terniflora		1–5%							
Artemisia barrelieri		1–5%	1–5%				1–5%		
Euzomodendron bourgeanum		1–5%						1–5%	iso
Salsola genistoides		1–5%		iso				iso	1–5%
Asphodelus microcarpus		iso							
Fulgensia fulgens		1–5%	1–5%		1–5%	1–5%	1–5%	1–5%	1–5%
Buellia spp.		6–10%	1–5%		1–5%	1–5%	1–5%	1–5%	1–5%
Brown crust		1–5%	iso		6–10%	6–10%	6–10%	6–10%	
Squamarina spp.			1–5%	1–5%	1–5%	1–5%	6–10%	6–10%	6–10%
Fagonia cretica				iso					
Psora decipiens			1–5%		iso				
Diploschistes dicapsis				1–5%	1–5%	6–10%	6–10%	6–10%	1–5%
Toninia spp.					iso				
Diplotaxis crassifolia					1–5%				
Teloschistes lacunosa							iso	iso	
Asparagus horridis							iso		iso
Anabasis articulata								6–10%	6–10%
Limonium insigne								iso	1–5%
Salsola papillosa									iso
Frankenia thymifolia									iso

Cover: ■ 6–10%   ▦ 1–5%   ▨ Isolated individuals

Frequency: □ Constant   ▭ 75–100%   ▭ 50–74%

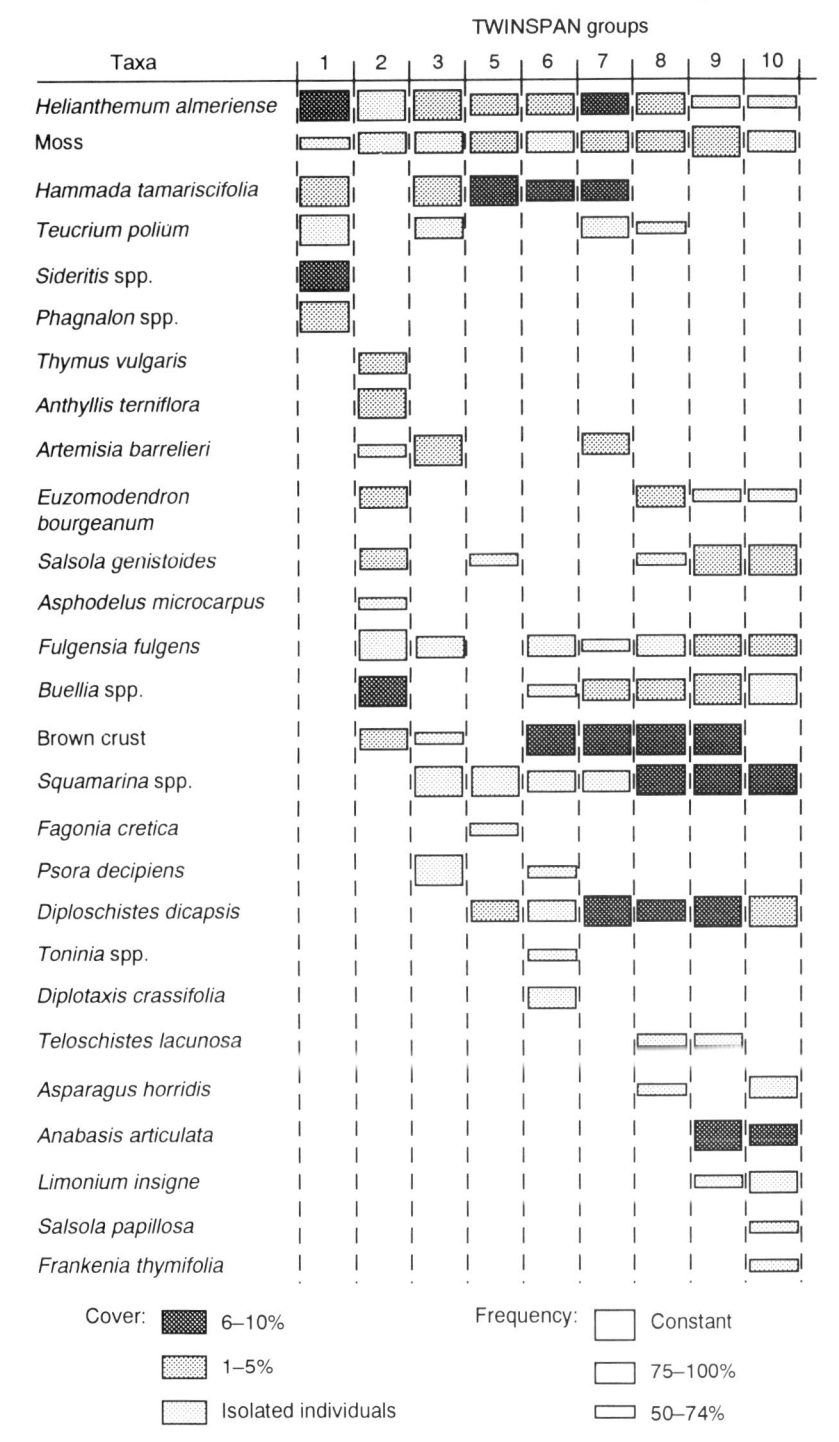

Figure 5.8 Frequency and cover values for the 27 most frequent taxa in the nine multi-member TWINSPAN groups

age. The shrub component is sparse and is characterised by *Euzomodendron bourgeanum*, *Salsola genistoides* and *Asparagus horridis*. Isolated individuals of *Serratula flavescens* occur on the steeper slopes. Crustose lichens achieve their highest cover values in this group with *Diploschistes dicapsis*, *Squamarina* spp., *Buellia* spp., *Fulgensia fulgens* and the indeterminate brown crust being constant or near constant members. The fruticose lichen, *Teloschistes lacunosa*, is a characteristic member of many examples of the group.

Group 9 (five quadrats) occurs on the lowermost D-age slopes and occasional isolated remnants of E age pediments. It is characterised by the presence of halophytes such as *Anabasis articulata* (constant), *Limonium insigne*, *Salsola papillosa* and *Frankenia thymifolia* (all occasional) and the constant presence of *Salsola genistoides*. These are sparsely distributed and the ground is largely covered by crustose lichens (*Diploschistes dicapsis*,

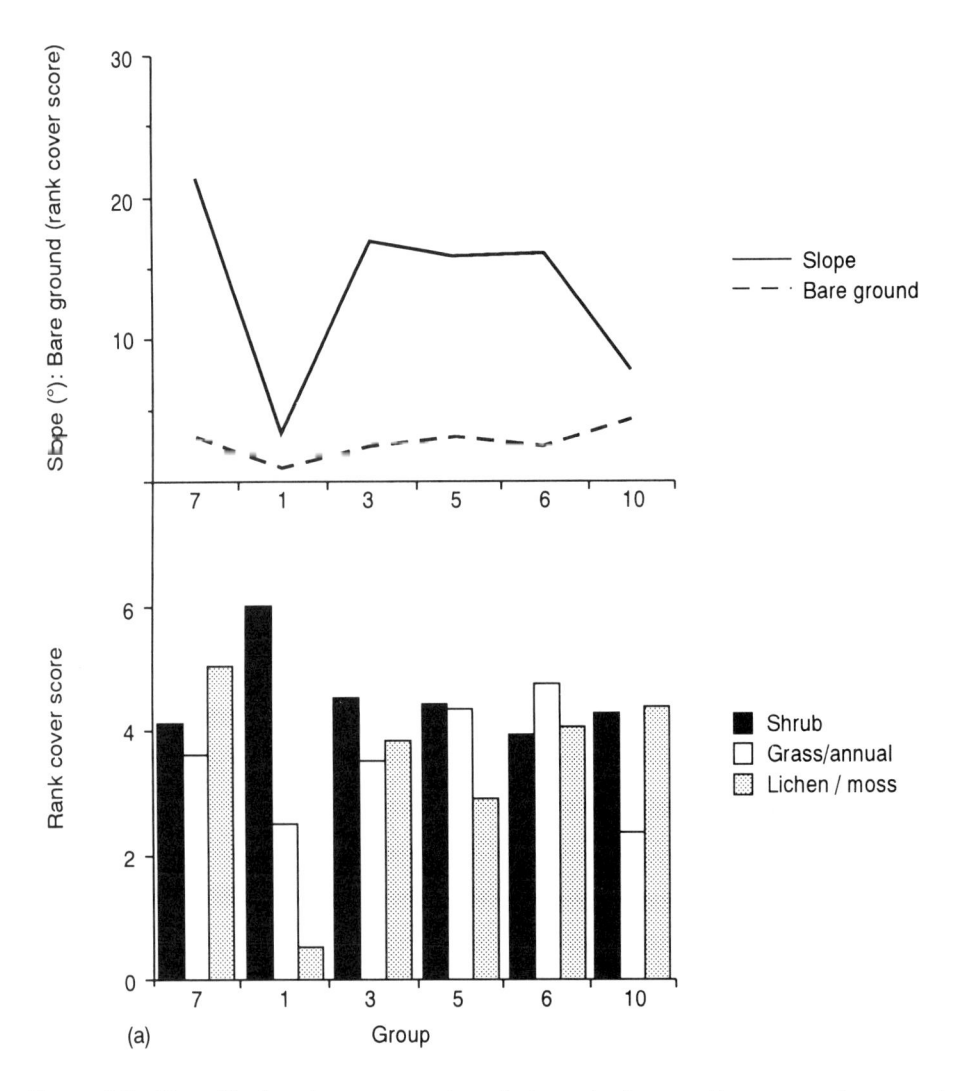

Figure 5.9 Mean life-form type cover values (columns), slope and bare ground cover values (lines) for the TWINSPAN groups located mainly on (a) pediments and (b) slopes

Squamarina spp., *Buellia* spp., *Fulgensia fulgens*), brown crust and occasional patches of *Teloschistes lacunosa*. Group 10 (six quadrats) occurs almost exclusively on the lowermost E-age pediments. It contains the same range of halophytes as group 9 and also *Asparagus horridis* and *Euzomodendron bourgeanum*. It is distinguished by lower cover values of crustose lichens (only *Squamarina* spp. retain similar cover), the virtual absence of brown crust and larger amounts of bare ground.

The general characteristics of the predominantly pediment sites and predominantly slope sites groups are displayed separately in Figure 5.9 as mean values for the life-form type cover scores in the member quadrats of each group (columns) and mean values for slope and bare ground cover (lines).

The pediment groups (Figure 5.9(a)) are arranged in sequence according to the age of

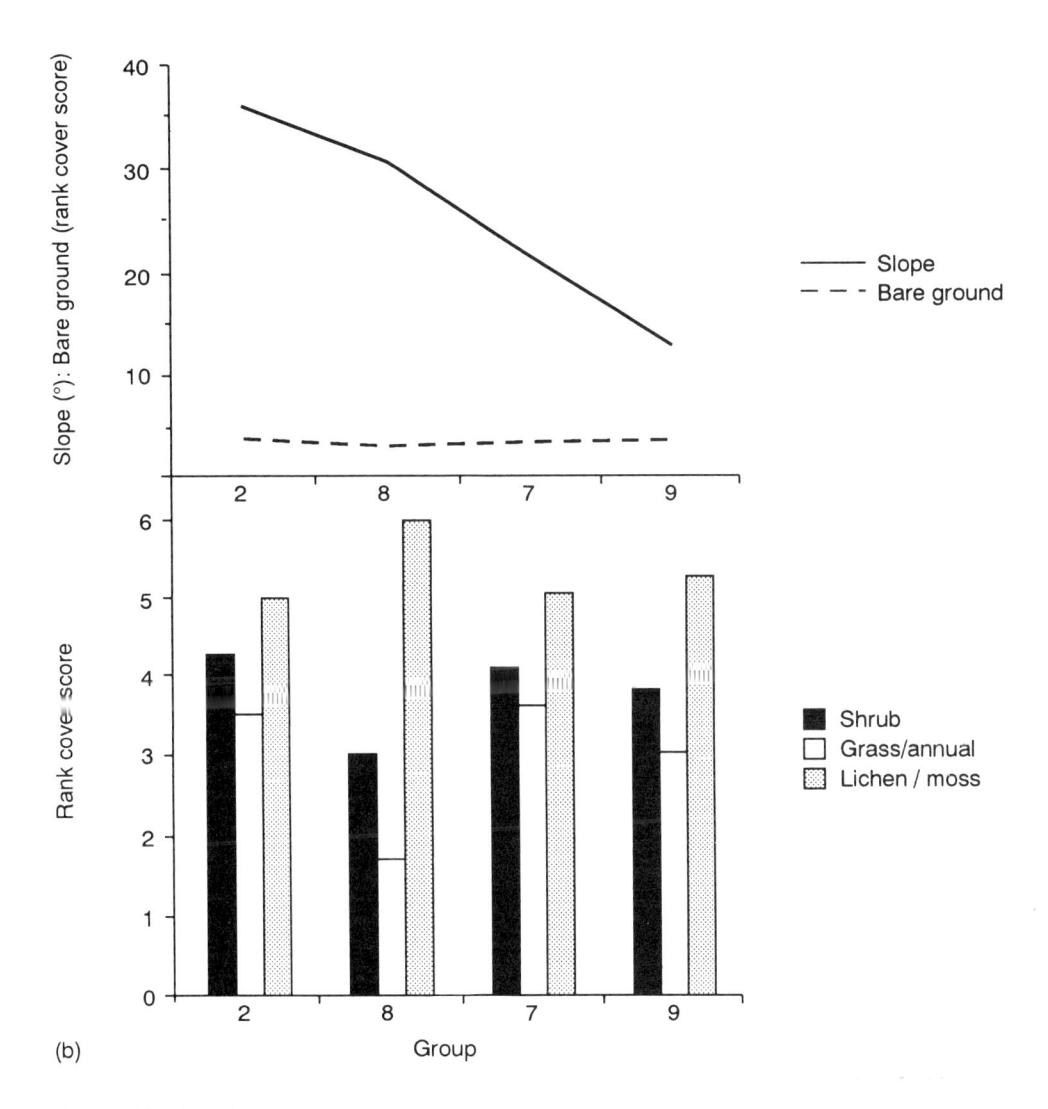

Figure 5.9 Continued

the majority of the surfaces that they represent. There is a sequence of development of vegetation from E-age sites, where lichens are co-dominant with, mostly halophytic, shrubs, through the various groups of C and D age where lichen cover decreases as first grasses and annuals and then shrubs rise to dominance. This sequence is associated with decreasing bare ground cover and a slight decline in slope values and it almost certainly reflects a change in soil moisture status from xeric to more mesic sites. Group 7, containing A/B-age pediments, represents a reversion of this sequence, with lichens again dominating in terms of cover values. The steeper slope gradients here partly reflect the inclusion of some C-age slopes in group 7, but the A/B-age pediment remnants are themselves more steeply sloping than their younger counterparts. The inclusion of a number of C-age slope sites in the group is itself indicative of the different surface conditions on A/B-age pediments, which are much older, more isolated, water-shedding and generally degraded, showing some evidence of truncated soil profiles (see below).

Figure 5.9(b) displays the slope groups in a similar fashion and includes the scarp slope sites (group 2) as well as the badland slopes. Less of a trend is evident with this smaller number of groups but two features that emerge are the increase in lichen cover with increasing slope gradient from groups 9 to 8 and also the similarities in vegetation structure between the badland groups 9 and 7 and the scarp group 2.

Soils

Soils have been examined in the field and characteristic soil profiles described and sampled for laboratory analysis. The Miocene marls weather to give soils of clay and silty-clay texture with abundant sub-angular sand-sized and larger rock fragments. The hue of all soils sampled was 10YR. Secondary iron oxide and organic matter contents are low. The soils are calcareous throughout, with the proportion of $CaCO_3$ varying between 14 and 32%. In most samples analysed gypsum was present as a trace. Higher quantities reached 42 m.e. 100 g^{-1} $CaSO_4$ in certain profiles. Soils on each surface were examined in the field and representative sites selected for sample collection from pits cut into surfaces of ages A, B, C, D, E and F. Another pit was located on freshly exposed marl (M) on an eroding badland slope. The respective sample codes are TP1 (A), TP2 (A), TP3 (B), TP5 (C), TP9 (E), TP10 (D), TP14 (M) and TP15 (F). Four profiles were analysed with four or five samples: TP1 (A), TP2 (A), TP5 (C) and TP9 (E). From the remaining profiles, samples were analysed only for surface (0–1 cm) and 'base' (depth varying from 14 cm for TP14 (M) to 90 cm for TP2 (A)).

Two sets of parameters were determined: the first included concentration of bases and electrical conductivity in a 1:5 soil/water extract. This was done in order to examine the leaching and upward migration of water and to evaluate the significance of sodium in surface soil stability (susceptibility to dispersion and crusting). The second set of parameters was chosen both for their importance in characterising the soil and in order to assess their potential as indicators of relative soil age. These parameters were organic carbon (Walkley–Black method); fixed inorganic phosphorus (digested with 1N HCl and determined colorimetrically); calcium carbonate (digested in 18% HCl and determined gasometrically); calcium sulphate (extracted in water and determined by the conductance method of Bower and Huss, in Hesse, 1971, p. 85); 'free' (pedogenetic/secondary) iron oxides: organic-bound iron (potassium pyrophosphate extractable), amorphous inorganic iron (extractable with acid ammonium oxalate) and crystalline iron oxides (extracted with sodium dithionite–citrate–

bicarbonate); and mineral magnetic parameters, as defined and following the methods given by Thompson & Oldfield (1986). Munsell colour notation and names are used. The results (except for gypsum, which in most samples was only a trace) are presented in Figures 5.10, 5.11 and 5.12.

In relation to erosion, stability and surface age the following features are pertinent. Leaching of bases, in particular sodium, has been effective in reducing quantities in the upper 60 cm of soils on E age and older surfaces. On surface F and in freshly exposed material, water-soluble sodium concentrations in the surface samples are significantly higher. Despite low sodium/calcium+magnesium ratios amongst the water-soluble bases, sodium is likely to have a significant effect upon dispersion and possibly swelling of clays on the younger geomorphic surfaces.

Organic carbon shows an expected increase towards the soil surface in most profiles including, rather surprisingly, those of the youngest geomorphological surfaces. Fixed inorganic phosphorus shows no consistent pattern with depth and no trend of increase with age. In fact, a decrease in the amount occurs with increasing age. Data for percentage calcium carbonate suggest a possible slight removal from the upper 20 cm and deposition in a peak of concentration between 20 and 40 cm, at least beneath surfaces of E age and older. Traces of gypsum occur in all profiles, but values of between 2 and 22 m.e. 100 g^{-1} CaSO$_4$ occur in the lower part (20 cm+) of soils of surface E and younger, except in TP14 (freshly exposed marl) where gypsum occurs at the surface. There are moderate concentrations of gypsum in the middle part of profile TP1(A).

Soil colour reflects changes in iron oxides through time. Soils are of 10YR hue throughout, but value and chroma increase to give a shift from dark grey (10YR4/1, in TP14) and greyish brown to yellowish brown (10YR5/6, in TP1(A)) with increasing age. Analyses of extractable iron oxides (conducted only on profiles TP1 (A) and TP9 (E)) show a higher proportion of crystalline compared with amorphous iron oxides in the surface A soil than that of surface E (Figure 5.11). Amounts of dithionite-extractable iron are low (< 0. 5%), this being one factor limiting the hue of the oldest soils to 10YR. The mineral magnetic parameters show considerable variation in magnetic mineral type and grain size in the basal samples of the three soils examined. The demagnetisation parameters, which show 'hard' and 'soft' magnetic responses, suggest a more significant presence of haematite by the higher 'hard' response (i.e. retaining remanence after withdrawal from a strong applied field) at the base of TP1(A) than in other parent material samples analysed. Haematite, the oxide most likely to be contributing to this response, may occur in a concentration (albeit very low) which is age dependent. Together with greater amounts of goethite, haematite may also account for the shift in colour observed in soils from the older surfaces. The difference between magnetic values for surface and basal samples of TP1(A), a difference not expressed in the younger profiles analysed, suggests the influence of soil development in the longer timescale. This surface 'magnetic enchancement' has been reported from a wide variety of soil-forming environments and has been interpreted as the result of 'fermentation' processes involving reduction–oxidation cycles (Le Borgne, 1955; Mullins, 1977). Amongst other possible enchancing factors, the influence of fire cannot be ruled out (Longworth & Tite, 1977; Maher, 1986), as burning is likely to have been common throughout the history of these badlands. Surface magnetic enchancement has not occurred in the surface D soil (TP10).

In comparison with profiles TP2(A), TP5(C) and TP9(E), profile TP1(A) shows higher values for the percentages of CaCO$_3$, CaSO$_4$ and the water-soluble bases, and electrical

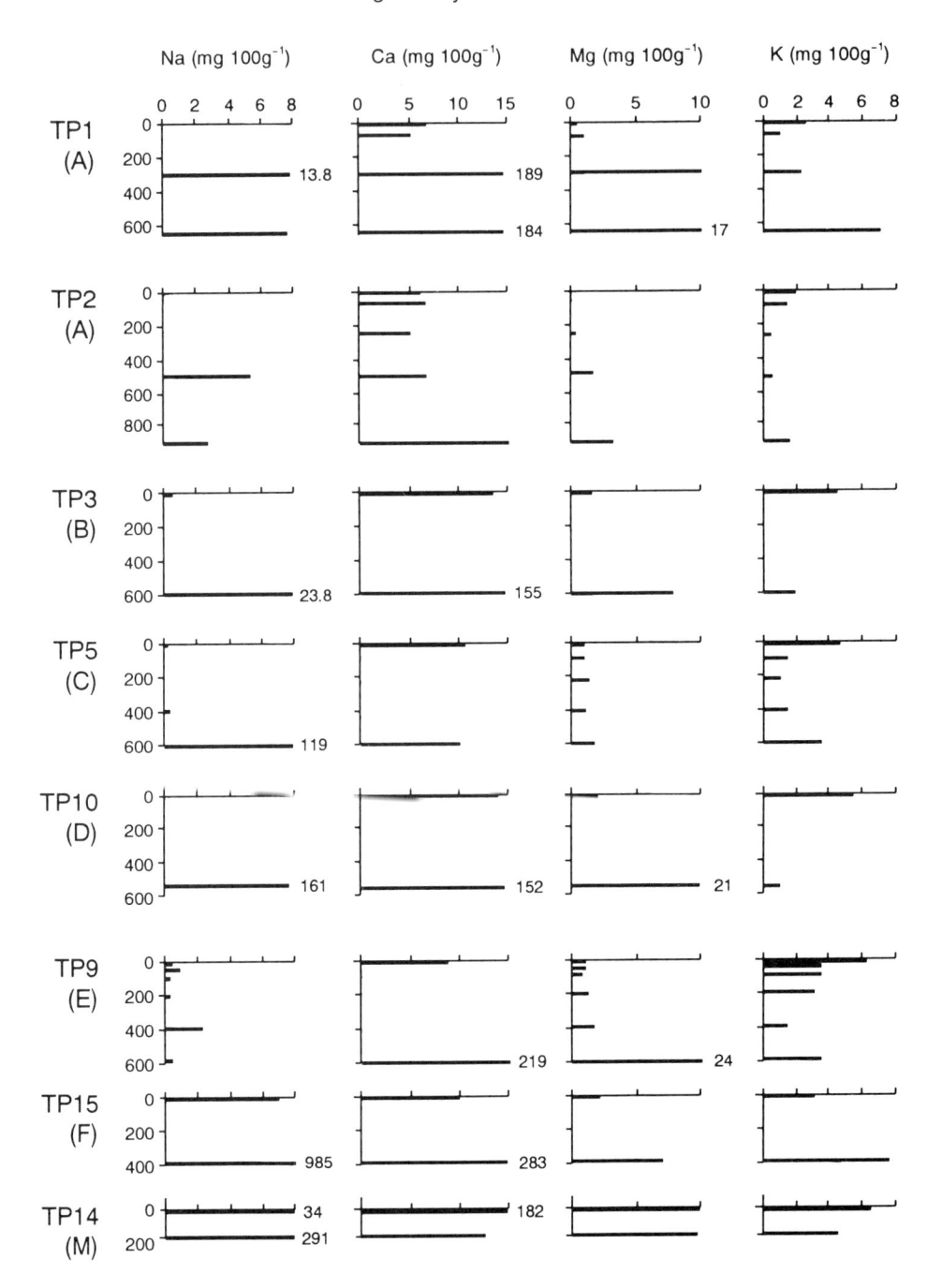

Figure 5.10 Soil chemistry. Na, Ca, Mg, K soluble in 5:1 water:soil, mg 100 g^{-1}. Na/Ca+Mg, based on m.e. 100 g^{-1}. Conductivity, mS cm^{-1}. Fixed P, % ×100. Organic carbon and free CaCO$_3$, %. Brackets indicate surface age

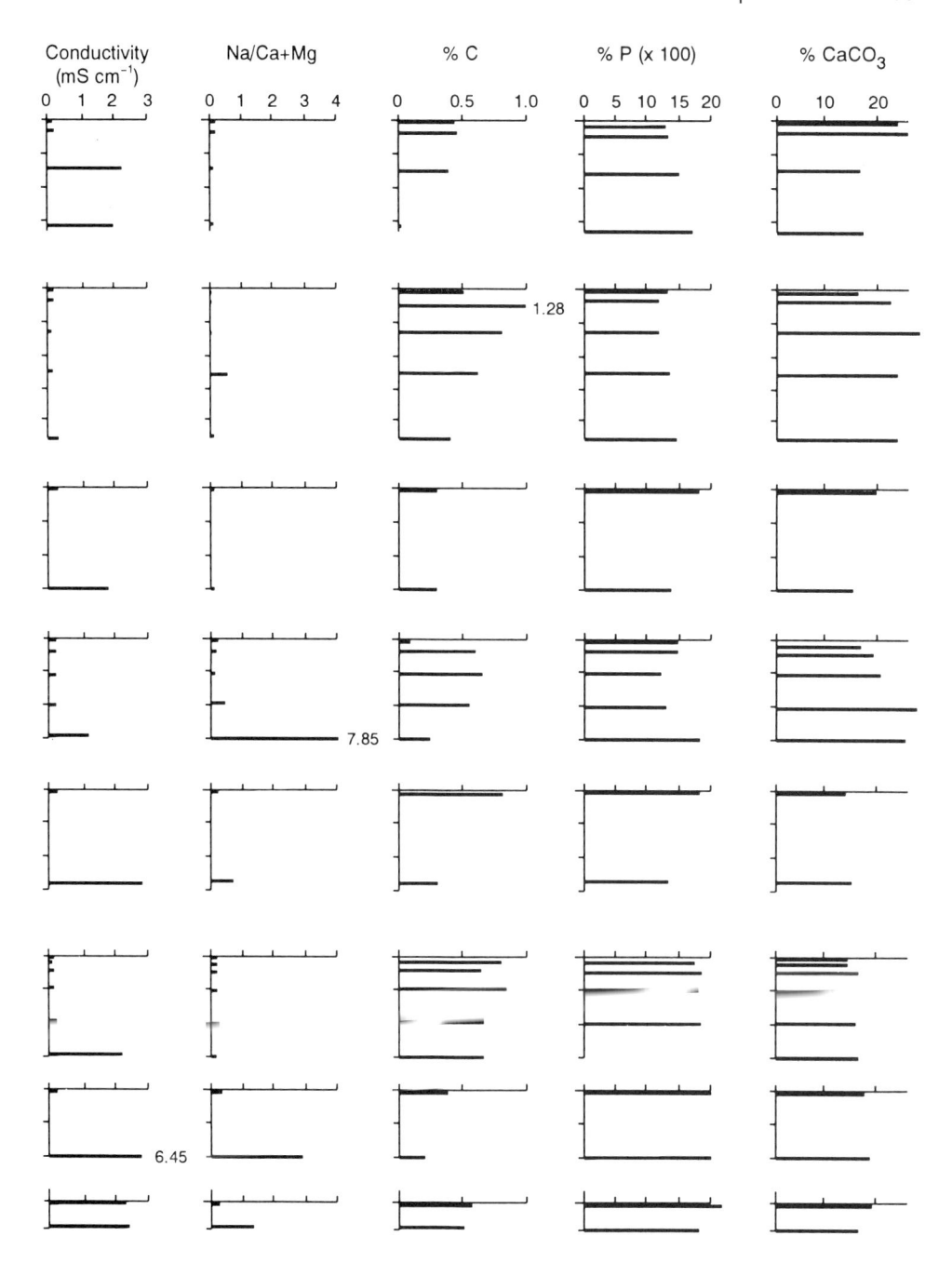

Figure 5.10 Continued

104 Environmental Change in Drylands

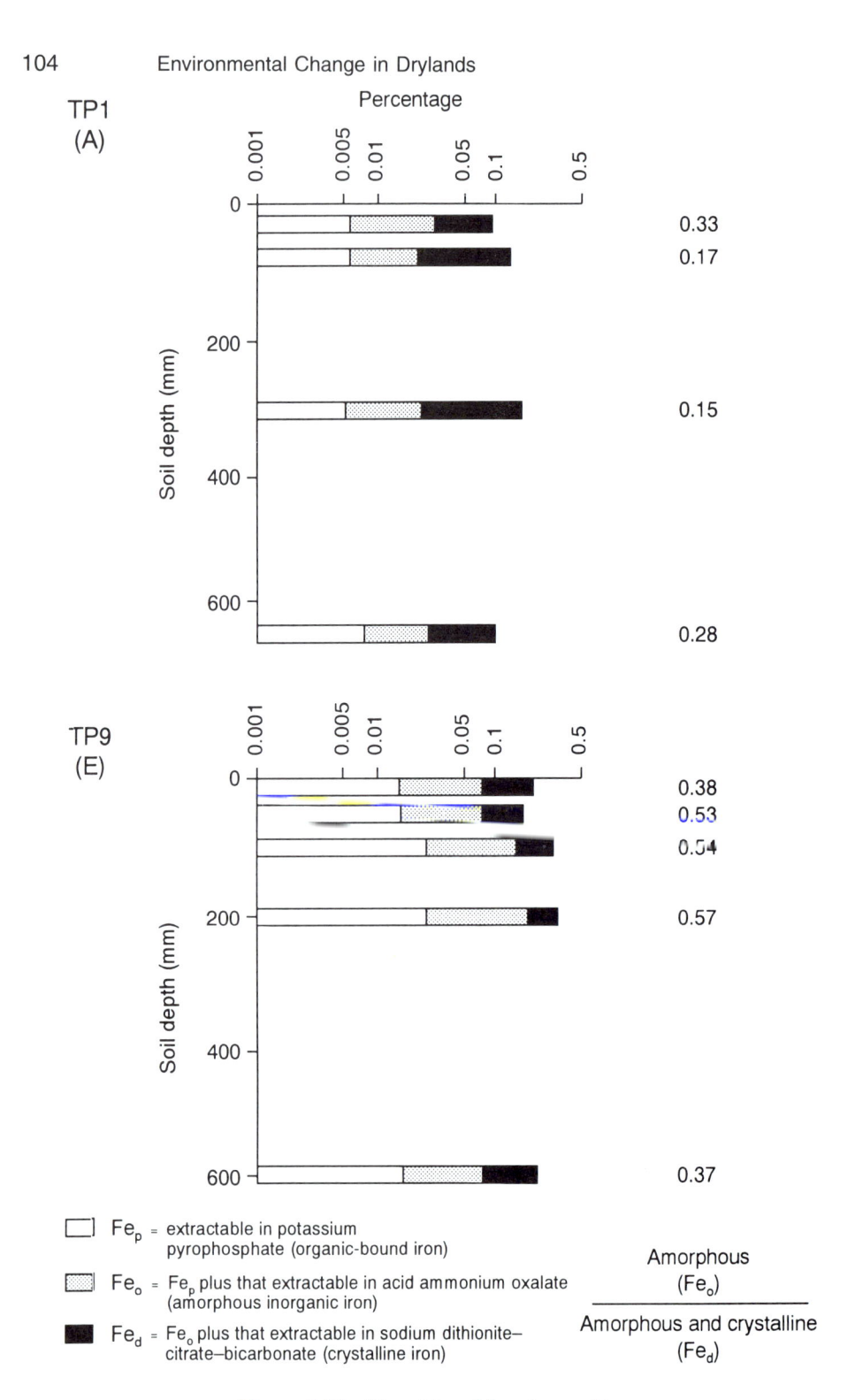

TP1 (A)

TP9 (E)

☐ Fe_p = extractable in potassium pyrophosphate (organic-bound iron)

▨ Fe_o = Fe_p plus that extractable in acid ammonium oxalate (amorphous inorganic iron)

■ Fe_d = Fe_o plus that extractable in sodium dithionite–citrate–bicarbonate (crystalline iron)

Amorphous (Fe_o)

Amorphous and crystalline (Fe_d)

Figure 5.11 Chemistry of free iron oxides

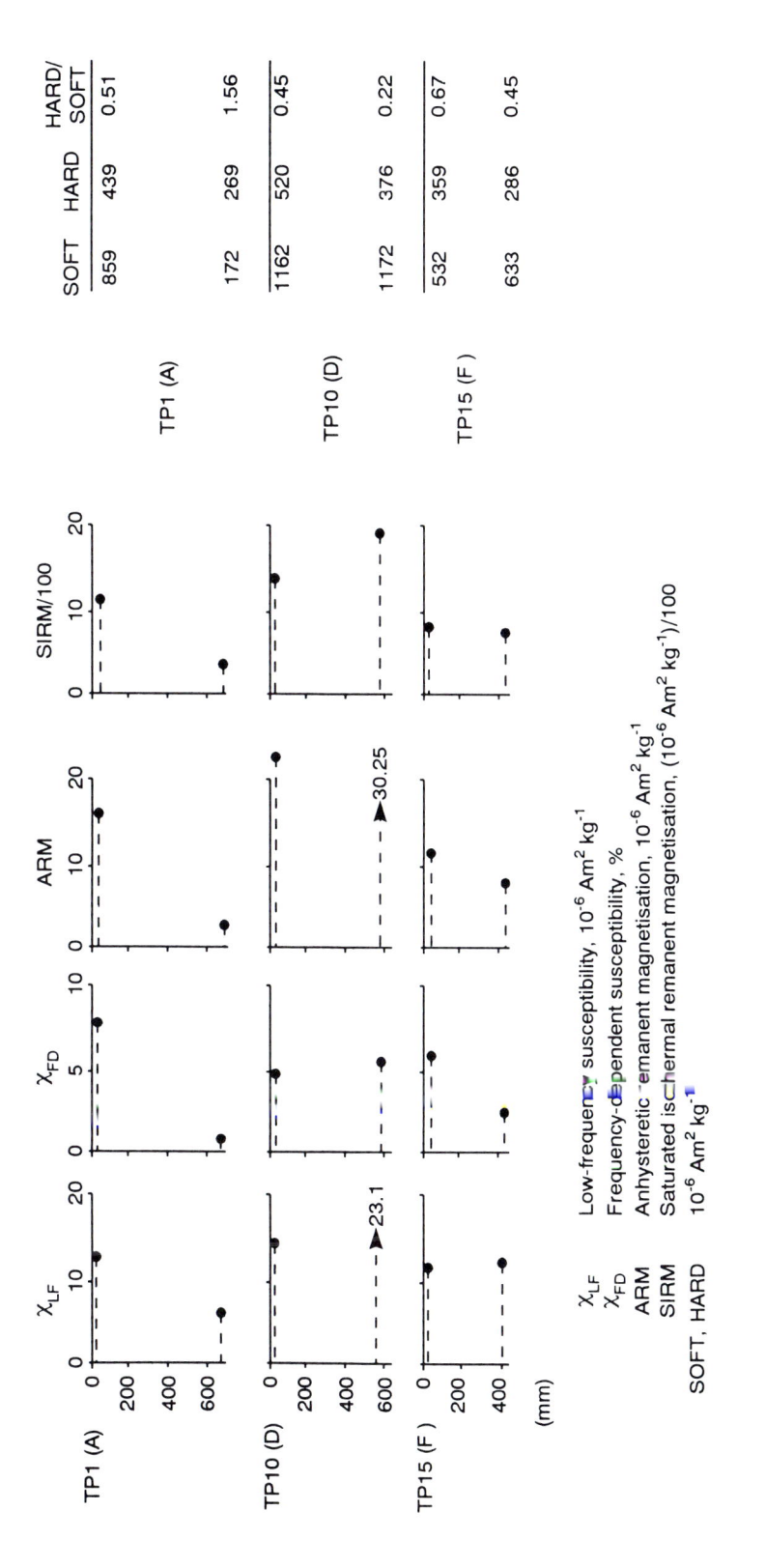

	SOFT	HARD	HARD/SOFT
TP1 (A)	859	439	0.51
	172	269	1.56
TP10 (D)	1162	520	0.45
	1172	376	0.22
TP15 (F)	532	359	0.67
	633	286	0.45

χ_{LF} — Low-frequency susceptibility, 10^{-6} Am^2 kg^{-1}
χ_{FD} — Frequency-dependent susceptibility, %
ARM — Anhysteretic remanent magnetisation, 10^{-6} Am^2 kg^{-1}
SIRM — Saturated isothermal remanent magnetisation, (10^{-6} Am^2 kg^{-1})/100
SOFT, HARD — 10^{-6} Am^2 kg^{-1}

Figure 5.12 Mineral magnetic properties

conductivity in the middle and upper part of the profile. This suggests the possibility of truncation of the soil profile by erosion on the flatiron slope where TP1(A) was dug. The depth curves for certain of the parameters suggest that any truncation has taken place following the redistribution of $CaCO_3$ to give the peak present in other profiles, but not at a time sufficiently long ago for there to have been redistribution of very soluble material typical of profiles older than surface D age.

Thus a number of the soil parameters measured are consistent with an age sequence that is in accordance with that derived from morphological criteria although, as with the vegetation analysis, the relationship is not a simple one.

STABILISATION MECHANISMS

The vegetation trends identified from Figure 5.9 have implications for natural stabilisation mechanisms, particularly as regards vegetation structure. Faulkner (1990) has discussed the protective role of vegetation in reducing effective rainfall intensity at the ground surface and the reduced likelihood of runoff generation when cover is dense. Alexander & Calvo (1990) have demonstrated the role of lichens, particularly crustose, in increasing runoff but decreasing sediment concentration in simulated rainfall plots on badlands. On the badland slope sites at Tabernas (Figure 5.9(b)), lichen cover is generally high and shows a slight increase with surface age, whilst shrub cover is lower and consists of isolated individuals (groups 7 and 8) or localised dense clumps (group 9). These structural patterns would lead to localised protection from erosion where clumps of shrubs occur and where lichen cover is complete but the increased runoff generated by the lichen crust would lead to localised flow concentration in intervening areas, giving rise to soil loss and a greater microtopographic amplitude. A similar picture would emerge on the youngest (E age, group 10) and oldest (A/B age, group 7 as above) pediment sites (Figure 5.9(a)). However, the increasing grass and annual, and shrub cover values in the C/D-age sites (groups 6 to 1) are likely to provide a reduction in both runoff and erosion rates through increased interception by a structurally more diverse and more complete canopy (c.f. the models of Thornes, 1990) and increased infiltration capacity due to greater litter contribution and increased root density (Faulkner, 1990).

Rainfall Simulation

In order to investigate the influence of vegetation type upon contemporary surface processes a series of rainfall simulation experiments was performed as part of, and building upon, a wider investigation of the response of badland surfaces in south-east Spain to simulated rainfall (Calvo *et al.*, 1991b). These experiments were carried out using a sprinkler-based rainfall simulator described in Calvo *et al.* (1988). The simulator operates over circular plots (*ca.* 0.24 m^2) at rainfall intensities of 60 mm h^{-1}. The duration of the experiments was 25–30 minutes and deionised water was used. This would be equivalent to an extreme, short-duration event. Runoff was measured in the field at 1–2 minute intervals and sediments were collected (also over 1–2 minute periods) at three times during each experiment: at the beginning of runoff, when the maximum runoff rate was reached and towards the end of the experiment when a constant runoff rate had been reached. Forty-seven experimental plots were used and these were distributed to reflect the range of surface characteristics and vegetation types in the area.

Runoff values show much similarity across the plots. In general the runoff rates are high, runoff starts quickly (within less than 7 minutes) and rapidly reaches a peak (within less than 12 minutes). These features are due to the presence of either a lichen crust or a mineral crust in most of the plots. Only in two sites where higher plants dominated and lichen crust was absent were runoff values low. Sediment concentrations and erosion rates were generally low in comparison to badland sites elsewhere in south-east Spain (Calvo et al., 1991b). The vegetated plots fall mostly within type A of Calvo et al. (1991b) which is characteristic of slopes with > 20% cover of lichens, stones or higher plants, and producing very low sediment concentrations and variable runoff. Plots from bare badland slopes fall within type D, producing low to moderate sediment concentrations, high runoff and thus moderate to high erosion rates. The value of 20% cover appears to represent an important threshold.

To examine vegetation–surface process relationships in more detail, the data from rainfall simulation plots located adjacent or close to a quadrat sample, and hence where vegetation type could be reliably established, were examined in more detail. This involved eighteen rainfall simulation plots in vegetation groups 6 to 10. Figure 5.13 displays the data in a similar fashion to that used in Figure 5.9. Average cover values for plant life-form types calculated from the quadrats in each group are shown as columns, and average slope, average bare ground cover (amongst the quadrats in a group) and average erosion value (from the rainfall simulation plots located on the vegetation type) are shown as lines. Examining the diagram from right to left, erosion values fall as lichen cover increases from group 10 to group 8, even though average slope is increasing. Erosion rises again in group 7 as lichen cover falls but grass and annual, and shrub cover rises. The two rainfall simulation plots in group 7 vegetation were located in a site dominated by large tussocks of *Stipa tenacissima* with much bare ground in the gaps between the clumps. Due to the small size of the plot ring relative to the tussocks, the rainfall simulations were performed in the inter-tussock gaps and thus the erosion values may be unrepresentative of group 7 vegetation as a whole. The five rainfall simulation plots from group 6 show a further decrease in average erosion values associated with increases in average cover by grasses and annuals, and shrubs. Clearly, the averaging of values in this exercise masks much local variability but two general trends emerge. Firstly, the variations in erosion rate display a similar pattern to those of bare ground cover (positive correlation significant at $p < 0.05$) as noted in other semi arid environments (Faulkner, 1990). Secondly, erosion decreases as vegetation succession proceeds.

DISCUSSION

The evidence examined suggests that vegetation and soil patterns within the Tabernas badlands do indeed reflect surface age and that a successional sequence of vegetation can be identified which is associated with a parallel sequence of soil development. Within this sequence, however, morphological controls (slope, aspect, hydrology) remain important, giving rise to much local variation. It would also appear that the vegetation exercises an influence on surface processes and thus a general model of natural stabilisation mechanisms can be proposed.

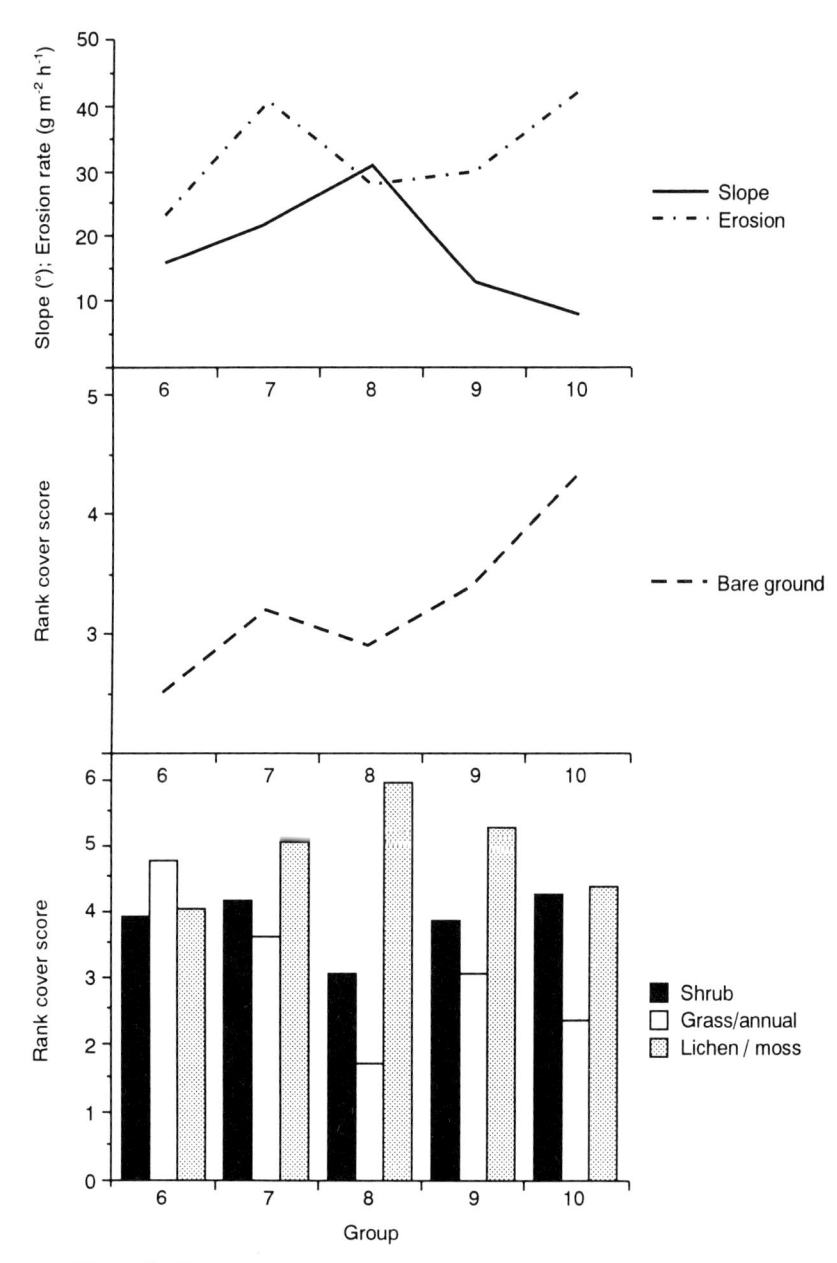

Figure 5.13 Mean life-form type cover values (columns), slope, bare ground cover and erosion rate values (lines) for TWINSPAN groups 6–10

Natural stabilisation would begin as basal incision ceases, and slopes are no longer undermined. At this stage, slope angles would be *ca.* 40°, but incipient pediment development at the slope base may reduce the basal angles to *ca.* 30–35°, causing a tendency for a reduction in erosion rate. Stones, derived either by erosion of a former pediment cover at the top of the slope or derived from within the material by erosion of thin interbedded stone bands or simply by the incomplete destruction of harder clasts of marl, would begin to accumulate at the slope base and, though not reducing runoff rates, would begin to reduce sediment concentrations and further reduce erosion rates. Previous work has clearly identified the role of a stone cover, which begins to become effective at a cover as low as 20%, in reducing sediment yield and erosion rate (Calvo et al., 1991a, and b).

With or without the development of a stone cover, colonisation by crustose lichens begins as soon as the erosion rate decreases sufficiently. This provides the beginnings of a biotic crust which leads to an increase in surface runoff rates but reduces sediment removal (Calvo et al., 1991a, b). This would be followed by colonisation by higher plants (grasses, herbs, shrubs), together with the beginnings of soil formation.

There is therefore a natural negative feedback mechanism reducing erosion rates on badland surfaces, following a reduction in basal incision rate or basal removal of eroded sediment. Its effectiveness is increased in environments where erosion yields large quantities of stones, lichen colonisation is effective in binding the surface or higher plants rapidly colonise the surfaces, causing interception and reducing the effective rainfall intensity as well as increasing the strength of the material by the development of a root system. By this stage pedogenic processes may have begun to influence erosion rates through the mechanisms related to leaching and increased infiltration capacity.

At Tabernas the first stages of the stabilisation sequence, stone accumulation and lichen colonisation, seem to occur at the active/E-age boundary; colonisation by higher plants and pedological change has begun in earnest by stage D/C. The overall timescale for complete stabilisation, whether in a geomorphological or in a pedological sense, would appear, in this environment, to be a long one—in the order of centuries rather than of decades, as has been suggested in more humid regions (Harvey, 1992). Beyond stage C, the vegetation sequence appears to revert as site characteristics become more important than age, but the soil chronosequence continues, albeit with some complication due to profile truncation.

The stabilisation sequence can thus reach quasi-equilibrium or it may be broken, initiating another erosion cycle. Such breaks can occur over the longer term by base-level change in response to tectonic or climatic factors. Over the shorter term they may be due to:

(i) hydrological or ecological change in response to human intervention through grazing or other disturbance;
(ii) in response to climatic fluctuations, or possibly;
(iii) by the occurrence of threshold-exceeding storm events.

ACKNOWLEDGEMENTS

The authors thank the Drawing Office and Photographic Section, Department of Geography, University of Liverpool, for producing many of the diagrams and Mr A. R. Henderson for undertaking the laboratory analyses. The authors also acknowledge the assistance, for some aspects of this work, of the Comission Interministerial de Ciencia y Tecnologia Project NAT89-1072-C06-04.

REFERENCES

Alexander, R. W. and Calvo, A. (1990). The influence of lichens on slope processes in some Spanish badlands. In Thornes, J. B. (ed.), *Vegetation and Erosion*, Wiley, Chichester, pp. 285–98.
Calvo, A., Gisbert, B., Palau, E. and Romero, M. (1988). Un simulador de lluvia portátil de facil

construcción. In Gallart, F. and Sala, M. (eds.), *Métodos y Técnicas para la Medición de Procesos Geomorfológicos*, Sociedad Española de Geomorfología, Monografía 1, Barcelona, pp. 6–16.

Calvo, A. and Harvey, A. M. (1992). Morphology and development of selected badlands in southeast Spain. *Earth Surface Processes and Landforms* (in press).

Calvo-Cases, A., Harvey, A. M. and Paya-Serrano, J. (1991a). Process interactions and badland development in S.E. Spain. In Sala, M., Rubio, J. L. and Garcia-Ruiz, J. M. (eds), *Soil Erosion Studies in Spain,* Geoforma Ediciones, Logrono, pp. 73–90.

Calvo-Cases, A., Harvey, A. M., Paya, J. and Alexander, R. W. (1991b). Response of badlands surfaces in southeast Spain to simulated rainfall. *Cuaternario y Geomorfología*, **5**, 3–14.

Campbell, I. A. (1989). Badlands and badland gullies. In Thomas, D. S. G. (ed.), *Arid Zone Geomorphology*, Belhaven Press, London, pp. 159–83.

Dargie, T. C. D. (1984). On the integrated interpretation of indirect site ordinations: a case study using semi-arid vegetation in southeastern Spain. *Vegetatio,* **55**, 37–55.

Faulkner, H. (1990). Vegetation cover density variations and infiltration patterns on piped alkali sodic soils: implications for the modelling of overland flow in semi-arid areas. In Thornes, J. B. (ed.), *Vegetation and Erosion*, Wiley, Chichester, pp. 317–46.

Geiger, F. (1970). Die Ariditat in Sudostspanien. *Stuttgarter Geographische Studien,* **77**, 173 pp.

Harvey, A. M. (1982). The role of piping in the development of badlands and gully systems in southeast Spain. In Bryan, R. and Yair, A. (eds.), *Badland Geomorphology and Piping*, Geobooks, Norwich, pp. 31–55.

Harvey, A. M. (1984). Aggradation and dissection sequences on Spanish alluvial fans: influence on morphological development. *Catena,* **11**, 289–304.

Harvey, A. M. (1987). Patterns of Quaternary aggradational and dissectional landform development in the Almeria region, southeast Spain: a dry-region, tectonically active landscape. *Die Erde* **118**, 193–215.

Harvey, A. M. (1992). Process interactions, temporal scales and the development of hillslope gully systems: Howgill Fells, northwest England. *Geomorphology,* **5**, 323–44.

Harvey, A. M. and Calvo, A. (1991). Process interactions and rill development on badland and gully slopes. *Zeitschrift für Geomorphologie Supplementband,* **83**, 175–94.

Hesse, P. R. (1971). *A Textbook of Soil Chemical Analysis*, John Murray, London, Hill, M. O. (1979a). DECORANA—a FORTRAN program for detrended correspondence analysis and reciprocal averaging. Section of Ecology and Systematics, Cornell University, Ithaca, New York.

Hill, M. O. (1979b). TWINSPAN—a FORTRAN program for arranging multivariate data in an ordered two-way table by classification of the individuals and attributes. Section of Ecology and Systematics, Cornell University, Ithaca, New York.

Hill, M. O., Bunce, R. G. H. and Shaw, M. W. (1975). Indicator species analysis, a divisive polythetic method of classification and its application to a survey of native pinewoods in Scotland. *Journal of Ecology,* **63**, 597–613.

Kirkby, M. J., Atkinson, K. and Lockwood, J. (1990). Aspect, vegetation cover and erosion on semi-arid hillslopes. In Thornes, J. B. (ed.), *Vegetation and Erosion*, Wiley, Chichester, pp. 25–39.

Kunkel, G. and Kunkel, M. A. (1987). *Florula del Desierto Almeriense*. Colección Investigacion 5, Instituto de Estudios Almerienses, Almeria.

Le Borgne, E. (1955). Susceptibilité magnétique anormale du sol superficiel. *Annales Geophys.,* **11**, 399–419.

Longworth, G. and Tite, M. S. (1977). Mossbauer and magnetic susceptibility studies of iron oxides in soils from archaeological sites. *Archaeometry,* **19**, 3–14.

Maher, B. A. (1986). Characterisation of soil by mineral magnetic measurement. *Phys. Earth Planet Interiors,* **42**, 76–92.

Malloch, A. J. C. (1988). VESPAN II. A computer package to handle and analyse multivariate species data and handle and display species distribution data. University of Lancaster, Lancaster.

Mullins, C. E. (1977). Magnetic susceptibility of the soil and its significance for soil science: a review. *Journal of Soil Science,* **28**, 223–46.

Poesen, J. (1986). Surface sealing as influenced by slope angle and position of simulated stones in the top layer of loose sediments. *Earth Surface Processes and Landforms,* **11**, 1–10.

Poesen, J., Ingelmo Sanchez, F. and Mücher, H. (1990). The hydrological response of soil surfaces to rainfall as affected by cover and position of rock fragments in the top layer. *Earth Surface Processes and Landforms,* **15**, 653–71.

Rohdenburg, H. and Sabelberg, U. (1973). Quartäre Kleinzyklen im Westlichen Mediterrangebiet und ihre Auswirkungen auf die Relief -und Bodenentwicklung. *Catena,* **1**, 71–80.

Rohdenburg, H. and Sabelberg, U. (1980). Northwestern Sahara margins: terrestrial stratigraphy of the Upper Quaternary and some palaeoclimatic implications. In Van Zinderen Bakker Sr, E. M. and Coetzee, J. A. (eds.), *Palaeoecology of Africa and the Surrounding Islands,* Volume 12, pp. 267–76, Balkema, Rotterdam.

Sabelberg, U. (1977). The stratigraphic record of late Quaternary accumulation series in southwest Morocco and its consequences concerning the pluvial hypothesis. *Catena,* **4**, 209–14.

Thompson, R. and Oldfield, F. (1986). *Environmental Magnetism.* Allen and Unwin, London.

Thornes, J. B. (1900). The interaction of erosional and vegetational dynamics in land degradation: spatial outcomes. In Thornes, J. B. (ed.), *Vegetation and Erosion,* pp. 41–53, Wiley, Chichester.

Weijermars, R., Roep, Th. B., Van den Eeckhout, B., Postma, G. and Kleverlaan, K. (1985). Uplift history of a Betic fold nappe inferred from Neogene–Quaternary sedimentation and tectonics (in the Sierra Alhamilla and Almeria, Sorbas and Tabernas basins of the Betic Cordillera, SE Spain). *Geol. en Mijnbouw,* **64**, 397–411.

6 Responses of Rivers and Lakes to Holocene Environmental Change in the Alcañiz Region, Teruel, North-East Spain

M. G. MACKLIN*, D. G. PASSMORE, A. C. STEVENSON, B. A. DAVIS
Department of Geography, University of Newcastle-upon-Tyne, UK

and

J. A. BENAVENTE
Taller de Arquelogía y Prehistoria de Alcañiz, Teruel, Spain

ABSTRACT

The chronology and pattern of Holocene environmental change in north-east Spain, as reflected in semi-arid river and lake sediment records, has been investigated in the Alcañiz region, lower Ebro basin. This paper presents initial palaeoenvironmental results from an endoreic saline lake (Salada Pequeña) and a nearby arroyo (Rio Regallo), evaluates the response of geomorphic and vegetation systems to environmental change and attempts to identify anthropogenic and climate-induced environmental degradation. Three Holocene alluvial fills (dated to 3840–500 BC, 410 BC–390 AD, post-Roman) separated by four phases of channel incision are evident. In broad terms, alluviation and arroyo infilling are associated with vegetation changes that imply a shift to drier conditions, while channel entrenchment appears to be linked to increased effective precipitation. Phases of accelerated lake sedimentation are also similarly coeval with arid conditions, although catchment erosion rates have been significantly increased by clearance and agricultural activities. With climatic and anthropogenic signals manifest in the region's Holocene lake and river sedimentary records, it is argued that the Ebro basin is likely to emerge as a key study area and harbinger of terrestrial ecosystem response to future climate change in southern Europe.

INTRODUCTION

The Ebro basin, north-east Spain, straddles the major climate and vegetation zones that characterise and separate northern Europe from the Mediterranean region. Moving south and south-west from the cool humid footslopes of the Pyrenees towards the seasonally dry steppe systems of the central Ebro basin, pronounced hydrological and vegetation gradients result in significant spatial variations in runoff and erosion. Soil and vegetation in semi-arid areas of the Ebro, similar to those elsewhere in the Mediterranean, also appear to be particularly vulnerable to human disturbance and degradation. This has been recently highlighted by soil loss and salinization following the dramatic increase in winter wheat pro-

*Present address: School of Geography, University of Leeds, UK.

duction, and irrigated agrosystems, since Spain joined the European Community. Although the link between recent land-use change and environmental degradation is well established, what is less clear is how it compares, both in nature and scale, with earlier periods of human disturbance and changes of Holocene climates. An evaluation of the impact of climate change on vegetation and geomorphic processes is especially timely in the light of the anticipated effects of global warming on hydrological and agricultural systems in southern Europe (Dickinson, 1986).

To investigate longer-term environmental change in the Ebro basin, a joint palaeoenvironmental research programme was initiated in 1990 (between the Department of Geography, University of Newcastle-upon-Tyne and the Taller de Arquelogía y Prehistoria de Alcañiz) in the Alcañiz region within the province of Teruel, 100 km south-east of Zaragoza (Figure 6.1). This area contains a series of endoreic saline lakes and alluviated and incised river valleys, which have been a focus of human occupation since the Palaeolithic (Benavente, 1987, 1988). It has been suggested that deforestation and early cultivation by Bronze Age and Iberian peoples led to soil erosion and accelerated sedimentation in these systems, transforming the lakes from productive freshwater ecosystems to unproductive *saladas* (saline lake systems) (Benavente, 1984) and also causing significant alluviation in river valleys (Gutierrez Elorza & Peña Monné, 1990; Burillo Mozata *et al.*, 1986; van Zuidam, 1975, 1976). Superimposed on this has been the effect of long-term climate change which is likely to have controlled lake size, volume and water quality. Alcañiz, with a present annual rainfall of 381 mm, lies in a transitional precipitation zone where moderate change in effective precipitation (as a result of changes in rainfall and/or temperature) can result in major changes in vegetation cover and yield of water and sediment which have probably induced metamorphosis of local river systems in the Holocene (cf. Knox, 1983).

The main aim of this study is to evaluate the response of geomorphic and vegetation systems to environmental change during the Holocene and attempt to separate the role of human from climatically induced environmental degradation. Sediments within these lakes and river basins contain a record of post-glacial environmental change and allow us to examine the relationships between climatic history, human activity and the sedimentary record. This paper explores the potential of the region's lakes and rivers to provide palaeoenvironmental information and presents initial results from two contrasting depositional environments: an endoreic saline lake — Salada Pequeña — and an arroyo — Rio Regallo.

This study is innovative in two respects:

1. To our knowledge it is the first in north-east Spain to utilise both alluvial and lacustrine pollen and geochemical records to document longer-term environmental change.
2. AMS ^{14}C and OSL (optically-stimulated luminescence, Smith *et al.*, 1990) dating techniques have been used to establish a provisional Holocene alluvial geochronology.

This complements artefact evidence, which in most of the Mediterranean has been the primary method of age control.

STUDY AREA

The Salada Pequeña and Rio Regallo, a south bank tributary of the Ebro River, are located in the lower Ebro valley some 5–10 km from the centre of Alcañiz in the north-western part of the province of Teruel (Figure 6.1). The region is underlain by lower Miocene clays,

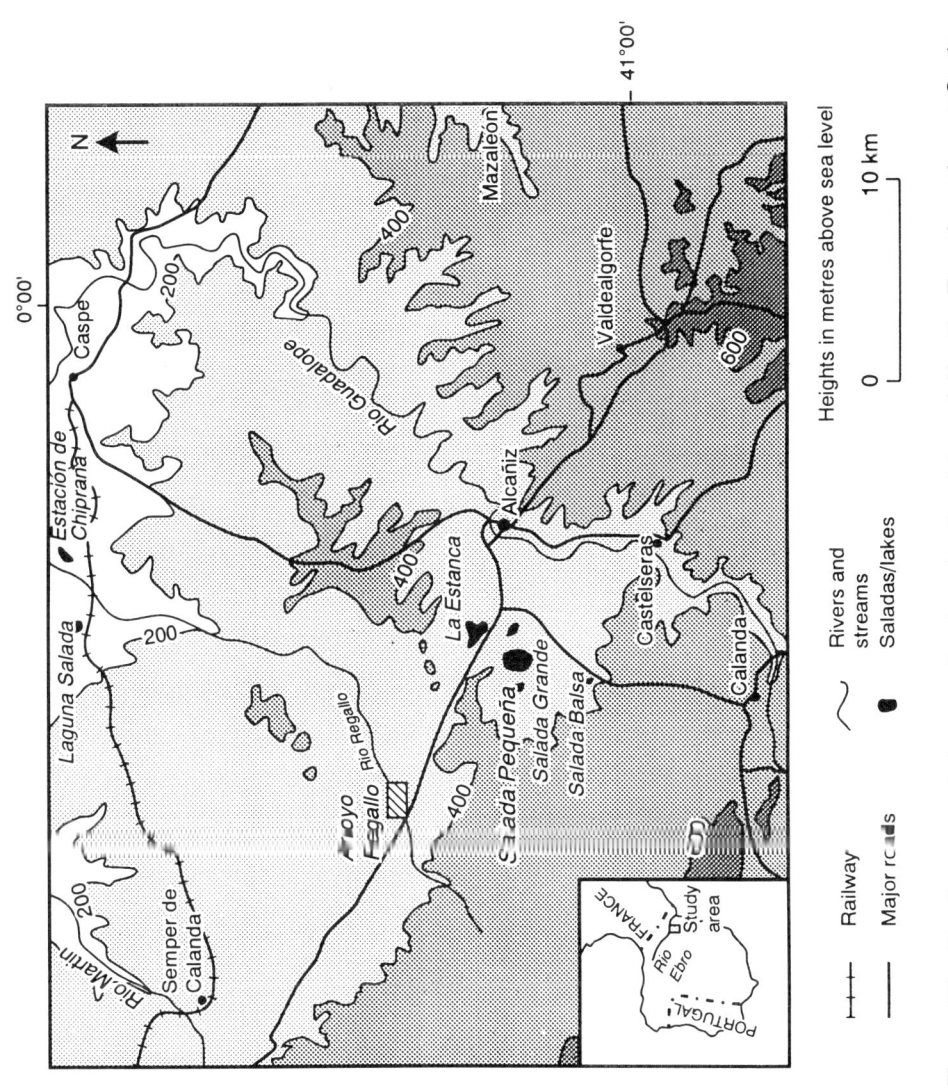

Figure 6.1 Physiography, drainage network and saline lakes of Alcañiz, Teruel, north-east Spain

sands and conglomerates with inter-bedded gypsum and limestone deposits (Ibañez, 1973, 1975, 1976). The natural vegetation is a meso-Mediterranean open woodland of *Quercus coccifera* (*Rhamno lycioidi–Querceto cocciferae sigmetum*). This has been extensively degraded to an *Artemisia* steppe. As a consequence of intense agricultural activity these vegetation types are now restricted to small rocky outcrops. The Salada Pequeña has internal drainage (*ca.* 1 km²) and, similar to other saline lakes in the 'Las Monegros', 'Cinco Villas' and lower Aragon, is contained in a basin formed by subsidence of limestone and gypsum which are strongly faulted and jointed. Lake sediments consist of structureless silty clays. The endoreic nature of the Salada Pequeña, together with leaching and weathering of halite and gypsum bands, ensures a very high lake water pH. This results in lake vegetation communities dominated by halophytes whose type is further controlled by frequency and duration of inundation.

The Rio Regallo (catchment area 350 km²) is a gravel-bed river whose headwaters rise on the lower northern slopes of the Iberian Cordillera. At the margin of the Iberian massif the Regallo valley floor widens and river morphology changes from a confined meandering pattern to a poorly defined network of shallow channels which alternate with short, entrenched reaches where channels are deeply incised into Quaternary alluvium or bedrock. Most of the Regallo alluvial valley floor in this piedmont reach, downstream to its junction with the Ebro, is presently cultivated. The Holocene sedimentation and erosion history of the Rio Regallo was investigated along a 1 km reach south of the *Puente del Regallo* (Figure 6.1), where channel entrenchment had produced a 7 m deep arroyo in alluvium infilling the valley bottom. A complex sequence of Holocene alluvial sediments, well exposed in the sides of the arroyo, were the focus of geomorphological, sedimentological and palynological investigations.

METHODS

A detailed morphological map was made of the alluvial terraces at the *Puente del Regallo* study reach with height relationships established by surveyed cross-profiles. Exposed sedimentary sequences were logged and sediment samples taken for subsequent grain size, geochemical and pollen analyses. In addition, a prominent palaeochannel on the south side of the valley floor was drilled using a percussion corer. Sediment cores were extracted from the Salada Pequeña using a Hiller corer. The deepest of these cores recovered 5.2 m of homogeneous, unstructured silty clays and was sampled at 5 cm intervals for physical, mineral magnetic, geochemical and biological analyses.

Pollen analyses were conducted on lake and river sedimentary sequences to provide regional, and local, vegetation and climatic histories. Geochemical and mineral magnetic analyses were undertaken to:

(i) identify periods of soil and sediment inwash into the lake basin;
(ii) determine the character and sources of eroded material within river and lake catchments; and
(iii) reconstruct the depositional chemical environment.

Organic content was estimated from weight loss following ignition at 450°C. Concentrations of sodium, magnesium, calcium, potassium, manganese and iron were determined by atomic absorption spectrophotometry (air/acetylene flame) following digestion in nitric acid. Magnetic susceptibility (χ_{LF}) and frequency-dependent magnetic susceptibility (χ_{FD}) were measured using a Bartington M.S.1 meter following standard operating instructions. Sediment samples were prepared for pollen analysis using standard procedures (Moore & Webb, 1978). At least 300 pollen grains were counted from each level.

RIO REGALLO QUATERNARY ALLUVIAL SEQUENCE

Five Quaternary alluvial units have been identified at the *Puente del Regallo* study reach and their sedimentary properties, height relationships, depositional environments and estimated ages are summarised in Table 6.1. The names used for alluvial units are informal and they are discussed below in order of apparent age.

Viejo Unit

The oldest alluvial fill in the study reach, the Viejo unit, forms a paired terrace lying 0.5–1.5 m above Holocene alluvium (Figure 6.2). It comprises limestone and sandstone sandy gravels which display numerous channel sloughs that suggest deposition in a braided river with a significant coarse bedload (Bryant, 1983) (Figure 6.2). Viejo sediments interdigitate with alluvial fans and cones developed along the valley floor-slope margin which appear to have been the principal source of frost-shattered gravel that makes up the bulk of the Viejo unit. This alluvial fill was probably deposited in Würm or earlier times (limited soil

Table 6.1 Quaternary alluvial units in the Rio Regallo

Alluvial unit age	Height of upper surface above present river bed (m)	Maximum observed thickness (m)	Coarse (C)/fine sediment (F) member ratio	Fluvial sedimentation style
Castellar, post-Roman	2.5	2.5	C > F	Coarse sediment arroyo inset fill
Altafulla, 410 BC–390 AD	3.5	2.0	C = F	Arroyo inset channel fill
Cerezuela, 3840–500 BC	7.0	5.6	C < F	Divided channel system aggrading by vertical accreition of fine-grained sediment
Moreno, Würm late glacial–early Holocene (?)	2.5	2.5	C > F	Incising, confined meandering gravel-bed river
Viejo, Würm (?)	8.5	3.5	C > F	Aggrading (?), low sinuousity, coarse sediment river system

118   Environmental Change in Drylands

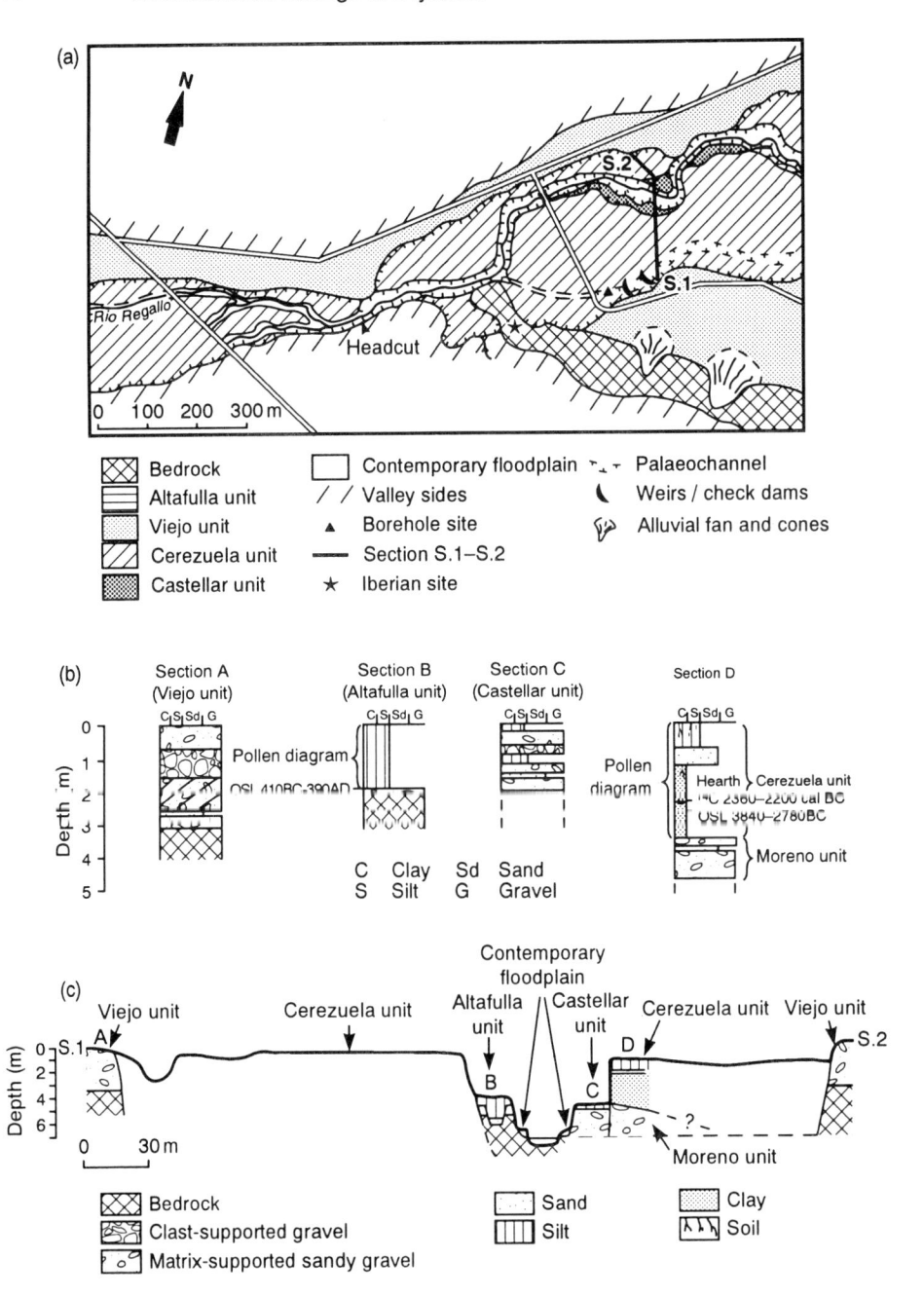

Figure 6.2 Puente del Regallo study reach: (a) terrace sequence and river channels, (b) sedimentary logs of alluvial units, (c) valley floor cross-section

and calcrete development suggest a more recent period of the late Pleistocene) under colder and possibly more arid conditions than present. Similar coarse gravel alluvial terraces are also well developed in the nearby Ebro and Guadalope rivers and are presumed to be Pleistocene cold (arid) stage fluvial deposits (Burillo Mozata *et al.*, 1986; van Zuidam, 1975, 1976).

Moreno Unit

Incision of the Viejo unit (up to 7 m) and Miocene sandstone bedrock underlying the valley floor probably occurred during the Würm late glacial–early Holocene transition. Relatively humid climatic conditions would have enabled vegetation to colonise and stabilise hillslopes, reducing the supply of coarse sediment to the river system. Coarse gravels termed the Moreno unit were probably deposited towards the end of this phase when channel incision had slowed or ceased. Inclined sets of gravels with dips orthogonal or sub-orthogonal to the valley axis are prominent in this unit and appear to have been generated by lateral migration of point bars in a high sinuosity, meandering gravel-bed river (cf. Gustavson, 1978). Several highly patinated flint artefacts (cores and flakes) of unknown age have been recovered from the upper part of this fill. The Moreno unit, unlike younger Holocene river gravels in the Rio Regallo, is strongly cemented by $CaCO_3$. This suggests a somewhat more humid climate than at present, with a seasonally rising water table and a high concentration of dissolved $CaCO_3$, but with relatively low sediment production.

Cerezuela Unit

Following the deposition of the Moreno unit, the Regallo valley bottom was infilled with red clays, silts and sands, between 3.5 and 5.6 m thick, termed the Cerezuela unit (Figures 6.2 and 6.3). This marked an important transformation of Holocene river sedimentation styles with a shift from mainly lateral accretion of coarse (bar and channel bed) material to predominantly out-of-channel vertical accretion of fine-grained alluvium. The Cerezuela unit is comprised of three lithofacies and in stratigraphic order these are:

(i) very uniform red silty clays;
(ii) bedded gravelly sands; and
(iii) buff-red pedogenically altered silts.

The red silty clays appear to have been derived principally from erosion of marl on adjacent valley sides and hillslopes, with silt coming from weathered gypsum. Both rock types are widespread in the Regallo catchment and are dissected by extensive (but currently inactive) gully systems that adjoin the valley floor. Dating control for the deposition of this lithofacies is derived from charcoal recovered from a hearth at a depth of 2.25 m, which has been AMS ^{14}C dated to *ca*. 2360–2200 BC (ETH-5697) (calibration after Stuiver & Becker, 1986), and also from an OSL sediment date at 2.5 m of 3840–2780 BC. These dates show that alluviation began sometime in the Neolithic and continued into the Bronze Age.

Pollen analysis (Figure 6.4) shows a progressive reduction in tree cover in the Regallo catchment during deposition of the silty clays, reflecting either prehistoric clearance and/or a period of aridification. Although *Vitis* and *Olea* are recorded in nearly all levels above the early Bronze Age hearth, and point to human activity nearby during this period, there is also evidence for drier conditions with a vegetation assemblage dominated by Chenopodiaceae and *Artemisia*. This pollen assemblage suggests deposition took place in a similar

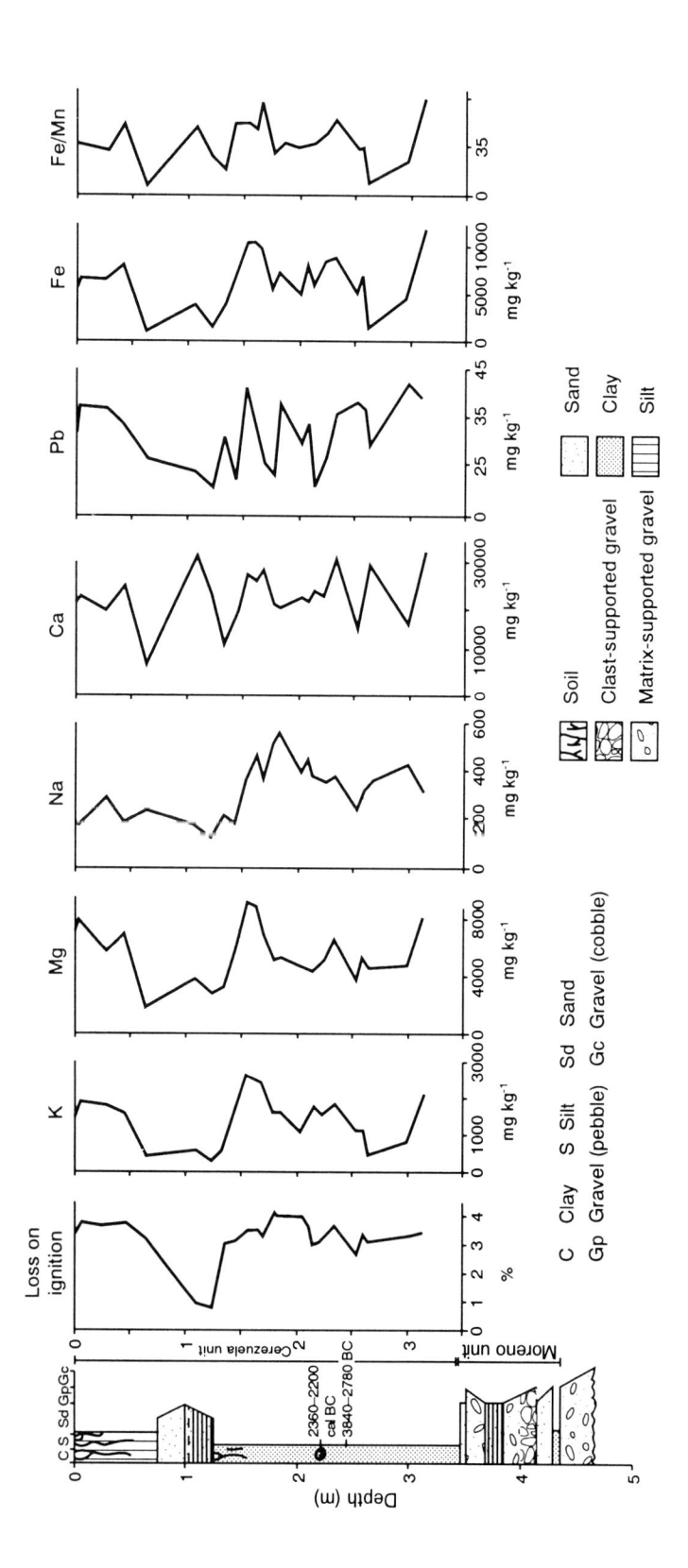

Figure 6.3 Cerezuela unit: loss on ignition and geochemistry

Figure 6.4 Cerezuela unit: pollen diagram

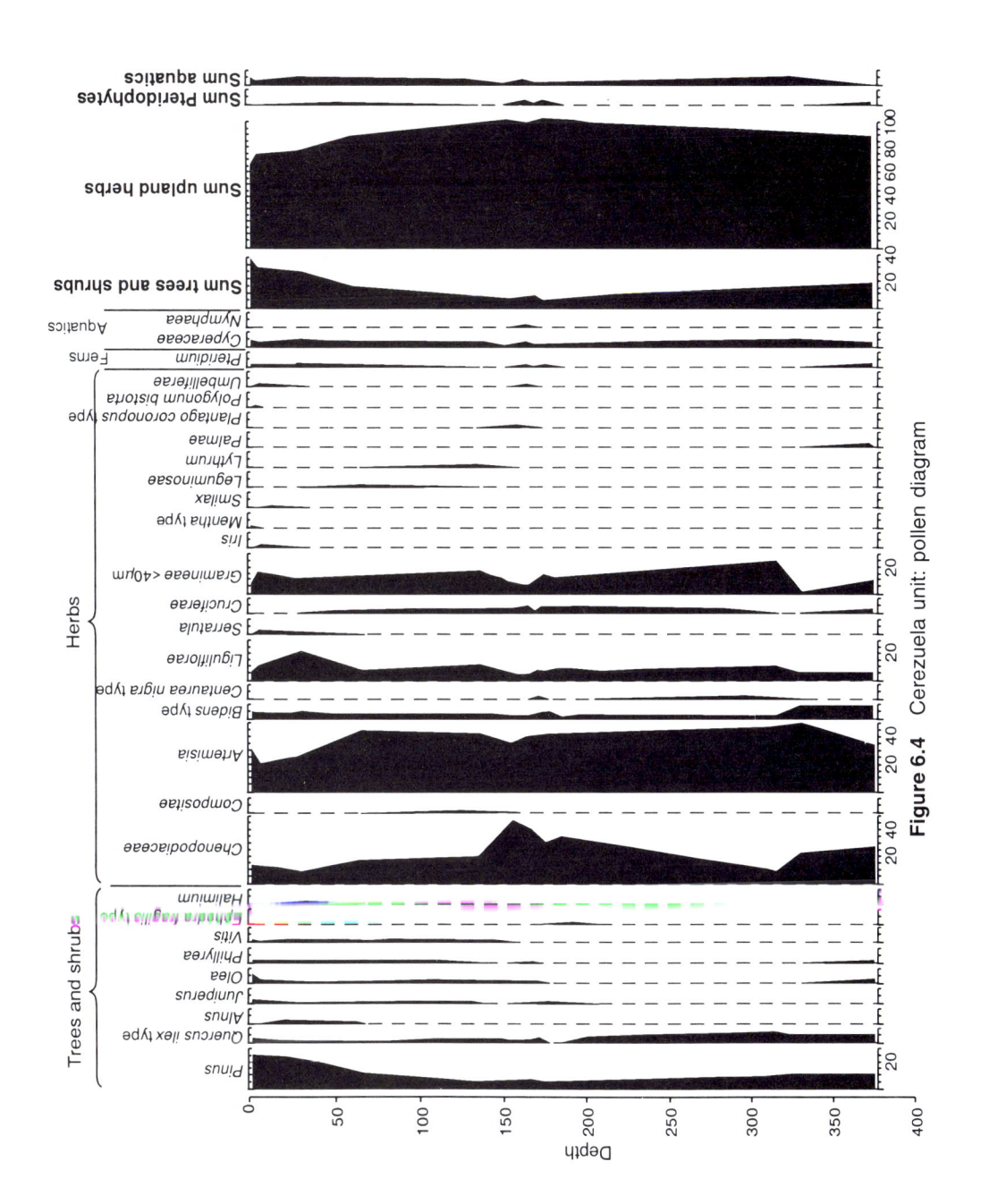

edaphic environment to that found today around saline lakes in the region. Pollen concentrations also show that rates of accretion progressively declined up-section, notably above the hearth, and a palaeosol is evident at the top of the silty clay unit. The palaeosol is truncated and overlain by laminated sands (with some granule and pebble-size clasts) (Figures 6.2 and 6.3) that reverse the overall fining-upward sequence. This unit would appear to indicate a significant increase in flood magnitude, though its stratigraphy provides no clue as to whether it was laid down during several floods or by a single major event. It is also geochemically distinct from overlying and underlying fine-grained alluvium in that it has higher calcium and lower iron contents (Figure 6.3), indicative of erosion of limestone or calcrete soils rather than marl or gypsum terrains in the catchment.

The upper part of the Cerezuela unit consists of buff-red silts (0.8 m thick) which have a non-erosional, gradational contact with the underlying gravelly sands. Pollen analysis indicates somewhat wetter conditions in this period than during deposition of the earlier silty clays, with a decline in halophyte and steppe plants concomitant with a significant regeneration of *Pinus* and, to a lesser extent, *Quercus* (Figure 6.4). Peak levels of *Olea* and a single cereal grain suggest cultivation in the Regallo catchment.

A tentative *terminus ante quem* for the Cerezuela unit comes from a series of check dams (or weirs), considered to be Iberian in age, built across a prominent meandering palaeochannel developed on the surface of the Cerezuela fill on the south side of the valley (Figure 6.2). These structures may have formed part of an extensive, now disused, valley floor water management system. Percussion drilling of the Cerezuela unit immediately upstream of one check dam proved 5.6 m of alluvium with slightly organic sediment at its base. Thus, during the Iberian period, and possibly earlier, the river in the study reach was probably divided into two anabranches, each with prominent leveés comprised of buff-red silts (Figure 6.2). One channel would have run along the line of the present arroyo and the second would have followed the course of the southern palaeochannel. It is likely therefore that the Cerezuela unit was actively accreting, albeit at a slowing rate, until Iberian times (*ca.* 500–100 BC).

Altafulla Unit

A second period of Holocene valley floor incision and refilling began with removal of earlier alluvium, followed by bedrock erosion and cutting of a trench, 4.5 m deep and 80 m wide, on the north side of the valley along which the present river runs. The Altafulla unit, similar to the later Castellar unit (see below), forms an inset fill which is only developed in the arroyo and is not present within the southern palaeochannel (Figure 6.2). This demonstrates that the latter channel system was cut off in the period between the deposition of the Altafulla and Castellar units. OSL dating (at a depth of 1.8 m, Figure 6.2) of the Altafulla fill brackets arroyo development more precisely to Iberian–Roman times and indicates that the river had begun to aggrade sometime after 410 BC–390 AD.

Pollen analyses of the Altafulla unit (Figure 6.5) show a similar vegetation assemblage and sequence to that evident in the Cerezuela fill. Halophyte and steppe plants, such as Chenopodiaceae and *Artemisia*, dominate the pollen assemblage, with the proportion of arboreal species increasing above 100 cm in the profile. In the upper part of the unit (*ca.* 50 cm) there is evidence of cereal and possibly *Olea* and *Vitis* cultivation.

Figure 6.5 Altafulla unit: pollen diagram

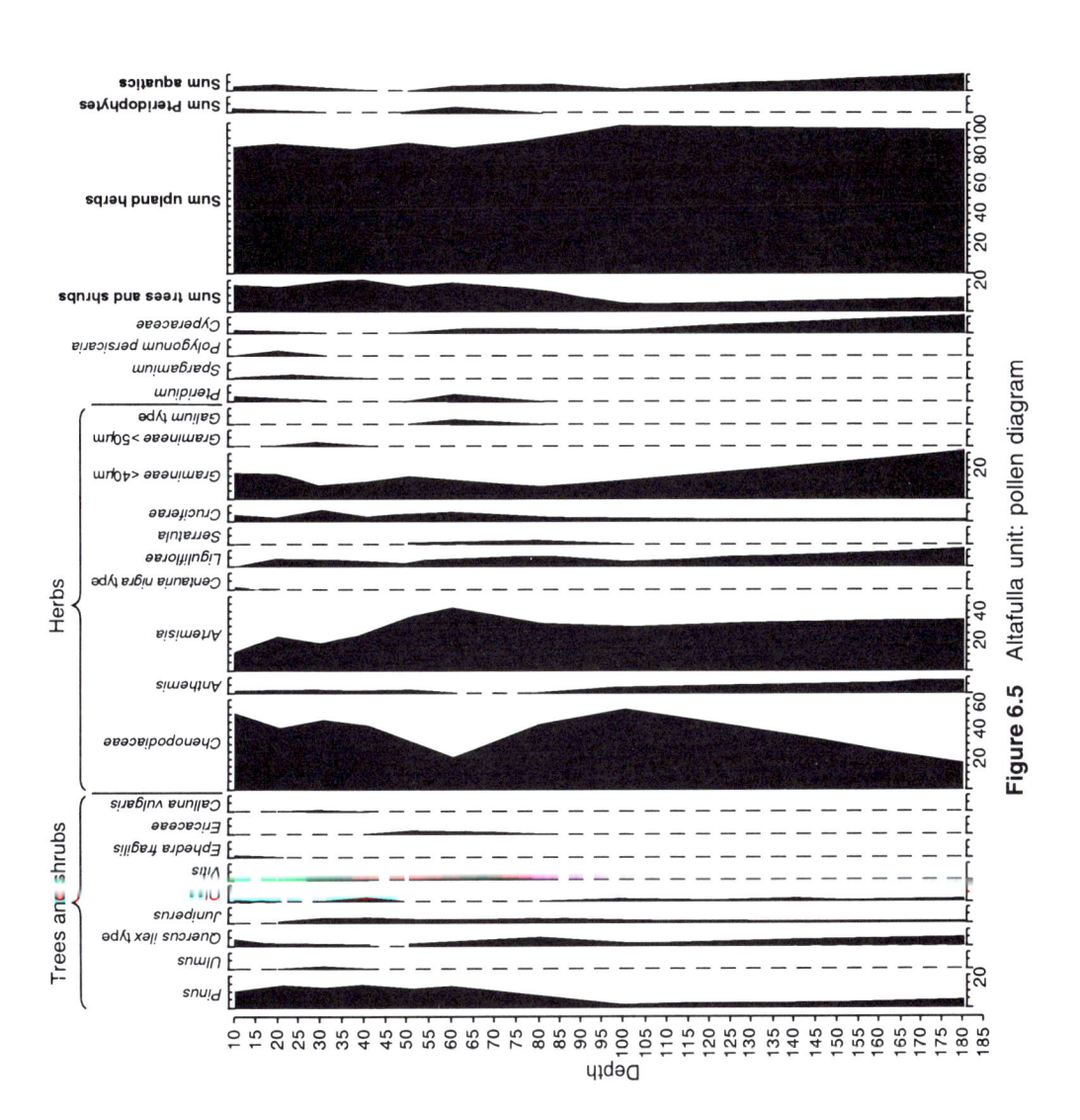

Castellar Unit

The Castellar unit, post-dating the Roman period, is the depositional phase of the third cycle of valley floor entrenchment and refilling, with sediment consisting mainly of gravel reworked from the Moreno and Viejo units. The higher proportion of gravel in the Castellar fill probably reflects both intrinsic hydraulic and sediment supply controls, as well as extrinsic hydroclimatological change. Progressive channel incision and confinement in the late Holocene appears to have resulted in deeper flows and higher shear stresses more capable of entraining and transporting coarser material.

The Castellar unit has itself been incised and forms a terrace 2.5 m above the present river bed. The current vertical tendency of the Regallo is of limited channel aggradation with dense stands of vegetation (mainly *Phragmites australis, Typha angustifolia* and *Tamarix africana*) within and adjacent to the river providing an effective trap for fine-grained sediment.

ENVIRONMENTAL CONTROLS OF HOLOCENE ALLUVIATION AND EROSION IN THE RIO REGALLO

Three Holocene alluvial fills separated by four phases of channel incision are evident in the Rio Regallo below the *Puente del Regallo*. Major erosion during the Würm late glacial to early Holocene period was followed by refilling of the valley bottom with fine-grained alluvium almost to the level of the late Pleistocene valley floor. ^{14}C and OSL dating, pollen analysis and archaeological evidence indicates significant alluviation during the Neolithic and Bronze Age, with rates of valley floor sedimentation slowing during the Iberian period. In broad terms, pollen analysis suggests that significant channel and/or valley alluviation is associated with vegetation changes that imply a shift to drier conditions.

OSL dating and archaeological evidence indicates that arroyo cutting began during the Iberian period. Check dams on the south side of the valley may therefore have functioned as erosion control devices, in an attempt to prevent a headcut eroding upstream, as well as impoundment structures for floodwater irrigation. Channel incision, as has been shown in many other semi-arid river systems (e.g. Graf, 1983), was probably initiated by a major flood or a series of flood events (perhaps represented by the prominent sand unit in the upper part of the Cerezuela unit) that evacuated alluvium and lowered downstream channel levels, causing a wave of incision to propagate upstream.

Such a major flood, or period of flooding, might or might not be coincident with land-use or climatic change, although pollen analysis does suggest somewhat wetter conditions towards the end of the Cerezuela unit immediately pre-dating river entrenchment. Unfortunately, to the authors' knowledge, there have been no investigations of historical rainfall patterns and arroyo activity in north-east Spain. Research in climatically analogous areas in the south-western United States (e.g. Balling & Wells, 1990), however, have demonstrated that arroyo incision is triggered by changes in precipitation patterns involving increased precipitation amounts and the number of high-intensity summer storms. Under these conditions increased runoff promotes channel erosion. Arroyo infilling and stability, on the other hand, occur as precipitation patterns change to fewer intense summer storms and less annual rainfall. If ephemeral streams in semi-arid north-east Spain have responded in a similar manner to changes in frequency, intensity and seasonality of precipitation events, periods dominated by alluviation and arroyo infilling would coincide with drought and channel entrenchment linked to increased effective precipitation. This hypothesis is supported in outline by palynological evidence from the Rio Regallo but obviously needs to be further tested by more detailed dating and pollen analyses. Würm late glacial to early Holocene period valley floor erosion, deposition of the Cerezuela fill and subsequent arroyo development may therefore reflect major hydroclimatological changes that could be of regional rather than local significance and may be recorded in other river and lake basins in the region. To evaluate this hypothesis we next consider the Holocene sedimentation history of the nearby Salada Pequeña.

HOLOCENE SEDIMENTATION HISTORY OF THE SALADA PEQUEÑA

On the basis of pollen, mineral magnetic and geochemical analyses, three major bio- and lithostratigraphic units can be recognised within the Salada Pequeña. These are identified at 520–210, 210–110 and 110–0 cm, and the palaeoenvironmental record of each unit is discussed below.

Unit 1: 520–210 cm

Pollen concentrations are too low in the early part of the unit (520–380 cm) for a reliable pollen diagram to be constructed. Concentrations rise to sufficient levels from 380 cm but still remain low and are oxidised. The pollen spectra (Figure 6.6), dominated by Chenopodiaceae/*Artemisia* and *Anthemis* pollen, indicate the development of extensive halophytic vegetation within the Salada and the presence of a drought-tolerant *Artemisia* steppe on the surrounding slopes. This combination of vegetation types, together with the absence of aquatic pollen taxa, the physical state of many of the pollen grains and the low tree pollen values, would suggest that the basin was rarely inundated for more than three months annually and climatic conditions were arid.

The geochemical and mineral magnetic measurements (Figure 6.7) further suggest that this arid phase was associated with strong erosional activity, peaking at 280 cm. Magnetic susceptibility (a measure of the concentration of magnetic minerals in a sediment sample; Oldfield, 1983) and trace element values rise up-profile between 420–280 cm before declining to 210 cm. Frequency-dependent magnetic (χ_{FD}) susceptibility values, however, are consistently low, which suggests that these sediments contain a relatively small component of eroded topsoil (Oldfield, 1983; Dearing *et al.*, 1985). This phase of catchment instability may therefore reflect erosion and inwash of catchment subsoils and bedrock rather than topsoil (with high χ_{FD} values) which may have been largely removed during the initial phase of instability (cf. the small χ_{FD} peak at 410–420 cm).

Unit 2: 210–110 cm

At the onset of this unit the halophytic and steppic indicators decline and are replaced by *Ruppia*. This suggests an increase in the inundation period of the Salada since *Ruppia* is only found in almost permanently flooded environments (Verhoeven, 1979). High pollen concentrations, the general good state of preservation of the grains and high manganese/iron

126

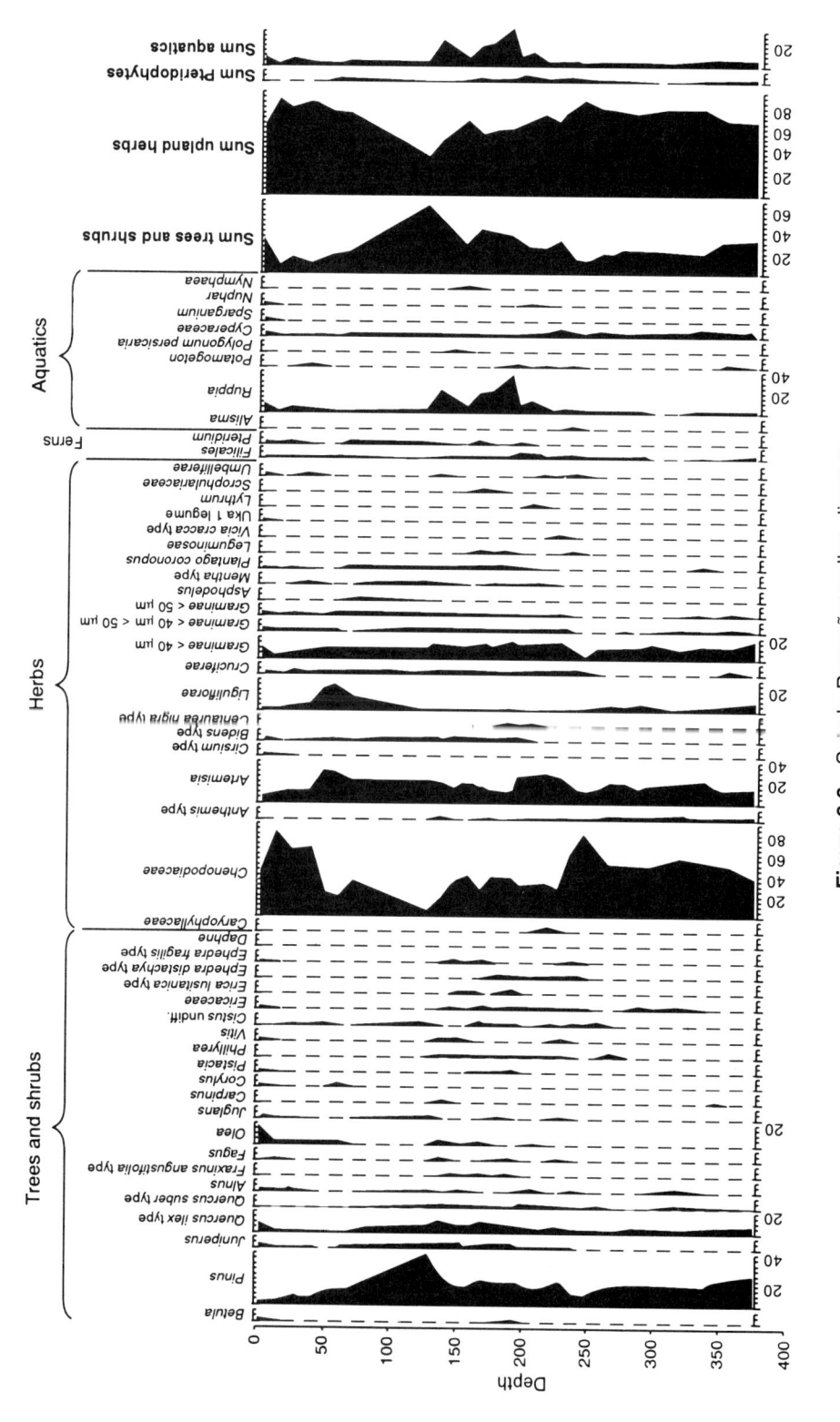

Figure 6.6 Salada Pequeña: pollen diagram

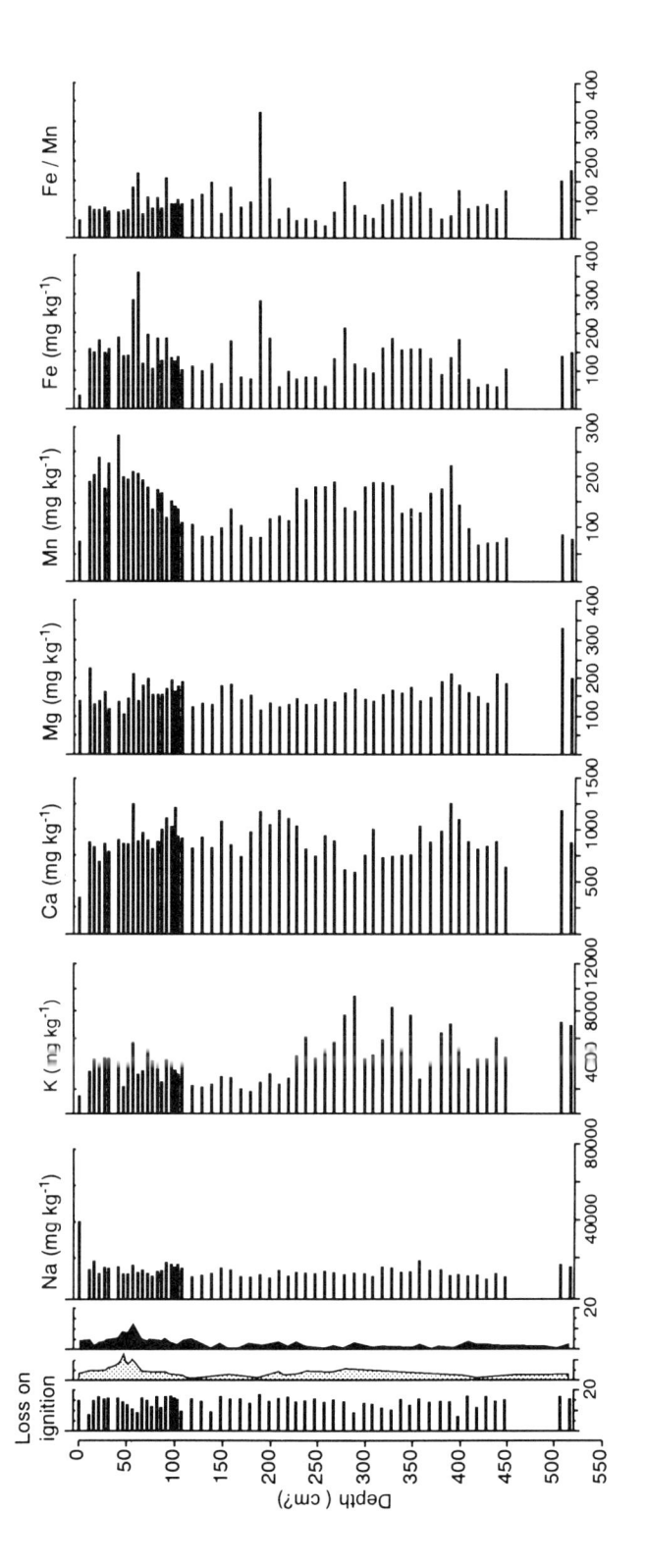

Figure 6.7 Salada Pequeña: loss on ignition, mineral magnetics and geochemistry

ratios confirms this general inference (Figure 6.7). A peak in cereal pollen (Gramineae > 60 μm) between 190 and 170 cm indicates cultivation in the basin at this time.

The *Ruppia* peak is interrupted by a short-lived depression at 160 cm, where values of Chenopodiaceae and *Artemisia* pollen rise. A similar transient rise in most trace metals, cations and mineral magnetic values occurs at this depth (Figure 6.7) and signals a short-lived arid phase accompanied by some limited catchment erosion and inwash. Since the Salada Pequeña basin is endoreic, with no obvious outlet apart from evaporation, temperatures must have been either lower or rainfall higher during this period. The low, stable geochemical and mineral magnetic record further suggest that erosional processes were generally reduced at this time with cultivation appearing to have little effect on sediment yields from the catchment.

Unit 3: 110–0 cm

A return to arid conditions occurs as the *Ruppia* phase is abruptly terminated by a section of the core containing little and badly degraded pollen. Pollen concentrations rise and preservation improves again from 70 cm, above which the spectra are dominated by halophytic and drought taxa like Chenopodiaceae, *Artemisia* and *Liguliflorae*. This, together with the absence of large amounts of *Ruppia*, suggests a return to more drier conditions. However, the limited presence of a few grains of Ruppia would suggest that this phase was not as arid as that recorded by the basal zone. Tree pollen values, e.g. *Pinus*, decline throughout and reflect continuing deforestation within the basin as agricultural practices intensified. The planting of extensive *Olea* plantations is only recorded from the uppermost sediments of the core.

Although prominent peaks in χ_{LF} and χ_{FD} (Figure 6.7) suggest markedly accelerated erosional activity in the catchment during this period this does not necessarily imply that climatic conditions were drier than in the bottom zone of the core. It is more probably a reflection of intense anthropogenic activities disturbing the catchment vegetation cover, combined with the presence of erodible and relatively well-developed topsoil (with high χ_{FD} values) inherited from the preceding wetter, stable period.

The development of the present salt crust across the Salada surface is reflected by the rise of sodium values towards the top of the core (Figure 6.7).

LAKE RESPONSE TO HOLOCENE ENVIRONMENTAL CHANGE: IMPLICATIONS FOR ARCHAEOLOGY

It can be seen that changes in vegetation, land use and erosion regimes have occurred throughout the period covered by the core. In the absence of a firm chronological framework for the Salada Pequeña, the implications of these changes for the local prehistoric populations can only be speculated upon at present. However, it does appear that the local climate was more favourable for human occupation from 210 to 110 cm than at any other time during the period spanned by the core.

It appears that increasing aridity led to catchment instability and accelerated erosion and that this was far more severe in the period represented by the upper part of the core, possibly as a result of greater catchment disturbance by human activity, than had previously occurred. Conversely, the apparent inverse relationship between increased humidity and lake sediment accumulation rates (indicated by variations in pollen concentration) demonstrates the crucial role of vegetation within semi-arid environments in preventing erosion (cf. Thornes, 1987).

TOWARDS A CORRELATION OF HOLOCENE RIVER AND LAKE HISTORIES IN THE ALCAÑIZ REGION

Correlations between river and lake sequences in the Alcañiz region must remain speculative until a more secure dating framework is established. Pollen evidence, however, does suggest that the relatively humid *Ruppia* phase (unit 2) in the Salada Pequeña may correspond in part with the uppermost sediments of the Cerezuela unit, as well as with arroyo cutting in the Rio Regallo shortly before 410 BC–390 AD. Accelerated sedimentation in the Salada Pequeña (represented by units 1 and 3) under drier conditions could therefore correspond with valley alluviation in the Rio Regallo during the Neolithic and Bronze Age (Cerezuela unit) and arroyo infilling in the post-Roman period (Altafulla and Castellar units). Confirmation of these links and inferences drawn about the relationships between human occupation, climate, sedimentation and erosion must await the provision of a geochronology from a further programme of ^{14}C and OSL dating. Nevertheless, significant and extensive alluviation in the area between *ca.* 3840 BC and 390 AD suggests the possibility of burial and preservation of archaeological material in valley bottomlands with good stratigraphic and environmental context. Clearly, in the light of preliminary results from the Rio Regallo, other entrenched Holocene river valleys in the region would repay systematic archaeological survey and palaeoenvironmental investigations.

CONCLUSIONS

This study conclusively demonstrates that saline lakes and incised ephemeral streams in north-east Spain provide suitable sedimentary sequences for palaeoenvironmental reconstruction. Climatic and anthropogenic signals are manifest in both environments and provide an excellent opportunity for disentangling the various potential causes of environmental disruption and degradation during the Holocene in a semi-arid environment. Although, as also suggested by Gutierrez Elorza & Peña Monné (1990), applications of new dating techniques (e.g. AMS ^{14}C and OSL) will ultimately improve the chronology and correlations of longer term environmental change, at present the main obstacle to progress is the lack of comparable integrated river and lake studies elsewhere in north-east Spain. This is especially true for the last 100 years or so, where river and lake histories can be directly related to daily weather records and documented land-use change. The Ebro basin is a zone of climatic transition whose vegetation and geomorphic systems appear to be particularly sensitive to recent climate change. It is likely therefore to emerge as a key study area and harbinger of terrestrial ecosystem response to future climate change in southern Europe.

ACKNOWLEDGEMENTS

The authors would like to thank Dr S. T. Patrick, Dr S. Juggins and Mr N. Rhodes for help with the fieldwork, Mr W. Stelling for laboratory analysis and Mr I. Bailiff for OSL dates. The work was funded by a Research Committee grant and a small grants award from the University of Newcastle-upon-Tyne to ACS and MGM and by Taller de Arquelogía de Alcañiz.

REFERENCES

Balling, R. C. and Wells, S. G. (1990). Historical rainfall patterns and arroyo activity within the Zuni River drainage basin, New Mexico. *Annals of the Association of American Geographers*, **80**, 603–17.

Benavente, J. A. (1984). Cambios geomorfológicos y distribución del hábitat prehistórico: Una aplicación en los focos endorreicos del Bajo Aragon. *Arqueología Espacial*, **2**, 53–74.

Benavente, J. A. (1987). *Arqueología en Alcañiz*. Diputación General de Aragón, Departamento de Cultura y Educatión.

Benavente, J. A. (1988). Las Lagunas de origen endorreico como focos de attraction del poblamiento antiguo: El ejemplo de 'La Estanca' de Alcañiz (Teruel). *Kalathios*, **7–8**, 45–61.

Bryant, I. D. (1983). Facies sequences associated with some braided river deposits of late Pleistocene age from southern Britain. In Collinson, J. D. and Lewin, J. (eds.), *Modern and Ancient Fluvial Systems: Sedimentology and Processes*, Special Publication of the International Association of Sedimentologists, Volume 6, pp. 267–75.

Burillo Mozata, F., Gutiérrez Elorza, M., Peña Monné, J. L., Sancho Marcén, C. (1986). Geomorphological processes as indicators of climatic changes during the Holocene in the North-East of Spain. In Lopez, G. (ed.), *Quaternary Climate in Western Mediterranean*, Madrid.

Dearing, J., Maher, B. A. and Oldfield, F. (1985). Geomorphological linkages between soils and sediments: the role of magnetic measurements. In Richards, K. S., Arnett, R. R. and Ellis, S. (eds.), *Geomorphology and Soils*, George Allen and Unwin, London, pp. 245–66.

Dickinson, R. E. (1986). How will climate change? The climate system and modelling of future climate. In Bolin, B., Jäger, J. and Döös, B. R. (eds.), *The Greenhouse Effect, Climatic Change and Ecosystems. Scope 29*, Wiley, Chichester, pp. 206–70.

Graf, W. L. (1983). The arroyo problem — palaeohydrology and palaeohydraulics in the short term. In Gregory, K. J. (ed.), *Background to Palaeohydrology: A Perspective*, Wiley, Chichester.

Gutierrez Elorza, M. and Peña Monné, J. L. (1990). Upper Holocene climatic change and geomorphological processes on slopes and infilled valleys from archaeological dating (N.E. Spain). In Imeson, A. C. and De Groot, R. S. (eds.), *Landscape Ecological Impact of Climatic Change on the Mediterranean Region*, Universities of Wageningen, Utrecht and Amsterdam, pp. 1–18.

Ibañez, M. J. (1973). Contribución al estudio del endorreismo de la depresión Ebro: el foco endorreico al W y SW de Alcañiz (Teruel). *Geographica*, **1**, 21–32.

Ibañez, M. J. (1975). El Endorreísmo del Sector Central de la Depresión del Ebro. Cuadernos de Investigación Geografía e Historia, Logroño, pp. 35–48.

Ibañez, M. J. (1976). *El piedemonte Ibérico Bajoaragones*, Estudio Geomorfológico, Madrid, pp. 337–8.

Knox, J. C. (1983). Responses of river systems to Holocene climates. In Porter, S. C. and Wright Jr, H. E. (eds.), *Late Quaternary Environments of the United States*, Volume 2, Longman, London, pp. 26–41.

Moore, P. D. and Webb, J. A. (1978). *An Illustrated Guide to Pollen Analysis*, Hodder and Stoughton, London. Oldfield, F. (1983). The role of magnetic studies in palaeohydrology. In Gregory, K. J. (ed.), *Background to Palaeohydrology*, Wiley, Chichester.

Smith, B. W., Rhodes, E. J., Stokes, S., Spooner, N. A. and Aitken, M. J. (1990). Optical dating of sediments: initial quartz results from Oxford. *Archaeometry*, **32**, 19–31.

Stuiver, M. and Becker, B. (1986). High-precision decadal calibration of the radiocarbon timescale, AD 1950–2500 BC. *Radiocarbon*, **28**, 863–910.

Thornes, J. B. (1987) The Palaeo-ecology of erosion. In Wagstaff, J. M. (ed.), *Landscape and Culture: Geographical and Archaeological Perspectives*, Blackwell, Oxford, pp. 37–55.

van Zuidam, R. A. (1975). Geomorphology and archaeology: evidences of interrelation at historical sites in the Zaragoza region, Spain. *Zeitschrift für Geomorphlogie*, **NF19**, 319–28.

van Zuidam, R. A. (1976). *Geomorphological Development of the Zaragoza Region, Spain*. ITC, Enschede.

Verhoeven, J. T. A. (1979). The ecology of *Ruppia*-dominated communities in Western Europe: I. Distribution of *Ruppia* representatives in relation to their autecology. *Aquatic Botany*, **6**, 197–268.

7 The Palaeolimnological Record of Environmental Change: Examples from the Arid Frontier of Mesoamerica

S. E. METCALFE
School of Geography and Earth Resources, University of Hull, UK

F. A. STREET-PERROTT
Environmental Change Unit, University of Oxford, UK

S. L. O'HARA
Department of Geography, University of Sheffield, UK

P. E. HALES
School of Geography, University of Oxford, UK

and

R. A. PERROTT
Environmental Change Unit, University of Oxford, UK

ABSTRACT

Like the Sahel, Ethiopia and Rajasthan, the highlands of central Mexico lie on the northern margin of the tropical summer rain belt. The sediments of closed-basin lakes record long-term variations in water level resulting from fluctuations in rainfall. In central Mexico, since the advent of sedentary maize agriculture *ca.* 3600 yr BP, episodes of accelerated erosion due to the clearance of pine–oak forests have been exacerbated by prolonged spells of drought. This paper focuses on mineral-magnetic, geochemical and palaeoecological evidence from La Hoya de San Nicolás de Parangueo (Guanajuato), La Piscina de Yuriria (Guanajuato) and Lago de Pátzcuaro (Michoacán).

INTRODUCTION

The development of the high civilisations of Mesoamerica, which occurred in the late Preclassic to Postclassic periods (*ca.* 3000–500 yr BP, see Table 7.1), was based on the domestication and cultivation of maize (*Zea mays*). The volcanic uplands of central Mexico, the Neovolcanic Axis (NVA), provided an attractive focus for settlement and agricultural exploitation with their lake basins and fertile alluvial soils. The NVA crosses Mexico at about 19°N, in the area of summer precipitation resulting from the monsoonal circulation at the western end of the Bermuda High. There is a strong northward decline in precipitation from the highlands (1000 mm yr^{-1}) towards the sub-tropical desert (< 400 mm yr^{-1}). In common with other regions lying close to the northern margin of the tropical summer rain belt, it is to be expected that the central highlands of Mexico would be sensitive to changes in climate associated with variations in the strength and location of atmospheric circulation systems such as the Bermuda High.

Many of the basins in the NVA contain, or contained, lakes which have been hydrologically closed for part or all of their histories (Figure 7.1). The sediments of these lakes retain records of variations in water level and changes in catchment stability and vegetation. Results are presented here from three lake basins forming a north–south transect across the present-day precipitation gradient. The lakes are: La Hoya de San Nicolás de Paranqueo (Guanajuato), La Piscina de Yuriria (Guanajuato) and Lago de Pátzcuaro (Michoacán) (Figure 7.1). Discussion is based on mineral-magnetic, geochemical and palaeoecological evidence for both climatic change and human disturbance. The interrelationships between these factors will be assessed.

Relatively little is known of the Preclassic and Classic cultures of Michoacán and southern Guanajuato. The best-known Preclassic group was the Chupícuaro, whose main site (now flooded by Presa Solis) was occupied between ca. 500 BC and AD 400–500 (Porter, 1956). Very little evidence of later Classic occupation has been found, although recent excavations in northern Michoacán (Zacapu Basin) suggest that it was very densely populated (C. Arnauld, personal communication). In the Postclassic, the militaristic Purépecha (or Tarascans) dominated the region and retained control until the Spanish Conquest in AD 1521.

Armillas (1969) proposed that the northern frontier of these agriculturally based groups coincided with a climatic boundary between forests to the south and desert to the north. He also suggested that this agricultural frontier moved north and south in response to increases, or decreases, in summer precipitation amounts. Thus, the lake basins chosen for this study, which lie along the precipitation gradient, might be expected to retain records of both climatic change and varying intensities of anthropogenic activity.

LA HOYA DE SAN NICOLAS DE PARANGUEO, GUANAJUATO (20°23′N, 101°17′W)

La Hoya de San Nicolás de Paranqueo is a small basaltic explosion crater lying at the southern end of a group of such craters in the Valle de Santiago, Guanajuato. Annual rainfall is between 600 and 800 mm, making it the driest of the study sites. As a result of agricultural activity, there is little natural vegetation in the area today with thorn scrub on the lower slopes and *Quercus* (oak) woodland at higher elevations. The basin occupies about 2.4 km^2 and has no outlet; its floor lies at about 1700 m a.s.l. The lake apparently desiccated around 1979 (Brown, 1984), although this cannot be linked directly to climatic drying. Little is known of the Preclassic or Classic occupation of the area. In the Postclassic, however, Valle de Santiago supported many farming communities, representing the northern frontier of the Purépecha empire.

A 5.85 m core was taken from the dry lake floor by R.B. Brown in 1979. The pollen record from the core was examined by Brown (1984) and the carbonate content, magnetic susceptibility, major cations, total P, C/N ratios and charcoal content were analysed by Steininger (1988). Unfortunately, diatom analysis of this core proved impossible due to the absence of preserved valves. The core has been dated by ^{14}C, with fifteen dates ranging

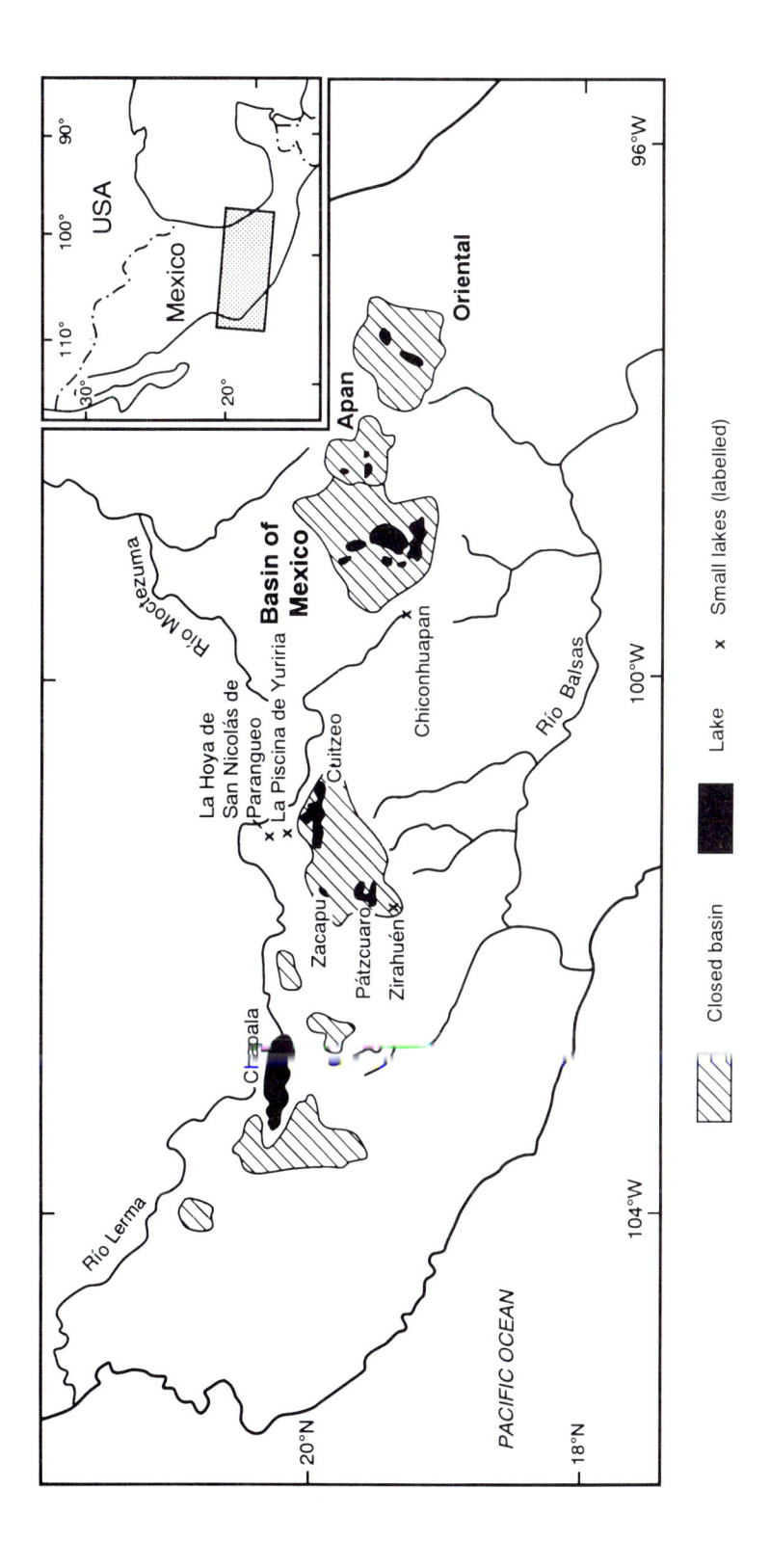

Figure 7.1 Basins of the Neovolcanic Axis, Mexico, showing the location of the study sites

Table 7.1 Chronology of the major cultural periods in Mesoamerica

Years BP	Period	(Major centres) West/Central Mexico
0	Hispanic	
250		
500		Tzintzuntzan
750	Post classic	Tenochtítlan
1000		
		Tula
1250		
1500	Classic	Teotihuacán
1750		
2000		Cuicuilco
2250		
2500		
2750		
3000	Preclassic (formative)	
3250		
3500		
3750		
4000		

from 10710 ± 410 yr BP (A-1963) at 5.30–5.35 m, to 280 ± 75 yr BP (A-2895) at 0.68–0.75 m. The uppermost five dates in the core are, however, in a reversed sequence, possibly due to the progressive erosion (and subsequent deposition) of soils from the inner basin slopes. The last reliable radiocarbon date is probably that of 1100 ± 90 yr BP (A-2641) at 1.18/-1.186 m

Detailed results from the core can be found in Brown (1984), Steininger (1988) and Metcalfe *et al.* (1989). Figure 7.2 summarises the main features of the record. Mass-specific magnetic susceptibility (χ) and total iron and aluminium are used to represent inputs of eroded soil due to disturbance in the lake catchment (Thompson & Oldfield, 1986). The *Pinus* (pine) pollen curve reflects the gross changes in surrounding vegetation caused by climatic change and human disturbance. Levels of > 70% pine suggest the presence of pine trees around the site (Palacios Chavez, 1977; Ohngemach & Straka, 1983). The curve for Chenopodiaceae–Amaranthaceae (cheno-ams) again may reflect either climatic change or human impact, as grain amaranth may have been cultivated as a maize substitute (Gorenstein & Pollard, 1983). The presence of *Zea mays* is, however, a clear indicator of human activity in the basin. The same disturbance indicators are used for the other sites in this study. Given the absence of diatoms from the lake sediments, there is little unambiguous evidence for changes in climate, but two major phases of catchment disturbance are indicated. In spite of the many ^{14}C dates, only a low resolution record may be gleaned due to a coarse original sampling interval (for pollen) and the uncertainties caused by the inverted dates in the upper part of the core. The first phase of erosion is indicated by a decline in *Pinus* (pine) pollen (from > 70 to 20%) occurring just after 3210 ± 120 yr BP. At the same time, values of *Quercus*, cheno-ams and Gramineae (grasses) increase (Brown's pollen zone III). Catchment disturbance is indicated by peaks in Al, Fe, total P and charcoal (290–260 cm) which occur slightly after the decrease in pine pollen in the core (*ca.* 3000–2800 yr BP). The pollen of *Zea mays* becomes more abundant above 240 cm (*ca.* 2700 yr BP),

135

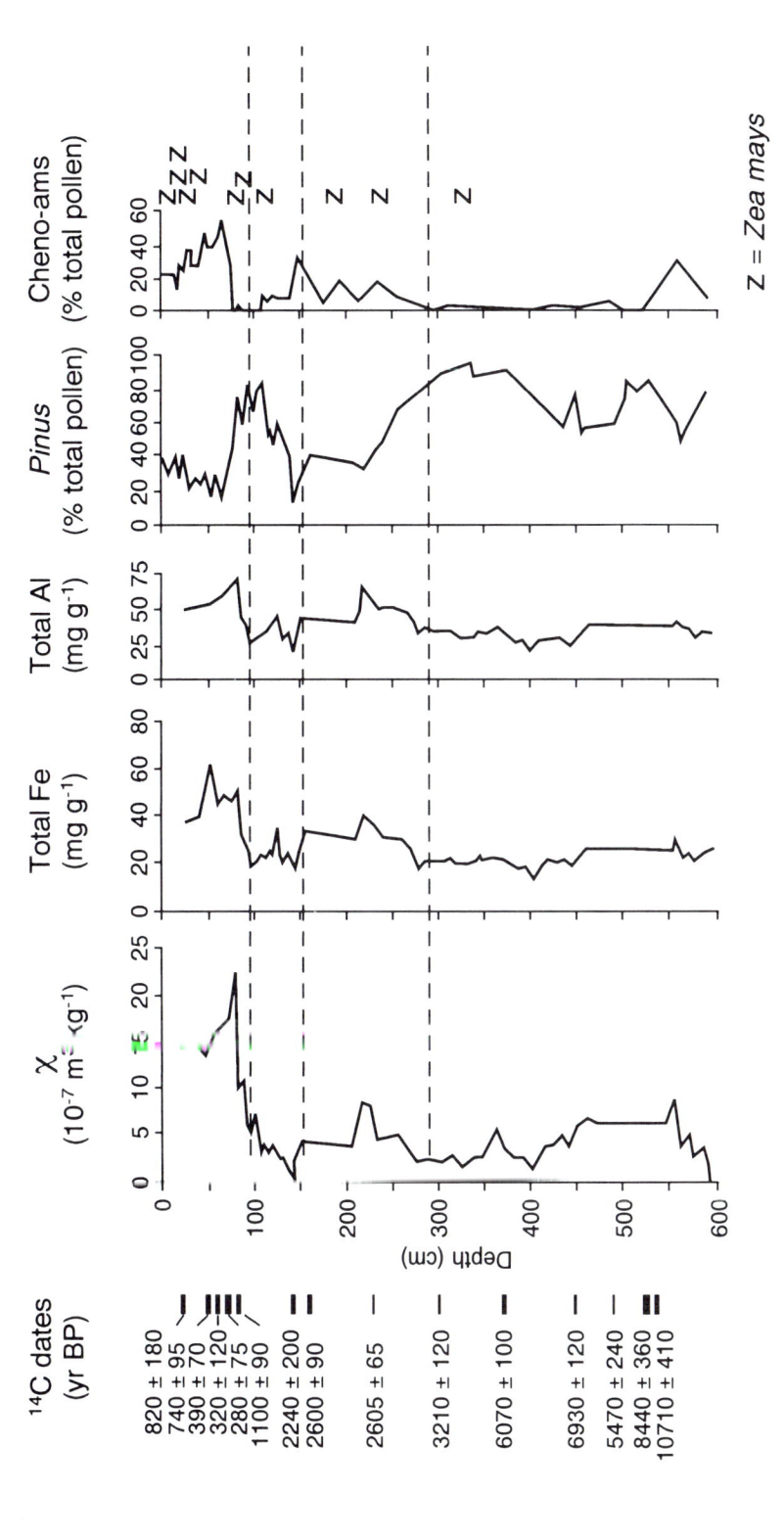

Figure 7.2 Selected indicators of catchment disturbance, La Hoya de San Nicolás de Parangueo

although isolated grains are recorded lower in the core (Brown, 1984). The record of this first phase of disturbance ends just above 150 cm (*ca.* 2350 yr BP). Increasing stability is then indicated by a strong recovery in *Pinus* pollen. It appears that there was little disturbance in the catchment of La Hoya de San Nicolás de Parangueo during the Classic period.

Renewed disturbance is recorded above 90 cm with lower pine, higher cheno-ams and peaks in χ, Al and Fe (Brown's pollen sub-zones IIa and Ib). The very high values of cheno-am pollen in the lower part of pollen zone I (50–75 cm) may reflect the cultivation of grain amaranth within the crater by local peoples (Gorenstein & Pollard, 1983). The uppermost part of the core exhibits a slight reduction in inputs of eroded soil, which may have resulted from reduced population pressure or/and a wetter climate. The poor dating of the top of the core does not allow the timing of the second disturbance phase to be established with any precision although it seems likely to have been initiated in the Postclassic period when the surrounding area was apparently densely populated (Brown, 1984).

LA PISCINA DE YURIRIA, GUANAJUATO (20°13′N, 101°08′W)

La Piscina de Yuriria is a small basaltic explosion crater in the south of the state of Guanajuato. Annual rainfall is *ca.* 700–800 mm. The modern vegetation consists of sub-tropical thorn bush/scrub in the lower areas and sparse *Quercus*-dominated woodland above 2200 m a.s.l. The basin covers an area of about 0.75 km^2 and the elevation of its floor is about 1740 m a.s.l. In 1981, it held a saline, alkaline lake of Na–CO$_3$–Cl type, about 2 m deep, fed by springs and surface runoff. This desiccated between 1985 and 1987, although a small pool of water was present in 1991. Recent lake level changes are probably associated with groundwater extraction for irrigation. Before the Conquest, Yuriria was a garrison town set up to repel attacks on the Purépecha from the Mexica (Aztecs) to the east and the Chichimecs (nomads) to the north.

Two cores have been taken from La Piscina de Yuriria: core 1 (4 m) long in 1981 and core 2 (14 m) long in 1982. Results from the shorter core are summarised here; analyses of core 2 are still in progress.

Core 1 has yielded 8 AMS ^{14}C dates, obtained on carbonate, ranging from 4100 ± 200 yr BP (RIDDL-63) at 375–380 cm to 580 ± 80 yr BP (OxA-1964) at 74–75 cm. A stratigraphically inconsistent date of 840 ± 90 yr BP (OxA-1963) was obtained from 44–46 cm. The core has been analysed for major cations and total P, C/N ratios, stable isotopes in bulk carbonate and magnetic susceptibility (Metcalfe *et al.*, 1989; Hales, 1991). The diatom record is discussed by Metcalfe (1990) and Metcalfe & Hales (in press). Pollen analyses by R.A. Perrott are in progress. In this paper, the results of these different techniques are compared.

The main features of the record from core 1 are summarised in Figure 7.3. In this case, two palaeoclimatic indicators are also plotted: the sample scores on diatom axis 1 resulting from an analysis using CANOCO (ter Braak, 1986) (see below) and the Sr/Ca ratios in the bulk sediment.

Two major diatom assemblages alternate in the core. The first assemblage, dominated by *Navicula elkab*, *N. halophila* and *Nitzschia* spp.(NN), is very similar to the flora of the lake in 1982. The second grouping comprises *Anomoeoneis sphaerophora*, *A. costata*, *Rhopalodia gibberula* and cysts of *Chaetoceros muelleri* (ARC). The species in the second group are largely benthic, found mainly in water of very high salinity, alkalinity and chloride

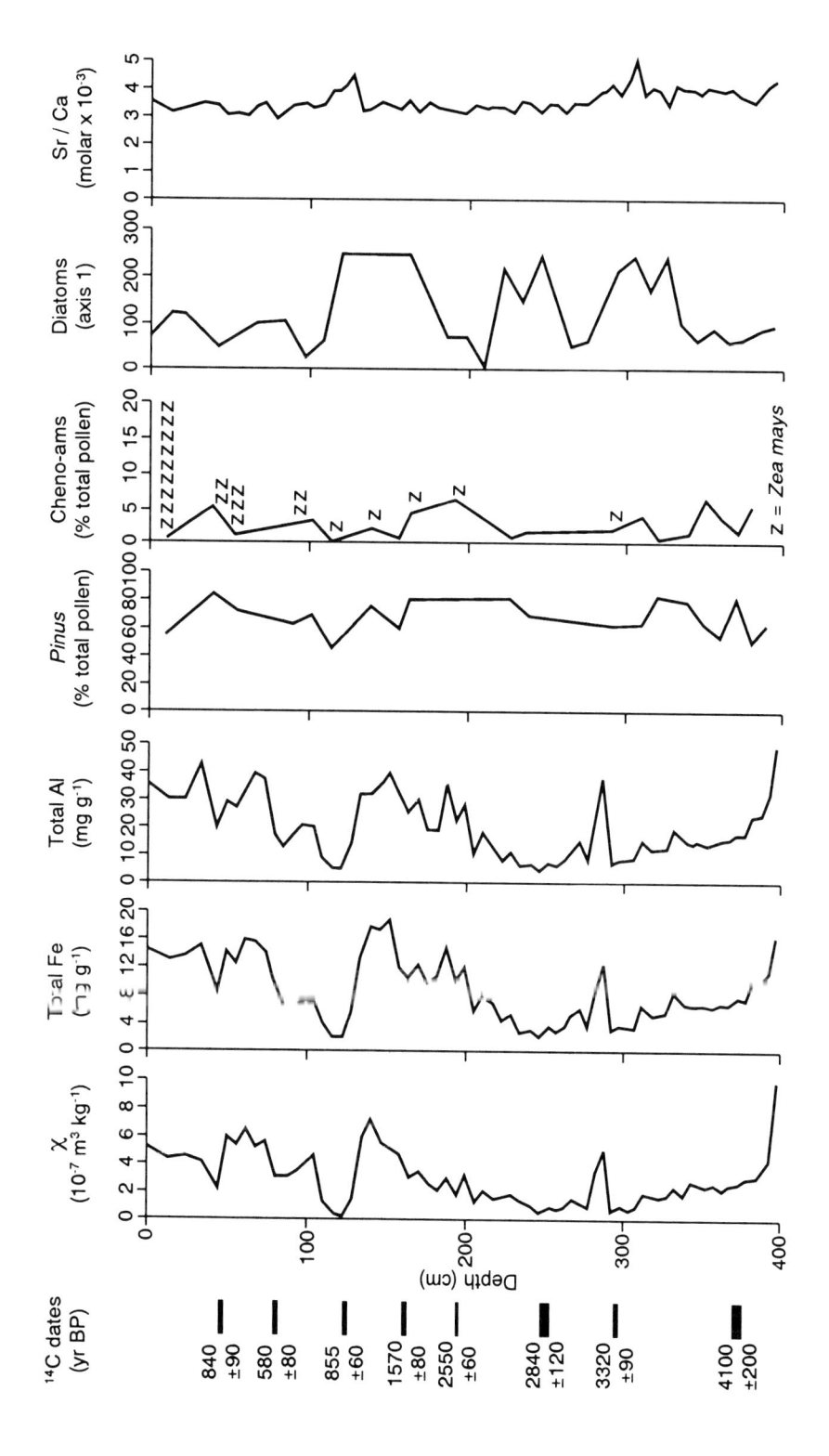

Figure 7.3 Selected indicators of catchment disturbance and climatic change from core 1, La Piscina de Yuriria

content. A shallower and chemically more concentrated lake is indicated. A shift from the first assemblage (NN) to the second (ARC) is believed to reflect increasing aridity and evaporative concentration. Analysis of the core samples using CANOCO confirmed the visual zonation. High scores on axis 1 reflect the predominance of the ARC grouping and of other taxa believed to indicate very shallow water or algal flat conditions. The Sr/Ca ratio in inorganic and biogenic calcite is positively correlated with the Sr/Ca ratio in the host water and hence with salinity (Chivas et al., 1985; Gasse et al., 1987). In La Piscina de Yuriria, strontium and calcium are concentrated in the HCl-soluble carbonate fraction of the sediments (N. Barber, personal communication). The major trends shown in the bulk Sr/Ca curve are currently being verified by microchemical analyses of hand-picked ostracods using ICP-AES (H. Davies, personal communication).

Four phases of disturbance are recorded in core 1. Only the end of the first phase, occurring before 4250 yr BP, is represented at the base of the core. In the absence of *Zea mays* pollen, this phase is attributed to dry climatic conditions.

The second disturbance phase was short-lived (*ca.* 3300–3000 yr BP). Given the presence of *Zea mays* at 290 cm, it is interpreted here as the first impact of maize cultivation in the early Preclassic. This phase is marked by peaks of χ, Fe, Al, Ti, P and other elements typical of tropical volcanic soils. *Pinus* percentages were still low due to a preceding period of dry climate (see below) as well as to disturbance.

It is difficult to date the onset of the third (late Preclassic–Classic) disturbance phase precisely. A rise in χ, Fe, Al and other soil-related elements was, however, underway by 2550 ± 60 yr BP. The rate of sediment accumulation increased considerably. Disturbance culminated around 1700–1000 yr BP (*ca.* AD 300–1000) during the Classic. The presence of *Zea mays* pollen in several levels indicates farming activity during this phase.

The disturbance indicators then drop sharply to a minimum around 855 ± 60 yr BP (OxA-1965). The fourth, complex phase of soil erosion began about 800 yr BP (*ca.* AD 1250) in the Postclassic. Its onset may have coincided with the arrival of the Purépecha in the area; however, the inverted ^{14}C dates preclude accurate dating.

Figure 7.3 also summarises the evidence for marked variations in climate since 4250 yr BP. As noted above, the base of the core appears to reflect the closing of a dry period. Three later periods of drought are registered by diatom assemblages, yielding high scores on axis 1 of the CANOCO analysis. They are dated *ca.* 3700–3300, *ca.* 2950–2700 and *ca.* 2300–800 yr BP respectively. The first of these, which apparently resulted in the retreat of pine from the local area, is associated with a peak in the Sr/Ca ratio centred on 3450 yr BP. The second does not feature in the Sr/Ca curve. However, a layer rich in carbonised material (fine-grained charcoal?) occurs around 240 cm (*ca.* 2800 yr BP). *Pinus* percentages are also low, at least in the early part of the second phase.

The third episode was the most intense, associated with the largest decrease in *Pinus* (culminating at 120–121 cm, *ca.* 850 yr BP) and high levels of *Juniperus* (juniper) pollen. The patchiness of the diatom record between 185 and 125 cm, in conjunction with high Sr/Ca and dewatered sediments, suggests that the lake desiccated between about 1400 and 900 yr BP. The last 800 years (the Postclassic) have been characterised by a generally wetter climate, although two minor episodes of drought are evident from the diatom curve (Figure 7.3).

LAGO DE PATZCUARO, MICHOACAN (19°55′N, 101°39′W)

Lago de Pátzcuaro, the southernmost of the lakes discussed, lies in the humid temperate highlands of Michoacán. This distinct C-shaped lake, which is situated in a lava-dammed valley, covers an area of 126 km^2 and drains a catchment of 927 km^2 (Chacón Torres, 1989). Mean annual precipitation in the basin is *ca.* 1040 mm. The level of the lake has fluctuated considerably over both the long and short term (Watts & Bradbury, 1982; O'Hara, 1991), but there is no evidence to suggest that the lake has ever desiccated completely. The present vegetation in the basin is highly degraded, but the upper slopes still support pine–oak woodlands, although these are mainly secondary.

Today, as in the Postclassic period, the basin is the heartland of the Purépecha culture. Immediately prior to the Conquest, the capital of the Purépecha Empire with its temple complex was located at Tzintzuntzan on the north-east shore of Lago de Pátzcuaro. It is estimated that at the time of the Spanish Conquest the population of the basin was between 60 000 and 105 000 (Gorenstein & Pollard, 1983). Due to the lack of archaeological evidence prior to the arrival of the Purépecha, the settlement history can only be inferred from the presence of *Zea mays* in the pollen record (Watts & Bradbury, 1982). There has been a series of palaeoenvironmental investigations within the basin from the 1940s onwards (e.g. Deevey, 1944; Hutchinson *et al.*, 1956; Watts & Bradbury, 1982). Stratigraphical evidence for two episodes of accelerated soil erosion within the last 2300 years was reported by Street-Perrott *et al.* (1989).

A series of 20 short cores, together with the upper 3.5 m of a 14.2 m core previously analysed by Watts & Bradbury (1982) (referred to here as the Mastercore), have provided the basis for a detailed study of the late Holocene sediments of Lago de Pátzcuaro (O'Hara *et al.*, 1993). AMS dating of ostracode tests has been used to provide chronologies for the cores collected by O'Hara, whilst dates on the Mastercore are conventional dates on bulk sample. The main environmental indicators of the Mastercore are summarised in Figure 7.4.

Four periods of increased catchment instability have occurred over the last 4000–5000 years. As in the case of La Piscina de Yuriria, the earliest phase of disturbance in the Basin of Pátzcuaro pre-dates the first appearance of *Zea mays* in the pollen record (*ca.* 3600 yr BP). This erosional event coincides with a decline in the percentage of *Alnus* (alder) and an accompanying increase in cheno-ams, which was interpreted by Watts & Bradbury (1982) as signifying either the removal of riparian alder by early agriculturalists or a slight shift to drier climatic conditions. Relatively high Sr/Ca ratios in the HCl–H$_2$O$_2$ soluble fraction (Engstrom & Wright, 1984) (Figure 7.4) support the latter hypothesis.

A second, short-lived erosional event, identified by small peaks in χ, Fe and Al, occurred between 3600 and 3200 yr BP. It coincided with the first appearance of *Zea mays* in the pollen record and, as with La Piscina de Yuriria, may have been triggered by the introduction of sedentary agriculture to the basin.

Intense erosion within the Basin of Pátzcuaro began about 2500 yr BP. Since that time, there have been two distinct erosional episodes: dated 2500–1200 yr BP (erosional phase 3) and 850 yr BP to present (erosional phase 4) respectively. During the first of these, erosion was largely confined to the northern slopes of the basin where widespread gullying occurred (O'Hara, 1991). Although there is no archaeological evidence dating from this time, it is possible that this region came under the influence of the Chupícuaro during the late Preclassic–early Classic period. The onset of the most recent (fourth) erosional episode may have been associated with the arrival of the Purépecha *ca.* 900 yr BP. Land degradation

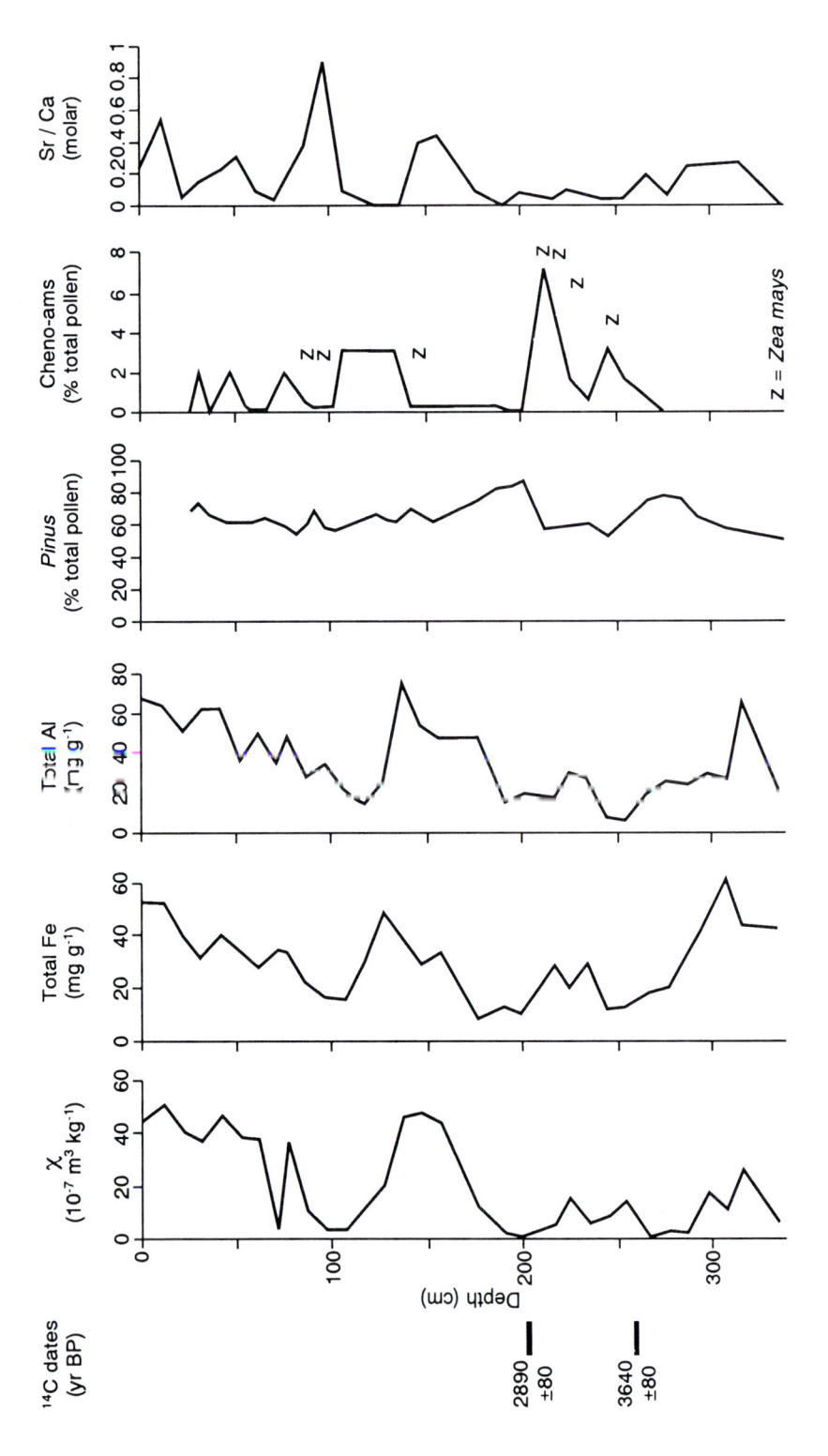

Figure 7.4 Selected indicators of catchment disturbance and climatic change from the Mastercore, Lago de Pátzcuaro

occurred throughout the basin, resulting in the deposition of a thick clay unit in the lake. There is no evidence to suggest that the influx of sediment increased after the arrival of the Spanish.

Fluctuations in the late Holocene climate can be inferred from the Sr/Ca curve presented in Figure 7.4. As discussed above, both the pollen and Sr/Ca ratios at the base of the core indicate relatively dry conditions. A small peak in Sr/Ca dated to *ca.* 3640 ± 80 yr BP (QL-1342) indicates a short-lived drier episode at that time. The upper 190 cm of the core display, several peaks in the Sr/Ca curve, all of which are taken to indicate periods of drought. A date of 2890 ± 80 yr BP (QL-1341), 10 cm below the first of these (190–140 cm), provides a maximum age for the onset of this event. The largest peak in the Sr/Ca ratio occurs at 96 cm (Figure 7.4). Although this severe dry period is not dated at this site, correlation with other cores indicates that it occurred at approximately 1100–1200 yr BP. The severity of this drought is indicated by the remains of shallow-water reeds in a number of cores, indicating a marked drop in lake level (O'Hara, 1991). The two small peaks in the Sr/Ca curve above 60 cm point to spells of drought during the Postclassic/Hispanic period, which may correlate with the two minor dry episodes seen in the diatom record from La Piscina de Yuriria (Figure 7.3); without better dating control, however, the timing of these events cannot be determined.

DISCUSSIONS AND CONCLUSIONS

Palaeolimnological analyses of sediments from three basins situated on a north-south transect across central Mexico have revealed a clear pattern of anthropogenic disturbance and climatic change. Four episodes of disturbance have been identified; their timing and relative intensities in the basins studied are summarised in Figure 7.5.

An early period of catchment disturbance is indicated by the records from La Piscina de Yuriria and Lago de Pátzcuaro. This occurred before 4000 yr BP and pre-dates any clear evidence for the cultivation of *Zea mays*. Accordingly, this erosional episode cannot be attributed to settlement and forest clearance but probably reflects a drier climate. A drought episode, centred on 4600 yr BP, has previously been reported from the Basin of Mexico (Flores, 1986) and from Lake Chiconahuapan in the Upper Río Lerma Basin (Metcalfe *et al.*, 1986, 1991) (Figure 7.1), indicating a regional, rather than a local, event.

The second disturbance episode dates to the Preclassic, between *ca.* 3600 and 3100 yr BP. This corresponds to the appearance of maize (*Zea mays*) pollen in all three sequences. Taken at face value, the radiocarbon dates for the onset of disturbance suggest that there was a northward migration of maize cultivation during an interval of moister conditions in the early Preclassic.

The third and more severe episode spanned the late Preclassic to Classic periods, *ca.* 2500–1100 yr BP, and was probably associated with the Chupícuaro culture. Little is known of this group, but they apparently had an economy based on permanent agricultural and craft centres with small pyramids (Porter, 1956). Charcoal remains in the colluvium surrounding Lago de Pátzcuaro suggest that extensive burning of woodland on basin slopes resulted in widespread erosion of topsoil (Street-Perrott *et al.*, 1989). The third phase of erosion appears to have begun during a relatively moist interval, although the driest basin (La Hoya San Nicolás de Parangueo) remained largely unoccupied; towards the end, however, the climate deteriorated rapidly. La Piscina de Yuriria desiccated and the level of

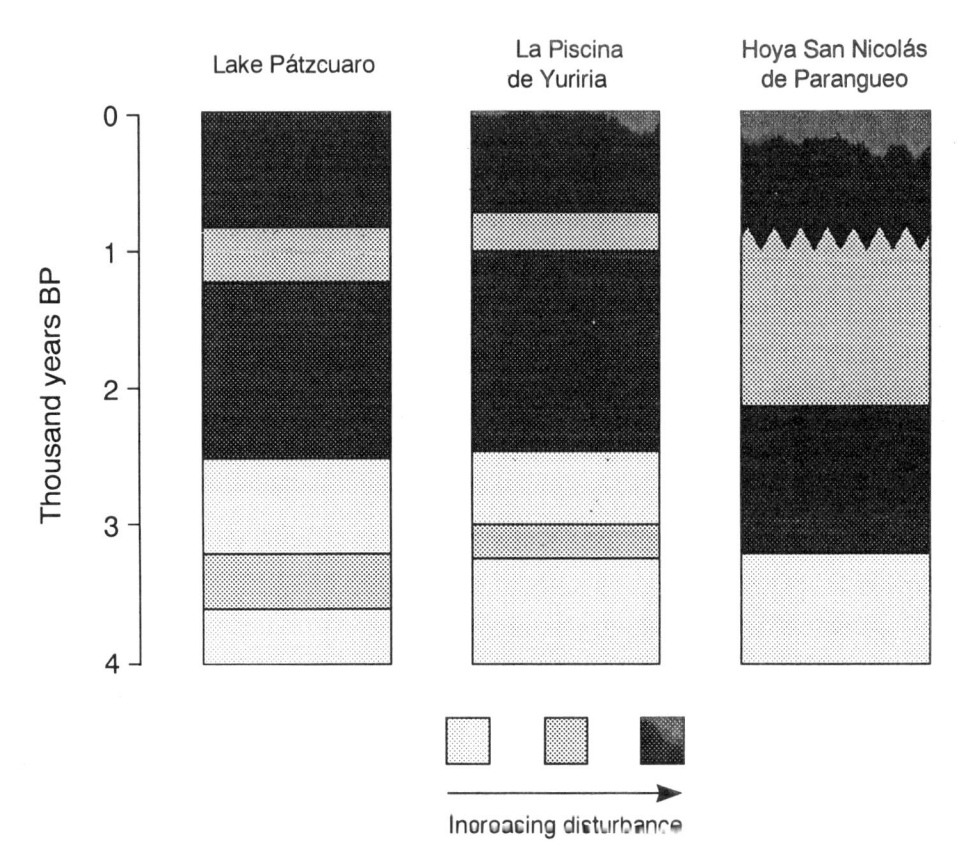

Figure 7.5 Chronology and intensity of disturbance for Lago de Pátzcuaro, La Piscina de Yuriria and La Hoya San Nicholás de Parangueo

Lago de Pátzcuaro fell significantly. Decreasing rainfall probably exacerbated the impact of land clearance. This intense drought during the late Classic (1400–900 yr BP) can be correlated with the desiccation of Lake Chiconahuapan, Upper Río Lerma, which is bracketed by dates of 1380 ± 50 and 870 ± 50 yr BP (Metcalfe *et al.*, 1986, 1991). The drought may have been a factor in the collapse of Teotihuacan (Basin of Mexico) and, possibly, of the Classic Maya cities. This finding contradicts the suggestion of Armillas (1969) that the late Classic and early Postclassic were wetter than present, allowing a northward expansion of the agricultural 'frontier', which was followed by drying and retreat in the twelfth and thirteenth centuries. Brown (1984) indicates a similar timing of events, although pointing to environmental degradation as the key factor. It should be noted, however, that the chronological control of the work reported here and elsewhere by the authors is much tighter than that available to either Armillas or Brown.

In the Postclassic, wetter conditions returned. The period of Purépecha and possibly Spanish dominance was marked by intense erosion (fourth erosional episode). High population densities during the Purépecha period, especially in the Basin of Pátzcuaro, account for the onset of erosion well before AD 1521. The severity of Postclassic disturbance and the difficulty of obtaining reliable radiocarbon dates have, however, made it hard to identify any specific impact resulting from the Spanish conquest in 1521 (with the introduction of draught animals and iron implements). In the cores from La Hoya de San Nicolás de Parangueo and La Piscina de Yuriria, the fourth episode is marked by very high percentages of *Zea mays* and by high cheno-am (possibly grain amaranth) pollen in the former. Further south, in the Pátzcuaro basin, human activity included large-scale burning of pine woodland (Street-Perrott *et al.*, 1989). Gully formation may have resulted (O'Hara, 1991). The inwashing of soil material contributed to the pronounced eutrophication of Lago de Pátzcuaro over the last 900 years.

The lake basins of the NVA have been important centres of occupation for at least the last 3500 years (longer in the Basin of Mexico). However, the region lies close to the arid frontier of Mesoamerica. As a result, it is vulnerable to prolonged episodes of drought associated with a failure of the summer rains. This vulnerability is at its greatest on the northern margin of the region where drought conditions (as in the early 1980s) are most severe. Our multidisciplinary studies of lake cores suggest that there has been a complex

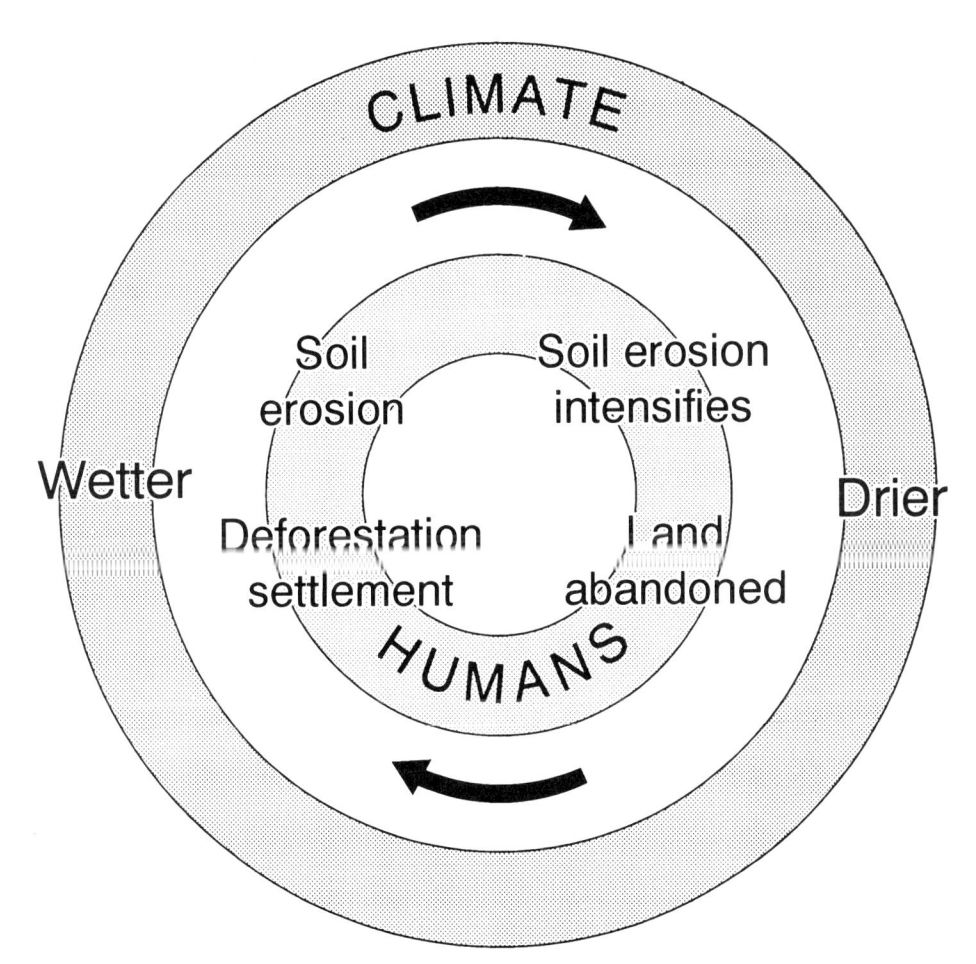

Figure 7.6 Possible relationship between climate, human occupation and disturbance at the arid frontier of Mesoamerica

relationship between climate, human occupation and land degradation in this region (Figure 7.6). During episodes of plentiful summer rainfall, settlement is established (or intensifies) and increased population pressure may result in deforestation and soil erosion. In the absence of excessive anthropogenic disturbance, the catchment may stabilise as precipitation becomes more abundant. As the climate dries, soil erosion initially intensifies as the vegetation cover is disrupted. However, with further drying, the land is abandoned in the drier areas and catchment stability is again restored.

ACKNOWLEDGEMENTS

The authors would like to thank Nick Barber, Ben Brown and Francine Steininger for their contributions to this study. Financial support was provided by the NERC (PEH, SEM, SLO'H), CONACYT and St Hilda's College, Oxford (SLO'H), the Faculty of Anthropology and Geography, University of Oxford (FAS-P and RAP); radiocarbon dating was funded by the NERC and SERC.

REFERENCES

Armillas, P. (1969). The arid frontier of Mexican civilization. *Transactions of the New York Academy of Sciences, Section of Anthropology,* **31**, 697–704.
Brown, R. B. (1984). *The Paleoecology of the Northern Frontier of MesoAmerica.* Unpublished doctoral thesis, University of Arizona, Tucson.
Chacón Torres, A. (1989). *A Limnological Study of Lake Pátzcuaro, Michoacán, Mexico.* Unpublished doctoral thesis, University of Stirling.
Chivas, A. R., De Deckker, P. and Shelley, J. M. (1985). Strontium content of ostracods indicates lacustrine palaeosalinity. *Nature,* **316**, 251–3.
Deevey, E. S. (1944). Pollen analysis and Mexican archaeology: an attempt to apply the method. *American Antiquity,* **10**, 135–49.
Engstrom, D. R. and Wright Jr, H. E. (1984). Chemical stratigraphy of lake sediments as a record of environmental change. In Haworth, E. Y. and Lund, J. W. G. (eds.), *Lake Sediments and Environmental History: Studies in Palaeolimnology and Palaeoecology in Honour of Winifred Tutin,* Leicester University Press, Leicester, pp. 11–67.
Flores, A. (1986). Fluctuaciones del Lago de Chalco, desdehace 35 mil años al presente, In Lorenzo, J. and Mirambell, L. (eds), *Tlapacoya: 35,000 Años de Historia del Lago de Chalco,* Colleción científica, serie prehistoria, Instituto Nacional de Antropología e Historia, México, pp. 109–56.
Gasse, F., Fontes. J. C., Plaziat, J. C., *et al.* (1987). Biological remains, geochemistry and stable isotopes for the reconstruction of environmental and hydrological changes in the Holocene lakes from north Sahara. *Palaeogeography, Palaeoclimatology, Palaeoecology,* **60**, 1–46.
Gorenstein, S. and Pollard, H. (1983). *The Tarascan Civilisation: A Late Prehispanic Cultural System.* Vanderbilt University Publications in Anthropology 28, Nashville.
Hales, P. E. (1991). Inorganic geochemistry of tropical lake and swamp deposits. Unpublished DPhil thesis, University of Oxford.
Hutchinson, G., Patrick, R. and Deevey, E. S. (1956). Sediments of Lake Patzcuaro, Michoacán, Mexico. *Geological Society of America Bulletin,* **67**, 1491–504.
Metcalfe, S. E. (1990). *Navicula elkab* O. Müller—a species in need of redefinition?. *Diatom Research,* **5**, 419–23.
Metcalfe, S. E., Street-Perrott, F. A., Perrott, R. A. and Harkness, D. D. (1986). Environmental changes during the Late Quaternary in the Upper Lerma Basin, Estado de México, In Ricard, M. (ed.), *Proceedings of the 8th International Diatom Symposium, Paris 1984,* pp. 471–82.
Metcalfe, S. E., Street-Perrott, F. A., Brown, R. B., Hales, P. E., Perrott, R. A. and Steininger, F. M. (1989). Late Holocene human impact on lake basins in Central Mexico. *Geoarchaeology,* **4**, 119–41.

Metcalfe, S. E., Street-Perrott, F. A., Perrott, R. A. and Harkness, D. D. (1991). Palaeolimnology of the Upper Lerma Basin, Central Mexico: a record of climatic change and anthropogenic disturbance since 11,600 yr BP. *Journal of Paleolimnology,* **5**, 197–218.

Metcalfe, S. E. and Hales, P. E. (in press). Holocene diatoms from a Mexican crater lake—La Piscina de Yuriria. In Kociolek, P. (ed.), *Proceedings of the 11th International Diatom Symposium, San Francisco 1990.*

O'Hara, S. L. (1991). Late Holocene environmental change in the Basin of Pátzcuaro, Michoacán, Mexico. Unpublished DPhil. thesis, University of Oxford.

O'Hara, S. L., Street-Perrott, F. A. and Burt, T. P. (1993). Accelerated soil erosion around a Mexican highland lake caused by prehispanic agriculture. *Nature,* **362**, 48–51.

Ohngemach, D. and Straka, H. (1983). Contribuciones para historia de la vegetación y del clima en la región de Puebla-Tlaxcala. Análisis polínicos en el Proyecto Mexico. El Proyecto México de la Fundación Alemana para la Investigación Científica, 18.

Palacios Chavez, R. (1977). Lluvia de polen moderno en diferentes habitats del Valle de México. *Boletin de la Sociedad Botanica de Mexico,* **36**, 45–69.

Porter, M. N. (1956). Excavations at Chupícuaro, Guanajuato, Mexico. *Transactions of the American Philosophical Society,* **46**, 517–637.

Steininger, F. M. (1988). Environmental changes in the Hoya San Nicolás de Parangueo Basin, Guanajuato, Mexico: a palaeolimnological study. Unpublished MSc thesis, University of Oxford.

Street-Perrott, F. A., Perrott, R. A. and Harkness, D. D. (1989). Anthropogenic soil erosion around Lake Pátzcuaro, Michoacán, Mexico, during the Preclassic and late Postclassic-Hispanic periods. *American Antiquity,* **54**, 759–65.

ter Braak, C. (1986). Canonical correspondence analysis: a new eigenvector technique for multivariate direct gradient analysis. *Ecology,* **67**, 1167–79.

Thompson, R. and Oldfield, F. (1986). *Environmental Magnetism,* George Allen and Unwin, London.

Watts, W. and Bradbury, J. P. (1982). Paleoecological studies at Lake Pátzcuaro on the west-central Mexican plateau and at Chalco in the Basin of Mexico. *Quaternary Research,* **17**, 56–70.

8 Lacustrine Sedimentation in a High-Altitude, Semi-Arid Environment: The Palaeolimnological Record of Lake Isli, High Atlas, Morocco

H. F. LAMB, C. A. DUIGAN*
Institute of Earth Studies, University of Wales, Aberystwyth, UK

J. H. R. GEE
Department of Biological Sciences, University of Wales, Aberystwyth, UK

K. KELTS
Limnological Research Center, University of Minnesota, Minneapolis, USA

G. LISTER
Geological Institute, ETH Zentrum, Swiss Federal Institute of Technology, Zürich, Switzerland

R. W. MAXTED†
Institute of Earth Studies, University of Wales, Aberystwyth, UK

A. MERZOUK
Département des Sciences du Sol, Institut Agronomique et Vétérinaire Hassan II, Rabat, Morocco

F. NIESSEN
Geological Institute, ETH Zentrum, Swiss Federal Institute of Technology, Zürich, Switzerland

M. TAHRI
Département des Sciences du Sol, Institut Agronomique et Vétérinaire Hassan II, Rabat, Morocco

R. J. WHITTINGTON
Institute of Earth Studies, University of Wales, Aberystwyth, UK

and

E. ZEROUAL
Institut de Géologie, Université de Neuchatel, 11, rue Emile-Argand, CH-2007, Switzerland

---

**Present address*: Countryside Council for Wales, Bangor, Wales.
†*Present address*: National Rivers Authority, Huntingdon, UK.
Publication No. 250 of the Institute of Earth Studies, University of Wales, Aberystwyth, UK (Lamb, Duigan, Maxted)
Publication No. 452 of the Limnological Research Center, University of Minnesota, Minneapolis, USA (Kelts)

Environmental Change in Drylands: Biogeographical and Geomorphological Perspectives.
Edited by A. C. Millington and K. Pye. © 1994 John Wiley & Sons Ltd

ABSTRACT

Lake Isli is an endorheic 95 m deep mesosaline lake located 260 m a.s.l. in the Moroccan High Atlas, close to the southern limit of Mediterranean forests. The lake currently receives detrital sediments from intermittent erosion of beach terraces formed during earlier, high lake-level episodes. Lake levels lower than the present are indicated by a series of submerged terraces revealed by side-scan sonar. Cores from the profundal zone show vertical alternation of proximal turbidite and slump facies and distal turbidite facies, which are probably associated with lake-level variation. An older, more extensive change in the lake level (± 50 m) is indicated by shallow-water authigenic calcites, containing pyrite framboids, littoral diatoms and steppe pollen, ^{14}C dated to 34 830 ± 400 yr BP in the basal unit of a core from a deep submerged terrace.

INTRODUCTION

The High Atlas mountains are a forbiddingly arid and inhospitable terrain, in which subsistence agriculture is subject to the unpredictable and forceful impact of drought, flash flooding and harsh winter conditions. The landscape is dominated by expanses of bare rock and soils, barren of all but the most sparse vegetation cover. To what extent this is a natural product of the semi-arid mountain climate, transitional between Mediterranean winter westerlies and Saharan aridity, or of overexploitation by grazing herds of sheep, goats and camels can only be guessed at in the absence of documentary or palaeoenvironmental data.

Among the few potential sources of such data for the High Atlas is an unusual pair of lakes near the village of Imilchil in the eastern High Atlas. This paper summarises the preliminary results of a project aimed at deciphering the sedimentary record of environmental change from these lakes, concentrating here on Lake Isli, the largest of the pair. In spite of extensive littoral erosion during episodes of lake-level change and the inherent instability of the steep-sided lake basin, the sediments provide evidence for past variations in lake level, water chemistry, lake biota and regional vegetation. In this case at least, the geomorphic response to climate change dominates to the extent that it may obscure details of the environmental record, especially that of human impact.

Other lakes in the less arid Middle Atlas region are currently under investigation (Lamb *et al.*, 1989; El Hamouti *et al.*, 1991; Roberts *et al.*, Chapter 9 in this volume). However, there have been few previous palaeolimnological investigations of high-altitude lakes in semi-arid regions; Lake Qinghai at 3200 m on the Tibetan plateau is an example, but it is very much larger than Lake Isli (Lister *et al.*, 1991). Thus, one justification for this investigation is the opportunity to obtain a temporal perspective on sedimentary processes in an unusual lake set in a mountainous, semi-arid environment where rates of surface erosion are high.

SITE DESCRIPTION

At 32°13′N, 5°32′W and 2270 m elevation (Figure 8.1), Lake Isli is located at the southern limit of Mediterranean forest. The position of this ecotone reflects the southward limit of precipitation-bearing winter westerlies, so it should be possible to infer the regional climatic history from the pollen stratigraphy of the lake sediments. Also, because Lake Isli is a

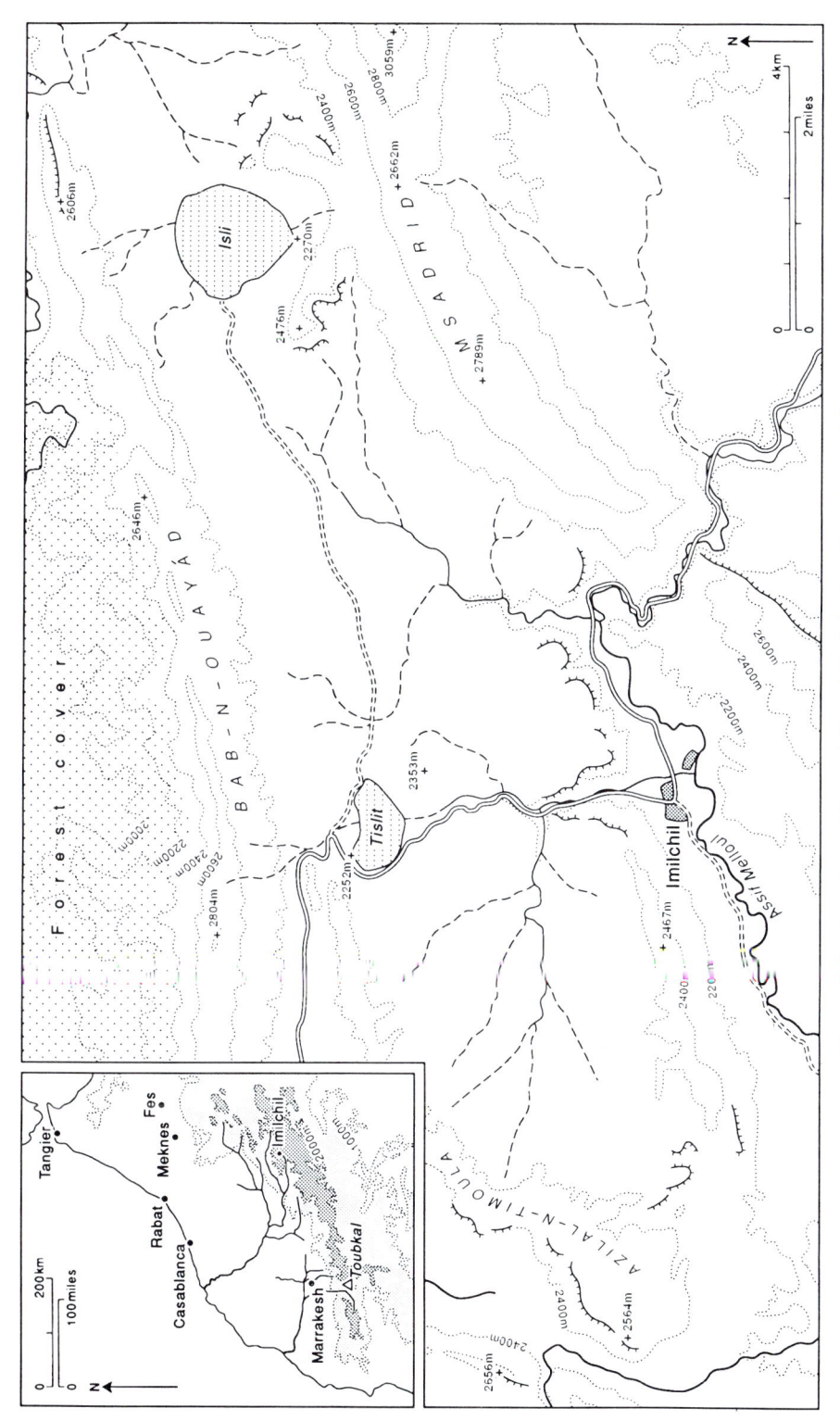

Figure E.1 Location of lakes Isli and Tislit in the High Atlas of Morocco

closed basin without any current surface outflow, and with a small catchment/surface area ratio (2.5), it should have been sensitive to past changes in the balance between precipitation and evaporation, fluctuating in water level like a giant, long-term rain-gauge. However, the extent of groundwater influence on the lake is unknown. Water chemistry may also have varied, affecting in turn the lacustrine biota and the nature of minerals precipitated from the water column. Evidence of past higher lake levels is present in the form of at least three raised shorelines. The highest of these, about 50 m above the present water surface, is associated with a probable former outlet, visible on aerial photographs as a channel-like feature leading from the lowest point of the lake's perimeter north-eastwards to the dramatic 600 m deep Agga-n-Ouanine gorge. Lacustrine marls are exposed at several places on these terraces.

Almost 2.15 km² in area and more than 95 m deep, Lake Isli is one of the largest and deepest natural lakes in North Africa. It is probably of structural origin, formed within a small 'pull-apart' basin in a trans-tensional regime. Its north and north-western sides are delimited by faults and most of the surrounding Middle Jurassic rocks dip towards the lake at angles between 1° and 45°. Because of its considerable depth, the lake is unlikely to have dried out completely except in conditions of extreme aridity. From a biogeographic standpoint, the presence of a population of apparently indigenous trout (*Salmo trutta*) in the lake is of interest (Morgan, 1982); they may have been isolated since the lake level fell below its outlet. There is no marginal vegetation, and only small amounts of *Chara* and *Potamogeton* are present. The depth, salinity and relatively small surface/volume ratio of the lake suggest that meromictic conditions conducive to the preservation of laminated organic sediments might occur, but limnological data show that this is not the case. In May 1990, the water column had a uniform O_2 content of 11.4 mg l^{-1}, declining to 9.0 mg l^{-1} O_2 at 13 m depth. This reflects the stratified temperature profile of 14.5°C at the surface, falling to 6.5°C below 40 m depth. Water conductivity was ca. 2.6 m3 cm^{-1} throughout the profile, with slightly higher surface values reflecting evaporative concentration. The surface waters have a pH of 9, and a Mg/Ca ratio of 18. In brief, the lake is mesosaline, temperature-stratified in summer, and the hypolimnion is sufficiently oxygenated to support benthic invertebrates.

Mean annual precipitation at Imilchil is 244 mm, with a standard deviation of 119 mm. Extremes include 471 mm in 1971, and no recorded precipitation in 1980 or 1981. Intense spring and summer rainstorms cause flash floods and severe surface erosion. Snowfall occurs for about 16 days at 2000 m a.s.1, increasing to 40–45 days on the highest summits. The mean maximum summer temperature is 30°C and the mean minimum winter temperature is −3.5°C (Tahri, 1991).

CATCHMENT AND LAKE MARGIN EROSION

Present-day land use of the area is sheep and goat pasture, with arable farming concentrated on the margin of the Assif Melloul River. Cereals are cultivated on the southern slopes of the Lake Isli catchment. The nearby forests are exploited for fuel, in spite of their distance from human settlement (principally the village of Imilchil), and must have declined as a result. Soil erosion is clearly a problem, and must be exacerbated by grazing pressure on the sparse vegetation. Application of Wischmeier and Smith's (1962) soil erosion model, following detailed survey of soils, slopes and vegetation cover of the 535 ha Lake Isli catchment, suggests a potential sediment yield of 28 T ha^{-1} yr^{-1}, equivalent to 15 000 T yr^{-1} (10 700 m^3 yr^{-1}, assuming a sediment density of 1.4 g cm^{-3}; Tahri, 1991). The lake sediments are thus a potential record of past variation in the intensity of soil erosion and the effect of human disturbance.

However, the sedimentary record of past variations in sediment yield is likely to be dominated by clastic input arising from reworking of littoral sediments. Unconsolidated beach gravels and sands, sometimes containing mollusc and ostracod shells, border Lake Isli; these have been exposed by recent lowering of water level, and are being eroded and redeposited by wave action. According to local inhabitants, the lake level has fallen by about 4 m during the last 20 years, which has been a dry period in Morocco. More than 50 ravines up to 16 m deep (most are 2–6 m deep) are cut into the terraces and beach deposits on all sides of Lake Isli, delivering snow meltwater and storm runoff to the lake. The volume of sediment derived from these gullies is crudely estimated from field measurements of the gully dimensions as 30 000 m^3. Assuming that this input occurred largely during the last 20 years, the recent rate of sediment accumulation over the 1.4 km^2 profundal plain of the lake may be estimated as approximately 1 mm yr^{-1}. An estimate based on the potential catchment sediment yield of 10 700 m^3 yr^{-1} is rather higher: 7.6 mm yr^{-1}. A still higher estimate of 11 mm yr^{-1} was obtained from measurements of material accumulated in a series of eight traps set at 10 m vertical intervals in the water column; this rate presumably reflects reworking of littoral and slope sediments, surface erosion and autochthonous sedimentation as well as gully erosion. Measured sediment accumulation rates from the profundal core Isli 90–5 range between 1 and 4 mm yr^{-1} (Table 8.1).

These estimates suggest that the input of clastic material associated with a relatively small change in lake level is substantial. Similar erosion must have occurred with previous changes in the lake level, and has almost certainly obscured any record of erosional input due to human impact. For anthropogenic erosional impact to be apparent in the sedimentary record, it would have to have taken place at a time of stable lake level. Even the longer-term record of climate change is likely to be obscured by reworking of older sediments and by dilution of autochthonous sediments by detrital influx.

FIELD AND LABORATORY METHODS

A 3.5 kHz pinger seismic transceiver, connected to graphic and magnetic recording systems, was used to make an initial survey of sediment distribution in the lakes. Greater penetration was achieved with a 1000 J sparker and hydrophone set, and the surface configuration of the sediments was determined with side-scan sonar. The results of the seismic survey

Table 8.1 AMS ^{14}C dates for Lake Isli cores 90–5 and 90–6

Core	Depth (cm)	Age ^{14}C years	Material
Isli 90–5	33.8	295 ± 60	Plant remains
Isli 90–5	493–494	1705 ± 55	Plant remains
Isli 90–5	668.5–669.5	2860 ± 60	Plant remains
Isli 90–5	789.4–790.4	3620 ± 70	Plant remains
Isli 90–6	649.7–652.7	34 850 ± 410	Ostracod valves

(detailed below) were used as a guide to optimum core location. A Kullenberg piston corer, operated from the Zurich laboratory's purpose-built raft (Kelts et al., 1986), was used to obtain ten cores up to 10 m in length from Lake Isli. The cores, still within their plastic core liners, were cut into sections, of 1 m length where possible, and capped for transportation. Each core section was split longitudinally, and one half archived in cold storage. The other half was photographed, logged and sampled at approximately 10 cm intervals, depending on stratigraphic variability. Sediment samples were weighed, then freeze-dried to determine water, organic and inorganic carbon contents with a coulometer. XRD determinations of sediment mineralogy were made on 56 samples of core Isli 90–6. Determinations were made on total sediment samples (non-oriented powder) and on insoluble residue, < 2 μm and 2–16 μm fractions (oriented preparations). Total sediment samples were prepared according to the method of Kubler (1987), by crushing to a homogenous powder in an agate hand mortar, and then pressed on to a sample holder at 20 bars. These non-oriented samples were analysed by a SCINTAG XDS-2000™. Semi-quantitative mineral composition was calculated from external standards, using the BASIC program Mac-dosage™. Pollen extraction was by standard procedures (Faegri & Iversen, 1989). Pollen grains (both determinable and indeterminable) were classified according to the preservation categories of Birks (1970). Algal preparations were treated with 10% HCl only, and the presence of chitinous remains recorded.

SIDE-SCAN SONAR AND SEISMIC SURVEY DATA

The steep (20–40°) marginal slopes of Lake Isli fall to a central, 1.4 km² profundal plain, 80–95 m deep (Figure 8.2). A side-scan survey shows that the marginal gullies continue below water down the lake slope, amalgamating as they do so (Figure 8.3). They give rise to fans which appear to spread sediments over much of the lake floor. In places, linear ledges parallel to the shoreline suggest past lower lake levels; some of these cut across the sub-aqueous gullies, so the latter are not only of recent origin. The side-scan data, summarised in Figure 8.4, also show an extensive slide at the north-east margin, and rotational slumps on the upper northern lake slope.

Sparker seismic reflection profiles reveal that the sedimentary infill of the fault-controlled basin is more than 100 m deep. However, only the top 25 m of this is stratified, presumably lacustrine, sediment. Material below this shows diffraction hyperbolae, suggesting that most of the infill is colluvial, and includes boulders. The pinger seismic equipment penetrated only the uppermost 10–15 m, and shows the presence of several reflectors, the main continuous one being at about 5 m depth. These data allow identification of the thinnest, and therefore oldest, sequences of sediments, away from the influence of turbidite fans. Near the lake's south-eastern margin, thin sediments are draped over a bedrock terrace. Core Isli 90–6 was taken here, at 60 m depth, in the hope of penetrating to older sediments.

CORE DATA

The cores consist of alternating proximal and distal laminated turbidite facies, with a massive proximal slump facies, the top of which is apparently equivalent to the main continuous seismic reflector. Core Isli 90–6 shows three main units. The basal grey unit (unit I, 716–

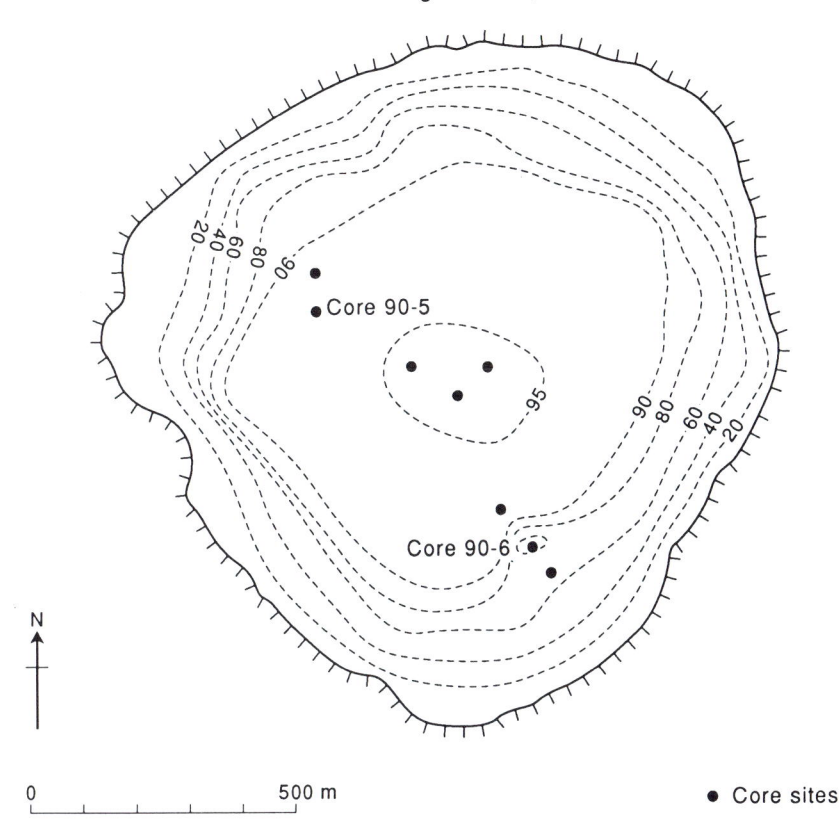

Figure 8.2 Lake Isli: generalised bathymetry (depths in m) and core locations

565 cm) is composed of typical shallow-water lacustrine carbonates, including tube-like incrustations of macrophyte plants and *Chara oogonia*. The middle unit (unit II, 565–265 cm) is made up of deformed (possibly by the coring operation), massive, but occasionally laminated, sediments. Banded sediments in the uppermost unit (unit III, 265–0 cm) are thinly bedded turbidites. Units I and II are apparently not represented in the other cores.

The sedimentary units of core Isli 90–6 are distinguishable by their mineral content (Figure 8.5). The basal unit I shows high pyrite abundance (visible in diatom preparations as framboids), up to 15% calcite, including some high-magnesium calcite, and low allochthonous phyllosilicate content. Bulk sediment digests reveal a high strontium content. Dolomite, probably of diagenetic origin, is present in the lowest sample, where bulk sediment digests show the highest magnesium content (3%). A sharp boundary at 565 cm is defined by replacement of pyrite by hematite, and by decreased calcium and strontium contents. The phyllosilicate content is highest in this middle unit (unit II), in which calcite is the predominant carbonate mineral. The uppermost unit (unit I, 265–0 cm) contains abundant aragonite, but less calcite; the hematite content is higher than that of unit II.

Preservation of diatoms in the core is poor (Figure 8.6). Although more than 40 taxa were recorded, their concentrations were low, presumably because of poor preservation (fracturing and dissolution). There are significant numbers present only at two levels (568 and 612 cm) in unit I, where the main taxa present (*Cocconeis placentula* Ehrenberg, *Cyclot-*

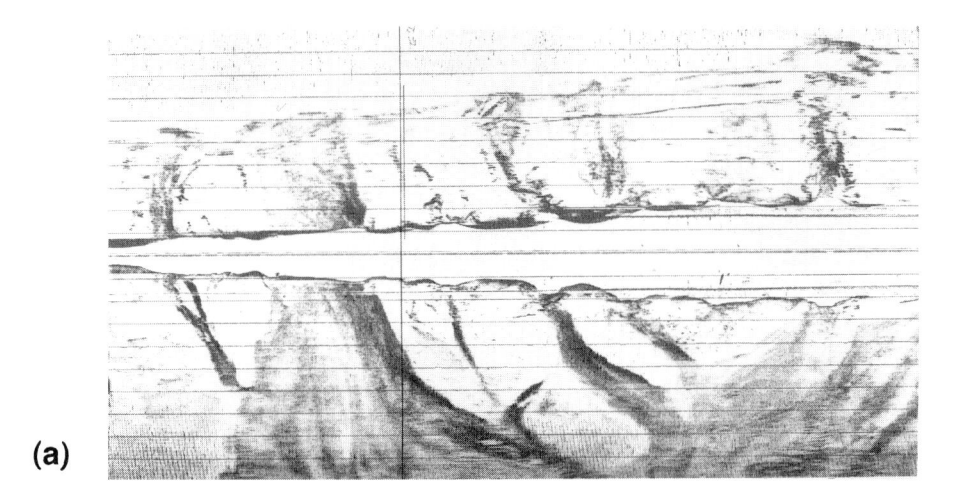

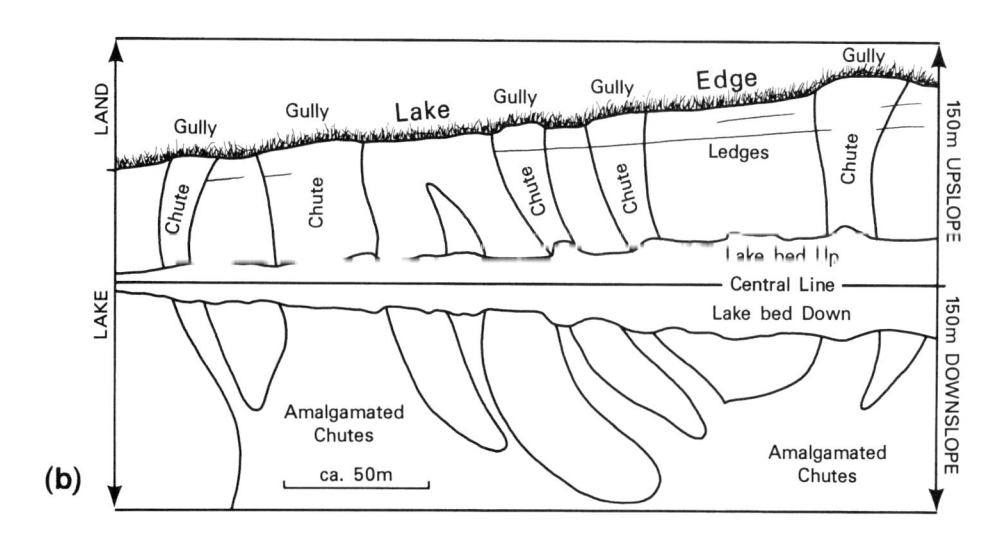

Figure 8.3 (a) Side-scan sonar record of part of the eastern slope of Lake Isli and (b) interpretative diagram of the record. Lake-marginal gullies continue sub-aqueously, merging downslope into fans which spread out on to the lake bed. Sub-parallel benches on the upper slopes indicate former lower lake levels

ella krammeri Haokansson, *Gomphonema gracile* Ehrenberg, *Staurosirella leptostauron* Ehrenberg and *S. pinnata* Ehrenberg) are periphytic, shallow-water forms. Desmids (*Staurastrum planctonicum* and *S. pingue*) are common at the base of unit III. These planktonic desmids occur in association with the remains of *Daphnia* (Crustacea) post-abdomens. *Chydorus sphaericus* headshields, valves and/or post-abdomenal claws were found in four samples from the basal unit.

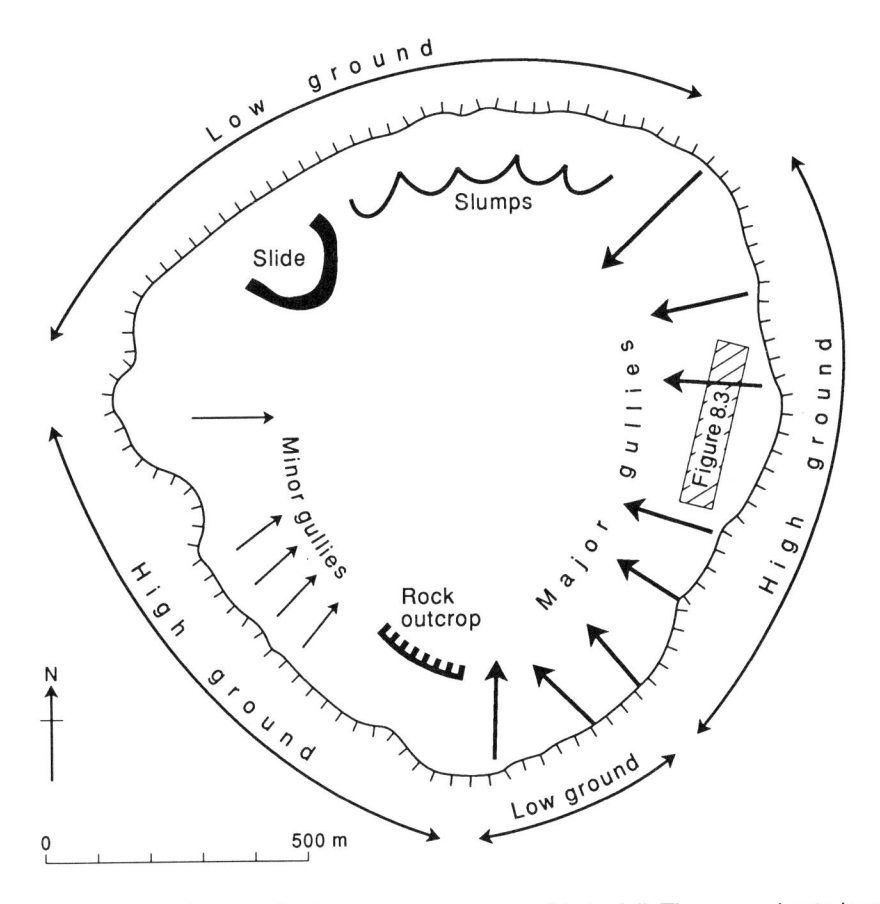

Figure 8.4 Schematic map of side-scan sonar survey of Lake Isli. The approximate location of the record shown in Figure 8.3 is indicated

Pollen is relatively abundant in the core. Numerical zonation procedures (constrained cluster analysis; Grimm, 1987) using the fifteen most abundant taxa (Figure 8.7) divide the sequence into three pollen assemblage zones at precisely the same levels as the lithostratigraphic boundaries. The basal unit is dominated by pollen of Chenopodiaceae, *Artemisia* and Gramineae. A similar pollen assemblage, but with increased Gramineae replacing much of the *Artemisia*, is present in unit II. These grains are frequently broken and degraded. The uppermost pollen assemblage zone is dominated by pollen of tree and shrub taxa (*Pinus*, *Quercus*, *Juniperus*) that are abundant in the nearby forests today. Pollen of the aquatic macrophytes *Potamogeton* and *Myriophyllum* is also present in unit III.

INTERPRETATION

Because the data available at this early stage of the investigation are sparse and sometimes apparently contradictory, the interpretations presented here should be regarded as working hypotheses, which will almost certainly be modified in the light of further analyses. The

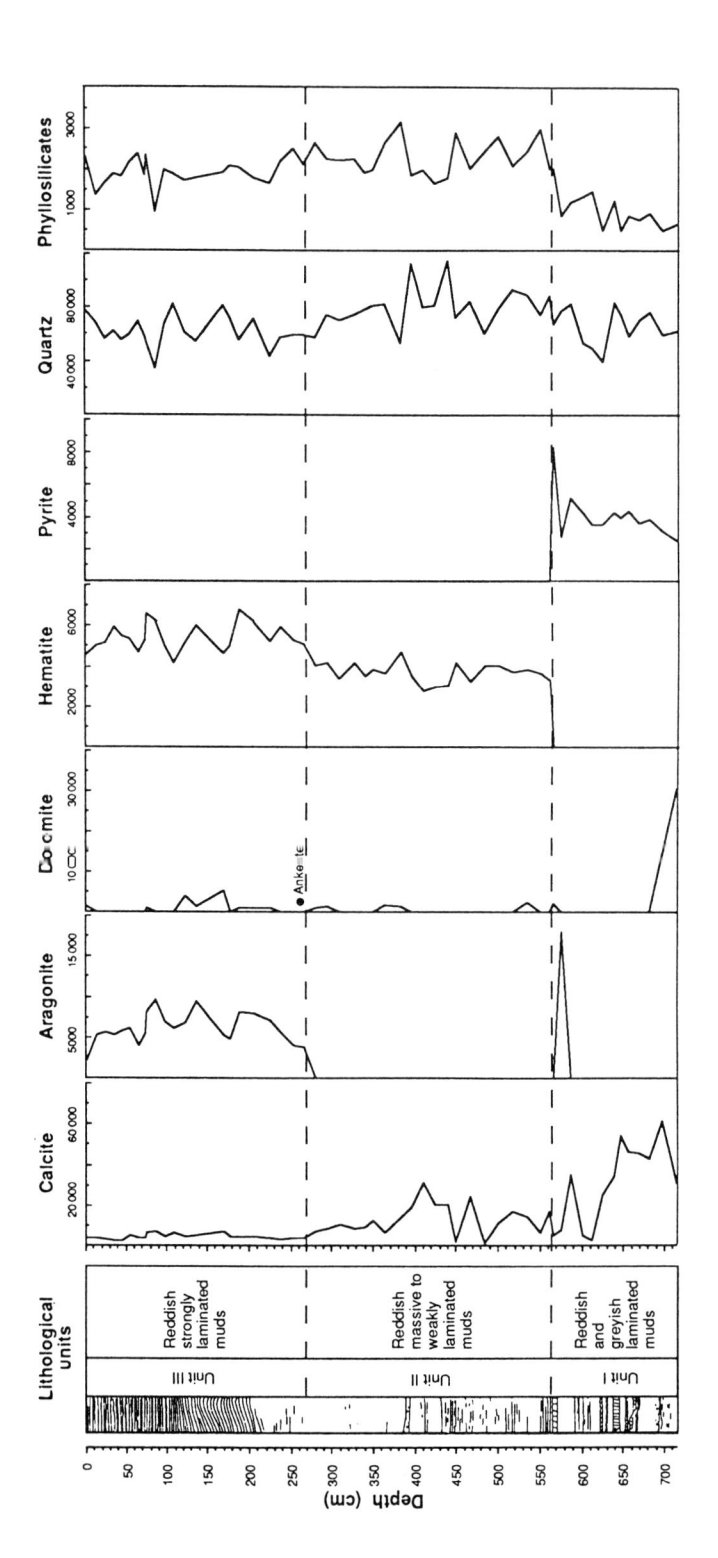

Figure 8.5 Semi-quantitative mineralogy of core Isli 90–6. Mineral abundance expressed as counts per minute, from X-ray diffraction determinations of non-oriented samples of bulk sediment

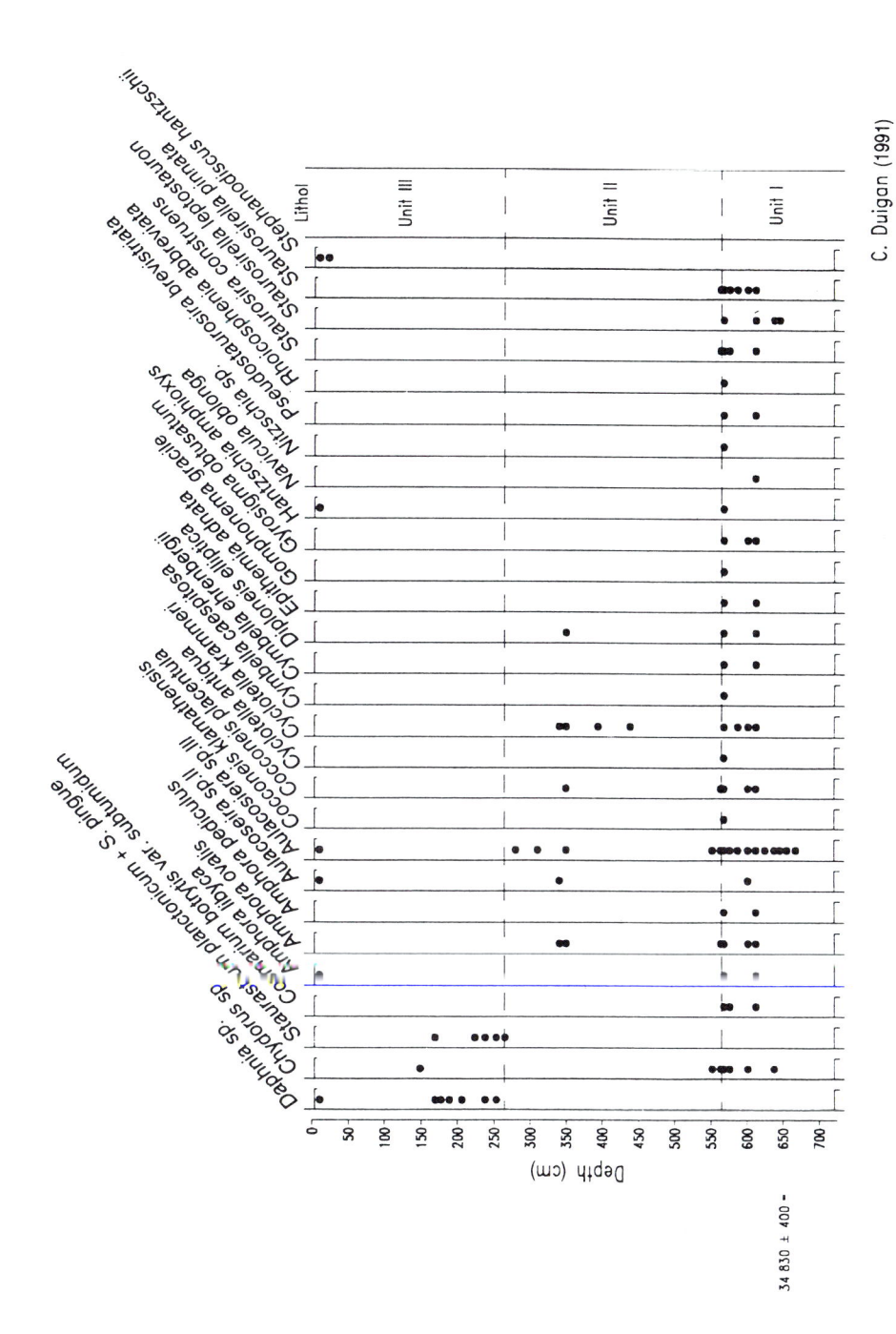

Figure 8.6 Summary algal and zooplankton stratigraphy of core Isli 90–6: presence of selected taxa

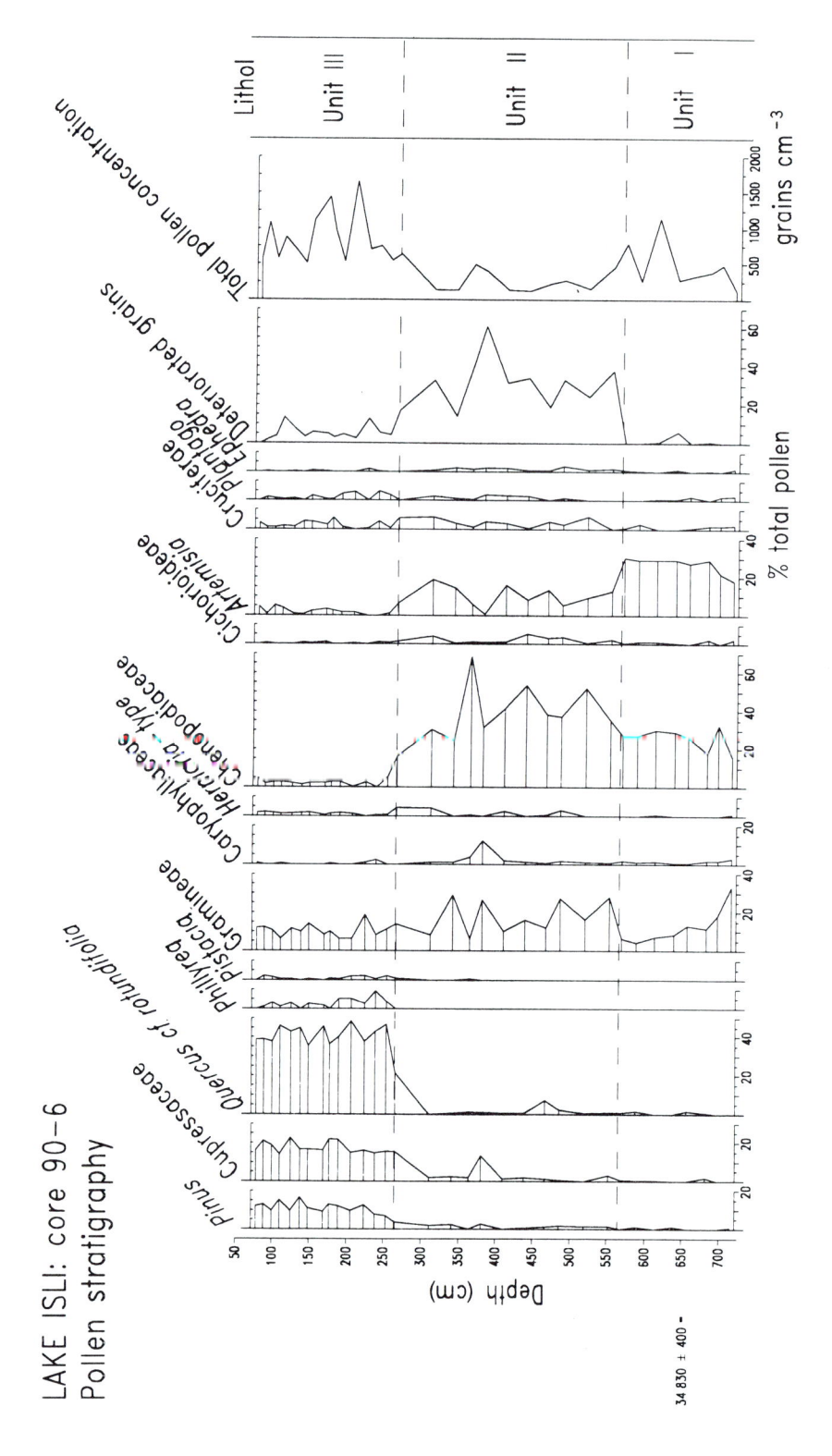

Figure 8.7 Summary pollen stratigraphy of core Isli 90–6. Deteriorated grains are mostly broken or degraded Gramineae and Chenopodiaceae

^{14}C dates thus far obtained (Table 8.1) show that sediments have accumulated on the profundal plain at rates of 1.02–3.84 mm yr^{-1}. The rates are probably highly variable at annual and decadal timescales, as suggested by the variable type and thickness of laminae, and by accumulation rate estimates from sediment yield, gully erosion and sediment trap data. The ^{14}C age of 34 850 ± 410 yr BP on ostracod valves from the base of core Isli 90–6 is one of the oldest from lacustrine sediments in North Africa, but there is almost certainly at least one major hiatus in the core, as explained below. Because carbon exchange with the atmosphere is probably relatively rapid in this alkaline system, and because there are few carbonate rocks in the catchment, these dates may be accepted, at least until additional ^{14}C analyses become available.

There is an apparent conflict between palaeoecological and mineralogical interpretations of the lake environment represented by the basal unit (unit I) of core Isli 90–6. On the basis of the occurrence of periphytic diatoms and chydorid remains, shallow freshwater conditions may be inferred for this earliest phase of the lake's history. However, the relatively high magnesium and strontium contents of the authigenic calcite matrix points to a saline environment. The presence of pyrite is an indication of high dissolved sulphate formed by evaporative concentration and deposited in a reducing environment associated with organic-rich sediments. The dolomite is suggestive of playa mud-flat conditions with early diagenetic, micritic dolomite formation. Thus, the mineralogical data imply that the lake water was more saline than at present, which might be expected from the location of these shallow-water sediments some 90 m below the former lake outlet. The apparent contradiction may be a function of rapidly changing or ephemeral conditions, in that the diatoms may derive from short episodes of freshwater within a period of overall salinity. More detailed sedimentary analysis is required to test this hypothesis. The pollen contained in unit I (Chenopodiaceae, *Artemisia* and Gramineae) is indicative of regional treeless steppe vegetation, and implies an arid climate, but little can be inferred about temperature, because these taxa are characteristic of both cold and warm steppe environments.

The sharp boundary at 565 cm in core Isli 90–6 probably represents a sedimentary hiatus. Because the sediments of the middle unit (unit II) have no counterpart in cores from the lake centre, the unit is almost certainly older than 3000 yr BP. The scarcity of laminae and the presence of black FeS spotting (possibly caused by reducing conditions inside worm burrows) suggest that the unit may be bioturbated. Hematite replaces pyrite, reflecting the change to detrital sediments of a low organic, low-sulphate content. The lack of intact diatoms and poor preservation of a pollen assemblage similar in composition to that of the lower unit suggest either that the sediments have been redeposited from the lake margin or slopes, or that they were deposited at a low rate (lower preservation potential), with bioturbation. By analogy with recent erosive input, any redeposition was probably associated with changing lake levels. Since the level was initially low, then a substantial rise in lake level, or an unstable period of several lake-level fluctuations, may be inferred.

The uppermost 265 cm of core Isli 90–6 (unit III) record lacustrine and catchment environments similar to those prevailing today. A weak correlation between the stratigraphic profiles of organic carbon contents of cores Isli 90–5 and 90–6 suggests that core 90–6 may have accumulated at half the rate of the profundal cores. Aragonite and calcite occur together in this unit; normally, lake sediments contain one or the other, not both. It may be that the calcite is derived from evaporative concentration of surface waters in the catchment and that the aragonite indicates evaporative concentration of the lake water; aragonite formation is usually associated with high aqueous Mg/Ca ratios (Eugster & Kelts, 1983).

The stratigraphic change is accompanied by preservation of *Daphnia* remains and of *Staurastrum*, desmids commonly associated with eutrophic conditions in extant communities. In this case, they may have responded to increased lake salinity, but it is unclear why they were present, or at least preserved, for only a short interval. Forests of pine, oak and juniper, with typically Mediterranean shrubs such as *Pistacia* and *Phillyrea*, became established on the northern slopes of the Atlas range (but probably not within the lake catchment) at this time. This suggests that the change to present climate is one of increased moisture relative to the periods represented by the middle and basal units. To accommodate both increased humidity and evaporative concentration of a higher-level lake, the present climate must be both more humid and warmer than the earlier periods represented. These preliminary interpretations do not take into account the complexities of change in climate seasonality during the last 30 000 years.

CONCLUSIONS

The sediments of Lake Isli are largely allochthonous, eroded from the lake catchment and reworked from the littoral zone. The dominant sedimentary process in the lake is regular episodic deposition by turbidity flows, probably initiated by intensive catchment runoff after rainfall and snowmelt, and by littoral wave action and water-level fluctuation. Lake-level fluctuations of several metres amplitude probably occur frequently, but more substantial changes, including a rise of at least 50 m, have occurred during the last 30 00 years. Because of the semi-arid mountainous catchment and the steep slopes of the lake basin itself, these changes are accompanied by strong detrital input to the lake, together with resedimentation by turbidity currents, slumps and slides. Minor lake level variations may thus be recorded in the laminated turbidite sequences, but distinguishing these from laminations caused by storm runoff events or by internal sediment movements will be difficult. The high rate of littoral erosion makes it less likely that anthropogenic influence on soil erosion can be inferred from the sedimentary record.

Evidence for past higher lake levels includes raised shorelines, a dry outlet channel and lacustrine marl outcropping above the present shoreline. Side-scan sonar surveys reveal evidence for past low levels of the lake. In the absence of better ^{14}C control and more precise stratigraphic data, conclusions about environmental change in the area must be treated with caution. About 30 00 years ago, the lake was shallow and saline, and the regional vegetation was a treeless steppe. These conditions suggest that the late Pleistocene climate of the area was cool but arid. This was followed, after an indeterminate period represented only by a sedimentary hiatus, by a phase of instability or rising lake level. The lake appears to have reached its present mid-level state, with slightly saline water and nearby oak, pine and juniper forests, at about 3000 yr BP. Since annual evaporation in the Atlas Mountains is ten times the precipitation, and since the lake catchment/surface area ratio is small, this relatively stable condition suggests that the lake is fed by groundwater, which may complicate attempts at modelling past hydrological regimes.

The only comparable lacustrine sedimentary record from Morocco is that of Tigalmamine, in the Middle Atlas some 100 km to the north (Lamb *et al.*, 1989; El Hamouti *et al.*, 1991; Roberts *et al.*, Chapter 9 in this volume). The High Atlas record sheds no further light on the surprising absence of *Cedrus atlantica*, the Atlas cedar, from the Middle Atlas before about 6000 years ago. Tigalmamine provides evidence for abrupt lake-level regressions at

2000–3000 year intervals during the Holocene. The High Atlas lakes may have been more sensitive to such hydroclimatic events, but the high rate and unstable nature of their sediment accumulation make it necessary to devise new approaches to interpreting the record of this unusual sedimentary environment.

ACKNOWLEDGEMENTS

This project was funded by grants from the Leverhulme Trust, the National Geographic Society, the Swiss National Science Foundation, the Royal Society and the British Ecological Society. The authors acknowledge Professor A. J. Brook, Dr Francoise Gasse and Dr W. L. Kovach for desmid identifications, comments on a preliminary diatom report and computing assistance respectively. ^{14}C dates were supplied by Professor W. Wolfli, Dr G. Bonani and others at the Mittelenergie Physik Institut, ETH-Honggerberg. The authors thank the Director of the Institut Agronomique et Vétérinaire Hassan II and the technical staff of the participating institutes, especially David Kelly, Lorraine Morrison, Franz Goenner, Kurt Ghilardi, Houssain Korkot and Omar Nakroum, and also the people of Imilchil for their hospitality.

REFERENCES

Birks, H. J. B. (1970). Inwashed pollen spectra at Loch Fada, Isle of Skye. *New Phytologist,* **69**, 807–20.

El Hamouti, N., Lamb, H. F., Fontes, J.-C. and Gasse, F. (1991). Changements hydroclimatiques abrupts dans le Moyen Atlas marocain depuis le dernier maximum glaciaire. *Comptes Rendus de l'Académie des Sciences, Series II,* **313**, 259–65.

Eugster, H. P. and Kelts, K. (1983). Lacustrine chemical sediments. In Goudie, A. S. and Pye, K. (eds.), *Chemical Sediments and Geomorphology,* Academic Press, London, pp. 321–68.

Faegri, K. and Iversen, J. (1989). *Textbook of Pollen Analysis,* 4th edition, Wiley, Chichester.

Grimm, E. C. (1987). CONISS: a FORTRAN 77 program for stratigraphically constrained cluster analysis by the method of incremental sum of squares. *Computers and Geosciences,* **13**, 13–35.

Kelts, K., Briegel, U., Ghilardi, K. and Hsu, K. (1986). The limnogeology-ETH coring system. *Schweiz Z. Hydrol.,* **48/1**, 104–15.

Kubler, B. (1987). Cristallinité de l'illite, méthodes normalisées de préparation, méthodes normalisées de mésures. Cahiers de l'Institut de Géologie de Neuchatel, Suisse. Série ADX-2.

Lamb, H. F., Eicher, U. and Switsur, V. R. (1989). An 18,000 year record of vegetation, lake-level and climatic change from the Middle Atlas, Morocco. *Journal of Biogeography,* **16**, 65–74.

Lister, G. S., Kelts, K., Chen, K. Z, Yu, J, -O. and Niessen, F. (1991). Lake Qinghai, China: closed basin lake levels and the oxygen isotope record for ostracoda since the latest Pleistocene. *Palaeogeography, Palaeoclimatology, Palaeoecology,* **84**, 141–62.

Morgan, N. C. (1982). An ecological survey of standing waters in northwest Africa III: site descriptions for Morocco. *Biological Conservation,* **24**, 161–82.

Roberts, C. N., Lamb, H. F., El Hamouti, N. and Barker, P. (1993). Abrupt Holocene hydroclimatic events: palaeolimnological evidence from north-west Africa (Chapter 9 in this volume).

Tahri, M. (1991). Cartographie des sols et de leur érosion dans les bassins versants des lacs d'Imilchil (Haut Atlas central). Diploma thesis, Institut Agronomique et Vétérinaire Hassan II, Rabat.

Wischmeier, W. H. and Smith, D. D. (1962). Rainfall erosion. *Advances in Agronomy,* **14**, 109–48.

9 Abrupt Holocene Hydro-Climatic Events: Palaeolimnological Evidence from North-West Africa

N. ROBERTS
Department of Geography, Loughborough University of Technology, Leicestershire, UK

H. F. LAMB
Institute of Earth Studies, University of Wales, Aberystwyth, UK

N. EL HAMOUTI
Department of Geology, Faculty of Sciences, Université Mohammed I, Oujda, Morocco

and

P. BARKER*
Department of Geography, Loughborough University of Technology, Leicestershire, UK

ABSTRACT

Superimposed on long-term, orbitally induced variations in the precipitation/evaporation ratio at the arid margin are shorter-lived climatic oscillations of 10^2–10^3 year duration. ^{14}C-dated lake-level chronologies from northern inter-tropical Africa indicate that a number of abrupt hydroclimatic events occurred during the early Holocene. Recent investigations of more northerly, extra-tropical lake basins such as Tigalmamine in Morocco reveal that these short-lived regressional events took place throughout the Holocene. Unlike twentieth century hydroclimatic fluctuations, the arid events appear to have occurred synchronously on both sides of the Sahara. Potential causal mechanisms are discussed, including Broecker et al.'s (1990) quasi-cyclic oceanic salt oscillator model for North Atlantic deep-water (NADW) formation.

INTRODUCTION

One of the primary influences on biotic, geomorphic and human activities in semi-arid lands is the quasi-periodic alternation between phases of wetter and drier climate. There is currently considerable scientific interest in the causes of these wet–dry climatic shifts, including how they may be affected by any future greenhouse gas-induced change in climate. Climatic oscillations have taken place on arid margins over a range of timescales, with the recent Sahelian drought providing a notable example within the period of observational record. In this paper the past climatic behaviour of the southern (Sahelian) and northern (Maghrebian)

---

*Present address: Department of Geography, University of Lancaster, UK.

Environmental Change in Drylands: Biogeographical and Geomorphological Perspectives.
Edited by A. C. Millington and K. Pye. © 1994 John Wiley & Sons Ltd

margins of the Sahara are compared, with particular reference to palaeolimnological evidence for abrupt Holocene arid events.

CLIMATE CHANGE ON THE MARGINS OF THE SAHARA

In discussing 'shrinking' or 'expanding' deserts, a number of models spring to mind. At a simple level, it is possible to identify a 'concertina model', in which the northern and southern limits of the Sahara expand or contract simultaneously, a parallel shift model, with margins moving in the same direction at the same time, and a 'no-model' model, in which the margins behave independently of one another and hence exhibit no temporal correlation. Over the observational time period, both the Maghreb and the Sahel have experienced climatic oscillations and both suffered droughts during the early 1980s. However, a comparison of a time series of yearly precipitation index values for 12 meteorological stations in Morocco and 20 stations in western sub-Saharan Africa shows no correlation for the period since 1941 (correlation coefficients of +0.11/+0.16; Lamb & Peppler, 1991) (Figure 9.1). In some respects this is not surprising given that the pronounced rainy seasons on each side of the Sahara are out of phase with each other, deriving respectively from the passage of winter cyclonic depressions and the summer migration northwards of convectional tropical storms. In terms of causation, drought years in the Sahel exhibit a correlation with a distinctive pattern of sea surface temperature (SST) anomalies, with the global ocean warmer than usual in equatorial and southern tropical waters and cooler than usual in the northern tropics (Folland et al., 1986). Droughts in Morocco correlate, not with SST anomalies, but with the so-called North Atlantic oscillation, which is an index based on the difference in atmospheric pressure between the Azores and Iceland, and is related to temperature anomalies between Greenland and southern Scandinavia (Lamb & Peppler, 1991). In essence, when the Azores High is weakened, westerly depressions take a more southerly track and North African rainfall is enhanced. Over an observational timescale, therefore, the climates of the northern and southern margins of the Sahara have been largely decoupled. There is no evidence to support the 'concertina' model of simultaneous expansion or contraction by opposing desert edges.

Over longer timescales, larger magnitude climatic changes have affected the Sahara and its surrounding lands. During the Quaternary, conditions alternated between climatic extremes both drier and wetter than at the present day. During the former, sand dunes (now inactive) extended more than 400 km to the south across a broad swathe of sub-Saharan Africa. During the latter, savanna-like conditions characterised the whole of the Sahara desert. ^{14}C and Th/U dating of lacustrine deposits, often from lakes no longer extant, indicate that the last major period of enhanced rainfall was during the early Holocene, ca. 10 000–5000 yr BP (Street-Perrott & Roberts, 1983; Gasse et al., 1987). At some sites 'wetting-up' began earlier than this during the terminal Pleistocene (ca. 12 500 yr BP on), while at others it may have continued until ca. 3000 yr BP (Gasse et al., 1987; Fontes & Gasse, 1991). Biotic indicators and an apparent south–north moisture gradient indicate that this was brought about primarily by a northward shift in the zone of tropical summer rains (Ritchie & Haynes, 1987; Street-Perrott et al., 1990; Lézine et al., 1990). In some sectors of the northern edge of the modern desert (e.g. southern Tunisia) the zone of enhanced precipitation reached almost to the Mediterranean Sea (Fontes & Gasse, 1991; Petit-Maire et al., 1991). However, further east there is no evidence for such a contraction in the northern

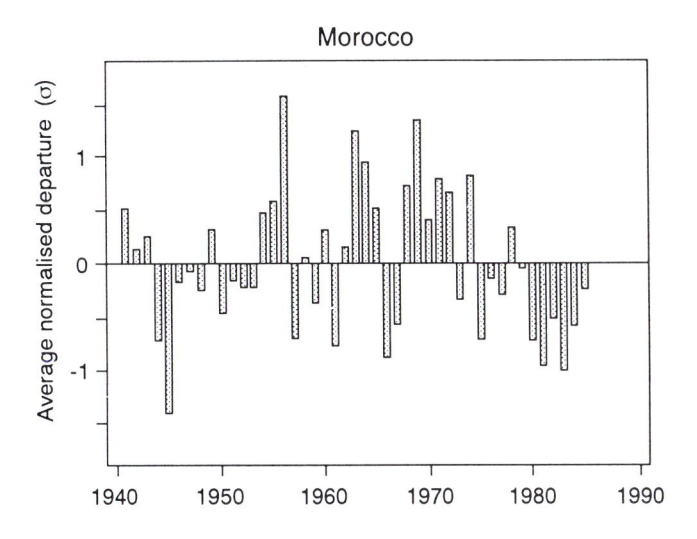

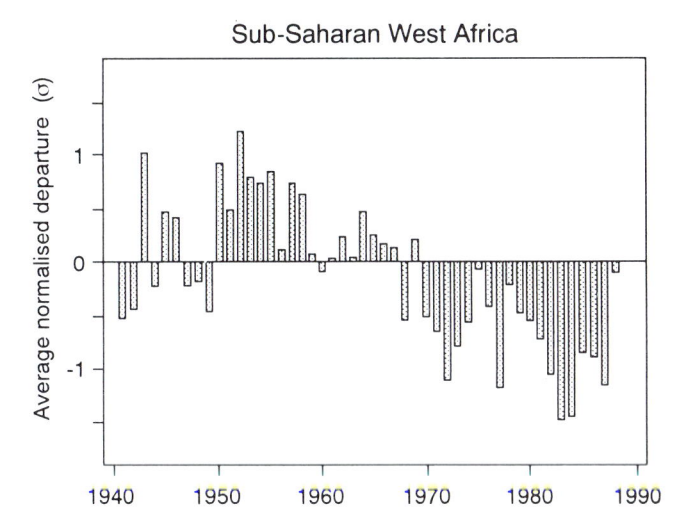

Figure 9.1 Time series of annual precipitation for Morocco (1941–85) and Sub-Sahara West Africa (1941–1988) (reproduced by permission of Cambridge University Press from Lamb & Peppler, 1991)

desert limit, and indeed in the Near East the northern edge of the Saharo-Arabian arid zone was drier than at present during the early Holocene (Roberts & Wright, 1993).

Time series of effective moisture (precipitation minus evaporation, or P–E) from low-latitude lake basins and deep-sea cores indicate a link, probably lagged, to the 23 kyr precessional cycle during at least the last 0.3 m yr^{-1} (Kutzbach & Street-Perrott, 1985; Dupont & Hooghiemstra, 1989; Clemens *et al.*, 1991). By contrast, the climate of the circum-Mediterranean lands shows a close match with the ice volume record dominated by a 100 kyr periodicity (van der Hammen *et al.*, 1971; Rossignol-Strick & Planchais, 1989). As with the observational time period, therefore, Quaternary climates on each side of the

Saharo-Arabian arid zone tend to support the 'no-model' model of climatic change. There is no obvious linkage between temporal variations in palaeoprecipitation derived from westerly depressions with that of monsoonal origin.

ABRUPT ARID EVENTS

In addition to this first-order sequence of water balance changes, a number of second-order oscillations are evident in many tropical African lake-level histories. These occurred over sub-Milankovitch timescales and are recorded by abrupt lake recessions, indicating a negative hydrological budget and arid climate (Street-Perrott & Roberts, 1983). Unlike the lagged glacial and vegetational responses to climatic change, most non-outlet lakes are known to respond rapidly to hydroclimatic changes. Abrupt events have a duration of a thousand years or less (possibly a lot less), but an intensity comparable to the first-order hydrological changes of the late Quaternary. In recognizing palaeoclimatic events such as these, care must be taken to avoid making spurious correlations between sites (cf. Rognon, 1987; Fontes & Gasse, 1989). In fact, a statistical distribution of ^{14}C-dated lake regressions shows a clear clustering, with early Holocene arid events centring on 10.2 and 7.4 kyr (Street-Perrott & Roberts, 1983, Fig. 8). There was also a widespread fall in lake levels after 5 kyr, but it has until recently been difficult to know whether this is truly an abrupt event, as it coincided with a first-order orbitally induced climate change. The non-outlet lakes which experienced these regressions are among the best-studied, the best-dated and the hydrologically most responsive, and they stretch geographically across northern inter-tropical Africa from Turkana (Kenya–Ethiopia, 4°N), Ziway-Shala (Ethiopia, 8°N) and Abhé (Ethiopia–Djibouti, 11°N) in the east to Bosumtwi (Ghana, 6°N) and Chad (13°N) in the west (Figure 9.2). Although the 10.2 kyr event is synchronous with the European Younger Dryas, it has generally been assumed that the abrupt arid events were a low-latitude phenomenon associated with a temporary weakening of the monsoonal circulation. They were not believed to have affected the northern (Mediterranean) margins of the Saharan and Arabian deserts.

This last issue may require some revision in the light of recent palaeolimnological investigations. The evolution of the southern basin of the Dead Sea (31°N) shows marked mid- to late Holocene water-level oscillations (Frumkin et al., 1991), while a sequence of older lake beds at Sebkha Mellala (Algeria, 32°N) clearly records dry events at 10.2 and 7.4 kyr (Gasse et al., 1990). However, perhaps the most interesting new record comes from Tigalmamine (33°N) in the Middle Atlas of Morocco.

MOROCCAN LAKE-LEVEL HISTORY

Tigalmamine comprises a series of small solution lakes orginally investigated for a pollen-based history of the surrounding cedar–oak forest (Lamb et al., 1989). Subsequent analysis of other palaeolimnological indicators has revealed that water levels have been subject to major Holocene fluctuations (El Hamouti, 1989; El Hamouti et al., 1991; Benkaddour, 1993). Analysis of both central and marginal cores from the main lake at Tigalmamine shows a number of shallow-water phases interrupting longer periods of relatively deep water.

A 16 m long sediment core taken from near the lake centre at 16 m water depth has a basal ^{14}C age of ca. 10 500 yr BP. Four conventional and seven AMS ^{14}C dates provide a consistent core chronology, with no evidence for breaks in sedimentation or gross changes

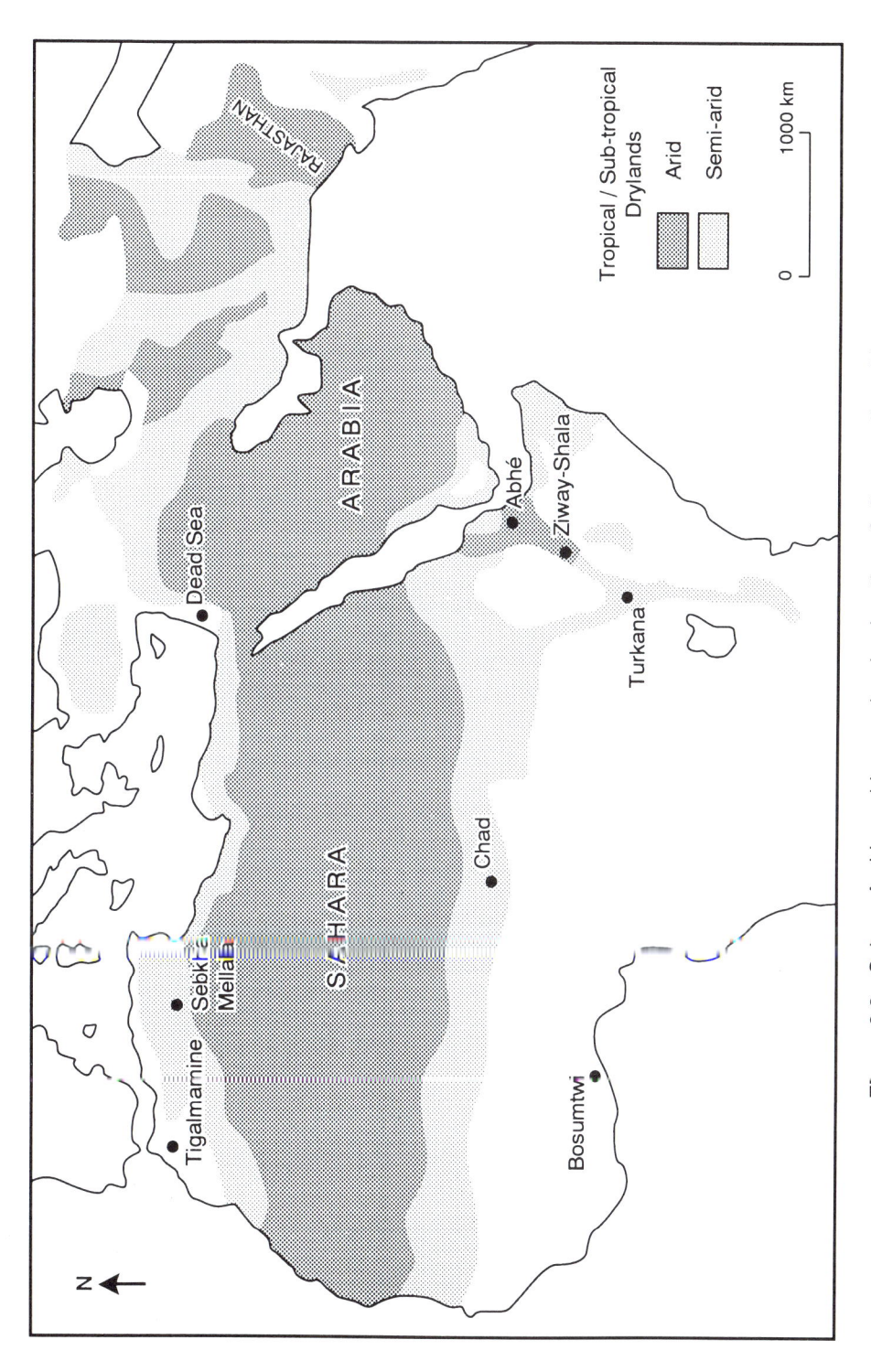

Figure 9.2 Saharo-Arabian arid zone showing location of sites mentioned in text

in the sediment accumulation rate over a Holocene timescale. On the other hand, changes in the nature of sedimentation are indicated by the presence of a number of coarser-grained layers with gastropods, ostracods and *Chara* oogonia. The diatoms at these and two other levels in the core show an increase in periphytic taxa such as *Cocconeis placentula* and *Gomphonema angustatum* (El Hamouti, 1989; El Hamouti *et al.*, 1991). Between these phases the diatoms are dominated by deep-water, planktonic forms of the genus *Cyclotella*. Oxygen-isotope analyses on ostracod carapaces and authigenic carbonates are consistent with the interpretation of these as regressional stages in the lake-level history (Benkaddour, 1993). Altogether five such stages appear to be evident in the central core, and they have associated ^{14}C ages of 10 400, 7760–7300, 4800–4400, 3500 and 2500–2000 yr BP (Figure 9.3).

A 21 m long core from the marsh at the northern edge of the main lake contains a more complex stratigraphic record. This is because the lake marginal zone was liable to desiccation and erosion during periods of lowered water level. In identifying temporal breaks in the core record, a sound chronology is critical. At first, it was thought that sedimentation was continuous, since paired dates bracketing the sharpest lithologic boundaries showed no apparent time gap (Lamb *et al.*, 1989). However, further dating and comparison with the more complete record from the lake centre suggests a number of probable hiatuses in the marginal core (El Hamouti *et al.*, 1991). The most obvious of these is at a depth of 17 m, below which is a calcareous silt of late Pleistocene age; thus, the earliest part of the Holocene is missing. The Holocene part of the marginal core has eight conventional and three AMS ^{14}C dates, and they suggest the existence of further gaps, notably related to the regressional event at 3500 yr BP. Sedimentary changes and fluctuations in the plankton/periphyton ratio of diatoms confirm the existence of Holocene water-level changes, but interpretation of the littoral record is complicated both by its semi-continuous nature and also because late Holocene sedimentation at the lake margin has been dominated by a hydroseral succession from lake marl to peat.

In correctly reconstructing the history of climate at Tigalmamine, it is obviously important to understand the local and regional hydrological system. The lake receives surface inputs from an inflowing stream and from runoff, and there are overflow channels between the lakes and below the lowest of them (this is currently active only in winter). However Tigalmamine is a karstic system and lake levels are primarily determined by the variations in the regional groundwater table. Twentieth century groundwater fluctuations have caused water-level changes in several Middle Atlas lakes, including Azigza and Sidi Ali, related to inter-annual variations in precipitation (Flower & Foster, 1992). The Holocene regressions at Tigalmamine are marked by an increase in diatoms typical of more saline waters (e.g. *Mastogloia* spp.), suggesting some evaporative concentration rather than opening or closing of underground outlets (El Hamouti, 1989). It is consequently likely that these regressive events are of climatic origin and that they reflect the same short-duration arid phases recorded in tropical African lake basins around 10.2 and 7.4 kyr (Figure 9.4).

POSSIBLE CAUSES OF ABRUPT ARID EVENTS

The Tigalmamine record is particularly significant because the lake's overall water level history during the Holocene does not correspond with those from Africa's monsoonal sector. Like Mediterranean Europe, the precipitation signal in the Middle Atlas of Morocco has responded to the location and intensity of westerly depressions rather than to tropical rain-

169

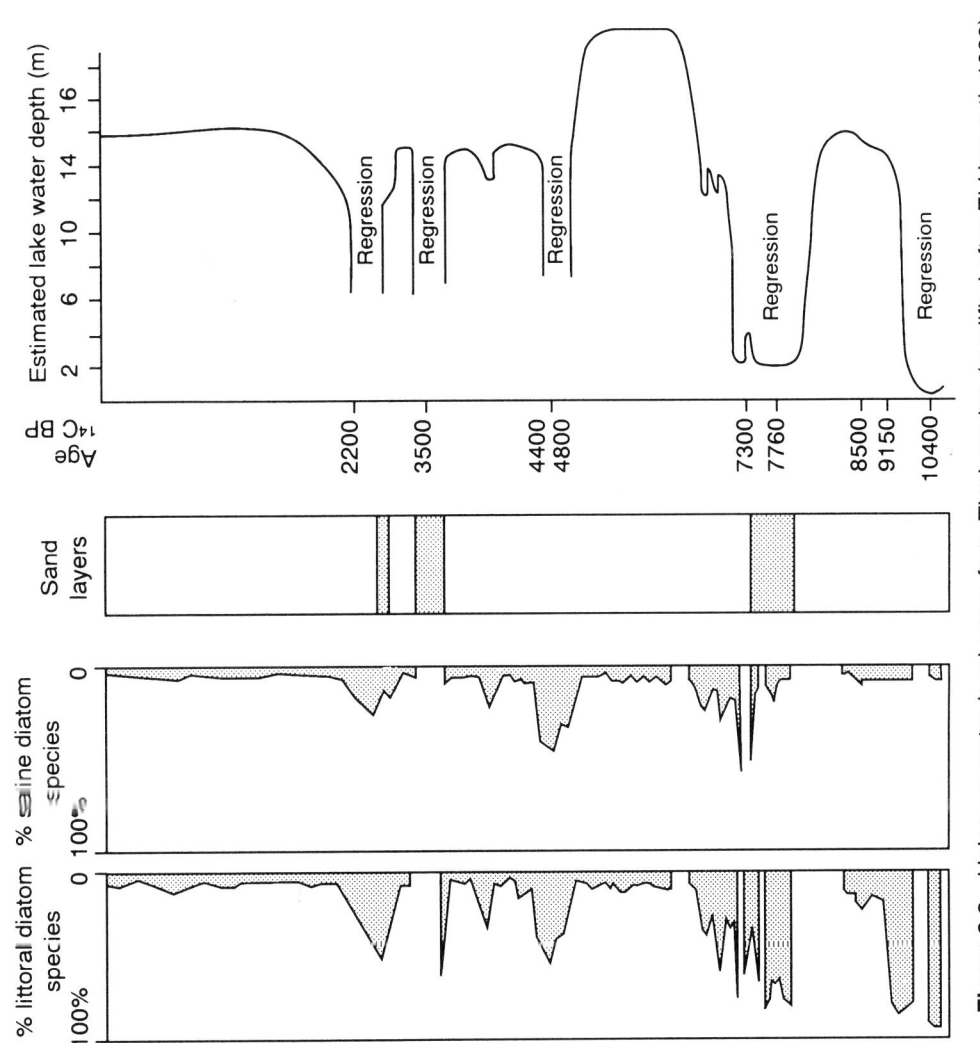

Figure 9.3 Holocene water-level curve from Tigalmamine (modified after El Hamouti, 1989)

170 Environmental Change in Drylands

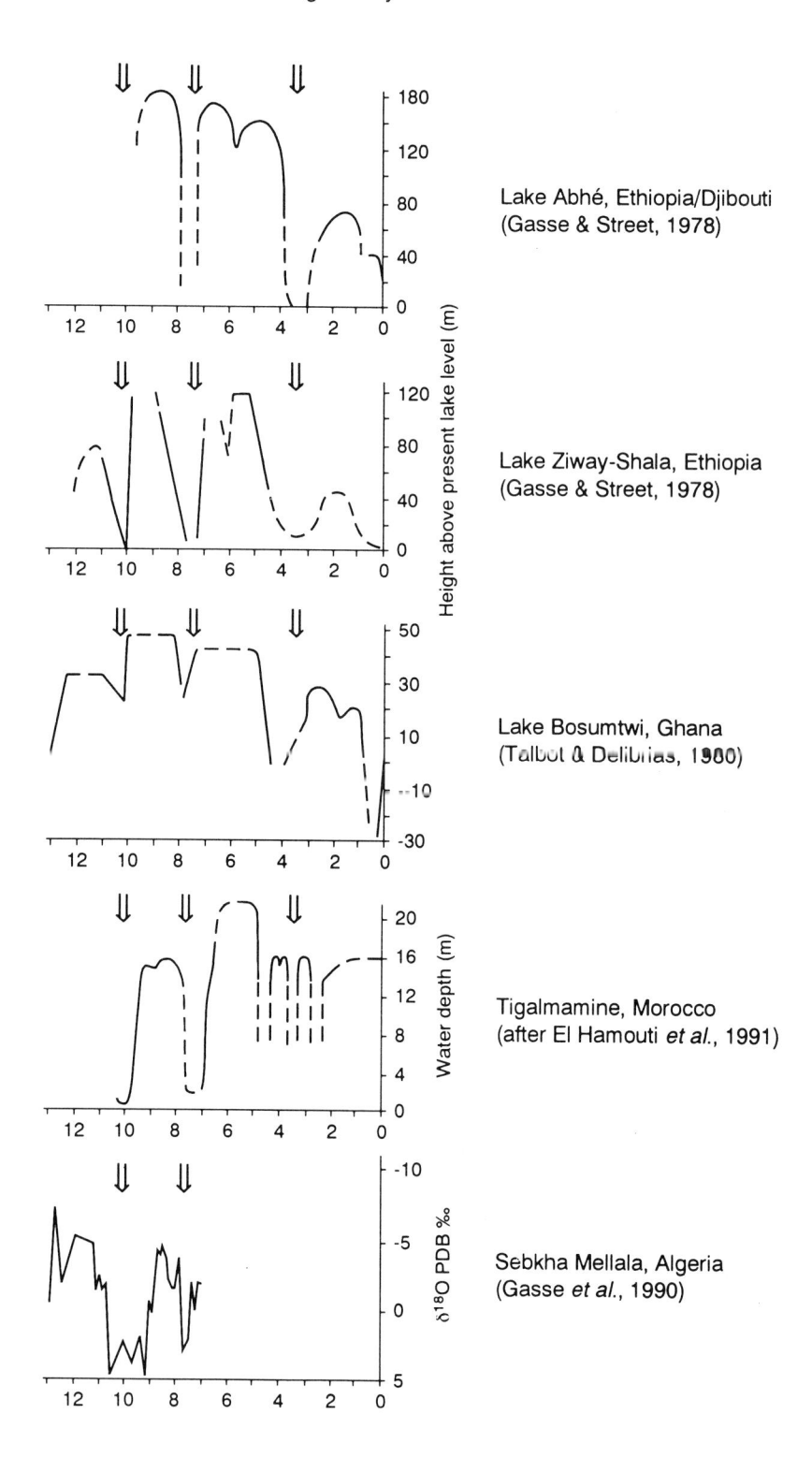

Lake Abhé, Ethiopia/Djibouti
(Gasse & Street, 1978)

Lake Ziway-Shala, Ethiopia
(Gasse & Street, 1978)

Lake Bosumtwi, Ghana
(Talbot & Delibrias, 1980)

Tigalmamine, Morocco
(after El Hamouti et al., 1991)

Sebkha Mellala, Algeria
(Gasse et al., 1990)

fall. Furthermore, because of its different response to orbital forcing, background lake levels in the Middle Atlas have remained high during the second half of the Holocene, unlike those south of the Sahara. Because of this, regressional events are stratigraphically visible throughout the Holocene, and not only during its first part. In many African lacustrine sequences any late Holocene water-level falls would not have been recorded because the basin was already dry or at a very low level. Tigalmamine, however, shows three probable palaeoclimatic events during the second half of the Holocene at *ca.* 4.5, 3.5 and 2.3 kyr. Existing evidence, while less definitive than that for earlier events, does suggest that these too may have been experienced elsewhere in Africa (Talbot, 1982); for example, an event at *ca.* 3.5 kyr is recorded in many sub-Saharan lake-level sequences (Figure 9.5). Sites elsewhere in the tropics tentatively indicate that a similar frequency of arid events may also have characterised the late Pleistocene (de Deckker *et al.*, 1991). This crude chronology suggests that arid episodes have occurred approximately every 2–3 kyr during the late Quaternary, implying a periodicity for sub-Milankovitch events that is not regular but probably not random either.

Sub-Milankovitch climate changes, such as those that may have caused the lake-level falls at Tigalmamine, were too rapid and too abrupt to be explained by orbital forcing. A number of possible explanations for these short-duration events are currently under discussion, including agencies such as major emissions of volcanic dust and North Atlantic deep-water (NADW) formation (Berger & Labeyrie, 1987; Street-Perrott & Perrott, 1990). Of these palaeoclimatic oscillations, the Younger Dryas event (11–10 kyr) has so far received the greatest attention. It seems likely that this event formed an integral part of the process of deglaciation in Europe and North America, with rapid ice-sheet decay resulting in desalination of North Atlantic water (Broecker, 1990). The evidence from lakes in Africa and elsewhere suggests that the Younger Dryas stadial was felt well beyond north-west Europe, with many areas experiencing a negative hydrological balance and an arid climate. A link with deglaciation and the suppression of NADW formation may also have applied to the 7.4 kyr arid event which was broadly coincident with the collapse of the Laurentide ice sheet over Hudson Bay.

An alternative model for the suppression of NADW formation has been put forward by Broecker *et al.* (1990). According to this salt oscillator model the thermohaline circulation of the oceans is driven by natural oscillations in the salinity budget of the Atlantic. Broecker *et al.* (1990) suggest a 2.0–2.5 kyr periodicity for these oscillations, although this would have been variable dependent on external parameters such as freshwater influx from the continents. The ocean–atmosphere link provided by NADW formation indicates that this would in turn produce a similar frequency of climate events. There is an obvious similarity between Broecker's quasi-cyclic suppression of the oceanic salt conveyor and the Holocene arid hydroclimatic events recorded at Tigalmamine and other African lake basins, not only in terms of frequency (every 2–3 kyr), but also because that periodicity has been somewhat irregular.

Figure 9.4 Comparison of African lake-level curves: Abhé and Ziway-Shala (reproduced from Gasse & Street, 1978, by permission of Elsevier Science Publishers BV); Bosumtwi (reproduced from Talbot & Delibrias, 1980, by permission of Elsevier Science Publishers BV); Tigalmamine (modified after El Hamouti *et al.*, 1991); Sebkha Mellala (from Gasse *et al.*, 1990. Reprinted with permission from *Nature*. Copyright © 1990. Macmillan Magazines Limited)

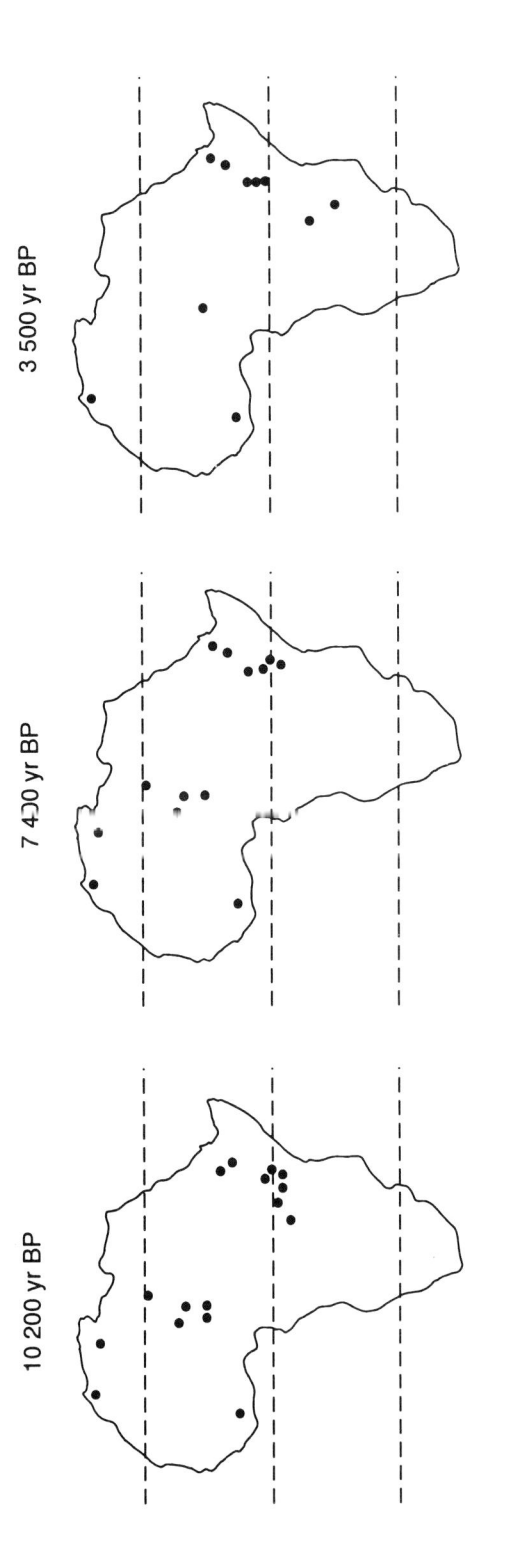

Figure 9.5 Sites showing regressional events at 10 200, 7 400 and 3 500 yr BP (modified from Street-Perrott & Roberts, 1983)

CONCLUSIONS

The climatic histories of the semi-arid zone on each side of the Sahara show little in common over an observational timescale; nor is a clear relationship apparent over Quaternary timescales. However, lakes on both sides record abrupt arid events too short and too rapid to be explained by orbital forcing — events that are more consistent with a 'concertina' model of climate change. The synchroneity of these events in both monsoonal and Mediterranean sectors suggests a common cause, probably different from mechanisms operating over longer or shorter time periods, and involving a temporary decrease in ocean-to-continent moisture flux. The 2–3 kyr periodicity of regressional events in African lakes is consistent with Broecker's quasi-cyclic oceanic salt oscillator for NADW formation.

Small lakes with a simple basin morphology contain a relatively clear palaeoclimatic signal and have a water balance responsive to these short-lived arid events. This offers the potential for precise resolution of their age, duration and intensity and an assessment of whether water levels fell and rose symmetrically or at different rates. Additionally, a site such as Tigalmamine provides a more complete record of late Holocene events than lake basins in tropical Africa, because background water levels have remained high. As these hydroclimatic events are abrupt in character, this further permits the investigation of the lag times and differential response of components of the regional vegetation to climate change via the pollen record in the same lake sediment cores. These problems are currently under investigation at Tigalmamine via detailed stratigraphic analysis of those parts of sedimentary sequences that correspond to regressional lake phases.

ACKNOWLEDGEMENTS

The authors are grateful to NERC (grants 5536 and GST/02/542) and to the Royal Society for financial support for the Middle Atlas research work, to the Ministry of Waters and Forests in Morocco for fieldwork permission, to the University Mohammed I at Oujda and to the Agronomy Institute at Rabat (especially Dr Aziz Merzouk) for valuable assistance in Morocco. The authors also thank J.-C. Fontes, F. Gasse, A. Benkaddour and A. van der Kaars for contributions to and comments on this work.

REFERENCES

Benkaddour, A. (1993). Changements hydrologiques et climatiques dans le Moyen-Atlas marocain; chronologie, géochime istopique et elementaire des sediments lacustres de Tigalmamine. Thèse Docteur en Science, Universitié de Paris-Sud.

Berger, A. and Labeyrie, W. H. (eds.) (1987). *Abrupt Climate Changes, Evidence and Implications*. Reidel, Dordrecht.

Broecker, W. S. (1990). Salinity history of the northern Atlantic during the last deglaciation. *Paleoceanography*, **5**, 457–67.

Broecker, W. S., Bond, G. and Klas, M. (1990). A salt oscillator in the glacial Atlantic? 1. The concept. *Paleoceanography*, **5**, 469–77.

Clemens, S., Prell, W., Murray, D., Shimmield, G. and Weedon, G. (1991). Forcing mechanisms of the Indian Ocean monsoon. *Nature*, **353**, 720–25.

de Deckker, P., Corrège, T. and Head, J. (1991). Late Pleistocene record of cyclic eolian activity from tropical Australia suggesting the Younger Dryas is not an unusual climatic event. *Geology*, **19**, 602–5.

Dupont, L. M. and Hooghiemstra, H. (1989). The Saharan-Sahel boundary during the Brunhes chronozone. *Acta Botanica Neerlandica*, **38**, 405–15.

El Hamouti, N. (1989). Contribution à la reconstitution de la paléohydrologie et de la paléoclimatologie du Maghreb et du Sahara au Quaternaire supérieur à partir des diatomées. Thèse Docteur en Science, Université de Paris-Sud.

El Hamouti, N., Lamb, H., Fontes, J.-Ch. and Gasse, F. (1991). Changements hydroclimatiques abrupts dans le Moyen Atlas marocain depuis le dernier maximum glaciaire. *Comptes Rendus Acad. Sci. Paris,* **313**, sér. II, 259–65.

Flower, R. J. and Foster, I. D. L. (1992). Climatic implications of recent changes in lake level at Lac Azigza (Morocco). *Bull. Géol. Soc. Française,* **163**, 91–6.

Folland C. K., Palmer, T. N. and Parker, D. E. (1986). Sahel rainfall and worldwide sea temperatures. *Nature,* **320**, 602–7.

Fontes, J.-Ch. and Gasse, F. (1989). On the ages of humid Holocene and Late Pleistocene phases in North Africa — remarks on 'Late Quaternary climatic reconstruction for the Maghreb (North Africa)' by P. Rognon. *Palaeogeography, Palaeoclimatology, Palaeoecology,* **70**, 393–8.

Fontes, J.-Ch. and Gasse, F. (1991). PALYDAF (palaeohydrology in Africa) program: objectives, methods, major results. *Palaeogeography, Palaeoclimatology, Palaeoecology,* **84**, 191–215.

Frumkin, A., Magaritz, M., Carmi, I. and Zak, I. (1991). The Holocene climatic record of the salt caves of Mount Sedom, Israel. *The Holocene,* **1**, 191–200.

Gasse, F. and Street, F. A. (1978). Late Quaternary lake-level fluctuations and environments of the northern rift valley and Afar region (Ethiopia and Djibouti). *Palaeogeography, Palaeoclimatology, Palaeoecology,* **24**, 279–325.

Gasse, F., Fontes, J.-Ch., Plaziat, J. C., et al. (1987). Biological remains, geochemistry and stable isotopes for the reconstruction of environmental and hydrological changes in the Holocene lakes from northern Sahara. *Palaeogeography, Palaeoclimatology, Palaeoecology,* **60**, 1–46.

Gasse, F., Tehet, R., Durand, A., Gibert, E. and Fontes, J.-Ch. (1990). The arid-humid transition in the Sahara and the Sahel during the last deglaciation. *Nature,* **346**, 141–6.

Kutzbach, J. E. and Street-Perrott, F. A. (1985). Milankovitch forcing of fluctuations in the level of tropical lakes from 18 to 0 kyr BP. *Nature,* **317**, 130–4.

Lamb, P. J. and Peppler, R. A. (1991). West Africa. In Glantz, M. H., Katz, R. W. and Nicholls, N, (eds.), *Teleconnections Linking Worldwide Climate Anomalies,* Cambridge University Press, Cambridge, pp. 121–89.

Lamb, H. F., Eicher, U. and Switsur, V. R. (1989). An 18,000 year record of vegetation, lake level and climate change from Tigalmamine, Middle Atlas, Morocco. *Journal of Biogeography,* **16**, 65–74.

Lézine, A.-M., Casanova, J. and Hillaire-Marcel, C. (1990). Across an Early Holocene humid phase in western Sahara: pollen and isotope stratigraphy. *Geology,* **18**, 264–7.

Petit-Maire, N., Burollet, P. F., Ballais, J.-L., Fontugne, M., Rosso, J.-C. and Lazaar, A. (1991). Paléoclimats holocènes du Sahara septentrional. Dépôts lacustres et terrasses alluviales en bordure du Grand Erg Oriental à l'extrême-Sud de la Tunisie. *Comptes Rendus Acad. Sci. Paris,* **312**, sér. II, 1161–6.

Ritchie, J. C. and Haynes, C. V. (1987). Holocene vegetation zonation in the eastern Sahara. *Nature,* **330**, 645–7.

Roberts, N. and Wright Jr, H. E. (1993). Vegetational, lake-level, and climate history of the Near East and Southwest Asia. In Wright Jr, H. E., Kutzbach, J. E., Webb III, T, Ruddiman, W. F., Street-Perrott, F. A. and Bartlein, P. J. (eds), *Global climates since the last glacial maximum,* University of Minnesota Press, Minneapolis.

Rognon, P. (1987). Late Quaternary climatic reconstruction for the Maghreb (North Africa). *Palaeogeography, Palaeoclimatology, Palaeoecology,* **58**, 11–34.

Rossignol-Strick, M. and Planchais, N. (1989). Climate patterns revealed by pollen and oxygen isotope records of a Tyrrhenian sea core. *Nature,* **342**, 413–16.

Street-Perrott, F. A. and Perrott, R. A. (1990). Abrupt climatic fluctuations in the tropics: the influence of Atlantic Ocean circulation. *Nature,* **343**, 607–12.

Street-Perrott, F. A. and Roberts, N. (1983). Fluctuations in closed-basin lakes as an indicator of past atmospheric circulation patterns. In Street-Perrott, F. A., Beran, M. and Ratcliffe, R. D. (eds.), *Variations in the Global Water Budget,* Reidel, Dordrecht, pp. 331–45.

Street-Perrott, F. A., Mitchell, J. F. B., Marchand, D. S. and Brunner, J. S. (1990). Milankovitch and albedo forcing of tropical monsoons: a comparison of geological evidence and numerical simulations for 9000 yr BP. *Transactions of the Royal Society of Edinburgh, Earth Sciences,* **81**, 407–27.

Talbot, M. R. and Delibrias, G. (1980). A new Late Pleistocene–Holocene water-level curve for Lake Bosumtwi, Ghana. *Earth and Planetary Science Letters,* **47**, 336–44.

van der Hammen, T., Wijmstra, T. A. and Zagwijn, W. H. (1971). The floral record of the late Cenozoic of Europe, In Turekian, K. K. (ed.), *The Late Cenozoic Glacial Ages*, Yale University Press, New Haven, pp. 391–424.

10 Aeolian Activity, Desertification and the 'Green Dam' in the Ziban Range, Algeria

J.-L. BALLAIS
Department of Geography, Université d'Aix-Marseille II, Aix-en-Provence, France

ABSTRACT

This paper attempts to explain desertification, particularly aeolian deflation and sand accumulation, in eastern Algeria by analysing a wide range of archaeological, geomorphological, historical, meteorological and sedimentological evidence. It is not possible yet to attribute specific variations in aeolian activity in the region to aridification or human activity because it has not been possible to establish a dated chronology of sediments. The importance of a detailed understanding of aeolian activity in the context of desertification control in the area is illustrated by the 'Green Dam' and it is argued that the 'Green Dam' has only been partially successful due to both geomorphological and socioeconomic reasons.

INTRODUCTION

The purpose of this study, which focuses on the Ziban Range (Figure 10.1), is to gain further knowledge about the causes of desertification in the eastern Algeria. The aim is not to choose between the two main current theories of desertification, i.e. climatic aridification or human mismanagement, but to provide information that could help to solve local problems. An increase in the number of detailed, local studies is indeed the only real way by which the part played by theoretical presupposition can be substaintially reduced. Such an approach has an added value as it involves an examination of the environmental history of the region over the past 2000 years (Shaw, 1981; Leveau, 1986).

From such a point of view, the Ziban Range is of considerable interest as it forms an obstacle to the pattern of winds that blow from the Hodna Basin towards the Grand Erg Oriental and transport sediment. Moreover, it is located on the northern borders of the North African desert, an area that has been cultivated for thousands of years, and therefore has a long history of environmental management.

CONTEMPORANEOUS WIND FLOW DIRECTIONS

Previous work (Ballais *et al.*, 1979; Ballais, 1984a; Rebillard & Ballais, 1984) has established that a dominant wind flow exists which entrains sand on the south-west rim of Chott el Hodna and carries it over at least 250 km to the Grand Erg Oriental (Figure 10.1). This

Environmental Change in Drylands: Biogeographical and Geomorphological Perspectives.
Edited by A. C. Millington and K. Pye. © 1994 John Wiley & Sons Ltd

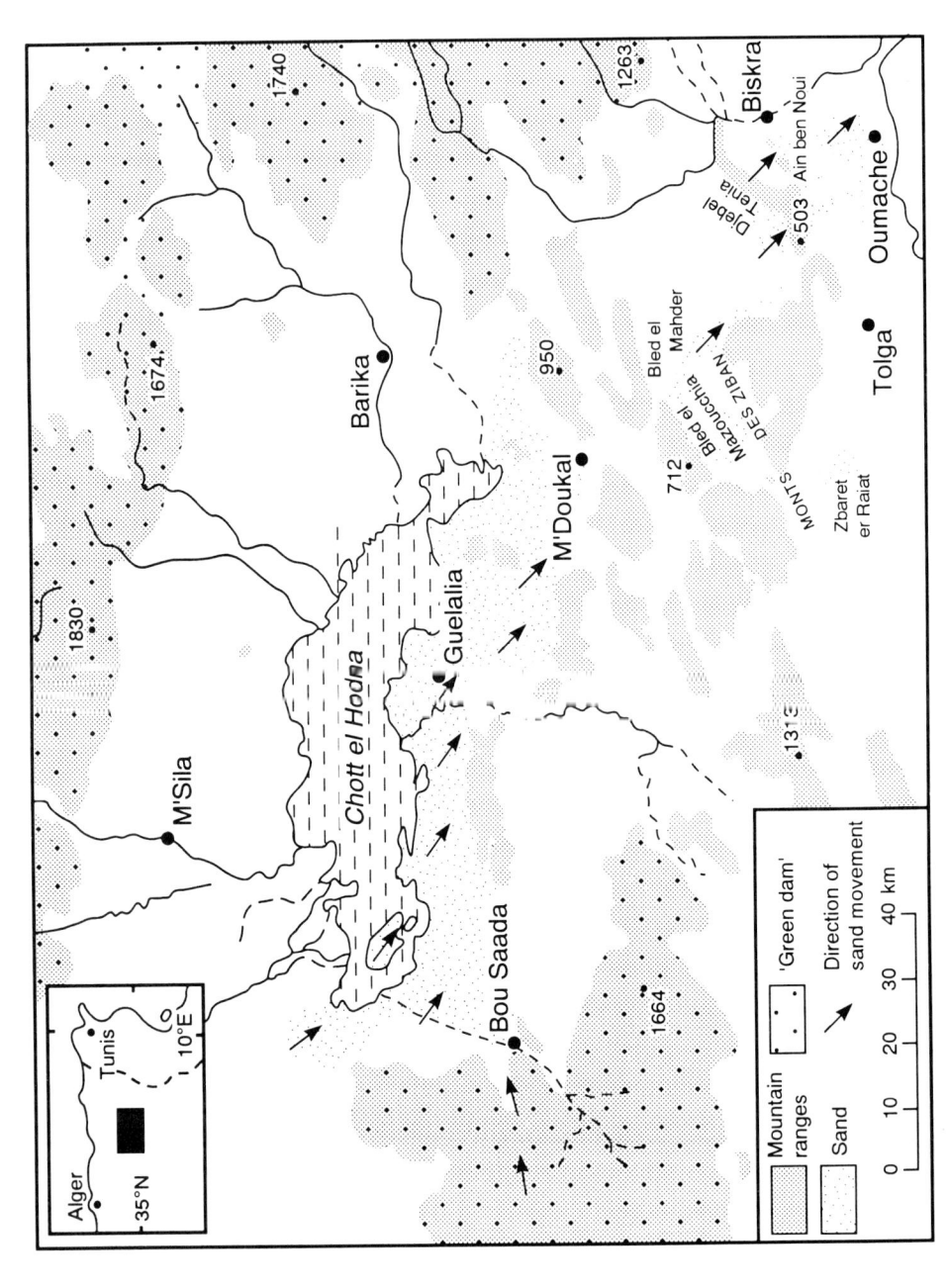

Figure 10.1 The Ziban Range: location map. 1, mountain ranges and hills; 2, the 'Green Dam'; 3, wind-blown sands; 4, direction of wind-blown sand

flow crosses the Ziban Range (Figure 10.2), then carries on southwards along the Wadi Biskra, crosses the Wadi Djedi and then continues on between Chott Merouane and Chott Melrhir. All along this route there is evidence of localised sand accumulations (wind-blown sand veils, sand-strewn slopes and dunes) which become highly mobile during sand storms (Baradez, 1949). These sand deposits move through passes (e.g. Wadi el Abiod, to the south of M'Doukal) and the depressions between the minor ranges (e.g. Wadi Matraf Kebir) which they sometimes succeed in crossing (e.g. Aïn Ben Noui).

The presence of this flow is linked to the dominant north-west winds in the region which, for instance, is the most frequent wind direction at Biskra (Seltzer, 1946): 40% of the winds at 7:00 h, 33% at 13:00 h and 36% at 18:00 h (Table 10.1). They are also the most effective in terms of sand entrainment, particularly in Autumn when they blow over ground that has dried out during the long summer drought (Sary, 1976).

QUATERNARY WIND FLOW DIRECTIONS

These same studies (Ballais et al., 1979; Ballais, 1984a; Rebillard & Ballais, 1984) have also shown that the present-day wind flow pattern has occurred many times, at least since the beginning of the middle Pleistocene. Indeed, during each dry phase, wind-blown sand has been trapped on the northern slopes of the hills and small mountains, and in particular the Djebel Tenia. In this area, sand accumulations have been modified to varying degrees before being truncated by the colluvio-alluvial aprons on which these wind-blown sands were once again deposited. At least six phases of wind-blown deposits (formations 11, 9, 7, 6, 4 and 2 in Figure 10.3) can be identified.

Until the Neolithic at least, these depositional sequences could have only been formed by natural processes. Subsequently, this was not the case and it is this realisation that has led to a reconsideration of some of the geomorphological evidence.

ARIDIFICATION AND DESERTIFICATION OVER THE PAST 4000 YEARS

Before attempting to draw any conclusions, a clear distinction has to be made between, on the one hand, natural morphogenesis and, on the other, historical changes in land use.

Natural Morphogenesis

This can be traced in two different locations: (i) the Wadi Matraf Kebir valley and (ii) the northern piedmont of the Djebel Tenia. Following the stabilization of the lower Holocene dunes (Figure 10.4) and formation of a fossil soil dated at 6350 ± 100 yr BP (dated by J.-C. Fontes), in the Wadi Matraf Kebir valley, these deposits were covered, firstly, by sub-angular pebbles and, secondly, by fluvial silts that were the lateral facies of a brown soil. On top of these deposits lie beds of fine gravels and colluvium, alternating with pockets of wind-blown sand. This sequence is breached by the wadi to a depth of 3–4 m, and the low terrace is contemporary with this breaching. Other examples of very recent incision can be seen throughout the Maghreb, particularly in the Ziban Range (Baradez, 1949; Guey, 1939), the Aurès (Ballais, 1984b) and the Nemencha (Ballais, 1976).

Figure 10.2 Dunes on the southern slopes of the Djebel Tenia (Aïn ben Noui)

Table 10.1 Direction of wind at Biskra

Month	7:00 h									13:00 h									18:00 h								
	N	NE	E	SE	S	SW	W	NW		N	NE	E	SE	S	SW	W	NW		N	NE	E	SE	S	SW	W	NW
January	22	7	3	2	0		6	59		11	5	3	10	3	5	12	51		10	5	3	10	3	6	7	56
February	16	8	6	5	1		6	56		7	4	5	14	4	6	9	51		7	4	4	14	4	6	9	52
March	15	6	8	12	2		5	48		6	4	5	19	7	8	9	42		10	4	7	18	6	5	7	43
April	11	5	8	18	2		6	47		6	3	7	22	7	7	10	38		7	4	7	20	6	7	9	40
May	8	5	3	27	5		7	31		4	3	10	28	9	11	10	25		5	4	8	22	12	7	9	33
June	8	6	12	31	8		6	24		4	4	7	29	15	12	10	19		4	5	7	26	16	11	7	24
July	4	8	16	40	6		5	16		2	3	7	33	15	22	6	12		3	3	4	32	15	20	7	16
August	8	9	14	38	8		4	16		2	4	6	28	16	19	11	14		3	4	5	26	18	16	10	18
September	10	11	12	29	5		5	25		2	3	11	26	15	12	10	20		5	6	8	27	15	10	8	21
October	18	8	9	14	2		2	43		6	3	3	24	9	11	8	31		6	4	9	23	8	10	6	34
November	18	5	7	8	1		6	53		6	3	8	20	5	7	11	40		7	3	7	21	5	6	7	44
December	21	7	4	3	1		5	57		10	4	5	9	4	4	10	34		11	5	4	10	3	4	8	55
Year	14	7	9	19	3		5	40		5	1	7	22	9	10	10	33		7	4	6	21	9	9	8	36

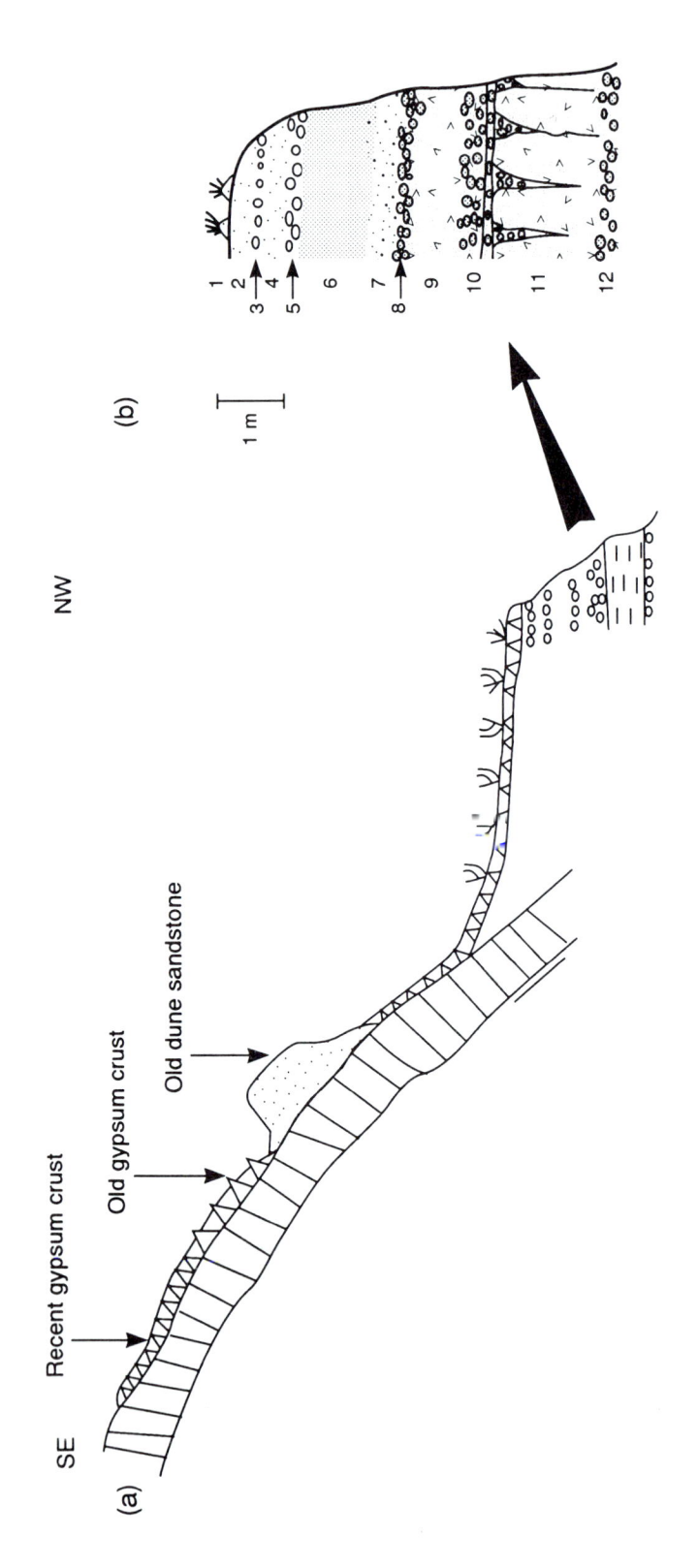

Figure 10.3 Sections along the northern piedmont of the Djebel Tenia. 1, 2, 4, 6, 7, 9, 11: aeolian formations; 3, 5, 8, 10, 12: colluvio-alluvial formations; A, general section; B, detailed section

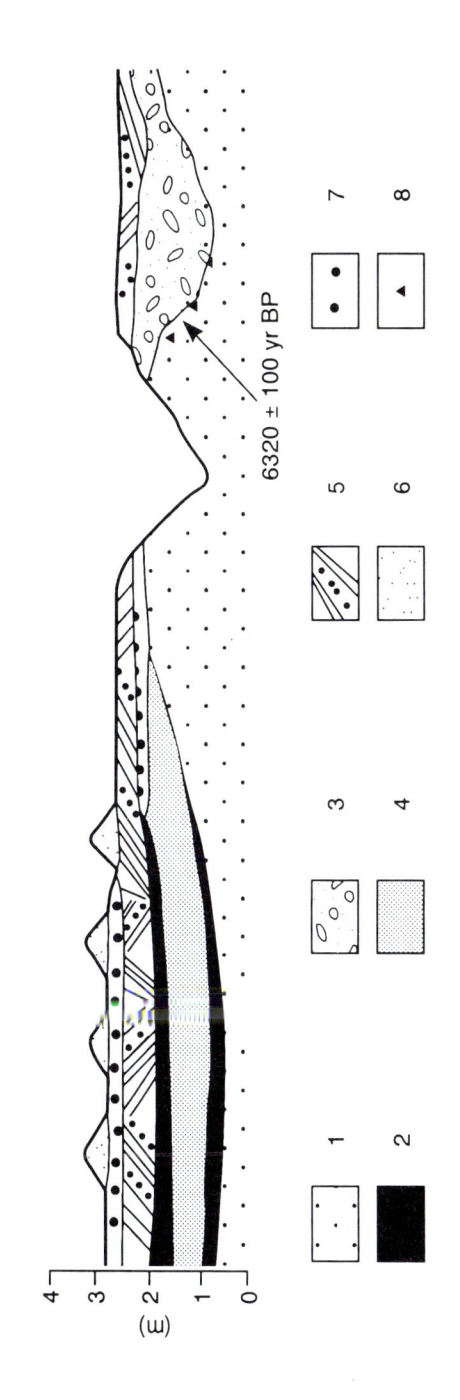

Figure 10.4 Section along the lower terrace in the Wadi Matraf Kebir. 1, lower Holocene dune; 2, palaeosols; 3, sub-angular pebbles; 4, brown silts; 5, alternating bands of wind-blown sand, fine gravels and colluvium; 6, contemporaneous wind-blown sand

On the northern piedmont of the Djebel Tenia (Ballais *et al.*, 1989), the basic Holocene sequence ends with a fossil soil (Figure 10.3: formation 4; Figures 10.5, 10.6 and Table 10.2: sample 127) which is 50 cm thick. The base is formed by up to 1.5 m of grey (10YR 6 or 7/3 or 4) quartziferous clayey-silty sands. These sands are well sorted (Figure 10.6) and richer in $CaCO_3$ and SO_3 than the previous upper Pleistocene sequence (samples 121 to 124, Table 10.2). The upper part of the lower half is characterised by a consolidated nature and a prismatic structure. Sands are finer textured (Figure 10.6) and contain high amounts of Fe_2O_3, $CaCO_3$, C, N, P_2O_5 and Na^+ and low SO_3 contents (Table 10.2). Finally, these sands are also characterised by low amounts of palygorskite and much well-crystallised smectite (Figure 10.5). This sequence is truncated by a thin discontinuous colluvial deposit made up of limestone pebbles from the slopes of the Djebel Tenia (formation 3 in Figure 10.3).

The second sequence is comprises beige (10YR 6 or 7/3 or 4), unconsolidated, structureless quartz sands (formation 2 in Figure 10.3; samples 132 to 137 in Figures 10.5, 10.6 and Table 10.2), which are 30–80 cm thick. These sands are also well sorted (Figure 10.6). They are poor in Fe_2O_3, $CaCO_3$, N, P_2O_5, Na^+ and Cl^- (Table 10.2); palygorskite is slightly more abundant and smectite slightly less abundant than in the previous sequence (Figure 10.5). The surface reveals an unevenly spread, white (10YR 8/3), dusty, gypsiferous 1–2 cm thick crust (sample 138). It has a high soluble salt content (Figure 10.5), much C and Na^+ and little $CaCO_3$, P_2O_5, Fe_2O_3 and Cl^- (Table 10.2). The clay minerals are the same as those in the lower sequence. This forms the contemporary grey, sub-desert soil which is approximatively 30 cm thick.

Both of these sequences, and the underlying upper (formation 6 in Figure 10.3; samples 121 to 124 in Figures 10.5, 10.6 and Table 10.2) and middle (formations 12 to 7 in Figure 10.3). Pleistocene accumulations are breached by wadis (Figure 10.3).

Despite the lack of either definitive or approximate dating in the second sequence, the following chronology and correlations can be assumed. Firstly, pedogenesis occurred at approximatively 6350 yr BP on the dune (Wadi Matraf Kebir) and on the lower sand sequence (Djebel Tenia). This was followed by gully erosion of the soils on the dune and the sands, which, in turn, was followed by the deposition of alluvium and colluvium. Subsequently there were colluvio-aeolian (Wadi Matraf Kebir) and aeolian (Djebel Tenia) accumulations. These deposits where then breached by wadis, and it is this fluvial activity that is still in progress.

Palaeoclimatic Interpretation

The fossil soil of the Djebel Tenia is a steppe soil typical of the so-called *arid floor* of the Mediterranean climate (Emberger, 1955). In general, in the *arid floor*, the mean annual rainfall is between 150 and 300 mm. A comparison with the existing soil at S'Gag (Aurès Range) (Ballais, 1984a), in the *upper sub-floor* of the Mediterranean arid floor suggests that this fossil soil can be attributed to *lower sub-floor*. At the present time, the mean annual rainfall at Biskra is 156 mm (Seltzer, 1946). During the Neolithic Era the soil was covered by a wooded steppe which prevented sand movement by aeolian saltation, although sand movement by aeolian suspension remained possible.

The current, extremely thin, gypsiferous soil co-exists with wind-blown deposits in an environment degraded by human activity. It can therefore be assumed that the movement of sand occurred in an environment very similar to that in the region at present time, i.e.

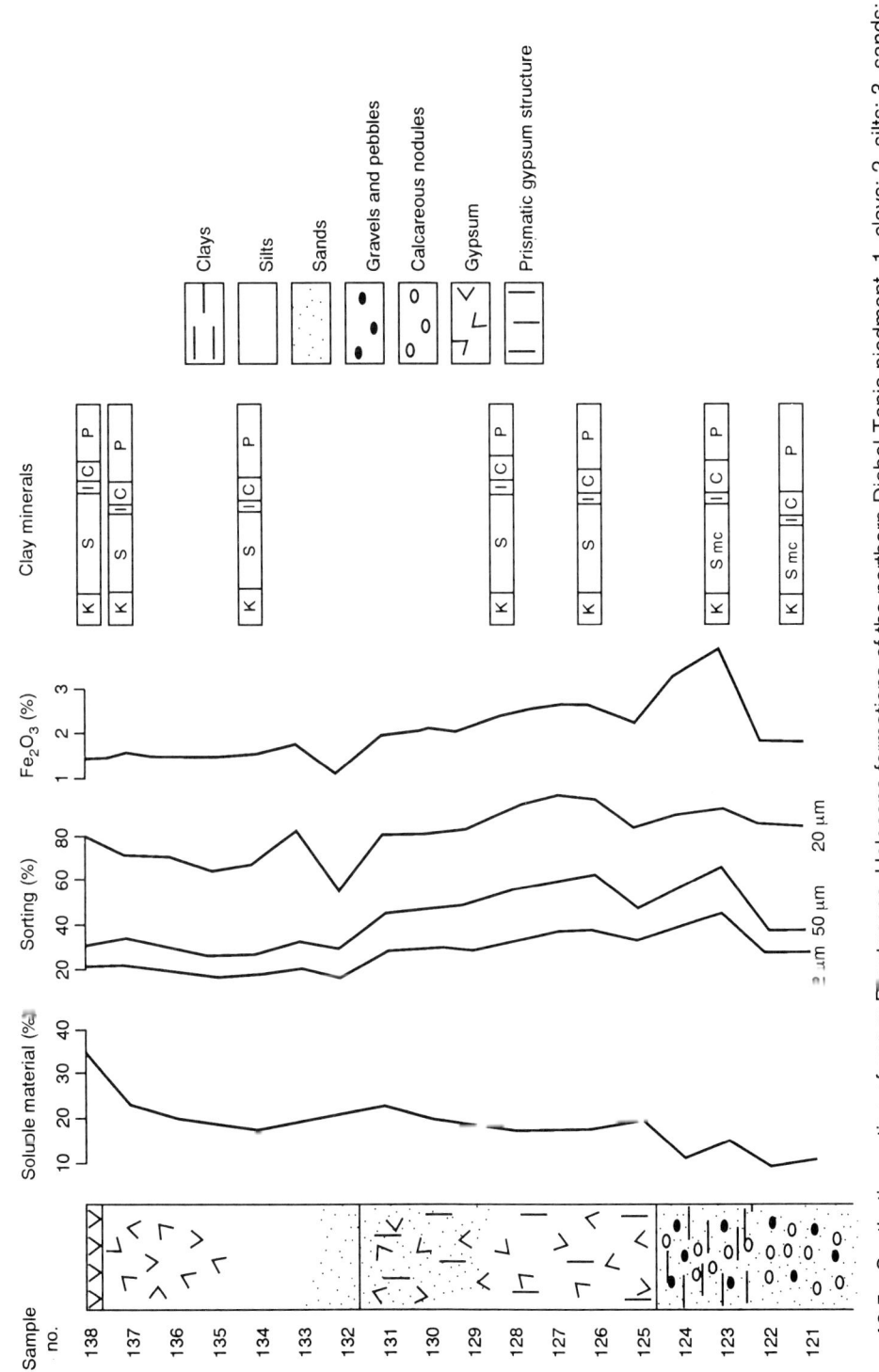

Figure 10.5 Synthetic section of upper Pleistocene–Holocene formations of the northern Djebel Tenia piedmont. 1, clays; 2, silts; 3, sands; 4, gravels and pebbles; 5, calcareous nodules; 6, gypsum; 7, prismatic structure caused by gypsum; K, kaolinite; S, smectite; Smc, poorly crystalline smectite; I, illite; C, chlorite; P, palygorskite

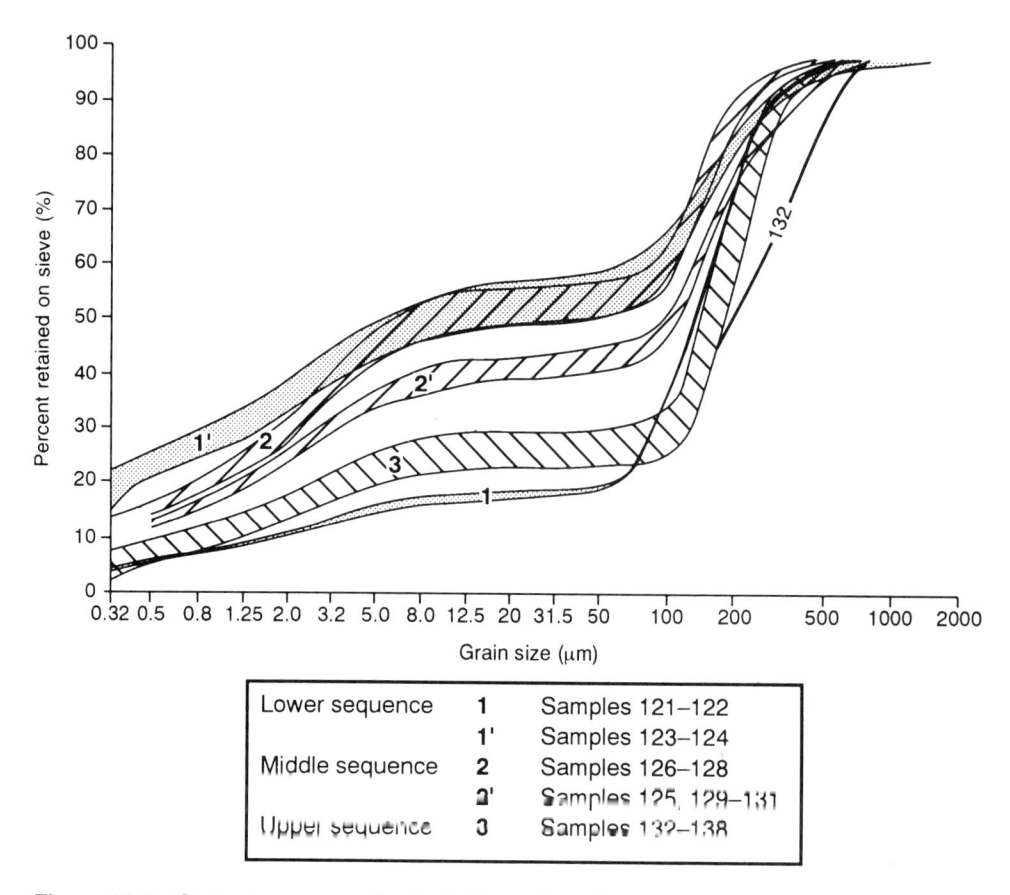

Figure 10.6 Grain-size curves of selected formations. Lower sequence: 1, samples 121 and 122; 1', samples 123 and 124; Middle sequence: 2, sample 126 to 128; 2', samples 125, and 129 to 131. Upper sequence: 3, samples 132 to 138

a vegetation which is typical of an *upper Saharan floor* climate, but which is limited by grazing and cultivation. When grazed, the steppe comprises very small bushes (20–40 cm high) with a cover of less than 20%, and as low as 10% on gypsiferous soils and gypcrete. This climate was more *Saharan*, more arid than the prevailing climate, particularly subsequent to the period of optimum humidity in the Holocene ($\approx$10 000–$\approx$3500 yr BP; Fontes & Gasse, 1989; Ballais, 1991). However, the currently active mobile sands are coarser (median size of 120–200 μm) than that of the lower sequence (median size of 85–100 μm) and comparable to those of the last sequence (median size of 135–185 μm) (Figure 10.6). It can therefore be assumed that since the Neolithic Era the amount of sand deposited after transport by suspension has decreased in relation to that transported by saltation.

The sheetwash phase is less problematical as it occurs between a phase of well-developed pedogenesis and the beginning of a new phase of aeolian sand accumulation, i.e. between humid and dry phases. During this phase sheetwash was prevalent on the very short slopes of the Djebel Tenia and continued onto the porous piedmont soils. It also assumes a dominance of high-intensity rains (compared to the phase that favoured pedogenesis) which were distributed more unevenly throughout the year than in the phase of pedogenesis and which

Table 10.2 Geochemical characteristics of the formations of northern Djebel Tenia Piedmont

Samples	CaCo$_3$	C	N	CaO	SO$_3$	Fe$_2$O$_3$	P$_2$O$_6$	Na$^+$	Cl$^-$
138	11.2	0.26	0.01	15.52	12.76	1.33	0.03	0.007	0.011
137	11.8	0.21	0.01	13.07	5.95	1.50	0.04	0.019	0.009
136	11.0	0.21	<0.01	11.30	6.64	1.42	0.04	0.011	0.018
135	10.1	0.24	<0.01	10.51	6.25	1.36	0.03	0.008	0.012
134	12.5	0.17	<0.01	11.45	5.77	1.38	0.04	0.008	0.019
133	12.7	0.24	<0.01	12.01	5.76	1.59	0.04	0.009	0.012
132	19.9	0.18	<0.01	16.03	7.88	1.01	0.03	0.011	0.007
131	21.4	0.18	<0.01	16.95	7.52	1.82	0.05	0.014	0.010
130	21.4	0.21	<0.01	16.61	6.57	1.91	0.05	0.014	0.026
129	22.6	0.20	<0.01	16.21	5.37	1.91	0.05	0.017	0.009
128	23.7	0.19	0.01	16.59	4.87	2.18	0.05	0.022	0.015
127	25.4	0.23	0.01	17.51	4.78	2.34	0.06	0.039	0.017
126	27.5	0.20	0.01	18.13	5.05	2.36	0.06	0.053	0.013
125	22.6	0.20	0.01	15.99	5.35	1.94	0.05	0.037	0.024
124	21.8	0.21	0.02	12.99	1.31	3.03	0.08	0.081	0.019
123	19.9	0.24	0.02	13.09	3.08	3.47	0.09	0.090	0.012
122	14.4	0.21	<0.01	9.64	1.13	1.44	0.03	0.033	0.035
121	14.2	0.21	<0.01	8.86	1.09	1.44	0.03	0.026	0.029

led to a decrease in vegetation cover. Overall, the pattern that emerges is one of aridification from a period during which the Holocene palaeosol was formed to the present-day gypsiferous soil.

The previous linear breaching phase caused by channel erosion at the end of the upper Pleistocene did not exceed more than a few decimetres and the only noteworthy, clearly visible, phase was prior to the formation of the upper Pleistocene deposits. Thus, the present-day breaching could have been the result of natural causes, as it occurred after the optimum climatic interglacial stage. However, it could also be the result of a change from *lower Saharan floor* climatic conditions to *upper Saharan floor* climatic conditions, i.e. a slight increase in precipitation.

Human Occupation and Land Use

From the Neolithic Era onwards evidence of human activity in the area is supported by the presence of transhumant sheperds in the Aurès (Roubet, 1979). However, too few sites have been identified (Grébénard, 1976) to confirm whether they had a significant influence on the natural environment.

Similarly, few traces remain of activity in the Protohistoric era (from about 4000 yr BP to about 220 BC, according to Camps, 1987). Although it is certain that agricultural techniques were developed during this period, in particular irrigation (Trousset, 1986). Many tumuli, especially some *bazinas* (Camps, 1987), have been observed on the crests around Bled el Mazoucchia and Bled el Mahder (Figure 10.1) as well as on the Djebel Tenia, although they have not been dated.

The Roman Era (first–fifth centuries AD) saw the gradual setting up of the *limes* (system of roads, fortifications and towns along the southern frontier of the Roman Empire) (Figure

10.7) near the Ziban Range, together with the development of rain-fed and irrigated agriculture (Baradez, 1949). It is therefore possible to obtain information concerning human modification of environment at this time, and therefore to date certain phases of morphogenesis.

With regard to the first point, two noteworthy studies have been carried out by Leschi (1943, 1957). The first study, which deals with the distribution of land between 198 and 201 AD, to the south of Chott El Hodna, shows that the area was divided, as it is today, between cultivated and grazing land. It shows that springs, which can still be identified today, already existed 6–8 km from the inscription studied by Leschi. The second study reveals that the Aqua-Viva 'Centenarium', founded in 303 AD, approximatively 6 km south of M'Doukal (Figure 10.7), was named after the nearby springs which, in 1943, were able to provide sufficient water to the Wadi Naïma-el Ahmar to maintain a steady flow. This was no longer the case in October 1987 when the underflow remained 1 m below the mean water bed. Some of the Roman wells are still operational; for example, the marabout at Sidi Abderhamane (south-west of the Gemellae castrum (Figure 10.7)) has plentiful water at a depth of 15–20 m (Baradez, 1949), as does the watering-place at Kherbet Djouala fort to the south-east of M'Doukal (Figure 10.7) (Leschi, 1943).

Baradez (1949) provides evidence of deflation from the Sallaouine, el Madher and Daya Bleds (to the north of Mesarfelta (Figure 10.7)) since Roman times, as well as the accumulation of more than 3 m of wind-blown sand on the ruins of the Gemellae camp, alluviation in depressions such as Bled el Madher and Bled Daya, and wadi entrenchment, commonly to 4–5 m depth and sometimes 12 m, since the Roman Era.

Thubunae (= Tobna) (Figure 10.7) and other areas of the region that was named 'Ziban Range' during the Mediaeval Era, and to which Hunirice exiled 4000 to 5000 Catholics, was a desert during the fifth and sixth Centuries (Courtois, 1954). This seems to indicate that there has been considerable environmental deterioration, although Thubunae was still flourishing in the Byzantine period, when the Zabi Justiniana citadel was built to the west of present-day Msila (Diehl, 1896).

There is little useful information concerning the pattern of land use during the height of the Islamic Era (eighth–ninth centuries). However, it appears to have been a prosperous period dominated by arable agriculture (cereals, cotton, olive and palm trees), and hydraulic installations (e.g. at Tobna and Msila) were used to irrigate some of the land. Furthermore, new towns were established in this period at Msila (925–6) and Achir, the Qal'a of the Bani Hammad (1007), to complete an already dense urban network (El Bekri, 1965; Al-Idrisi, 1983; Despois, 1953).

In the eleventh century it appears that, based on work carried out by El Bekri, the southern limit of olive cultivation reached the town of Bentîous, near Wadi Djedi. This latitude is similar to the southernmost limit of olive cultivation that prevailed during the Roman Era. There is an impression that the land was still fertile a century later, in particular to the north of the Chott El Hodna, although the olive groves seem to have completely disappeared (Al-Idrisi, 1985). However the verification of this change, its interpretation and its relationship to the arrival of nomadic Arabs has still to be established. Towards the middle of the eighth century, Msila (famed for its fruit), Tobna (a large-sized town known for its cotton) and Biskra (renowned for its excellent dates) were still prosperous (Ibn-Sayd, in Aboulféda, 1848). However, from that time onwards there is very little documentary information available, and only fleeting reference is made to the disappearance of Tobna and the decline of Msila in the seventeenth century (Despois, 1953).

Around 1860, the levels of biomass in the Hodna Basin and around the Wadis Djedi and

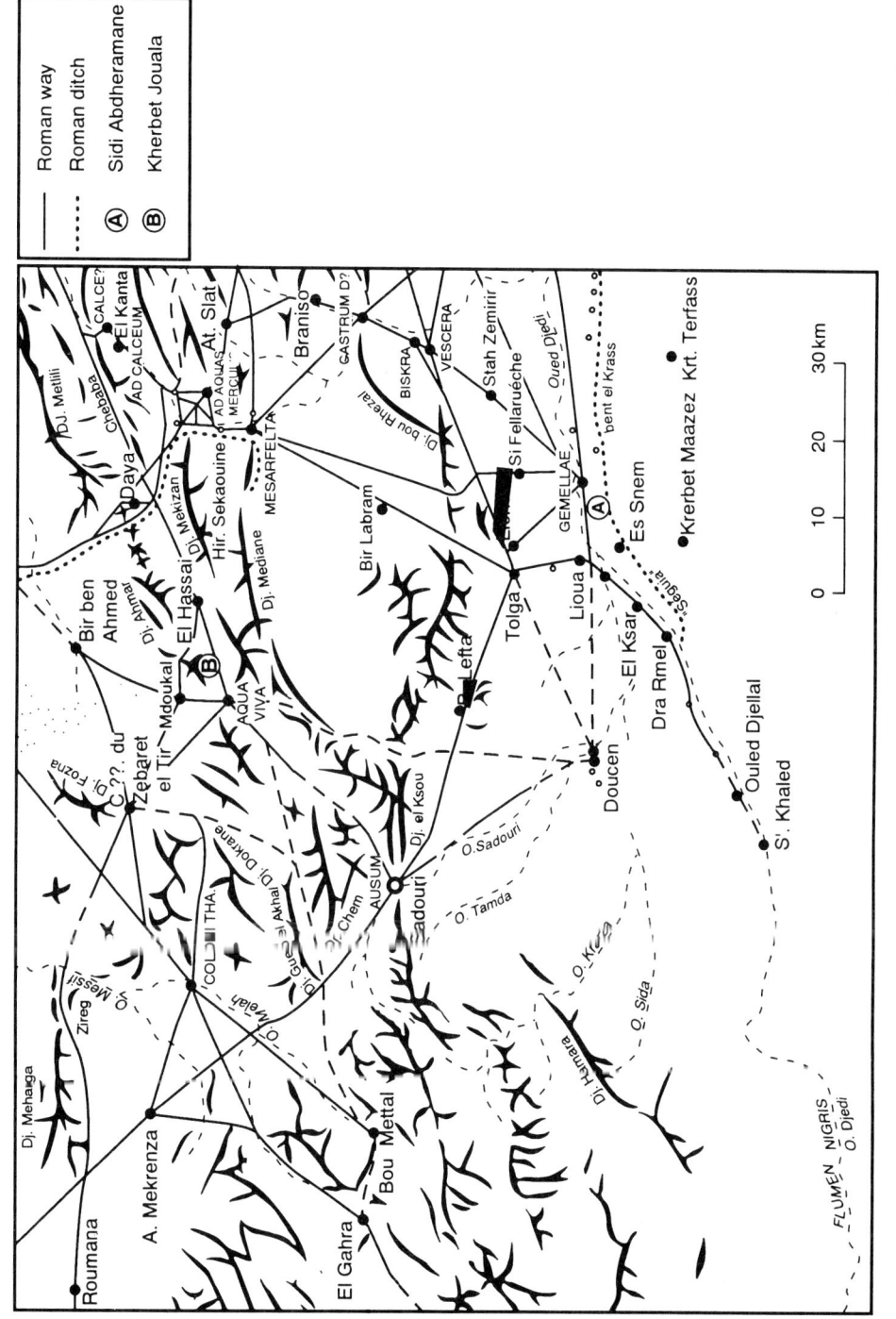

Figure 10.7 The *limes* (after Baradez, 1949). A, Sidi Abdheramane; B, Kherbet Jouala; 1, Roman way; 2, Roman ditch

Biskra were greater than at present, and the undergrowth was '... rife with wild boar ...' (Shaw, 1981). Towards the middle of the twentieth century the olives groves had disappeared from the land which separates Biskra and the Wadi Djedi had become '... salt land ...', frequented by cameleer nomads (Guey, 1939). Sand encroachment on to the steppe had been observed which filled up, within the space of several years, the 2–3 m deep excavations which had previously revealed the *fossatum* (ditch of Roman age which was a part of the *limes*) (Baradez, 1949). However, the wells still supplied the same brackish water (Guey, 1939) as they had in the eleventh century (El Bekri, 1965). The soils were suitable for arable agriculture providing it rained twice a year at appropriate times in the cultivation cycle. At least one French settler started to build a farm at the beginning of the century, which was abandoned at the outbreak of World War I.

Thus, on a regional scale, it is very difficult to reveal any significant variation in the chemical characteristics of the groundwater, the level of the water table or the amount of aeolian activity over several centuries. Furthermore, the land-use patterns during the historical period do not seem to be markedly different from those of central and eastern Maghreb as a whole.

Anthropogenic Interpretation

Generally speaking, anthropogenic activity can be divided into two distinct periods: firstly, one that runs from the end of Protohistory (about the third century BC) to the twelfth century, which was marked by improved methods of agriculture and involved considerable vegetation clearance in the depressions and valleys; secondly, a period that lasted at least until the end of the nineteenth century, which was dominated by animal husbandry and the decline of urban communities (Table 10.3).

Table 10.3 Morphoclimatic evolution during the Upper Pleistocene and Holocene

Morphogenesis		Palaeoclimatic interpretation	Anthropogenic interpretation
Wadi Matraf Kebir	Djebel Tenia		
Pedogenesis on dune	Pedogenesis on wind-blown sand	Arid climate, approximately 6350 yr BP	Neolithic
Gullying, alluviation	Gullying, colluvio-alluviation	Heavy rains, annual total reduced, approximately 4000 yr BP	
Colluvio-wind-blown accumulation	Wind-blown accumulation	Lower *Saharan* climate	Clearning, cultivation, Roman to Hammadite colonisation
Breaching	Pedogenesis and breaching	Upper *Saharan* climate	Decline in cultivation present-day

In the first period, land clearance destroyed the protection provided by the steppe vegetation in the Hodna Basin and at the same time increased topsoil erodibility. This exacerbated soil erosion and aeolian transportation. As a result there was increased wind-blown sand deposition on the northern piedmont of the flanking hills (last sequence in Figure 10.5). The current renewal of pedogenesis could therefore be linked to a decrease in aeolian deposition coinciding with a reduction of land cultivated and the development of animal husbandry, which has accompanied the population decline in the region between the eleventh and nineteenth centuries. The implication here is that animal husbandry reduces aeolian erosion, when compared to arable cultivation, at least as long as overgrazing is controlled.

The post-Roman, linear, incision of the wadis has, at the regional level, been linked to the hydrological effects that resulted from the extension of farming (Ballais, 1976, 1985, 1984b, 1986). It is possible to apply this hypothesis at a local level, in restricted areas, such as the Djebel Tenia and its piedmont, and the Bled el Madher. For example, at Bled el Madher, a Roman presence has been proven by the ruins of the fort at Bir Labrach (Figure 10.7) (Baradez, 1949). No proof of Roman cultivation has been found on the Djebel Tenia piedmont, but it probably was a pasture for sheep and goats at this time, as were the other piedmonts in the region. It must be noted that, in spite of the low altitude of the Djebel Tenia (503 m a.s.l.), its height is significant and is favourable to an important activity of the wadis. Furthermore, as the thalwegs are very well embedded in the limestone substratum, this allows the concentration of water. The last factor, which is favourable to incision downstream, is constituted by loose sandy sediments.

Conclusions

There remain two interpretations. Firstly, there is a paleoclimatic interpretation which, beginning with an arid Neolithic climate, was followed in succession by a *lower Saharan* climate and an *upper Saharan* climate. This aridification led to further deposition of wind-blown sand, followed by a slight increase in humidity which compensated for the anthropogenic desertification. Secondly, there is an anthropogenic interpretation which holds that a phase of clearance, followed by crop cultivation, led to renewed deposition of wind-blown sand. A subsequent decrease in cultivation limited this sand deposition.

Uncertainties with regard to a dated chronology make it impossible to choose clearly between a paleoclimatic or an anthropogenic hypothesis. Indeed, Table 10.3 should not raise false hopes. No accurate dating is available for the past 6000 years and if subsequent events have been established for each of these interpretations, there is no way of positioning the possible climatic variations on an historical timescale; the situation here is different from that in Tunisia (Brun & Rouvillois-Brigol, 1985; Ballais, 1991). Moreover, it is not possible to draw a parallel between these variations and those recently suggested with regard to the south-west Aurés (Rouvillois-Brigol, 1986). On the other hand, the climatic reconstructions proposed by Nicholson (1981) are based on very debatable arguments, i.e. her theory of coincidence of famine and dryness, or of plague and humidity in Algeria. However, it can be assumed that for some 4000 years, the climate in the Ziban Range has remained *Saharan* (*Saharan floor*), and has possibly fluctuated only by half of Emberger's floor, i.e. from the lower to the upper.

THE 'GREEN DAM'

Ideological and Scientific Assumptions

In 1975 the Algerian authorities, conscious of problems raised by the severe Sahelian drought (1968–73) and by numerous warnings with regard to the seriousness of desertification (e.g. Le Houérou, 1968), decided to establish an extensive 'Green Dam' across the country in the Saharan Atlas (Figure 10.8). This was modelled on Soviet and Chinese experimental projects and, in particular, the plan drawn up in the 1950s to build an extensive 'green wall' around the Gobi Desert and thus halt sand encroachment in China (Eckholm, 1977). The objectives in Algeria were similar (Sari, 1979). The idea of planting a dam which, in China, conjured up images of the Great Wall, was in Algeria linked to the colour green which, in Islamic countries, is the colour associated with the Prophet.

This dam was therefore designed to prevent sand from the Sahara from submerging the more northerly steppes, similar to the model drawn up for the Sahel where, because of the dominance of the north-east Harmattan, the wind-blown sand from the Sahara tends, particularly in dry years, to encroach upon the wooded Sahelian steppe.

A Tool Unsuited to its Intended Purpose

Unfortunately, as pointed out previously, throughout eastern Algeria, and probably at least in part of western Algeria, the dominant winds blow from the north-west and, although they move in the opposite direction during the dry season (Table 10.1), the resulting sand flow is in a south-east-south direction. More precisely, the 'Green Dam', planted on high mountains, completely surrounds the Hodna Basin (Figure 10.1). Therefore it does not act as a barrier to sand that is entrained in the Hodna Basin and is transported out of it between the hills of the Ziban Range (Figure 10.2). The location of this dam can therefore be seen to be more technocratic than scientific, as can the planting of the dam which was carried out, not by the local inhabitants, but by the National Liberation Army. It corresponds clearly to an instrumentalist view of the fight against desertification.

Therefore the question that must be asked is whether is the 'Green Dam' is useless. The answer is no as it (i) maintains the soil or, where there is no soil, helps to recreate it; (ii) provides the local population with wood for cooking, heating and building; (iii) reduces soil erosion; and (iv) eliminates ligneous perennials which favour desertification.

Nonetheless, work should be directed towards the steppes of the Hodna Basin. Action should be taken to prevent the sand from being moved by the wind, which means adopting appropriate methods of crop cultivation, protecting the steppe from overgrazing and reducing the consumption of ligneous plants, which are all generally accepted technical procedures (Le Houérou, 1975). Above all, sand should be anchored in the area in which it originates, to the south of the chott. Here again, the appropriate technical measures are widely known and are already being used in Algeria; e.g. at Djelfa station herbs, bushes and shrubs (e.g. *Eleagnus angustifolia*) are being planted between dunes (Figure 10.9); dune slopes are stabilised with a layout of plastic wire mesh together with herb planting; and grazing is banned (Figure 10.10). However, more detailed studies are also necessary to determine how much of the mobile sand from the Hodna Basin comes from the sand seas that stretch along the southern strip of the Zahrez Rharbi and Chergui, 100–150 km to the west, from which a large quantity of sand is blown through the gap made by the Wadi

Figure 10.8 The 'Green Dam' west of Bou Saada

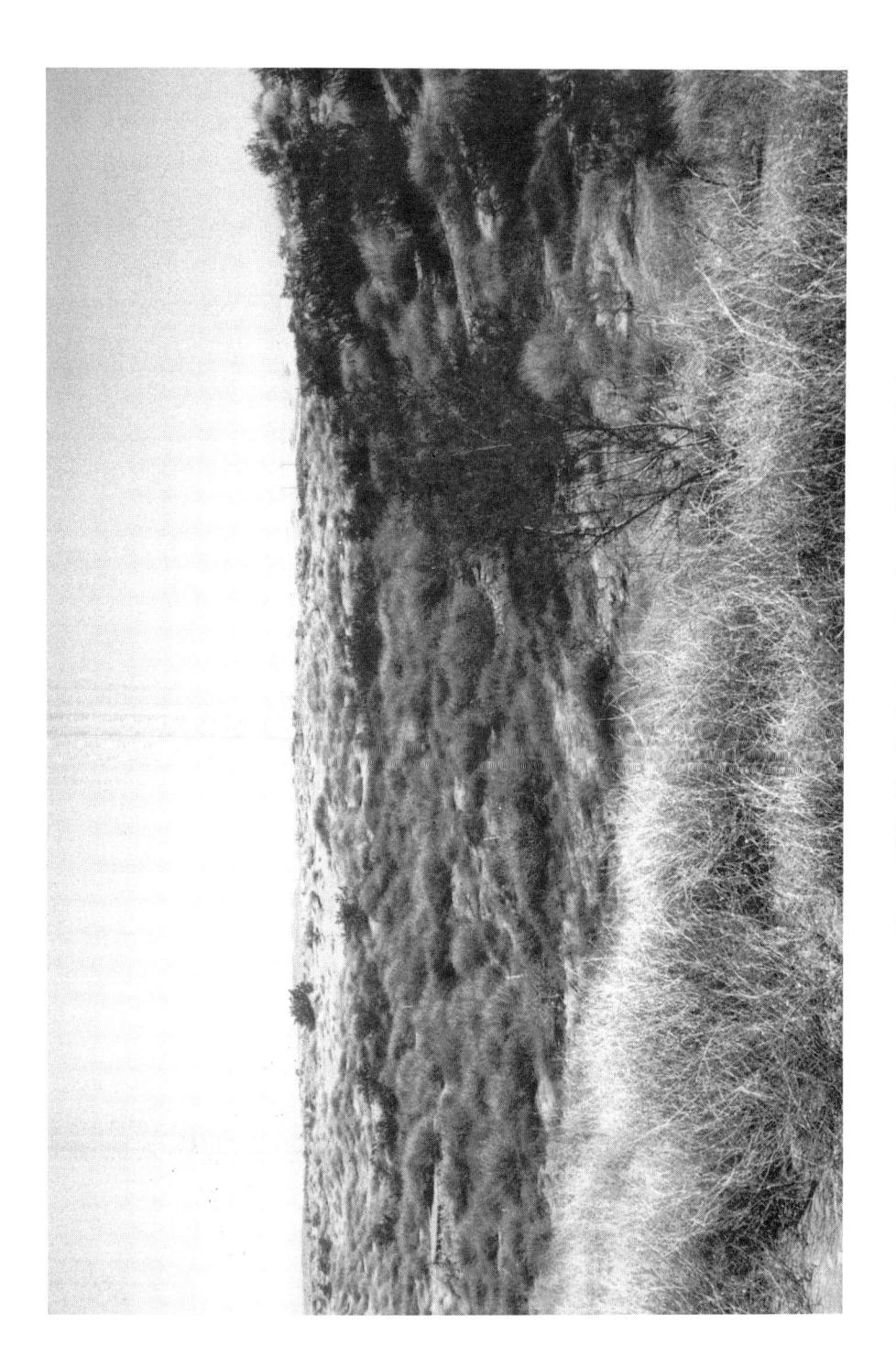

Figure 10.9 Planting between the dunes (Djelfa station)

Figure 10.10 Fixation of dunes with a layout of plastic wire mesh and planting (Djelfa station)

Figure 10.11 Wind-blown sand on the southern bank of Wadi Maïter, north-west of Bou-Saada

Maïtar (Figures 10.1 and 10.11). Finally, the fight against desertification has little chance of success if it does not gain, if not the enthusiastic support, at least the approval of the local population (Davidson, 1977; Whitney, 1977; Sari, 1979).

CONCLUSION

In the Ziban Range, on the northern borders of the Sahara Desert, the state of aeolian activity and desertification is very different from that of the Sahel, on the southern borders of the Sahara Desert, where is has been studied for over 20 years. As well as the differences due to the different land-use histories, aeolian activity appears to function in the opposite direction; i.e. from the Sahara to the steppes in the Sahel, and from the steppes to the Sahara in the Ziban Range. This major difference shows that it is essential to carry out a precise and thorough scientific study before each technical decision concerning the fight against desertification. In particular, the objective of such a study in this region should be to identify aeolian sand movement and control present-day aeolian activity.

REFERENCES

Aboulféda (1848). *Geography*, transl. and notes by M. Reinaud, Imp. Nat., Paris.

Al-Idrisi (1985). *Le Maghrib au 6ème siècle de l'Hégire*, translated by M. Hadj-Sadok, Publisud, Paris.

Ballais, J.-L. (1976). Morphogenèse holocène dans la région de Chéria (Nementchas-Algérie). *Actes Symp. Versants Pays Médit.*, Aix-en-Provence, pp. 127–30.

Ballais, J.-L. (1984a). *Recherches Géomorphologiques dans les Aurès (Algérie)*. ANRT, Lille.

Ballais, J.-L. (1984b). Grandes phases de modifications de l'environnement dans les Aurès (Algérie) au cours de la période historique. *Bull. Ass. Géogr. Fr.* , **499**, 73–6.

Ballais, J.-L. (1985). Modifications de l'environnement des Aurès (Algérie) au cours de l'Holocène. *Cah. Lig. Préh. et Protoh.* , new series **2**, 125–39.

Ballais, J.-L. (1986). Variations du milieu à l'Holocène dans les Aurès (Algérie). *Trav. ORSTOM*, **197**, 19–21.

Ballais, J.-L. (1991). Evolution holocène de la Tunisie saharienne et présaharienne. *Méditerranée*, **4**, 31–8.

Ballais, J.-L., Marre, A. and Rognon, P. (1979) Périodes arides du Quaternaire récent et déplacement des sables éoliens dans les Zibans (Algérie). *Rev. Géol. dyn. et Géogr. phys.*, **21**, 97–108.

Ballais, J.-L., Dumont, J.-L., Le Coustumer, M. N. and Levant, M. (1989). Sédimentation éolienne, pédogénèse et ruissellement au Pléistocène supérieur-Holocène dans les Ziban (Algérie). *Rev. Géom. Dyn.*, **2**, 49–58.

Baradez, J. (1949). *Fossatum Africae*, Arts et Métiers Graphiques, Paris. Brun, A., and Rouvillois-Brigol, M. (1985). Apport de la palynologie à l'histoire du peuplement en Tunisie. In *Palynologie Archéologique*, CNRS, Paris, pp. 213–26.

Courtois, C. (1954). *Victor de Vita et son Oeuvre. Etude Critique*, Alger.

Davidson, B. (1977). Mass mobilization for national reconstruction in the Cape Verde Islands. *Economic Geography*, **53**, 393–6.

Despois, J. (1953). *Le Hodna (Algérie)*. PUF, Paris.

Diehl, C. (1896). *L'Afrique Byzantine. Histoire de la Domination Byzantine en Afrique*, Leroy, Paris.

Eckholm, E. (1977). The shrinking forests. *Focus*, **28**, 12–16.

El Bekri, A. O. (1965). *Description de l'Afrique septentrionale*. translation by de Slane, Adrien-Maisonneuve, Paris.

Emberger, L. (1955). Une classification biogéographique des climats. *Trav. Inst. Bot.*, **7**, 3–43.

Fontes, J.-C. and Gasse, F. (1989). On the age of humid Holocene in late Pleistocene phases in North Africa; remarks on 'Late Quaternary climatic reconstruction for the Maghreb (North Africa)' by P. Rognon. *Palaeogeogr., Palaeoclimatol., Palaeoecol.*, **70**, 393–8.

Grébénard, D. (1976). Le Capsien des régions de Tébessa et d'Ouled-Djellal, Algérie. Université de Provence, Aix-en-Provence.

Guey, J. (1939). Note sur le limes romain de Numidie et le Sahara au IVème siécle. *Mél. Archéol. et Hist. Ecole franç. Rome*, **61**, 178–248.

Le Houérou, H.-N. (1968). La désertisation du Sahara septentrional et des steppes limitrophes. *Ann. alg. Géogr.*, **6**, 5–30.

Le Houérou, H.-N. (1975). Peut-on lutter contre la désertisation? In *La Désertification au Sud du Sahara*, Nouakchott, pp. 158–63.

Leschi, L. (1943). Le 'Centenarium' d'Aqua-Viva près de M'Doukal (Commune mixte de Barika). *Rev. Afr.*, **394–395**, 5–22.

Leschi, L. (1957). Une assignation de terres en Afrique sous Septime Sévère, In *Etudes Africaines*, Arts et Métiers Graphiques, Paris, pp. 75–9.

Leveau, P. (1986). Occupation du sol, géosystème et systèmes sociaux. Rome et ses ennemis des montagnes et du désert dans le Maghreb antique. *Annales ESC*, **6**, 1345–58.

Nicholson, S. E. (1981). Saharan climates in historic times. In Faure, H. and Williams, M. A. J. (eds.), *The Sahara and the Nile*, Balkema, Amsterdam, pp. 173–200.

Rebillard, P. and Ballais, J.-L. (1984). Superficial deposits of two Algerian playas as seen on SIR-A, Seasat and Landsat coregistered data. *Zeitschrift für Geom.*, **28**, 483–98.

Roubet, C. (1979). *Economie Pastorale Préagricole en Algérie Orientale: Le Néolithique de Tradition Capsienne. Exemple: l'Aurès*, CNRS, Paris.

Rouvillois-Brigol, M. (1986). Quelques remarques sur les variations de l'occupation du sol dans le Sud-Est algérien. In *Hist. et Archéol. Afr. du Nord*, CTHS, Paris, 35–53.

Sari, D. (1979). Les tentatives de restructuration du monde rural en Algérie. *Méditerranée*, **1–2**, 65–72.

Sary, M. (1976). Géographie physique d'une Haute Plaine steppique algérienne: le Hodna. Thèse de 3ème cycle, Strasbourg.

Seltzer, P. (1946). *Le Climat de l'Algérie*. Carbonnel, Alger.

Shaw, B. D. (1981). Climate, environment and history: the case of Roman North Africa, In Wigley, T. M. L., Ingram, M. J. and Farmer, G. (eds), *Climate and History Studies in Past Climates and Their Impact on Man*, Cambridge University Press, Cambridge, pp. 379–403.

Trousset, P. (1986). Limes et frontière climatique. In *Hist. et Archéol. Afr. du Nord*, CTHS, Paris, pp. 55–84.

Whitney, J. (1977). Striking a blow against desertification: cooperative initiative in Chungwei county, China. *Economic Geography*, **53**, 381–4.

11 The Ambiguous Impact of Climate Change at a Desert Fringe: Northern Negev, Israel

A. YAIR
The Institute of Earth Sciences, The Hebrew University of Jerusalem, Israel

ABSTRACT

This paper reports on interdisciplinary research into the effects of climatic change on three contrasting arid environments in Israel: (i) the rocky Negev Highlands, (ii) the loessic Beer Sheva depression and (iii) the sandy Nizzana plain. Present-day hydrological processes were combined with the study of environmental processes over longer timescales (back to 100 000 years). The effect of climatic change on environmental processes is shown to vary with local site conditions. The responses are partly controlled by two variables: (i) changes in surface properties and (ii) the salt balance. However, ambiguities are also introduced because the scale of climate models cannot account for these local variations in surface properties and the fact that only annual average data are presented by these models. The result of a regional climatic change is a non-uniform ambiguous effect on hydrological processes.

INTRODUCTION

The influence of expected climate changes caused by the Greenhouse Effect upon arid and semi-arid areas has not yet been subject to thorough discussion. The evaluation of the potential impact of future climatic change in the low latitudes encounters several difficulties. Some of these difficulties are specific to semi-arid and arid areas; for example:

1. General circulation models are usually operated at the global scale. They have not been able to suggest unequivocally whether global warming will result in a decrease or increase in rainfall in the sub-tropical belts. Furthermore, they are not able to project changes at the regional or local scales. Such scales are of great importance in sub-tropical areas where non-climatic factors such as lithology, topography, soils and biological variables can exert, alone or in combination, a strong influence on the environment.
2. Most climatological models deal with annual averages of precipitation and temperature. They do not have the ability to deal with climatological variables which control water availability, such as the timing of storms during the wet season, the frequency, distribution and magnitude of extreme events responsible for catastrophic floods or prolonged drought periods.

Among the variables specific to semi-arid and arid areas, two factors have a pronounced effect on the environment. The first is the substantial change in the surface properties and the second is the salt balance. A climatic change at the desert fringe is not limited to purely climatological variables such as temperature and precipitation. Such changes may be accompanied by a rapid alteration of surface properties. Scientists in different disciplines working in North Africa and the eastern Mediterranean tend to agree that aeolian loess penetration into the desert fringe occurred during relatively wet periods whereas sand penetration took place under relatively dry conditions (Bruins & Yaalon, 1979; Goldberg, 1981; Coude-Gaussen *et al.*, 1983). An assessment of the possible effects of global climate change in arid and semi-arid areas should therefore consider two sets of variables: climatological conditions, such as rain amount, temperature and evaporation, as well as the modification of surface properties. Needless to say, the new surface conditions may result in a drastic change in the hydrological regime, greatly affecting water availability. It should be noted that the combination of the two sets of variables described above is specific to arid and semi-arid areas. In wetter areas, where the input of aeolian materials (sand or loess) is negligible, there is little or no effect on the pre-existing surface materials.

The second factor is the salt balance. In addition to water availability, salinity is the next-limiting factor of biological activity in semi-arid and arid areas. It has been well established (Yaalon, 1962) that salts incorporated in the soils of arid areas are mainly derived from ions dissolved in rain water and, to a lesser extent, from dry dust fall. The positive relationship between rain and salt amount means that salt input increases during wet periods. Salt input can be further increased by a parallel increase in dust or loess input. Such conditions raise a very important and intriguing question. As an increase in rainfall is assumed to have a positive effect on the environment, whereas that of salts has an opposite effect, which of them will have a predominant impact on the environment?

AIM OF THE PRESENT STUDY

This case study will review several inter-disciplinary investigations conducted in the Negev Desert during the last two decades. It will cover past and present-day hydrological processes in sandy, rocky and loessic environments and will draw attention to the following aspects:

1. A climatic change, representing a change at the regional scale, had a non-uniform ambiguous effect on hydrological processes within the region considered, being highly influenced by the local site-specific conditions that prevailed in the area prior to climate change.
2. Soil salinisation processes induced by the increased salt input during wet periods and their negative effect on the biota.
3. The need to re-examine the validity of existing aridity indices for management policy in semi-arid and arid areas.

This study focuses on the transition from drier to wetter conditions, a transition that was accompanied by loess penetration into the desert fringe.

DESCRIPTION OF THE STUDY AREA

The study area (Figure 11.1) forms part of the northern Negev Desert and extends from the semi-arid Beer Sheva depression in the north to the anticlinal Ramon ridge in the south.

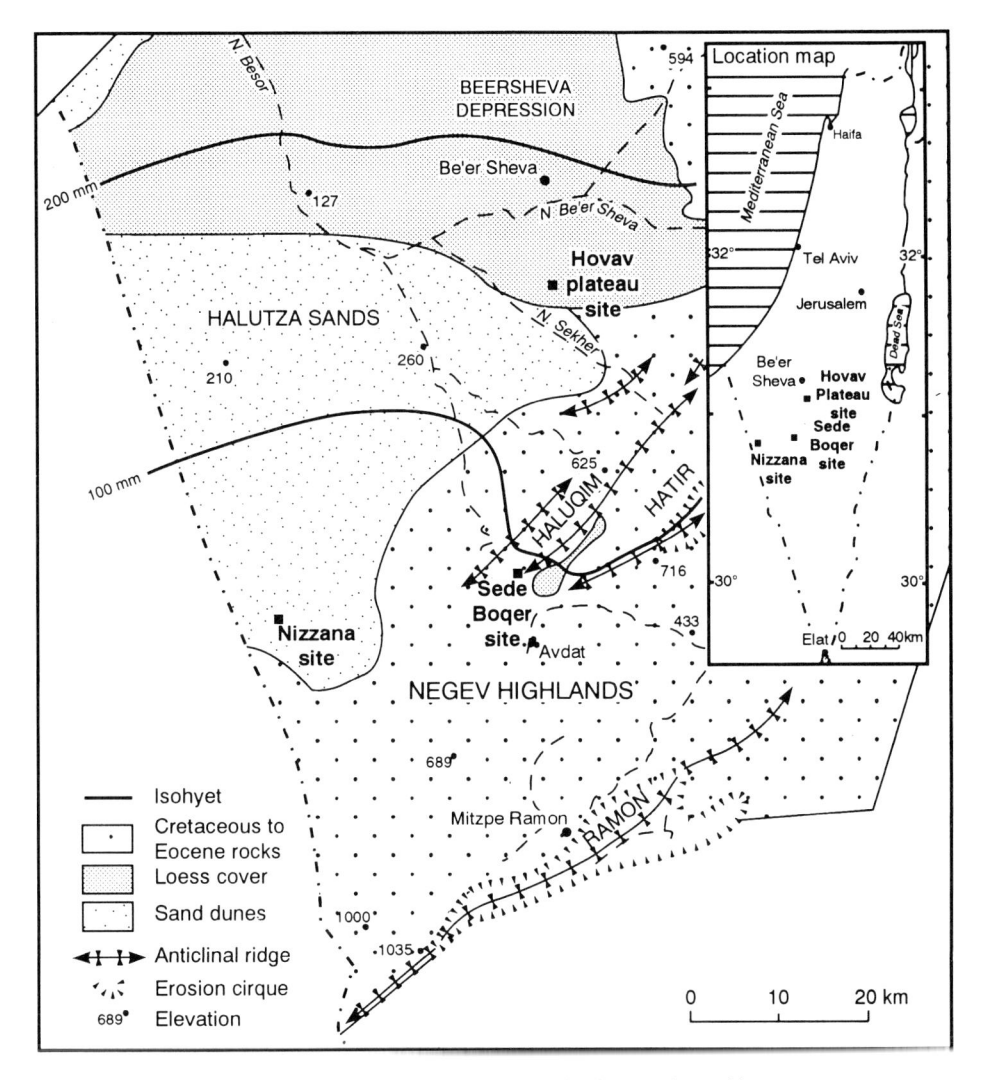

Figure 11.1 Physiographic units in the northern Negev

The topographic elevation increases southward from 200 m at the central part of the depression to some 1000 m at the summit of the Ramon ridge. The mean annual rainfall decreases gradually southward, being 200 mm at Beer Sheva and 73 mm at Mitzpe Ramon. The rainfall period is limited to the winter months which extends from October to April. Mean annual temperature is 19.2°C in the north and 18°C at Mitzpe Ramon. Annual potential evaporation is around 2000 mm. According to several different aridity indices calculated for this region (Yair & Berkowicz, 1989), the area is considered to be at the boundary of the semi-arid zone, with aridity increasing in a southerly direction (Figure 11.2). The study area is comprised of three main distinct physiographic units.

202 Environmental Change in Drylands

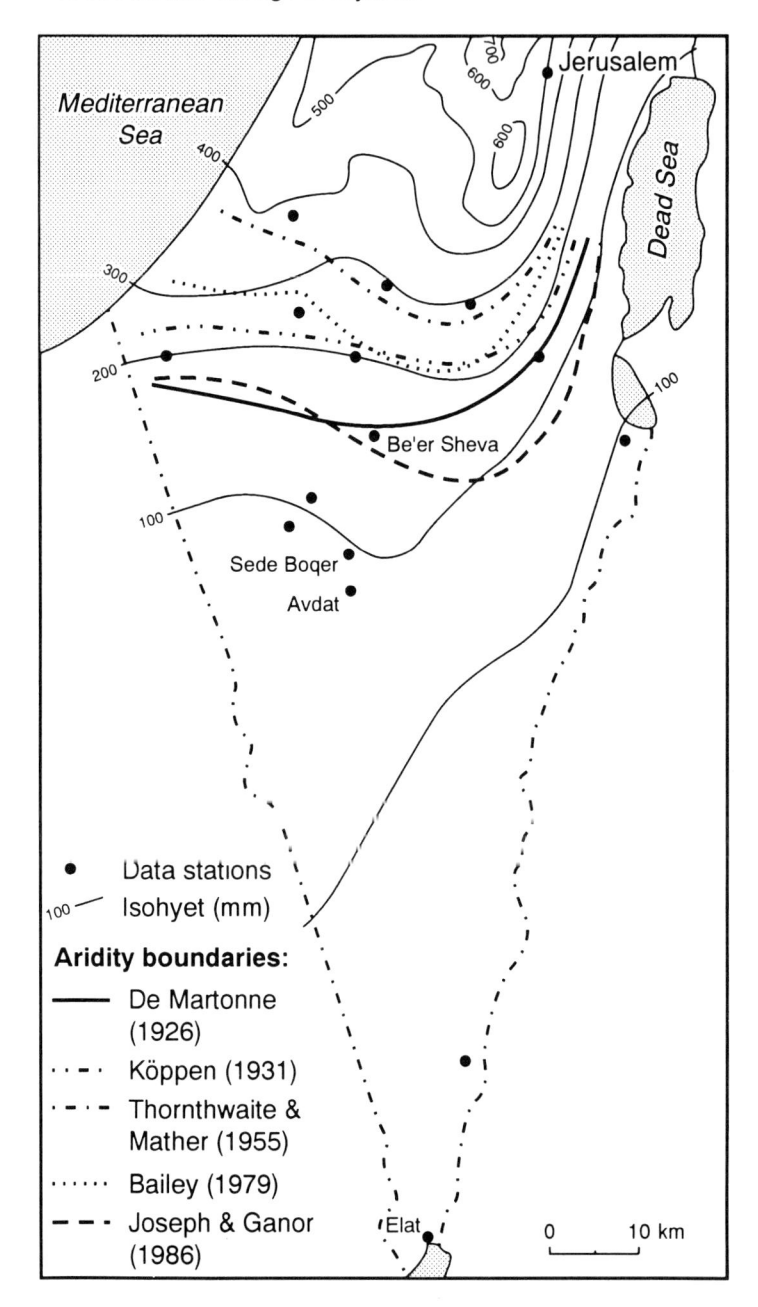

Figure 11.2 Aridity boundaries in Israel as calculated for five climate classifications (from Yair & Berkowicz, 1989. Reproduced by permission of Catena Verlag)

The Rocky Negev Highlands

This area consists of Cretaceous and Tertiary limestones and chalks and is characterised by a rugged topography of dissected plateaux and anticlinal ridges with extensive steep rocky slopes (Figure 11.3). Bare bedrock outcrops extend over 50–80% of the surface forming the upper to middle part of the hillslopes. A contiguous stony soil cover forms the lower colluvial part of the hillslopes. The valley bottoms are flat, with a relatively thick loess alluvial fill.

The Loess-Covered Area

A contiguous loess cover of up to 12 m is characteristic of the Beer Sheva depression (Bruins & Yaalon, 1979). The topography consists of gently rolling plains, with loess cover increasing in the downslope direction (Figure 11.4). Loess cover decreases towards the margins of the depression where rocky outcrops become more common. The loess cover overlies a conglomeratic unit of assumed Upper to Middle Palaeolithic age (Goldberg, 1981). On slopes, loess is in direct contact with the older marine Eocene or Cretaceous bedrock.

The Sandy Area

The sandy area occurs in the western part of the region at the border of the rocky and the loess-covered areas. It is characterized by longitudinal dunes, trending west–east, representing the easterly extension of the extensive Sinai sand field (Figure 11.5).

METHODOLOGY

Three complementary approaches were adopted for the study:

1. A detailed study of present-day runoff processes and their implications with respect to the spatial variability of runoff and water availability in the area under consideration.
2. A study of the validity of the conclusions derived from data collected on present-day processes for longer time scales, up to 100 000 years, and for larger areas at a regional scale.
3. Application of results obtained for the understanding of ancient agricultural systems used in the area.

FACTORS AFFECTING THE SPATIAL VARIABILITY OF WATER RESOURCES IN SEMI-ARID AND ARID AREAS

Runoff generation integrates a climatic factor—rainfall—and a non-climatic factor—surface properties. Pronounced differences in infiltration will influence the rate of transformation of rainfall into runoff, resulting in the spatial redistribution of water resources. Areas that respond quickly to rainfall can shed their water to an adjoining absorbing area. The water volume collected and stored in the soil may thus be out of proportion to the annual rainfall, thereby enhancing the creation of a locally improved environment. A proper under-

Figure 11.3 View of the rocky area at Sede Boqer experimental site

Figure 11.4 View of the loess-covered area on Ramat Hovav Plateau

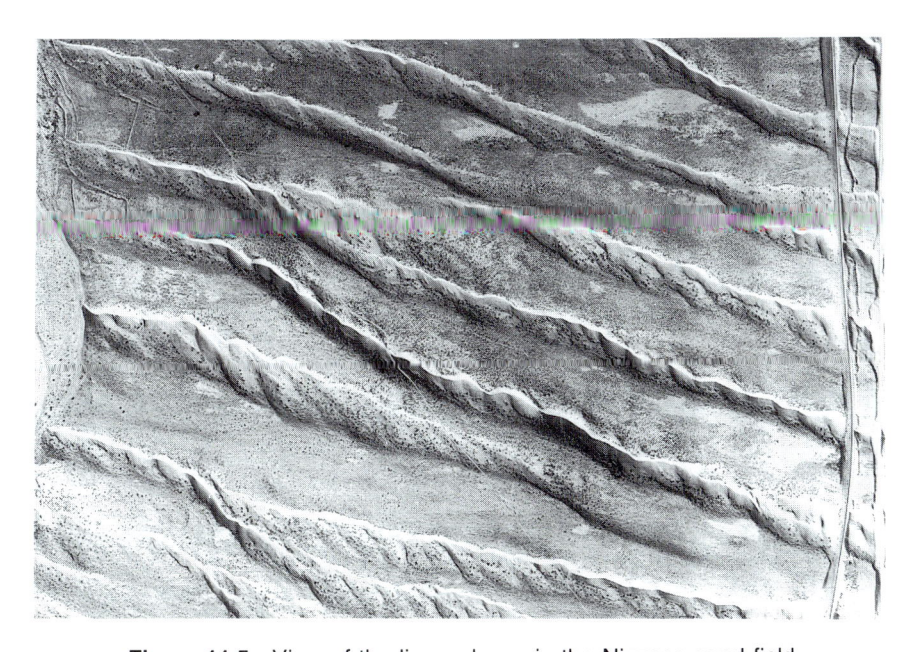

Figure 11.5 View of the linear dunes in the Nizzana sand field

standing of the spatial distribution of water availability in the northern Negev Desert must therefore be based on our understanding of rainfall–runoff relationships. A detailed analysis of these relationships has already been published (Yair & Lavee, 1985; Yair, 1990). Only a brief review of the salient points is provided here.

Effect of Surface Properties on Runoff

With respect to surface properties, the study area may be divided into two distinct units: rocky and soil-covered areas. Figure 11.6 represents the hydrological response of simulated rainfalls over these two surfaces. The simulated rainstorms represent rather extreme conditions of rain intensity and, in particular, duration. The data obtained clearly indicate that the colluvial loess soils are characterized by a high infiltration rate (about 17 mm h^{-1}) which remains so even after one hour of high-intensity rainfall. On the other hand, a very low infiltration rate, which drops to zero after about ten minutes, is characteristic of the dense, low-porosity crystalline limestone. These results are fully corroborated by hydrological data collected under natural rainfall conditions at the Sede Boqer experimental site located in the Negev rocky area (Figure 11.1) (Yair, 1983; Yair & Lavee, 1985).

Available data indicate that the threshold of daily rain necessary to generate runoff on rocky areas is in the order of 1–3 mm. Runoff occurs with any rainstorm having an intensity exceeding 5 mm h^{-1}. Soil-covered areas (having a lower bulk density, higher porosity and higher clay and silt contents) have a much higher absorption capacity, thus limiting runoff generation. The threshold for runoff generation over such areas exceeds 10 mm h^{-1} (Stibbe, 1974) and runoff does not occur for rain intensities below 10–15 mm h^{-1}. As low-intensity rainfall and low daily rainfall values are more common, runoff events are much higher over rocky than over soil-covered areas (Yair 1983).

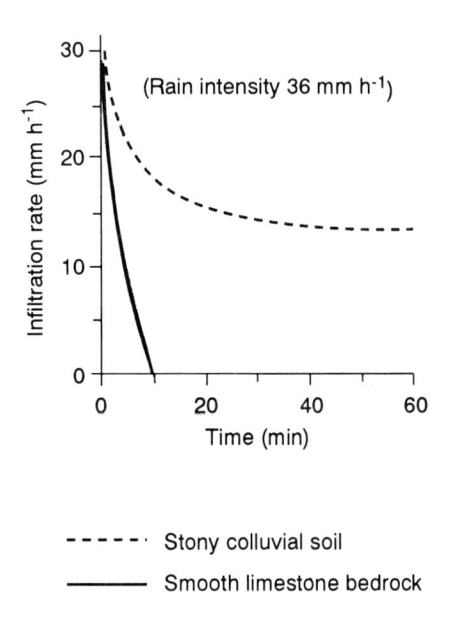

Figure 11.6 Hydrological response in rocky and soil-covered areas to simulated rainfall

Effect of Rainfall Properties on Runoff

Rainfall in the Negev is limited to the winter season. Most rainstorms belong to the category of frontal storms and are of low to medium intensity. Rainstorms with an intensity exceeding 30 mm h^{-1} represent only 5% of the total rainfall, whereas rainstorms with an intensity between 5 and 10 mm h^{-1} constitute about 77% of the total annual rainfall (Kutiel, 1978). This point is well illustrated in Figure 11.7. Even for a high-magnitude storm, which occurred in December 1980, the average rainfall for the entire storm did not exceed 3 mm h^{-1}, with very short peak intensities of 27 mm h^{-1} lasting only for a few minutes. The second important characteristic of these storms is that they consist of several separate showers, each of only a few millimetres. Consecutive showers are separated by time intervals, ranging from a few minutes to more than one hour, during which the soil can drain and dry.

The rainfall characteristics described above greatly accentuate the difference between the hydrological response of rocky and soil-covered areas. Due to the prevailing discontinuous low-to-medium intensity rainstorms coupled with the high infiltration capacity, the soil-covered areas are able to absorb all rainwater in most rainstorms. Furthermore, when runoff does occur the rate may be expected to be very low and highly discontinuous both in time and in space. A converse situation exists in the rocky areas. In this case, the quick response to rainfall results in runoff generation during most of the low-intensity, short-duration, rain showers (Yair & Lavee, 1985). The differential response to rainfall of rocky and soil-covered areas under natural rainfall conditions is well illustrated by the data presented in Table 11.1.

The minicatchments drain a loess-covered area some 15 km south of Beer Sheva, whereas data on the rocky slopes are derived from the Sede Boqer experimental site. Each of the minicatchments drains an area of about 500 m^2.

Runoff Response of Sandy Areas

When dealing with the effect of surface properties on runoff generation, rocky and sandy areas are usually considered at opposite ends of the spectrum. Whereas rocky areas respond quickly to rainfall, sandy areas are most often regarded as being shaped by wind activity and, due to their very high infiltration rate, as areas devoid of any runoff. In this respect it is quite interesting to consider the effect that a loess mantle deposited on a sandy sub-stratum would have on both aeolian and runoff processes. This possibility is provided by the longitudinal dunes at the Nizzana research site (Figure 11.5). The upper part of the sand ridges is composed of loose sand, and the silt and clay content is less than 2%. Well-developed ripple marks clearly indicate that the crests represent the active part of the dunes most responsive to wind activity. The lower flanks of the dunes, as well as most of the inter-dune area, however, are covered by a rather contiguous flexible topsoil crust 1–2 mm thick (Figure 11.8). The particle size composition of the crust is similar to that of aeolian loess; the silt and clay content represents 60–80% of the material. At a depth of only 1 cm below the surface both fractions drop to only 5% of the material. The mineralogical composition of the crust is also typical of aeolian loess. The mineral grains are held together by a fibrous mat of cyanobacteria (blue-green algae) and other biological elements. The composite biological and fine-grained crust is an efficient agent in stabilizing the underlying sandy material and in sealing the surface.

In order to study the hydrological role of the topsoil crust, sprinkling experiments were performed in the area (Yair, 1990). Runoff started over the topsoil crust only 3 minutes

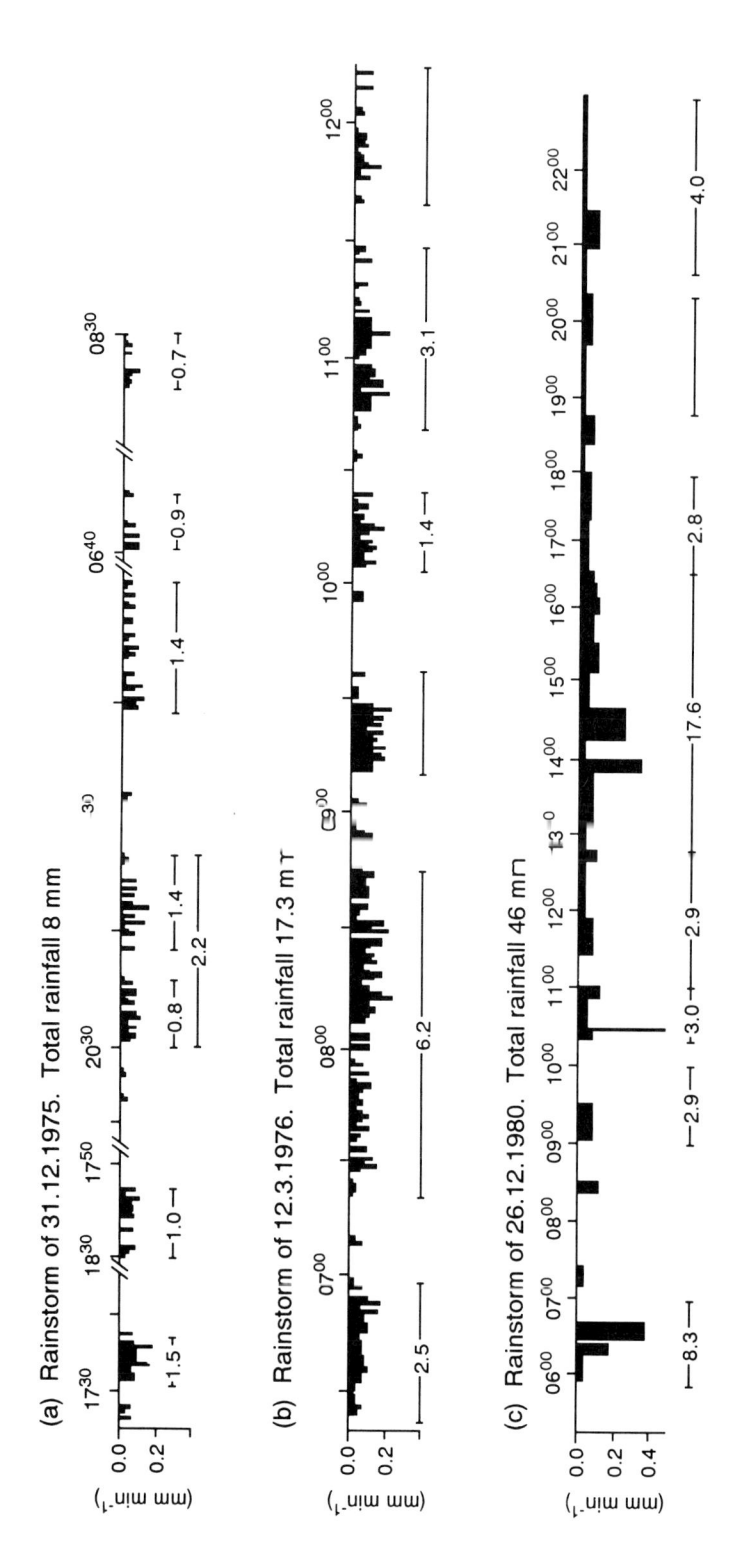

Figure 11.7 Temporal variations in rainfall during rainstorms at Sede Boqer experimental site

Figure 11.8 Biological topsoil crust at the flank of a dune

Table 11.1 Hydrological data for rocky and soil-covered hillslopes

Year	Rainfall (mm)	Number of flows	Runoff (mm)
Rocky hillslope			
1982–3	136	21	27.0
1983–4	72	5	6.5
Soil-covered hillslope			
1982–3	170	4	7.8
1983–4	72	1	2.0

after the beginning of rainfall applied at an intensity of 18.4 mm h^{-1} (Figure 11.9). The topsoil crust was then removed, exposing the underlying sandy material. Sprinkling was resumed on the previously wetted surface. Despite the fact that sprinkling intensity was increased to 54 mm h^{-1}, no runoff was observed for 42 minutes even though the accumulated rain amount reached 38 mm. Given that the Nizzana area has an average annual rainfall of 90 mm, the second sprinkling experiment represents extreme conditions in terms of both accumulated rain amount and, particularly, rainfall intensity. It clearly shows that no runoff should be expected in sandy areas devoid of a thin loessic topsoil crust.

EXTRAPOLATION TO LONGER TIMESCALES UNDER CHANGING CLIMATE CONDITIONS

The data presented so far refer to short-term hydrological events collected at the Sede Boqer and Nizzana research sites. The following section will deal with the validity of the con-

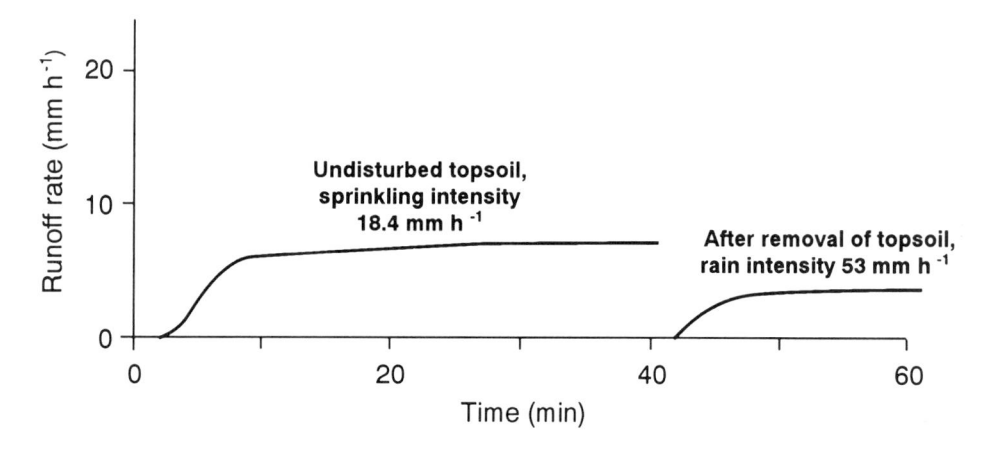

Figure 11.9 Effect of a biological topsoil crust on runoff generation

clusions based on data collected at present for longer time periods. Three timescales were selected. The first timescale, from tens of years to a few hundred, is represented by the perennial shrubs found in the area. The second timescale will refer to thousands of years and is represented by soil-forming processes which, in the arid environment under consideration, can be expected to be extremely slow. Finally, the geological timescale, covering a period of up to 100 000 years, with changing climatic conditions, is considered. This timescale is represented by the evolution of the drainage network following loess penetration into the area.

THE RECENT GEOLOGICAL EVOLUTION OF THE NORTHERN NEGEV

One approach to the assessment of the effect of global climate change on the environment is the study of the recent geological and pedological evolution of an area under changing climatic conditions. Arid and semi-arid areas, because of their sensitivity to slight changes in water resources, offer especially favourable conditions for investigation. This sensitivity is well established in the Negev. A large number of palaeosols, assumed to be of late Pleistocene age, have been identified at the northern fringe of the Negev Desert (Bruins & Yaalon, 1979). Similarly, a large number of sedimentary units, alternating loess and sand, have been identified in the Nizzana area located at the western Negev (Yair, 1990).

Evolution of the Rocky Negev

Analysis of a stratigraphic section typical of the Negev Highlands and their northern vicinity (Figure 11.10) shows that loess material overlies a conglomeratic unit deposited in former channels. Several palaeosols, indicative of a fluctuating regime of loess penetration, were identified within the loess mantle. Calcic nodules found in these palaeosols gave ^{14}C dates ranging from 20 900 to 6000 yr BP (Magaritz & Enzel, 1990). Older undated loess palaeosols also exist in the area. At those places where the loess was deposited on hillslopes, it is in direct contact with the older marine Cretaceous to Eocene formations (Figure 11.11).

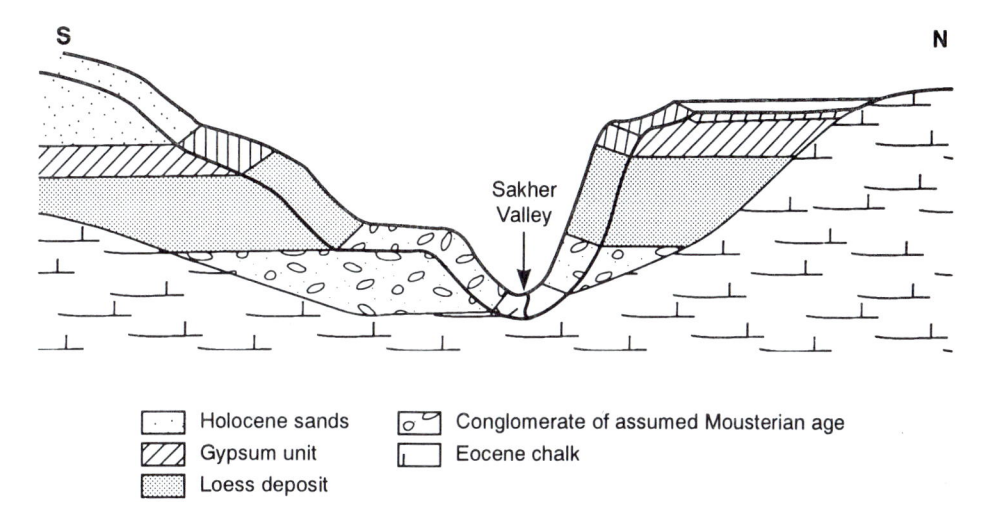

Figure 11.10 Stratigraphic section at Wadi Sekher, northern Negev (from Yair & Enzel, 1987. Reproduced by permission of Catena Verlag)

Prior to loess deposition the landscape therefore consisted of barren rocky hillslopes with very active channels capable of transporting coarse boulders; in other words, a regime of limited weathering, intense hillslope erosion processes and high-energy channel flows, all indicative of arid conditions, prevailed in the area prior to loess deposition.

Evolution of the Sandy Area

A detailed history of sand penetration into the western Negev is still unclear. Until recently it was assumed that dune penetration into the area started in the late Pleistocene, some 20 000 yr BP (Goldberg, 1986; Magaritz, 1986; Goring-Morris & Goldberg, 1990; Magaritz & Enzel, 1990). The dating was based on palaeolithic artifacts and ^{14}C analyses. A recent study (Yair, 1990) conducted in the Nizzana linear dunes revealed a complex multi-phase evolution characterized by a fluctuating regime of sand and loess deposition (Figure 11.12). An initial study based on thermoluminescence dating provided ages in the range of 43 000–6000 yr BP (Rendell et al., 1993). Stabilized, probably older, dunes exist in the area, pointing to an earlier sand penetration.

Climatological Context of Recent Geomorphological Evolution of the Northern Negev

As stated earlier, there is general agreement that aeolian loess was deposited under relatively wetter climatic conditions and sand under drier conditions. The wet period is usually connected with the Würm glaciation, during which time the climatic belts had shifted southwards. Issar & Bruins (1983) studied the isotopic composition of groundwater in the Negev and Sinai, as well as the clay content of the loess soils, and reached the conclusion that between 100 000 and 10 000 yr BP 'the amount of precipitation was 50–100% more than that of present'. The relatively wet period ended at the beginning of the Holocene when the tran-

212 Environmental Change in Drylands

Figure 11.11 Sharp contact between the loess mantle and earlier rocky outcrops

Figure 11.12 Alternating sandy and silty-clayey layers in the inter-dune corridors, Nizzana

sition to drier conditions resulted in the massive penetration of sand into the western part of the area.

The sedimentary sequences and the fluctuating climatic conditions described above offer a unique opportunity to examine the effects of a climatic change—in this case a transition from dry to wetter conditions, coupled with loess penetration, for two different pre-existing environments. The first is represented by the rocky Negev Highlands, where loess was deposited on previously rocky surfaces (Figure 11.11). The second is a sandy environment where loess was deposited on longitudinal dunes separated by wide inter-dune depressions (Figure 11.12).

ANALYSIS OF THE EFFECTS OF LOESS PENETRATION

Effects on Rocky Surfaces and Drainage Network Evolution

The relationship between annual rainfall, annual runoff and annual sediment yield is a frequently addressed topic in geomorphology (see reviews by Jansson, 1982; Walling & Kleo, 1979). Most authors assume that a positive relationship exists between these variables up to a threshold of about 300 mm of annual rainfall. The explanation proposed is that runoff energy available for erosion and sediment transport increases positively with the amount of annual rainfall up to 300 mm, at which point the vegetal cover increases infiltration and surface protection, thus limiting runoff and erosion processes.

Following the principles outlined above, a transition from a dry to a wetter period can be expected to increase runoff and erosion rates in areas where average annual rainfall is less than 300 mm. This outcome is not supported by the field evidence for the northern Negev (Figure 11.10). The deposition of a thick loess mantle on top of a conglomerate is enigmatic if one accepts that runoff and transporting energy are positively related to annual rainfall. The shift from the deposition of coarse to fine particles took place in the study area during a climate change for which an increase in precipitation has been assumed. Channels that were once able to transport coarse particles began to deposit, under wetter climatic conditions, fine silty and clayey particles. The deposition rate of fine grained materials was so rapid that entire valleys were completely buried under the loess mantle (Figure 11.11). Furthermore, the ensuing drastic decrease in the transporting capacity of the pre-existing drainage network led to a considerable decrease in the drainage density (Yair & Enzel, 1987). This apparent contradiction can be resolved if the conditions controlling runoff generation in arid rocky and soil-covered areas are well understood and if the change in the surface properties following the shift to wetter climatic conditions is considered. The runoff data presented earlier clearly show that runoff generation and runoff energy in the study area are strongly controlled by the extent of bare rocky surfaces.

Such data can be used to explain the decrease in runoff energy and transporting capacity of the drainage network following loess penetration into the northern Negev. Loess penetration during the wetter climatic period buried the rocky landscape with an extensive soil mantle and led to a drastic increase in infiltration and an accompanying decrease in runoff and erosion energy. Consequently, the removal of the incoming aeolian loess was inhibited and resulted in the general obliteration of the drainage network.

Soil Salinization Processes Induced by Loess Penetration

Soil-forming processes are strongly influenced by the amount of water available for weathering and leaching. Pedological models which relate annual rainfall to the intensity of leaching suggest that increasing annual rainfall leads to deeper leaching and thus to better conditions for biota (Figure 11.13). Such models are based on two assumptions. The first is that surface properties do not change significantly with increasing annual rainfall and the second assumes that precipitation represents the main source of water input into the soil. Both assumptions cannot be applied to the situation prevailing in the northern Negev, where runoff water seems to play a dominant role in water input into the soil and where surface properties are far from being uniform when passing from the rocky Negev Highlands to the loessial Beer Sheva plain.

On the basis of data presented, an alternative hypothesis is presented here. According to this hypothesis, water availability and therefore the depth of water infiltration and leaching are positively related, at the desert fringe, to the rock/soil ratio. Where this ratio is high, runoff from the upper rocky parts of the slopes will be high. Most of this water will infiltrate into the soil-covered part of the slopes, thus permitting a more efficient leaching of salts from the limited soil cover. Conversely, where this ratio is low (i.e. extensive soil cover), the positive effect of runoff water concentration from rocky areas will be inhibited, leading to limited leaching intensity.

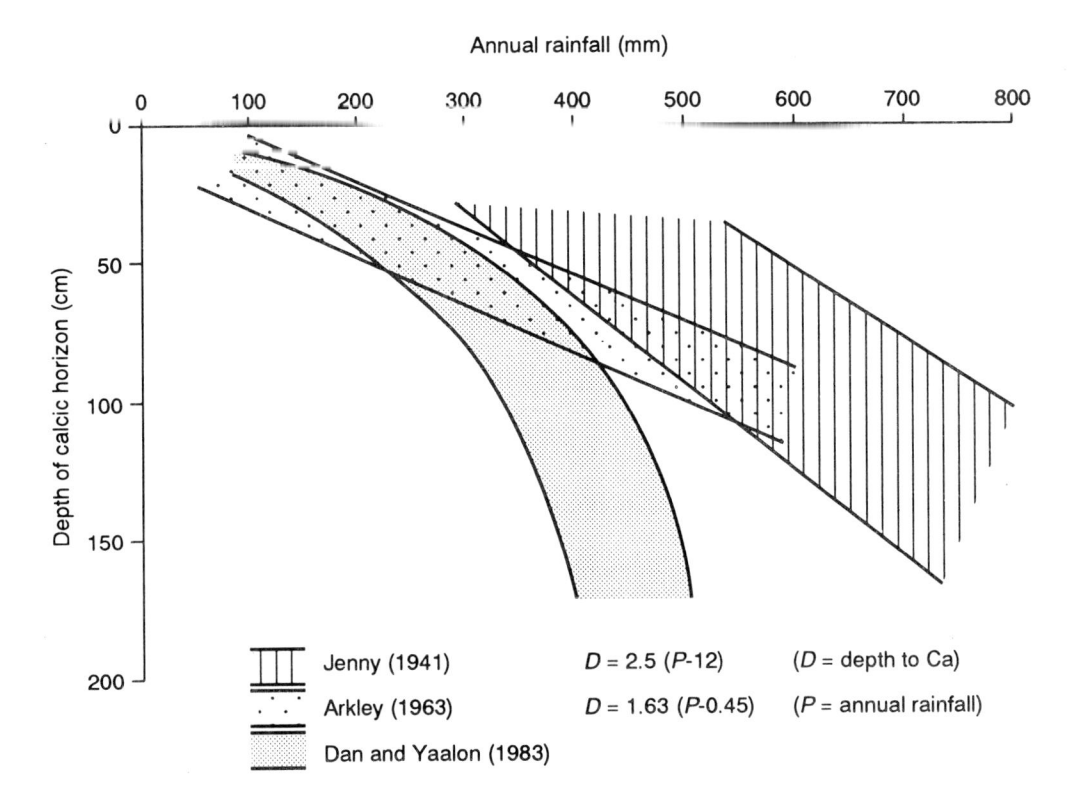

Jenny (1941)
Arkley (1963)
Dan and Yaalon (1983)

$D = 2.5 (P-12)$
$D = 1.63 (P-0.45)$

(D = depth to Ca)
(P = annual rainfall)

Figure 11.13 Relationship between annual rainfall and intensity of leaching processes

In this respect, one can advance the idea that the gradual covering of rocky surfaces with a soil material should result in a decrease in runoff with a parallel increase in aridity. The possibility of testing this hypothesis is provided by the natural conditions prevailing in the area. As stated earlier, loess material of aeolian origin started to penetrate in great amounts into the area, covering rocky surfaces devoid of any soil cover (Figure 11.11).

In order to test the above hypothesis, two hillslopes were selected (Wieder *et al.*, 1985). The first is located within the rocky Negev Highlands where average annual rainfall is 100 mm. The second is located at the southern margins of the Beer Sheva depression where the loess cover is extensive and average annual rainfall is about 180 mm. Both slopes are north-facing and are subdivided into an upper rocky and a lower colluvial section (Figure 11.14). The slope in the Negev Highlands is steeper than its counterpart to the north. The rocky slope at Sede Boqer is composed of massive limestone which forms a stepped topography. Smooth bedrock is exposed over 60–80% of the surface. The colluvial section is very stony, thickening in the downslope direction. The rocky slope at the Hovav site is composed of very densely jointed flinty chalk. The rock weathers into cobbles and gravels embedded in a quite contiguous, thin loess veneer. The colluvium is stoneless and thinner than that at Sede Boqer. Due to the proximity of the northern site to the Halutza sands (Figure 11.1), the sand content in the Hovav soils is higher than at Sede Boqer. Four pits, located at diagnostic positions, were dug down to the bedrock along each of the slopes (Figure 11.14). A detailed analysis of the data was reported in a recent paper (Yair, 1990).

The two toposequences presented display the same trends in terms of variation of soil properties along the slopes (Table 11.2). However, a striking difference is that the northern soils, located in a relatively wetter area, represent from a pedological point of view, an environment that is more saline than that of the climatologically drier Negev Highlands. The greater salinity of soils in wetter areas does not fit the pedological models which positively relate annual rainfall to the intensity of leaching. The alternative model, the rock/soil ratio effect, appears to be a better fit regarding water availability. The quick response to rainfall over the steep rocky slopes of the Negev Highlands results in the concentration of water from the large, rocky contributing area into a relatively limited adjoining soil-covered area. The process of water concentration leads to deeper wetting of the soil, down to a depth of 100 cm, beyond that allowed by direct rainfall. The deeply infiltrated water is an efficient agent in the evacuation of salts from the soil profile and, most importantly in the creation of a water reservoir at a depth little influenced by evaporation.

The opposite exists for areas with extensive soil cover. Because of the high absorption capacity of the soil, runoff frequency is low. When runoff does occur, its rate is also low. The positive effect of water concentration is inhibited, and the only source of water is often direct rainfall. Due to a high absorption capacity and compaction of the loess soils, the limited amounts of rainwater available at most rainstorms infiltrates to a shallow depth and are lost afterwards by evaporation during the lengthy intervals that separate consecutive storms. Under such conditions, the depth of water infiltration is limited to the top 20–40 cm (Figure 11.15), and leaching is confined to shallow depths, contributing to a gradual soil salinization process. The high evaporation losses from the whole wetted profile result in a low soil moisture content and an increased environmental aridity. An additional factor contributing to the high salt content in the northern area is the high salt input by rainfall. Studies of dissolved salts in rainwater show that concentrations are similar throughout the loess-covered areas and the Negev Highlands (Nativ *et al.*, 1983; Nativ & Mazor, 1987).

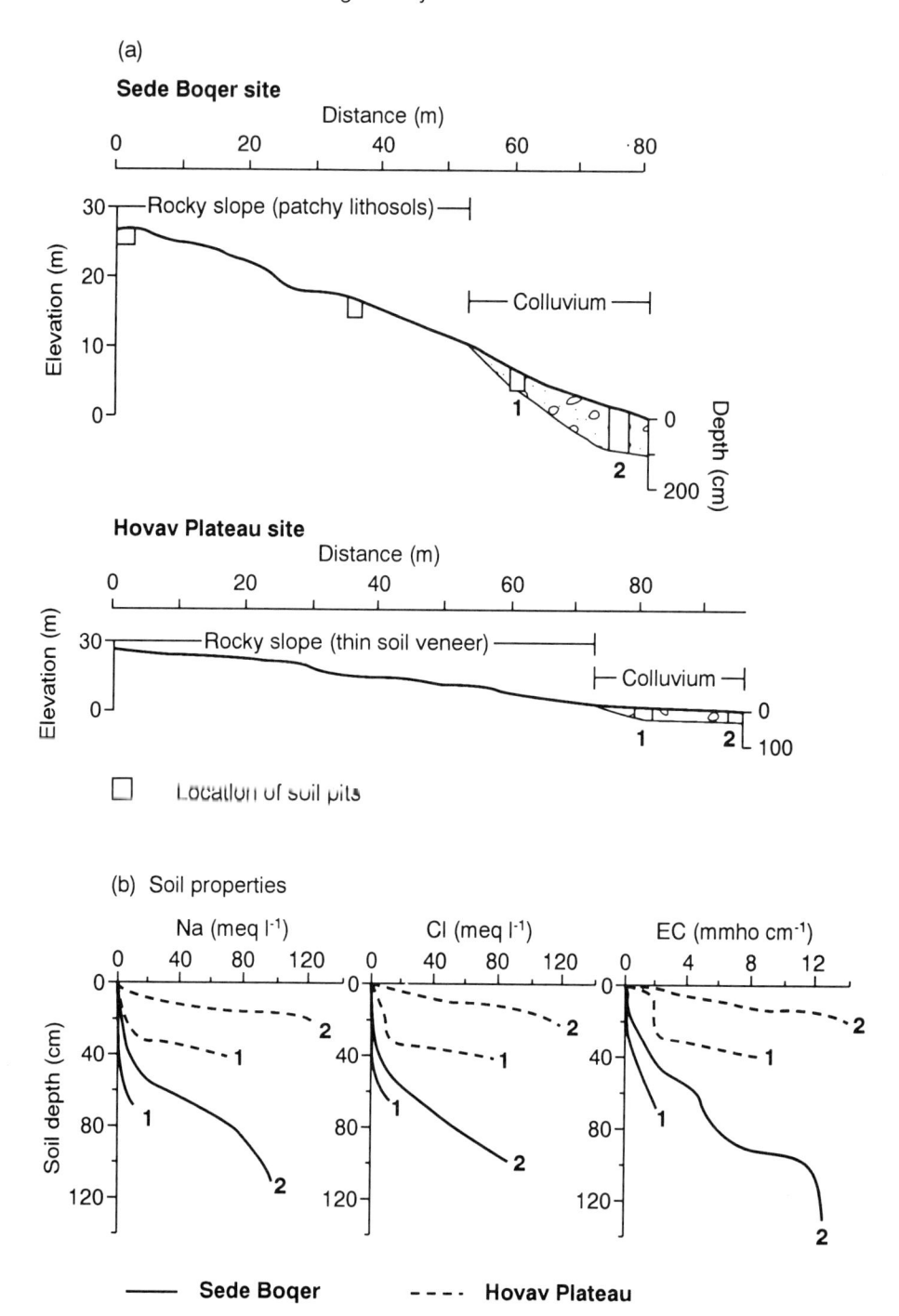

Figure 11.14 Soil properties along a rocky and a soil-covered slope. (Reproduced from Yair, 1983, by permission of Academic Press, London, Ltd)

The Ambiguous Impact of Climate Change at a Desert Fringe 217

Table 11.2 Soil Properties at Sede Boqer and at the Hovav Plateau (from Yair, 1983. Reproduced by permission of Academic Press, London, Ltd)

Location of soil pit	Depth (cm)	EC (25°) (mmho cm^{-1})	Na (meq l^{-1})	Cl (meq l^{-1})
Sede Boqer				
Top of slope	0–10	3.9	36.8	30.0
Rocky section	0–16	0.7	4.4	3.5
Upper colluvium	0–8	0.5	2.0	1.0
	8–20	0.4	2.1	1.1
	20–40	0.4	2.6	1.5
	40–65	1.4	11.3	13.0
Lower colluvium	0–12	0.3	2.0	1.0
	12–27	0.3	2.6	1.2
	27–43	1.1	10.1	7.5
	43–65	5.3	50.7	36.5
	65–80	6.1	68.1	49.5
	80–100	12.7	90.9	97.5
Hovav Plateau				
Top of slope	0–4	8.9	65.0	77.2
	4–16	19.8	187.5	176.5
	16–30	28.1	200.0	233.0
	30–54	26.0	225.0	229.6
Rocky section	1–10	1.7	8.4	9.9
Upper colluvium	0–4	2.0	11.3	11.0
	4–29	1.9	10.2	9.9
	29–42	8.2	71.1	77.6
Lower colluvium	0–6	3.2	16.8	24.0
	6–16	114.8	105.0	101.0
	16–24	13.9	125.0	120.0

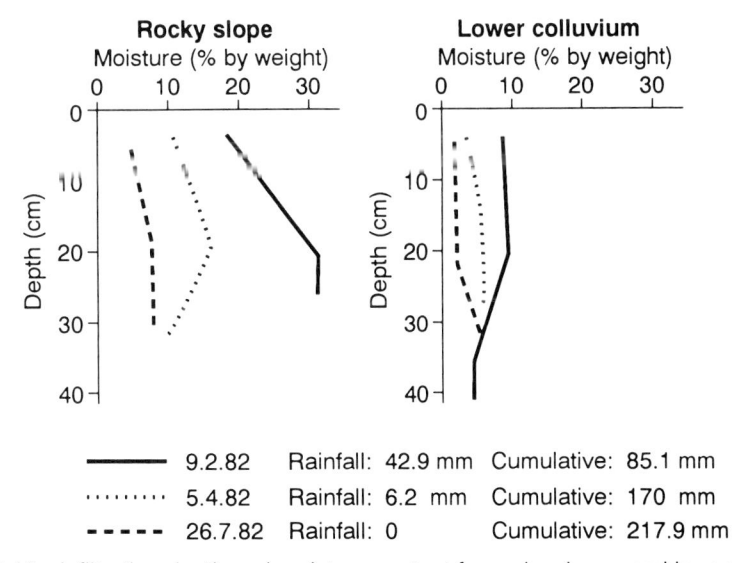

Figure 11.15 Infiltration depth and moisture content for rocky slopes and lower colluvium at the Hovav Plateau (from Kadmon et al., 1989. Reproduced by permission of Catena Verlag)

As rainfall totals are higher near Beer Sheva than near Sede Boqer, more salts accumulate in the former than in the latter area.

In order to better understand the significance of salt input by rainwater, it is worth while noting that the concentration of dissolved ions in rainwater in the study area varies from 50 to 300 mg l^{-1}. The average is about 100 mg l^{-1}, which represents 100 mg per mm of rain over 1 m^2. Some 20–25% of the dissolved ions are represented by NaCl (Yair et al., 1991). In an area receiving 200 mm annual rainfall, the net input of NaCl amounts to 4–5 g m^{-2} yr^{-1}, and over a period of 100 years is equivalent to 400–500 g. As most rainwater infiltrates to a shallow depth, the process of salt accumulation and soil salinization can be quite rapid. The process can be further enhanced by an increase in rainfall that is not accompanied by an important change in the depth of rainwater infiltration and by salts included in dry dust fall.

Effects of Loess Penetration on Biological Activity

The composition of the vegetal cover as well as that of snails and some burrowing animals is highly influenced by soil moisture and soil salinity in arid and semi-arid areas. Insofar as general principles are concerned, ecological models (Shmida, et al., 1986), in parallel with climatic and pedologic models, suggest a positive relationship between annual rainfall and various biological variables (Figure 11.16). The following tables (Tables 11.3 and 11.4) compare and contrast data obtained along the two slopes, at Sede Boqer and the Hovav Plateau, where the pedological studies were conducted. Table 11.3 presents some vegetation characteristics that can be related to the soil water regime and salinity. The relative importance of the perennial vegetation (expressed by cover or number of species) has a very strong positive correlation with the water regime in a desert environment. Similarly, the presence of species with Mediterranean chorotypes is indicative, in the same environment, of an improved water and low-salinity regime. In contrast, the dominance of Saharo Arabian species expresses real desert conditions (Danin et al., 1975; Yair & Danin, 1980).

An analysis of Table 11.3 highlights two main points:

1. The number of perennial species, as well as the water requirements of the species identified, are higher over the rocky than over the colluvial slope sections.
2. The most striking result is the difference in vegetation characteristics between the slopes. The vegetation cover, the number of woody plants and the relative importance of the Mediterranean chorotypes are much higher at the southern, climatologically drier area, and it is the northern, climatologically wetter area, that is dominated by Sahero-Arabian

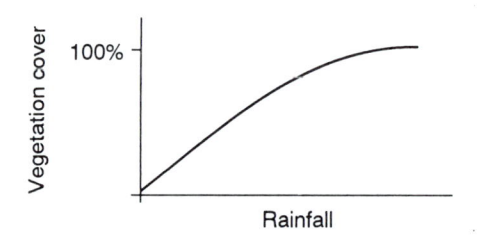

Figure 11.16 General relationships between vegetation cover and annual rainfall (modified from Shmida et al., 1986)

Table 11.3 Floristic comparisons between Sede Boqer and Hovav Plateau (from Kadmon et al., 1989. Reproduced by permission of Catena Verlag)

Parameter	Rocky sections		Colluvial sections	
	Sede Boqer	Hovav Plateau[h]	Sede Boqer	Hovav Plateau[h]
PVC (%)[a]	30[f]	10	10[f]	0
PS[b]	27[f]	11	n/d[i]	2
S–SS[c]	18[g]	7	14[g]	0
Smd[d]	22[g]	18	7[g]	0
Ssa[e]	33[g]	55	42[g]	100

[a] Perennial vegetation cover (%).
[b] Number of perennial species.
[c] Number of semi-shrub species.
[d] Percentage of perennial species belonging to the Mediterranean, Mediterranean Irano-Turanian, and Mediterranean Saharo-Arabian chorotypes (%).
[e] Percentage of perennial Saharo-Arabian species (%).
[f] After Yair & Danin (1980).
[g] After Olsvig-Whittaker et al. (1983).
[h] Hovav Plateau data after Kadmon et al. (1989).
[i] n/d = no data.

Table 11.4 Zoological comparisons between Sede Boqer and Hovav Plateau (from Yair & Shachak, 1987. Reproduced by permission of Rowman & Littlefield)

Species	Index	Sede Boqer	Ramat Hovav
Animal abundance (per 100 m²)			
Hemilepistus reaumuri	Families	25	1
Hystrix indica	Diggings	30	0.2
Trocoidea seetzeni	Individuals	260	6
Sphincterochila zonata	Individuals	20	20

Species		Sede Boqer	Ramat Hovav
Snail species richness			
Euchondrus albulus		+	
E. desertorum		+	
Sphincterochila zonata		+	+
S. prophetarum		+	
Trocoidea seetzeni		+	+
Granopupa granun		+	

species. This trend is especially pronounced over the colluvial slope sections of the two slopes (Kadmon et al., 1989).

Table 11.4 presents some zoological characteristics that are also related to water availability. Isopods (*Hemilepistus reaumuri*) can survive in the study area only at those sites where soil moisture at a depth of 40–50 cm is greater than 6% at the end of the hot and rainless summer. They are therefore more abundant in relatively wet sites. Similarly, porcupines (*Hystrix indica*) feed on bulbs of Mediterranean origin which are more abundant in wet and low-salinity sites. Once more, data obtained display a greater aridity in the northern, climatologically wetter area. Of special interest is the data concerning the snail species. Species diversity of snails is greater at Sede Boqer than at the Hovav Plateau. Furthermore, three out of the five snail species are of Mediterranean origin whereas those identified at the northern Hovav site, especially *Sphincterochila zonata*, are indicative of the driest conditions for snails (Yair & Shachak, 1987).

Application of Hydrological Results to Ancient Agricultural Systems

Extensive abandoned agricultural systems (Figure 11.17) have been found in the Negev Desert. According to archaeological findings (Aharoni *et al.*, 1960), some of the fields are more than 2500 years old. All researchers agree that the existence of sedentary agriculture in a desert area was made possible by ingenious water-harvesting techniques designed to collect overland flow from slopes adjoining the fields (Evenari *et al.*, 1982; Kedar, 1957; Yair, 1983). An examination of the distribution of ancient farms, compiled by Kedar (1957), reveals that most of the cultivated area is confined to the rocky Negev Highlands (Figure 11.18) where average annual rainfall is 75–100 mm. The highest density of small farms is found in the Avdat–Mitze Ramon area. The percentage of cultivated area decreases northward and is accompanied by increasing loess cover. Practically no vestiges of ancient runoff farms can be observed within the loessial rolling hills, though average annual rainfall is twice as much as in the Avdat area.

Figure 11.17 A sequence of ancient agricultural terraces on the Avdat Plateau

Figure 11.18 Areal distribution of ancient agricultural systems in the northern Negev

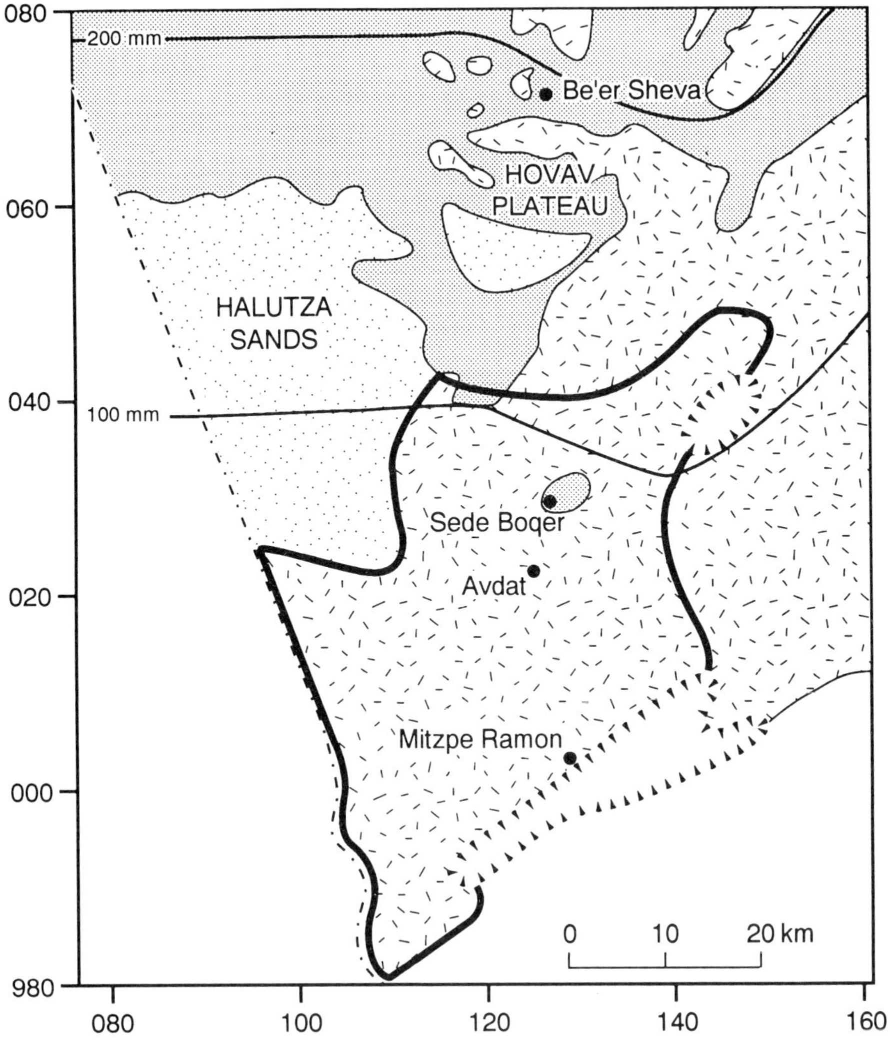
080
060
040
020
000
980
080
100
120
140
160
200 mm
100 mm
Be'er Sheva
HOVAV
PLATEAU
HALUTZA
SANDS
Sede Boqer
Avdat
Mitzpe Ramon
0
10
20 km

Loess cover

Sand dunes

Mainly rocky areas

Erosion cirques

Average annual rainfall (mm)

Limit of agricultural fields

The same pattern is encountered in the distribution of cisterns and springs. The cisterns represent an integral part of the ancient agricultural systems, with each farm unit having one or more cisterns. Some of the cisterns are located at a slight elevation above the main channel and fed by floodwater. The majority of cisterns, however, are found on the slopes where they were fed by conduits collecting hillslope runoff. All cisterns found in the Negev were carved by man in soft, chalky formations. These formations outcrop in the Negev Highlands as well as in the margins of the northern loessial plains. The loess area, despite the occurrence of soft chalk and relatively higher rainfall, has few cisterns whereas over 400 cisterns have been identified so far in the rocky Negev Highlands.

The hydrological processes discussed in the previous section can shed light on the spatial distribution of ancient agricultural systems. These processes stress the positive effect played by the rocky surfaces in arid areas (Yair, 1983). Rocky surfaces which respond quickly to slight rainstorms shed their water to adjoining, soil-covered, absorbing areas where the water volume collected and stored in the soil may be out of proportion to the annual rainfall. This process favours the creation of a locally improved environment. A converse situation exists in areas having an extensive soil cover. Because of the high absorption capacity of the soil, runoff frequency is low. When runoff does occur, the rate is also low. Under such conditions the positive effect of water concentration is inhibited and the only source of water is often only direct rainfall. The limited rainwater available during most rainstorms infiltrates to a shallow depth and is lost afterwards by evaporation during the lengthy interval that separates storms. A detailed mapping of hydrological installations reveals two facts (Evenari *et al.*, 1982; Kedar, 1957). The first is that the ratio of the runoff-contributing area to the cultivated areas was on average 20:1. The second is that most conduits collecting runoff extended over the rocky part of the slopes and not over the colluvial or alluvial soil-covered areas. The runoff coefficient of the contributing area was estimated as 15–20% of the annual rainfall. Under such conditions, each cultivated unit in the valley could be expected to receive an amount of runoff water equivalent to a rainfall of 300–400 mm, in addition to some 100 mm of direct rainfall. These amounts are sufficiently high to allow for deep wetting of the valley soils and to permit sedentary agriculture.

A completely different situation exists in the loess-covered area where, due to the high infiltration losses, the runoff coefficient is much lower than in the rocky areas (Yair, 1987). It is therefore clear that only a dissected area, with narrow valleys and steep rocky slopes, could offer conditions suitable for runoff farming in the desert. This combination does exist in the dissected plateaux of the Negev Highlands, but is absent in the loessial flat plains which extend south of Beer Sheva. Here, most of the area is covered by contiguous soil. The potential runoff-contributing areas—smooth, bare bedrock outcrops—are limited. In other words, the ratio of possible runoff-contributing area to cultivated area is far less than the minimum required for irrigated agriculture, thus explaining its absence from the northern, climatologically wetter area.

Effects on Sandy Surfaces

Hydrological Versus Aeolian Processes

The sedimentary evolution of the sandy area is displayed in Figure 11.12. It is characterized by alternating sandy and silty-clayey units. The horizontal and layered appearance of the fine-grained units point to water-lain deposits within the inter-dune corridors. It is possible to explain the fine-grained units by relating them to the effect of loess penetration into the area. The mantling of sand dunes by a thin loess cover can be expected to stabilize the dunes, thus enhancing the development of the biological topsoil crust. This crust, by limiting rainwater infiltration, plays a decisive role in runoff generation and in the transport of the fine-grained material from the dune into the inter-dune area where the material eroded from the dune is being deposited.

It appears that the mantling of the sandy area with a loess cover led to a basic change in the process shaping the landscape. The importance of aeolian processes decreased tremendously due to sand stabilization by the loess cover and the development of biological topsoil crusts. Simultaneously, the importance of runoff generation and erosion by runoff water greatly increased, resulting in the deposition of fine-grained units within the inter-dune corridors. A reverse trend can be expected when a transition to drier climatic conditions occurs. Loess input decreases while that of sand increases. Aeolian processes gain in importance and result in sand deposition within the inter-dune corridors.

Effect of Sand Penetration on Infiltration and Salinity

It was mentioned earlier that soil salinisation following loess penetration was accelerated due to the high water absorption capacity and restricted depth of water infiltration into the loess. The water that did penetrate was eventually lost by evaporation and led to salt accumulation at a shallow depth. This process was enhanced by an increase in salt input during a wet period. The opposite can be expected to occur during a drier period.

Whereas the depth of water infiltration into a loess-covered area is limited to 30–50 cm (Figure 11.19), it attains 150–300 cm in a sandy area (Figure 11.19). Sand penetration permits deeper water infiltration and salt leaching. As the capillary movement of moisture in sand is very low, there is in fact more water available for biological activity. This favourable setting is enhanced by the low input and the deep leaching of salts during a dry period.

SUMMARY AND CONCLUSIONS

The inter-disciplinary research presented in this paper covered climatological, hydrological, pedological and biological aspects. It combined the study of present day hydrological processes with the study of processes at longer timescales, during which time the area underwent several climatic changes. The results obtained draw attention to the complex relationships between climatic and environmental conditions at the desert fringe. The important points that deserve to be stressed are the following:

1. Climate is generally viewed as a zonal phenomena. This approach explains why climatologists contend that aridity is inversely proportional to annual precipitation and positively related to temperature. Similarly, ecologists and pedologists tend to assume that productivity, number of species and depth of leaching are positively related to annual rainfall. Such generalizations are certainly valid at the global scale when the whole of the arid and semi-arid areas are compared with more humid Mediterranean or tropical belts. However, when regional or local scales are considered, the relationship between climate and environmental conditions becomes problematic. Local factors such as topography, lithology and soils play a decisive role in the spatial redistribution of water resources—

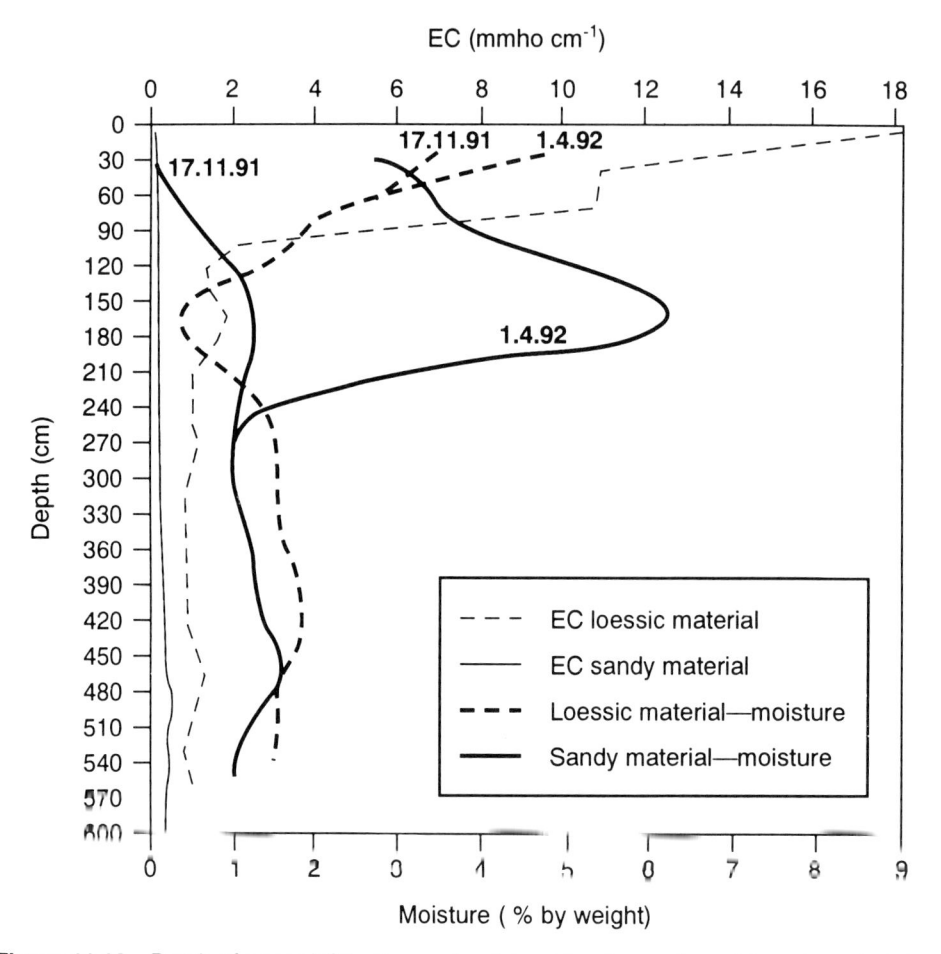

Figure 11.19 Depth of water infiltration and salt leaching in sandy and loessic materials at the Nizzana site

a role that in some cases is far more important than the absolute annual rain amount (Yair & Berkowicz, 1989). The influence of local factors can be expected to increase with decreasing annual rainfall.

2. The relationship between climate and environment in arid and semi-arid areas is even more problematic when climate change is considered. In the study areas, past climatic changes initiated, through aeolian inputs of loess or sand, an alteration in surface properties and in salt input. This paper has established that a period of increased rainfall can be accompanied by an increase in the extent of soil cover and, in parallel, an increase in salt input. The overall effect is an increased environmental aridity caused by the combined effect of reduced runoff, reduced water availability and increased soil salinity. An opposite effect can be expected during a drier period.

3. The data presented show that the effect of a climate change in such areas is highly controlled by the surface conditions prevailing in the area prior to climatic change. The covering of rocky surfaces by a porous loess cover with a higher water absorption capacity resulted in a drastic decrease in runoff and obliteration of the drainage network.

An opposite effect was observed over sandy areas. As the porosity of the fine-grained loess particles is lower than that of sand, the mantling of sandy areas by a thin loess cover led to runoff generation in an area that had previously not experienced any runoff.
4. Soil erosion is usually regarded as a negative process due to soil thinning and fertility losses that may affect crop yields. In rocky arid and semi-arid areas, however, where water availability is by far the controlling factor in agriculture, soil erosion can be regarded as beneficial. The removal of the soil uncovers the underlying bedrock, thus increasing the rock/soil ratio. Runoff is accordingly enhanced and can make the remaining soil cover more productive.
5. If one accepts the rock/soil ratio as an important component of available water in the desert, then it may also be used to explain landscape evolution on a geological timescale. A decrease in the rock/soil ratio would reduce runoff and the efficiency of sediment removal. Evidence for this can be found in the northern Negev during the late Pleistocene. Loess penetration into the area led, because of decreased runoff, to the burial and disintegration of the pre-existing, well-integrated drainage network (Yair & Enzel, 1987).
6. Climatologists use aridity indices to express the relationship between climatic variables and the environment. These indices tend to imply that the acuteness of aridity depends upon prevailing atmospheric conditions, such as annual precipitation, temperature and evaporation, and that aridity is inversely correlated to annual precipitation. This approach, correct as the global scale, does not appear to fit the relationship between climate and environment within arid and semi-arid areas. Here, more importance should be attributed to non-climatic factors such as surface properties that encourage or inhibit the positive effect of water concentration.

ACKNOWLEDGEMENTS

This study was conducted at the Nizzana and Sede Boqer Research stations operated by the Arid Ecosystems Research Centre (AERC) of the Hebrew University of Jerusalem. The financial support of the AERC is gratefully appreciated.

REFERENCES

Aharoni, Y., Evenari, M., Shahnan, L. and Tadmor, N. H. (1960). The ancient desert agriculture of the Negev. *Israel Exploration Journal*, **10**, 23–36 and 97–111.
Arkley, R. J. (1963). Calculation of carbonate and water movement in soil from climatic data. *Soil Science*, **96**, 239–48.
Bailey, H. P. (1979). Semi-arid climates: Their definition and distribution. In Hall, A. E., Cannell, G. H. and Lawson, H. W. (eds.) *Agriculture*.
Bruins, H. J. and Yaalon, D. H. (1979). The stratigraphy of the Netivot section in the desert loess of the Negev (Israel). *Acta Geologica Academiae Scientarum Hungaricae*, **22**, 161–4.
Coude-Gaussen, G., Olive, P. and Rognon, P. (1983). Datation de dépôts loessiques et variations climatiques a la bordure nord du Sahara Algéro-Tunisien. *Revue de Geologie Dynamique et Geographie Physique*, **24**, 61–73.
Dan, J. and Yaalon, D. H. (1982). Automorphine saline soils in Israel. *Catena Supplement*, **1**, 103–115.
Danin, A., Orshan, G. and Zohary, M. (1975). The vegetation of the northern Negev and the Judean desert of Israel. *Israel Journal of Botany*, **24**, 118–72.
de Martonne, E. (1926). Une nouvelle fonction climatologique: l'indice d'aridité. *La Meteorologie*, **2**, 449–58.

Evenari, M., Shanan, L. and Tadmor, N. H. (1982). *The Negev: The Challenge of a Desert*, 2nd edition, Harvard University Press, Cambridge, Mass.

Goldberg, P. (1981). Late Quaternary stratigraphy of Israel: an ecclectic view. *Colloques Internationaux du CNRS Prehistoire du Levant*, **598**, 58–66.

Goldberg, P. (1986). Late Quaternary environmental history of the southern Levant *Geoarcheology*, **1**, 225–44.

Goring-Morris, A. N. and Goldberg, P. (1990). Late Quaternary dune migration in the southern Levant: archeology, chronology, and paleoenvironments. *Quaternary International*, **5**, 115–37.

Heathcote, R. L. (1983). *The Arid Lands: Their Use and Abuse*, Longman, New York.

Issar, A. S. and Bruins, H. J. (1983). Special climatological conditions in the deserts of Sinai and the Negev during the latest Pleistocene. *Palaeogeography, Palaeoclimatology, Palaeoecology*, **43**, 63–72.

Jansson, M. B. (1982). Land erosion by water in different climates. Uppsala University, Report 57, Uppsala Sweden.

Jenny, H. (1941). *Factors of Soil Formation*. McGraw-Hill, New York.

Joseph, J. H. and Ganor, E. (1986). Variability of climatic boundaries in Israel—use of a modified Budyko-Lettsu acidity index. *Journal of Climatology*, **6**, 69–82.

Kadmon, R., Yair, A. and Danin, A. (1989). Relationships between soil properties, soil moisture and vegetation along loess-covered hillslopes, northern Negev, Israel. *Catena Supplement*, **14**, 43–57.

Kedar, Y. (1957). Water and soil from the Negev: some ancient achievements in the Central Negev. *Geographical Journal*, **123**, 179–89.

Köppen, W. (1931). *Die Klimate der Erde*, Berlin.

Kutiel, H. (1978). The distribution of rain intensities in Israel (in Hebrew). Unpublished M. Sc. thesis, Hebrew University of Jerusalem. Magaritz, M. (1986). Environmental changes in the Upper Pleistocene along the desert boundary, southern Israel. *Palaeogeography, Palaeoclimatology, Palaeoecology*, **53**, 213–29.

Magaritz, M. and Enzel, Y. (1990). Standing water deposits as indicators of late Quaternary dune migration in the northwestern Negev. *Climatic Change*, **16**, 307–18.

Nativ, R. and Mazor, E. (1987). Rain events in an arid environment. Their distribution and ionic and isotopic composition patterns, Maktesh Ramon Basin, Israel. *Journal of Hydrology*, **89**, 205–37.

Nativ, R., Issar, A. and Rutledge, J. (1983). Chemical composition of rainwater and floodwaters in the Negev Desert, Israel. *Journal of Hydrology*, **62**, 201–23.

Olsvig-Whittaker, L., Shachak, M. and Yair, A. (1983). Vegetation patterns related to environmental factors in a Negev desert watershed. *Vegetatio*, 54, 153–65.

Rendell, H. M., Yair, A. and Tsoar, H. (1993). Thermoluminescence dating of periods of sand movement and linear dune formation in the northern Negev, Israel. In Pye, K. (ed.), *The Dynamics and Environmental Context of Aeolian Sedimentary Systems*, Geological Society of London, Special Publication 77, pp. 69–74, Geological Society Publishing House, Bath.

Shmida, A., Evenari, M. and Noy-Meir, I. (1986). Hot desert ecosystems: an integrated view. In Evenari, M., Noy-Meir, I. and Goodall, D. W. (eds.), *Hot Desert and Arid Shrublands*, Elsevier, Amsterdam, pp. 379–88.

Stibbe, E. (1974). Hydrological balance of limans in the Negev. Project 304, Volcani Institute for Agricultural Research.

Thornthwaite, C. W. and Mather, J. R. (1955). The water balance. *Publications in Climatology* (Dexel Inst. of Technology), **8**, 1–104.

Walling, D. A. and Kleo, A. H. A. (1979). Sediment yields of rivers in areas of low precipitation—a global view. *International Association of Scientific Hydrologists Publication*, **128**, 479–92.

Wieder, M., Yair, A. and Arzi, A. (1985). Catenary relationships on arid hillslopes. *Catena Supplement*, **6**, 41–57.

Yaalon, D. H. (1962). On the origin and accumulation of salts in groundwater and soils of Israel. *Bulletin of the Research Council of Israel*, **11G**, 378–93.

Yair, A. (1983). Hillslope hydrology, water harvesting and areal distribution of some ancient agricultural systems, northern Negev, Israel. *Journal of Arid Environments*, **6**, 283–301.

Yair, A. (1987). Environmental effects of loess penetration into the northern Negev desert. *Journal of Arid Environments*, **13**, 9–24.

Yair, A. (1990). Runoff generation in a sandy area. The Nizzana sands, Western Negev, Israel. *Earth Surface Processes and Landforms,* **15**, 597–609.

Yair, A. and Berkowicz, S. M. (1989). Climatic and non-climatic controls of aridity: the case of the northern Negev of Israel. *Catena Supplement,* **14**, 145–58.

Yair, A. and Danin, A. (1980). Spatial variations in vegetation as related to the soil moisture regime over an arid limestone hillside, northern Negev, Israel. *Oecologia (Berl.),* **4**, 83–8.

Yair, A. and Enzel, Y. (1987). The relationship between annual rainfall and sediment yield in arid and semiarid areas. The case of the northern Negev. *Catena Supplement,* **10**, 121–35.

Yair, A. and Lavee, H. (1985). Runoff generation in arid and semiarid zones. In Anderson, M. G. and Burt, T. P. (eds.), *Hydrological Forecasting,* Wiley, Chichester, pp. 183–220.

Yair, A. and Shachak, M. (1987). Studies in watershed ecology of an arid area. In Berkovsky, L. and Wurtele, G. (eds.), *Progress in Desert Research,* Rowman and Littlefield, Totowa, New Jersey, pp. 145–93.

Yair, A., Karinieli, A. and Issar, A. (1991). The chemical composition of rainfall and runoff along arid hillslopes; northern Negev, Israel. *Journal of Hydrology,* **129**, 371–88.

12 The Environmental Consequences and Context of Ancient Floodwater Farming in the Tripolitanian Pre-Desert

D. D. GILBERTSON
Department of Archaeology and Prehistory, University of Sheffield, UK

C. O. HUNT
Department of Geographical and Environmental Sciences, University of Huddersfield, UK

N. R. J. FIELLER
Department of Probability and Statistics, University of Sheffield, UK

and

G. W. W. BARKER
School of Archaeological Studies, University of Leicester, UK

ABSTRACT

This paper reports on the biogeographical and fluvial geomorphological changes brought about by the adoption of floodwater farming in the Libyan pre-desert. Ancient floodwater farming systems, which were probably introduced in the first century AD, are described, particular attention being paid to their role in controlling runoff and in the crops that were grown. Floodwater farming profoundly changed the wadi floors and closed basins in the region. These changes have been examined by the textural and palynological analysis of sediments from the Grerat D'nar Salem. This analysis covers a range of archaeological, hydrological, sedimentological and palynological techniques.

INTRODUCTION

This paper has two related objectives. It summarises progress made in understanding the biogeography and geomorphology of past floodwater farming systems which were built on wadi floors and in plateau basins in one part of the desolate semi-arid rock desert, known as the Libyan Pre-Desert (Figure 12.1). Then it tries to place this understanding in the context of a longer picture of environmental change in the area that is derived from our initial studies of sediment cores from the infill sediments of a large shallow basin at Grerat D'nar Salem (Figure 12.1). To achieve these objectives the paper describes and integrates information from archaeological, hydrological, sedimentological and palynological studies

Environmental Change in Drylands: Biogeographical and Geomorphological Perspectives.
Edited by A. C. Millington and K. Pye. © 1994 John Wiley & Sons Ltd

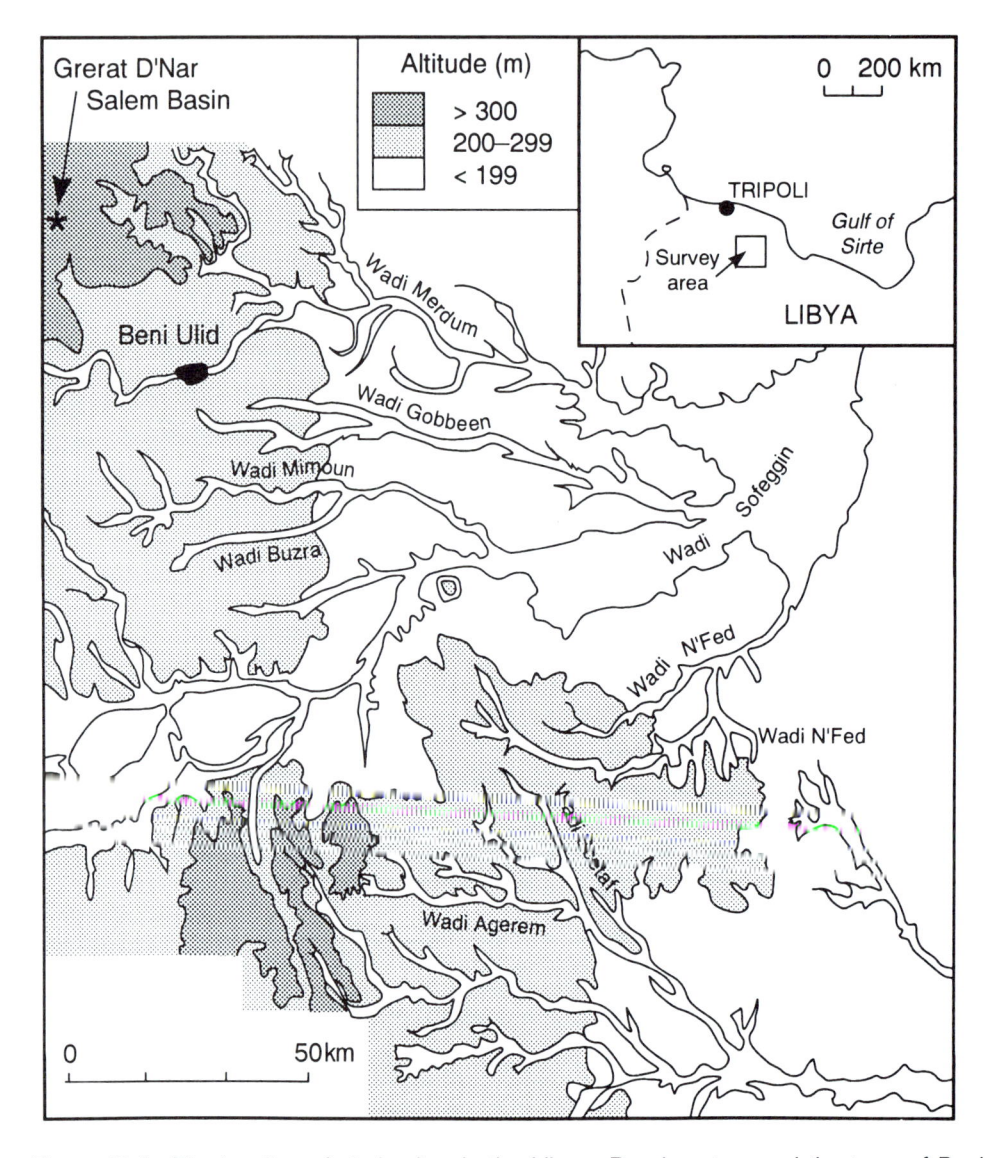

Figure 12.1 The location of study sites in the Libyan Pre-desert around the town of Beni Ulid. * Grerat D'nar Salem basin; contours at 200 and 300 m

derived from different types of sites and data sources around the pre-desert town of Beni Ulid (Figure 12.1).

The pre-desert is a broad zone whose northern boundary is between 100 and 500 km inland of the Mediterranean shoreline, and which merges further south into the 'true' desert of the Sahara in the modern state of Libya. This is an area of low and unreliable annual precipitation, which ranges from an average figure of approximately 70 mm at Grerat to less than 25 mm at the more southern floodwater farms such as Ghirza, some 300 km to the south-south-east (Brogan & Smith, 1984).

The introduction of floodwater farming in the region brought about a most profound environmental change on wadi floors and in basins on the plateaux. The farmers were sedentary local people who managed the floodwater produced by the one or two winter storms that occur (on average) each year in the Libyan Pre-desert. These people were 'floodwater farmers', using processes first described by Kirk Bryan (1929) from the American south-west. The ancient farming of semi-arid lands through harvesting rainwater is probably best known from the remarkable studies of Evenari *et al.* (1982) in the Negev, but it also occurs in one form or another in many parts of the semi-arid world (see papers and bibliographies in Bruins *et al.*, 1986; Gilbertson, 1986; and Reij *et al.*, 1988). Such ancient farming in Mediterranean or semi-arid lands is often viewed in one of two lights. It may be seen as a self-sustaining and a very desirable route to promote agricultural self-sufficiency in semi-arid areas. This is the thrust of the well-known work in the Negev and its extensions in various forms of runoff farming in other countries (Evenari *et al.*, 1982; Bruins *et al.*, 1986; Pacey with Cullis, 1986; and in the popular press, Connor, 1984). However, in other studies, sometimes more directly concerned with archaeological problems (typically framed in terms of 'the decline and/or collapse of civilisations *etc*'), agriculturally induced soil erosion, or the failure to maintain soil conservation measures such as terraces, are often postulated (e.g. Hughes & Thirgood, 1982) or shown in modern analogue (e.g. Ballais, 1990; Bonvallot, 1986) to have played a key role(s) in bringing about substantial economic or sociopolitical problems. The appropriateness of the explanation obviously depends upon the particular example and its context. However, behind these questions lies the issue of the extent to which ancient farmers were actually aware of the problems of erosion that they may have induced. Such questions were raised in the context of floodwater farming in the Libyan Valleys project by Gale & Hunt (1986), Gale *et al.* (1986) and Gilbertson & Hunt (1990).

All these issues have been addressed in the UNESCO Libyan Valleys Survey Project of the Universities of Leicester, Manchester and Sheffield, and the Department of Antiquities in Tripoli, the results of which have been progressively published over the last decade (of which the following are particularly important in the present context—Barker & Jones, 1984; Barker *et al.*, 1991; Dorsett *et al.*, 1984; Gale *et al.*, 1986; Gilbertson *et al.*, 1984; Hunt *et al.*, 1986, 1987; Laronde, 1986, 1989; Mattingly, 1988, 1989). Floodwater farming in the pre-desert was based upon the augmentation and management of stormwater using complex systems of walls whose location, construction and impact all merit detailed geomorphic and ecological investigations. Investigations of ceramic pottery assemblages recovered from the associated middens and more generally around the adjacent numerous ruined buildings have shown that the extensive and intensive use of the pre-desert occurred in the period from the first to the fifth century AD (by people who appear to have spoken Punic and are referred to as Romano-Libyan, a term which is used in the same sense as the more familiar Romano-British) and that this type of occupation and agriculture continued through subsequent periods to dates as late as the fifteenth or sixteenth century AD (Barker & Jones, 1981; Dore & Van der Veen, 1986; Gilbertson & Hunt, 1988). Away from the modern Libyan towns such as Beni Ulid (Figure 12.1), the contrast between the abundance and apparent richness of the modern and 'ancient' human landscape is very apparent. The life and habitation of the modern sheep and goat herder in a comparatively empty rock desert contrast vividly with the abundance and quality of buildings inherited from the work of its earlier occupants—these structures can be located over thousands of square miles of present rock desert.

ANCIENT FLOODWATER-BASED AGRICULTURE

The most obvious biogeographic impacts of these floodwater farmers are of course the direct and indirect biological transformations achieved by the development of the substantial arable and pastoral husbandry that developed in the region. At present, this is principally understood from detailed palaeoeconomic examination of the well-dated midden deposits associated with these settlements and farms in the region. The results of these studies are summarised in Table 12.1, and are described in more detail in Brogan & Smith (1984), Clark (1986) and Van der Veen (1981, 1985). These indicate the presence of a productive and mixed agricultural economy in which the clearest local variant was the increasing use of hunted-wild animal meat, such as gazelle, with increasing passage into the desert to the south. Unfortunately, these material-rich middens are only rarely inter-bedded with other deposits or features that would enable the information they yield to be integrated into broader pictures of biogeographic and geomorphic variations in space or through time.

At present detailed reconstructions of how and where the different crops were grown and animals managed depend upon three lines of evidence. The first is an observed, if unquantified, general association between the presence of olive or grape presses in the study area and the presence of low-angle alluvial fans descending to the wadi edge from the plateau.

Table 12.1 Farm products of the Tripolitanian pre-desert 1st to 15th Century ad. There were no fundamental differences between the agricultural economies of the Romano–Libyan and Islamic periods. Hunted animals become more important further south

	Century $_{ad}$ 1–5	10–16	
Plant crops			
Hordeum vulgare (hulled six rowed barley)	*	*	
Triticum (wheat)	*	*	
Pisum sativum (field pea)	*	*	
Lens culinaris (lentil)	*	*	
other pulses	*	*	
Ficus carica (fig)	*	*	
Vitis vinifera (grape)	*	*	
Phoenix dactylifera (date palm)	*	*	
Olea europea (olive)	*	*	
Prunus amygdalus (almond)	*	*	
Pistacia atlantica (wild pistacia)	*		
Animal products			
sheep/goat	*	*	dominant
gazelle	*	*	
bovid	*	*	
pig	*		
canid	*	*	
camel	*	*	
hare/rabbit	*	*	
equid	*	*	
antelope	*	*	

This suggests that the colluvium and fan conglomerates of many (but certainly not all) such alluvial fans may have been used for the production of these crops (Gilbertson *et al.*, 1984; Gilbertson & Hunt, 1988).

Walls, Erosion and Agriculture

The second approach has been to examine the nature of the floodwater management systems and their relationships with:

(i) the modern tree and cereal crop production in the wadi floors and
(ii) the observable patterns of erosion and deposition that are produced.

Figure 12.2 illustrates a typical floodwater farming system, in this case at the confluence of a number of wadis in the Wadi Gobbeen approximately 20 km south-east of the town of Beni Ulid (see Figures 12.1 and 12. 2, and Gilbertson *et al.*, 1984, for a full description). Although not all the walls located in the pre-desert are concerned with the management of floodwater (Gilbertson *et al.*, 1984; Barker *et al.*, 1991), this is the case in the present example. The operation of the system is readily understood by following the course of water through a storm episode. Intense rain (typically) in the one or two winter storms soon produces overland flow which runs down the gentle (1–2°) gradients of the basins on the plateaux. The water is intercepted by carefully laid-out lines of boulders forming walls 0.3–0.5 m tall which lead the runoff towards the wadi edge. Sometimes, these plateaux walls are fronted by an excavated channel, 2–3 m wide. This is the case with wall 1250 (illustrated in Barker & Jones, 1982, Fig. 16), which is unusual in that both wall and channel follow a route that has been excavated across a watershed. The wall and channel open on to a plateau-edge basin at which water is spread by a series of oblique walls with the effect that sediment is deposited there. Palaeohydrological estimates by Gale *et al.* (1986) indicate that the channel would have exceeded its capacity during storms in which greater than 25 mm of rainfall fell in one hour. The combination of the form of the system and the presence of numerous huts (1203) at the borders of the depression lead to the conclusion that crops of some type were being produced in the area. There is only very limited information if this was the case. The only relevant evidence is the discovery of Hunt *et al.* (1987) in the water conduits of Gasr Mm10 in nearby Wadi Mimoun (Figure 12.1) of the seeds and pollen of edible plants, especially cereals and grasses in sediments that must have been swept from such plateaux areas. Unfortunately this plateau-edge taphonomic situation is complex and must include problems introduced by the possible location of ancient crop-processing activities outside the large gasr, as well as the more familiar fluvial and aeolian taphonomic processes. All such plateau water is then directed down the steep wadi-edge cliff. At many other floodwater farms this wadi-edge water is directed through a series of water-storage cisterns or more probably via 1–2 m deep boulder-built ponds in wadi-edge gullies. These are seen to fill rapidly with sediments in a storm. These ponds contain wet sediment which may also have been of agricultural use, perhaps once functioning as a well-irrigated garden. Whenever the dip of the bedrock in the region (typically thinly bedded porcellaneous Cenozoic or Cretaceous limestones) at the wadi cliff is *into* the wadi, there are further walls (1207). These may run obliquely down that cliff edge, taking water into cisterns at the edge of the wadi floor, or (more commonly) they may direct water on to the wadi floor. It is important to note that they commonly introduce this water to the wadi floor at entrances that are located between one and ten metres downstream of a cross-

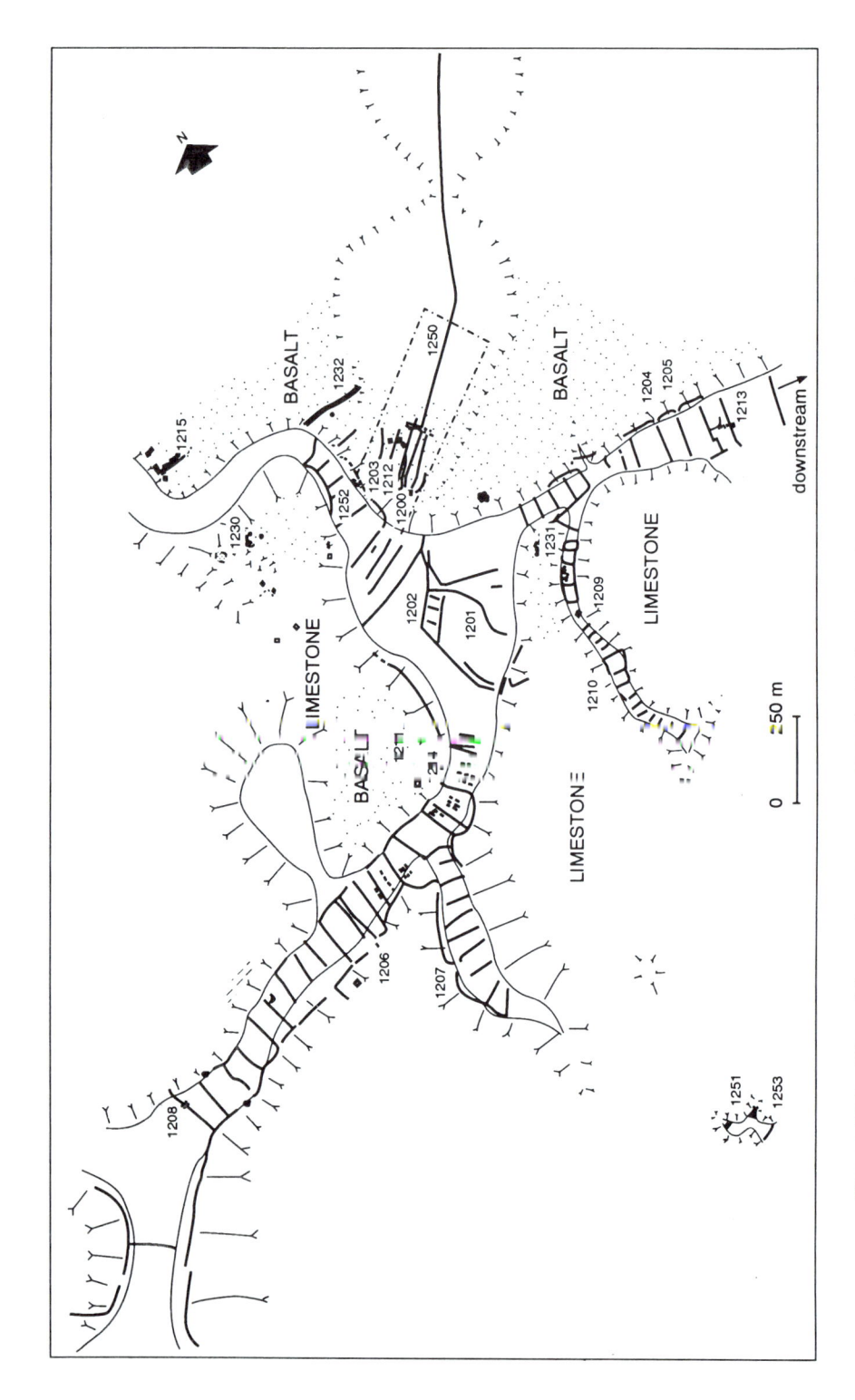

Figure 12.2 Plan of the wall systems associated with floodwater farming in the upper Wadi Gobbeen

wadi wall—never immediately upstream of a wadi wall (e.g. 1200). The wadi floors are characterised by numerous types of walls, of which the most common cross-wadi wall is a simple check-dam (1252), its 0.5–1 m height being made up of three or four courses of boulders. Confluences of large wadis are associated with much larger barrages which can be 1 km long, of up to 4 m wide and 3 m high (1202 is a smaller example), and sometimes built carefully with shaped and faced masonry. Boulder-built sluices or overflows, as extensively described from the Negev (Evenari *et al.*, 1982), are common. They serve to diminish soil erosion associated with waterfall effects, when and if floodwaters threaten to overtop a wall. Less common are large walls which run obliquely across the floodplain (parts of 1202). These appear to have a variety of purposes. Some appear to be protecting areas by diverting storm flow; typically they occur at confluences of wadis. At other sites, walls on the floodplain appear to serve as a means for abstracting water from the wadi floor and directing it into cisterns (via sediment traps) or into wadi-edge field systems fed by these diversion channels (see Figure 4 in Gilbertson *et al.*, 1984)—again parallels exist in the Mahgreb (El-Amani, 1983), in the Negev (Evenari *et al.*, 1982) and the American southwest and elsewhere (*e. g.* Bruins *et al.*, 1986; Nabhan, 1986a, 1986b).

In the pre-desert, three modern analogue situations were encountered which appeared to offer insights into the spatial aspects of the former agricultural system and its relationships between floods, walls, tree and cereal growth, and erosion and deposition. The cultivation of barley using floodwater farming techniques appears to be particularly successful in modern small side wadis which were not subject to significant gully formation (Figure 12.3). Secondly, it is obvious that olives and date palms can withstand significant floods—for example the floods responsible for erosion in Figure 12.4 reached halfway up the trunks of the olives. Similarly, these trees are clearly able to tolerate the gullying and instability that was generated by the interactions between the flood, the trunks and roots. Thirdly, a close examination of the inter-relationships between any of the cross-wadi walls in the main wadis (see Figure 12.5) reveals several features of importance. The upstream side (to the right) of the cross-wadi wall is characterised by ponded water and flat accumulating surfaces of fine-grained sediment, both of which have contributed to a relatively lush and even growth of pasture grasses in the illustration. The downstream side is different. The much greater proportions of undulating bare ground and linear spreads of boulders from damaged walls emphasise the numerous erosive gullies that have developed as a result of the 'waterfall' and sapping effects associated with the walls. Gullying immediately downstream of the cross-wadi walls is always promoted by the larger plateaux and wadi-edge wall systems which (almost) invariably introduce large quantities of high-velocity water to the wadi floor at this location (Figure 12.2). All of these features can be combined by juxtaposing Figures 12.3 and 12.4 on either side of a cross-wadi wall to produce the theoretical model shown in Figure 12.6, which suggests how floodwater farming might have been used to grow a mix of plant and animal products on the broader wadi floors to which the vast majority of floodwater was directed.

This model implies not only a sophisticated understanding of runoff hydrology (a recognition that is widespread in the literature on ancient floodwater management) but also points to both the understanding and manipulation of the erosional and depositional results of concentrating floodwater. Another consequence is the implication that relatively few of the 'simple' cross-wadi walls ought to have been concerned with the control of stock—especially sheep and goats. This prediction is difficult to test with the present palaeoecologi-

Figure 12.3 Modern cultivation of barley using floodwater farming techniques in a small side wadi in Wadi Merdum, near Beni Ulid, photographed one day after the same November storm responsible for the erosion in Figures 12.4 and 12.5

Figure 12.4 A plantation of olives and date palms immediately after a November rainstorm, indicating the extent to which these tree crops can withstand the erosional consequences of their cultivation in the Wadi Merdum at Beni Ulid

Figure 12.5 Floodwater interacting with a standard cross-wadi wall leads to the deposition of fine-grained sediment upstream of the wall (to the right of the wall). Gullies are produced downstream by waterfall and spring-sapping effects, which are frequently augmented by the introduction to the area immediately downstream of a wall of high-velocity plateau water by feeder walls on sloping ground of the wadi edge

cal evidence, although control is carried out in much of the world by active management by people, rather than by the passive management of enclosure boundaries.

Further detailed examination of Figure 12.5 indicates that a line of shrubs is associated with the lee of the cross-wadi wall. This is a very common feature throughout the area (e.g. see Figs. 8, 9 and 11 in Barker & Jones, 1982). The walls trap seed, enlarge the sub-surface seed bank, offer greater protection from insolation, grazing, disturbance by ploughing and subsequent erosion, as well as providing enhanced soil moisture levels. Cross-wadi walls that have become totally buried by mobile wadi-floor sediment may be readily detected in the field by the lines of trees and shrubs evident on the modern floodplain. In hydrological and geomorphological senses, the floodwater farming systems in the Libyan Pre-desert still function. As Figure 12.5 shows, they are one of the prime environmental agencies influencing the distribution of vegetation and the productivity of the plants and animals on the modern wadi floor, and determining the topography of the land surface with which they are associated.

ENVIRONMENTAL HISTORY OF THE LIBYAN VALLEYS PROJECT AREA: THE CORE FROM GRERAT D'NAR SALEM

Little is known of the more general Holocene vegetational or environmental history of the Libyan Pre-desert, before, during or after the major period of development of the floodwater

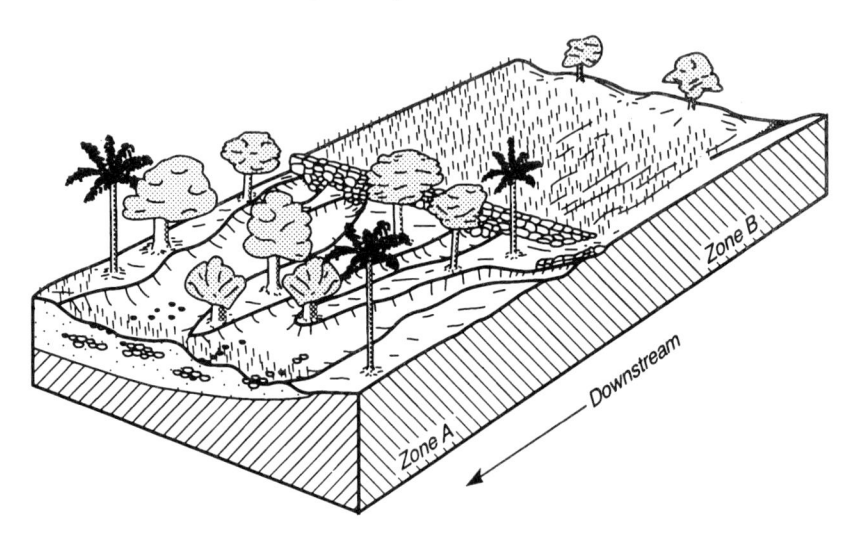

Figure 12.6 A sketch indicating the general characteristics of a hypothesised pattern of mixed agricultural land use that might have occurred within a sector of wadi floor characterised by cross-wadi walls. Water flow is from right to left. In Zone A, upstream of the cross-wadi wall ponded water leads to deposition and higher soil moisture contents favouring the growth of cereals, perhaps pasture crops, and providing stubble for autumn/winter grazing after harvesting. In Zone B, the effects of the tree crops, waterfalls and sapping combine with the erosive power of plateau water introduced immediately downstream of the cross-wadi wall to create a gullied landscape which is tolerable to tree crops. Shade and some pasture grasses are created for animals grazing in the summer. Animal excrement forms a nutrient input into the wadi floor sediments/soils

farms. In the last decade complex sequences of presumed Holocene superficial deposits have been recognised in this part of the pre-desert by Barker et al. (1983), Gilbertson and Hunt (1988), Gilbertson et al. (1984, 1987), Hunt et al. (1986, 1987), with an attempted lithostratigraphic correlation with coastal sequences being published by Anketell (1989) and Anketell et al. (in press). Unfortunately, at present, this work lacks chronostratigraphic precision and is not easily applied to questions of climatic change or human impact for the middle and late Holocene. Elsewhere in North Africa, more robust evidence has been obtained from a series of palaeoecological, stratigraphical and archaeological studies (for instance Maley, 1977; Sarnthein, 1978; Rossignol-Strick & Duzer, 1979; Pachur, 1980; Adamson et al., 1980; Rognon, 1980; Coude-Gaussen & Rognon, 1988; Petit-Maire et al., 1991). These studies suggest that, in general, climates during the earlier Holocene were moister than those which prevail today, although Schulz (1980) questioned the validity of some of the earlier interpretations of palaeobiological evidence for changes in humidity. This work suggests that many areas that are now desert have previously been relatively well-vegetated steppe grassland which was used by hunter-gatherers and then pastoral agriculturalists. This humid phase ended during the period 6000 to 3000 yr BP and the climate became more arid. The opportunity to place the present research on floodwater farming in the Libyan Pre-desert into the context of this broader picture of environmental change arose when a >2 m core was obtained with an Eijkelkamp auger from the sandy infill sediments in the 2–3 km broad, shallow limestone basin of Grerat D'nar Salem. This appears to be a large polje on a gently undulating plateau north-west of Beni Ulid (Figure 12.1). At present the area is used to grow barley, is ploughed to a depth of 0.2 or 0.3 m and is subject to occasional flooding. Unpublished surveys have shown that the margins of the basin were relatively densely occupied in Romano-Libyan times. Numerous contemporaneous buildings with olive presses, as well as walls, check-dams and cisterns associated with intensive and widespread floodwater farming, have been located (unpublished information).

The sediments obtained from the core were subjected to three types of analysis. The first was an attempt to use thermoluminescence-dating techniques to date important parts of the sediment sequence in the core. Unfortunately this approach failed. The second was to obtain simple chemical measurements and textural data to explore the geomorphic history of the site, and the third was to investigate the palynology of the core.

Stratigraphy of the Core

Two sedimentary units were observed in the field. The upper part of the core, unit 1, above 1.05 m, comprised light red-brown fine sands and silty sand, which had been disturbed by ploughing to a depth of 0.3–0.35 m. Unit 2 lay below this at a depth of 1.05 m to the base of the core at 2.10 m. The sediment was light red-brown sand with frequent small pebbles of the local limestone. Sample selection for sedimentological and pollen analysis was influenced by the gravelly nature of some parts of the core which prevented a fully regular sampling pattern.

Sedimentological Studies

The textural data were obtained using standard procedures by dry-sieving using 0.25 phi sieve intervals. The mass–size data were analysed by fitting log skew Laplace distributions—a procedure described originally in Fieller et al. (1984) and which has now been usefully applied to several data sets of terrestrial sands from semi-arid Africa (Fieller et al., 1990a, 1990b; Flenley et al., 1987). The results are summarised in Figures 12.7 and 12.8. The critical textural properties of the two stratigraphic units recognised in the field are readily understood from Figure 12.7. This displays simple plots of the weight of sand caught on each of the sieves, for samples from the base, mid-point and near-surface part of the Grerat core, together with a reference sample from modern climbing dunes accumulating at the side of the Wadi Merdum, some 15 km north of Beni Ulid (Figure 12.1, and see Flenley et al., 1987 and Gilbertson & Hunt, 1988, for full description of the samples and site). The basal sample at Grerat (#244) is made up of a wide range of fine to coarse sand and gravel-size particles; the material is coarse and poorly sorted, probably reflecting the transportation and mixing of a range of superficial materials by overland flow. Well-sorted fine sands typical of the wind-blown deposits in the region (#149 Beni Ulid) are totally missing in the samples from the lowest levels at Grerat. However, with passage upwards they become progressively more evident, especially above a depth of 1 m (e.g. #227), where the samples resemble the modern surface sands (#212). Figure 12.7 shows that at 1 m depth the poorly sorted coarser materials are now accompanied by a distinct windblown input of well-sorted, fine sands, which are also identical in size range to those currently accumulating as climbing dunes in the Wadi Merdum (#149, Figure 12.7). This evidence suggests that the sediments in the lowest part of the core accumulated in a depositional environment in which overland flow (perhaps floods) reworked a range of surface materials. However, the environment became increasingly influenced by aeolian processes, in addition to floods, and

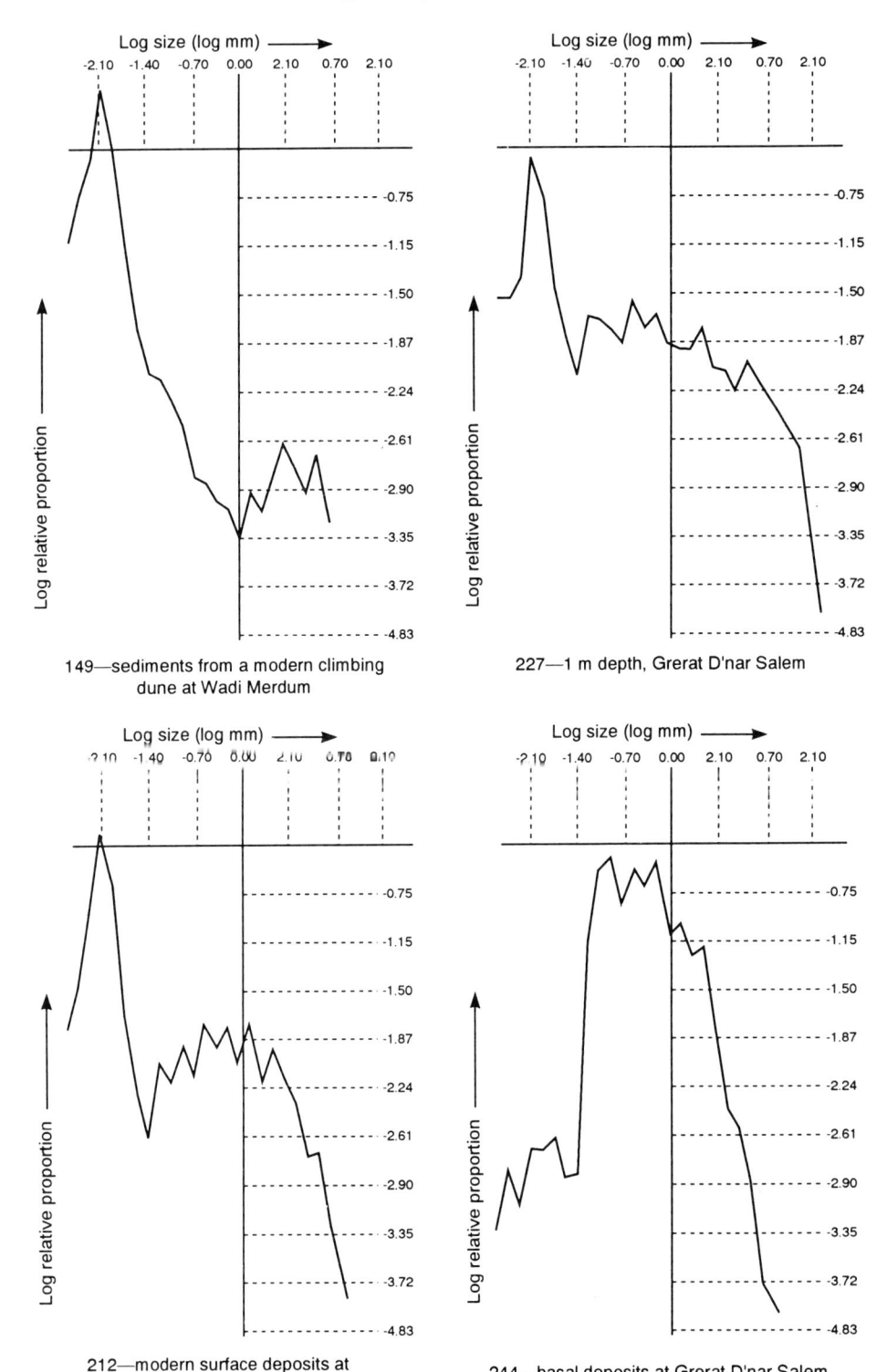

Environmental Change in Drylands

149—sediments from a modern climbing dune at Wadi Merdum

227—1 m depth, Grerat D'nar Salem

212—modern surface deposits at Grerat D'nar Salem

244—basal deposits at Grerat D'nar Salem

this mix of processes, essentially similar to those that prevail today, were well established by the period of time (as yet unknown) represented by the sediments between 1.3 and 1.1 m depth in the core. This changing mix is clearly identified in all the parameter plots (unpublished data), as illustrated by the simple depth plot of values of the principal modal grain size parameter μ which was derived from this statistical modelling of the grain size distributions (Figure 12.8).

Other aspects of the changing depositional environment are indicated by loss-on-ignition (LOI) values (after Ball, 1974) down the core. The uppermost deposits currently support crops of barley and are shown in Figure 12.8 to be low in oxidisable organic matter. The situation is essentially the same until unit 2 is reached, just below 1 m depth, which is seen as a unit to be notably richer in oxidisable organic matter, especially from 1.5 to 1.9 m. This suggests that significantly higher levels of biological activity were taking place when these materials were being deposited. Unit 2 can therefore be envisaged accumulating in wetter, more biologically productive conditions in the basin; the textural and LOI data suggest that unit 1 accumulated in a depositional environment akin to that which prevails today—characterised by aeolian as well as flood processes. In contrast the values for sediment conductivity obtained (see Gilbertson et al., in press for details) do not correspond to the stratigraphic or textural properties noted in the core. Below 0.85 m, there is a substantial increase in conductivity, pointing to the presence of a significant salt content in the lower parts of the sediment column. Salt-rich bedrocks, especially the limestones and anhydrites, are common in the area. They are associated with karstic features (Hunt et al., 1985) and are likely to be responsible for much of the salt in these sediments. This salt may also have agricultural significance because a farming technique involving the continued artificial augmentation of the quantity of floodwater reaching the area is likely to raise the level of a salt-rich groundwater to the level of the roots of tree, cereal or pasture crops.

Pollen Analysis

Samples of 5 ml were prepared by boiling in 5% KOH, then swirling the residue on a clock glass, following the procedures described in Hunt (1985). In the upper part of the core (above 0.8 m), pollen was abundant and aliquots of the residues, usually containing 200–300 grains, were counted. Below this depth, pollen was rare and all the pollen in the sample was counted. The results are shown graphically in Figure 12.9 where pollen counts are shown as bar graphs on the pollen diagram where the total pollen count was over 150 grains.

The pollen was variably preserved, with most of the samples showing signs of compression, corrosion and thinning, except for the topmost two samples which contained well-preserved pollen. Virtually every sample had slightly different preservation characteristics, and this, together with the differing pollen counts down the core, was taken as evidence for a stratified and substantially undisturbed sequence, which has not been subject to significant down-washing of pollen grains. In addition to the pollen, some samples contained organic-walled algal microfossils and a few contained diatoms and sponge spicules. Most samples from unit 1 contained very abundant fungal remains, principally spores and vesicu-

---

Figure 12.7 Plots of the logarithm of the weight of sediment against the logarithm of the sieve mesh for sediments from: #244 = basal deposits at Grerat D'nar Salem; #227 = deposits at 1 m depth at Grerat D'nar Salem; #212 = modern surface deposits at Grerat D'nar Salem; #149 = sediments from modern climbing dunes in Wadi Merdum

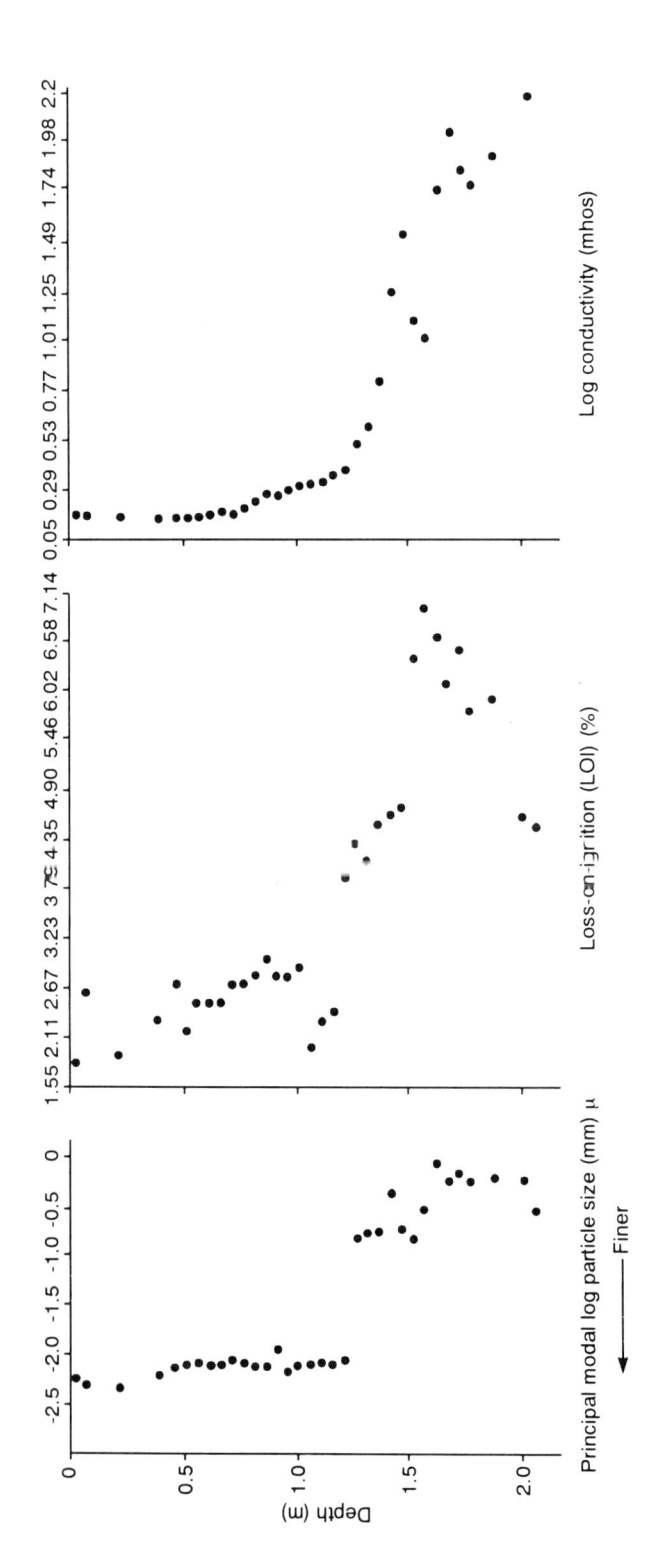

Figure 12.8 Sedimentological changes down a 2-m core from the alluvial basin at Grerat D'nar Salem

lar arbuscular micorrhyza. Fungal remains were infrequent in unit 2. The pollen assemblages in the core were variable, but could be divided into three assemblage biozones, which are described below.

Assemblage Biozone GR-A (Pollen Rare, Mostly Gramineae), 1.80–1.05 m depth

Only half of the samples in this biozone contained pollen. Where pollen was present, the biozone was characterized by extremely sparse, poorly preserved assemblages, usually containing at least 10%, and often as much as 40%, Gramineae, the remainder usually being herb pollen. Organic-walled microfossils, mostly Zygnemataceae, occurred in a few samples between 1.80 and 1.65 m and siliceous microfossils were detected in the same interval, and also between 1.25 and 1.10 m.

Assemblage Biozone GR-B (Gramineae–Compositae), 1.05–0.60 m depth

The biozone was characterized by samples containing at least 10% and often over 20% Gramineae and usually over 20% Compositae. *Artemisia* and Chenopodiaceae were usually present and Polypodiaceae, Polygonaceae, *Pistacia*, *Pinus* and *Quercus* were often noted. Organic-walled microfossils, mostly *Sigmopollis*, were often identified.

Assemblage Biozone GR-C (Compositae–Gramineae–Artemisia), 0.60–0 m depth

The biozone was characterized by samples containing over 30% Compositae, and usually 5–10% Gramineae and 5–10% *Artemisia*. Chenopodiaceae, *Plantago*, Cruciferae, Labiatae, *Ephedra* and *Pinus* were usually present. There was a small peak of *Olea* around 0.42 cm. Organic-walled microfossils, mostly *Sigmopollis*, were recorded in the lower half of the biozone.

Interpretation

The interpretation of 'non-standard' palynological sites such as this is difficult because unusual taphonomic pathways and preservational mechanisms must be considered. At present it is impossible to identify or quantify the importance of all the taphonomic and preservational mechanisms or artefacts that might be represented by these samples from Grerat D'nar Salem. Nevertheless, some broad interpretative statements can be advanced.

Assemblage Biozone GR-A

Pollen is very scarce in this biozone, so interpretation can be no more than tentative. The high incidence of Gramineae and other herbaceous pollen is perhaps indicative of an open, steppe-like landscape. The presence of centric diatoms in some samples might be interpreted as evidence for the intermittent occurrence of a water-body several metres deep persisting for more than several months. Centric diatoms are planktonic and floras typically take some months to develop. The poor preservation of pollen may also reflect the frequent wetting of the basin-floor sediments and consequent high microbial activity.

244

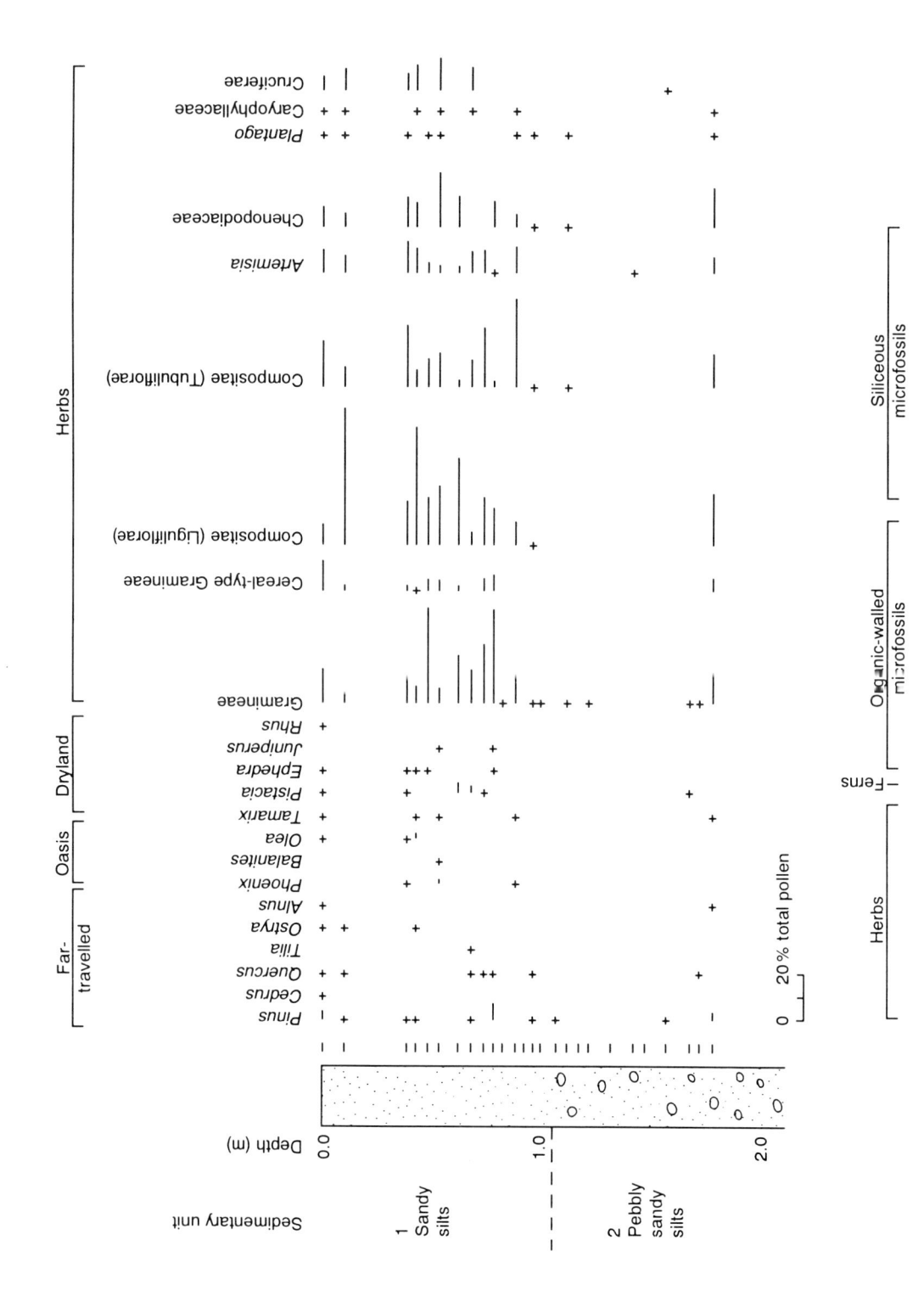

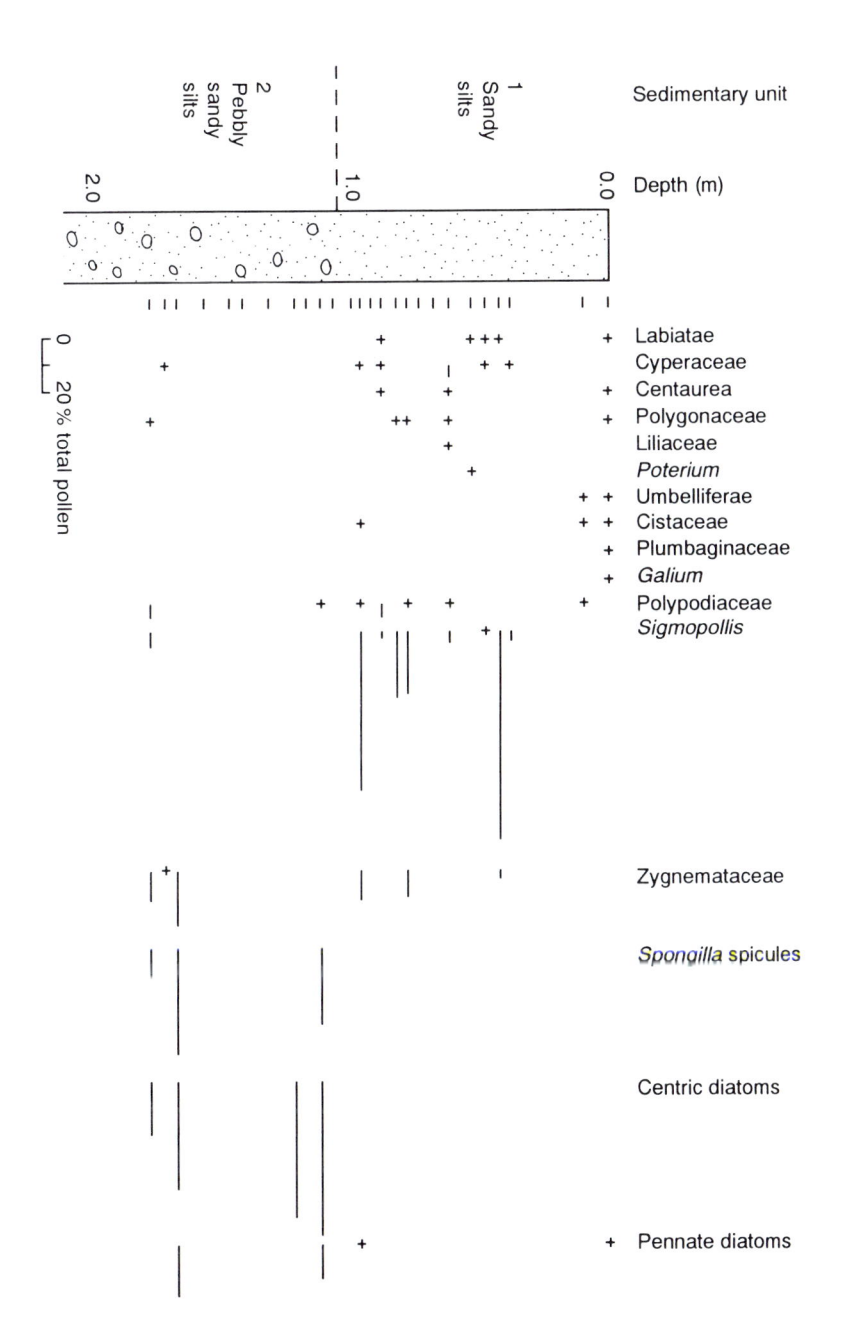

Figure 12.9 Palynology of a 2+ m core from the alluvial basin at Grerat D'nar Salem

Assemblage Biozone GR-B

The high incidence of Gramineae and other herbaceous pollen, together with some pollen of dryland taxa such as *Pistacia*, in this assemblage biozone can be used to suggest a grassy steppe landscape with occasional trees and shrubs. *Pistacia* is a low pollen producer, so that the 1 to 4 % of total pollen recorded from the samples may reflect a substantial number of trees. It typically grows along watercourses in the pre-desert today, but it is now rare. The cereal-type pollen may reflect cultivated cereals, but may alternatively be from wild plants. Polypodiaceae spores may indicate relatively humid conditions — ferns are classified as hygrophilous plants in studies of the palynology of arid localities in the Levant (Horowitz, 1979). The incidence of the far-travelled component is low enough to suggest a fairly dense vegetation cover around the polje (cf. Schulz, 1980). The organic-walled microfossils most probably reflect short-lived, seasonal flooding of the basin floor.

Assemblage Biozone GR-C

The fall in the general level of Gramineae pollen and the virtual disappearance of *Pistacia* and Polypodiaceae in this biozone, together with the rise in Cruciferae and the more frequent appearance of *Ephedra*, may reflect a general aridification of the area around Grerat D'nar Salem. These assemblages are more like the modern pollen rain of the interior deserts of Libya (Schulz, 1980) which are characterized by higher values for Cruciferae and Chenopodiaceae and lower values for Gramineae than is typical for steppe lands. The disappearance of the organic-walled microfossils may reflect the diminution of flooding events on the basin floor, which would be consistent with this hypothesis. An alternative explanation of the fall in Gramineae would be that the environment became significantly degraded, perhaps by soil erosion resulting from overgrazing by domesticated animals. A third possibility that cannot be neglected totally is that the taphonomic or preservational regime changed and that in fact the general character of the vegetation of the area was unchanged. At present, our published and unpublished evidence suggests that the appearance of *Olea* at 0.42 m reflects the start of 'Romano-Libyan' olive farming in the basin. At this level, the incidence of cereal and Gramineae pollen are also seen to decline, which raises the possibility that cereal cultivation might have been reduced or discontinued locally. The fall in Gramineae pollen in this sample might be the result of the common strategy of removing the ground cover from around olive trees in order to enhance infiltration, a practice that is widespread around the Mediterranean today.

Discussion of the Grerat D'nar Salem Core

The initial discussion of this core is obviously limited and speculative because of the absence of both reliable dating evidence and an appropriate body of sound taphonomic information with which to interpret the evidence. In particular, pollen is likely to have arrived by a number of taphonomic pathways which will have led to bias in the resulting assemblages. For example, plants on the basin floor at the time of sampling will have contributed pollen directly to the sediments. The regional pollen rain will have had an important input. The ephemeral rivers and overland flow that result from winter storms and flood the basin will have transported airfall-pollen, as well as pollen derived from eroding soils and surfaces in the catchment. These (intermittent) floods may have been of considerable taphonomic importance at the Grerat site because nowadays they are seen to transport and deposit significant quantities of sediment which will bury and hence allow the preservation of pollen. Schulz (1980) suggested that pollen will last no more than six months on the ground surface in the Libyan desert before it is destroyed by oxidation and microbial activity. Pollen preservation is unlikely to be 'perfect' in a normally well-aerated, base-rich environment with variable soil moisture content such as occurs at Grerat D'nar Salem. Differential pollen preservation is thus likely, with resistant taxa such as the Compositae, Polypodiaceae and *Pinus* likely to be preserved preferentially. The degree of humidity of the sediments on the basin floor will also have a significant effect on preservation. Pollen preservation is often good in very dry environments because microbial activity is limited by lack of water (see, for instance, the investigations of Fish, 1985, in the Sonoran Desert).

Nevertheless, despite these and other uncertainties, the Grerat core does appear to offer significant insights into the Holocene vegetational history of this area of the Libyan Predesert. The combination of the sedimentological and palynological studies suggests the following sequence of events. The deposits of unit 2 appear to have collected in the early Holocene when precipitation was sufficiently high and/or regular to support 'fairly deep', semi-permanent water bodies on the basin floor, with overland flow and/or stream discharges sufficiently high to erode and transport gravel-size materials hundreds of metres across the basin floor to the core site. The pollen of assemblage biozone GR-A indicates that vegetation of the surrounding basin at that time was probably a grassy steppe, though it cannot be reconstructed in any detail. The inferred intermittent dampness of the basin floor probably had the effect of increasing the biological richness of the site, hence increasing LOI values of the accumulating sediments, but simultaneously bringing about the poor preservation of the pollen assemblages through enhanced microbial activity and repeated oxidation/reduction cycles.

In time, the climate became less humid, the basin floor was flooded less often, leading to lower LOI values and the increased presence of aeolian processes which deposited wind-blown sand at the core site. Pollen from this phase—assemblage biozone GR-B—has preserved well because of the changing microbial and geochemical situation of the basin floor, and suggests a grassy steppe landscape with trees and/or shrubs along the watercourses around the basin. It is *possible*, but not certain, that there was some cereal cultivation near the site. The pollen and sedimentological data both indicate that the environment continued to become more arid, and it is impossible to rule out the possibility that the landscape became more degraded. It is possible that a loss of plateaux soils suggested by Hunt *et al.* (1986) may have taken place at this time. Flooding of the basin became ever more ephemeral, until by the depth of 0.4 m the surface environment could no longer support the previously notable algal flora (Figure 12.9). The pollen of *Olea* makes a sudden and significant first appearance at the depth of 0.4 m in the core. We suspect that this event most probably reflects the start of Romano-Libyan farming in the area. At the same time the incidence of cereal-type pollen appears to fall. There is, however, a small but distinct increase in LOI values, suggesting greater biological activity or even perhaps a manuring of the accumulating sediment surface of the time. These are interesting phenomena because the size and frequency of archaeological remains around the basin would suggest that the cultivation of barley and wheat can be anticipated to have increased at that time. This evidence suggests that the type of mixed olives–cereals–grasses–animal husbandry suggested in Figure 12.6 and Gilbertson *et al.* (1984) in the context of wadi floors did *not* occur at this broad basin site. It is possible that various taphonomic factors may be involved. However, one interesting possibility is that cereal cultivation was largely discontinued in such favourable localities, especially once the olive leaf canopy was established. The cereal crops may have been grown more extensively (and perhaps intensively) in other locations such as the 'deep soils' in plateaux basins as exemplified by sites 1212 and 1251/3 at the floodwater farm in Wadi Gobbeen (Figure 12.2). These basins are reported widely in the region by Dorsett et al. (1984), Gilbertson et al. (1984) and Gilbertson & Hunt (1988). Evidence of olive cultivation then disappears from the pollen record until the trace of *Olea* is noted in the pollen in the modern surface sediment—the nearest known olive plantations being 20 km distant. The sedimentological and palynological evidence is, however, clear on three critical points (subject to the constraints imposed by the limits of precision associated with the sampling intervals). Apart from the presence or absence of olive pollen, there has been no significant change between the olive-containing Romano-Libyan levels and the modern pollen rain at the surface. The inferred Romano-Libyan level was not only followed by arid conditions broadly similar to those that prevail today but was also accompanied and preceded by similarly arid conditions (with perhaps degraded landscapes). However, most recent research on cave deposits at nearby Beni Ulid suggest slightly wetter climates may have occurred in the Medieval period (Gale et al., 1993). A significant period of greater humidity, ephemeral water bodies and biological richness does seem to have occurred in the Libyan Pre-desert in the early Holocene, but it had ended well before this identified episode of olive farming in and around the basin.

DISCUSSION AND CONCLUSIONS

This study has involved differing bodies of information, ideas and theoretical 'retrodictions' which have been obtained from different types of depositional and biogeographical contexts, as well as from archaeological excavations and inferences in the Libyan Pre-desert. The attempted combination of such (often non-compatible) information is both a strength and weakness of this multidisciplinary and multilevel approach, with the result that the following first, integrated and chronologically based account of environmental change and human impact in the desert must be read with due caution.

In the large shallow basin of Grerat D'nar Salem there appears to have been a large, semi-permanent water-body during some part of the early/middle Holocene, which was surrounded by a landscape dominated by grassy steppe vegetation with some tree and shrub species, perhaps growing in wetter stream and wadi beds. The climate of the region was moister in the early/middle Holocene than occurs today. Sometime in the mid- or early/late Holocene, the region became more arid. The semi-permanent lake at Grerat disappeared and aeolian processes became more dominant in moving sediment into the basin. The adjacent vegetation continued to be a grassy plain, but there are possibilities that some cereal cultivation might have been practised. Floodwater farming appears to have developed in this already more-arid landscape sometime in the first century AD and led to a major change in the geomorphology and biogeography of the region. Cereal crops, grapes, olives, as well as date-palms, figs and almonds were grown by a sizeable sedentary population by concentrating floodwater into wadi floors and plateaux basins with well-designed and geomorphologically-sensitive wall systems. Examination of the few surviving modern analogues of these agricultural practices, as well as the erosional and depositional consequences of the many walls that occur on wadi floors, has indicated that it is possible to build a theoretical model that predicts that all the crops, known from excavations, could be grown as a mixed crop on the mobile sediments in the wadi floors. This model implies that intense mixed farming in these wadis would have been able to cope with the patterns of erosion and sedimentation that are consequences of this type of floodwater farming. This agricultural development was long-lived, which implies a sufficient robustness to cope with the range of environmental fluctuations and stresses that must have occurred with the passage of time. Floodwater farming must have brought about a substantial change in the pattern and richness of plant and animal life in the area. At the Grerat D'nar Salem basin site, the available palynological information indicates that the ancient cultivation there is better viewed as a monoculture rather than the mixture of plant and animal crops postulated for the more geomorphologically mobile wadi floors. Studies at Grerat have also shown that the 'excessive' addition of runoff waters on to the basin, which is itself associated with salt-rich bedrocks, might have caused the rise of salt-rich groundwaters with deleterious consequences for tree, cereal or pasture crops. No evidence of any other environmental change has been found that might have caused the abandonment of these floodwater farms over such large areas. It is probable that issues concerned with individual's aspirations, their societies, economics or political situations were of greater significance than 'environmental change or problems' in this context. The evidence from the Grerat site suggests that the modern environment may not be very different in its 'climate' to that experienced before, during and after the major episode of olive farming in the basin. Nevertheless, as well as recognising the minimal nature of the evidence for late Holocene geomorphologically or biogeographically significant environmental change in the region, suggested by the Grerat study, it is also important to recognise that in a hydrological sense, much of the wall engineering is still operating, gathering, augmenting and concentrating floodwaters. The result is that much of the distribution and productivity of the modern vegetation (and presumably fauna) in the wadis and plateaux, and the scale and character of the fluvial geomorphological processes that prevail, are *still* the direct consequences of the work of ancient floodwater-farmers.

REFERENCES

Adamson, D. A., Gasse, F., Street, F. A. and Williams, M. A. J. (1980). Late Quaternary history of the Nile. *Nature,* **288**, 50–5.

Anketell, J. M. (1989). Quaternary deposits of nothern Libya—lithostratigraphy and correlation. *Libyan Studies,* **20**, 1–29.

Anketell, J. M., Ghellali, S. M., Gilbertson, D. D. and Hunt, C. O. in press. The Quaternary wadi and floodplain sequences in Tripolitania, north west Libya, and their implications in landscape development. In Lewin, J., Macklin, M. and Woodward, J. (eds.), *Mediterranean Quaternary River Environments,* Balkema, Rotterdam.

Ball, D. F. (1974). Loss-on-ignition as an estimate of organic matter and organic carbon in non-calcareous soils. *Journal of Soil Science,* **15**, 84–92.

Ballais, J.-L. (1990). Terrasses de culture et jessours du Mahgreb oriental. *Méditerranée,* **3**, 51–3.

Barker, G. W. W. and Jones, G. D. B. (1981). The UNESCO Libyan Valleys Survey 1980. *Libyan Studies,* **12**, 9–48.

Barker, G. W. W. and Jones, G. D. B. (1982). The UNESCO Libyan Valleys Survey 1978–1981: palaeoeconomy and environmental archaeology in the pre-desert. *Libyan Studies,* **13**, 1–34.

Barker, G. W. W. and Jones, G. D. B. (1984). The UNESCO Libyan Valleys Survey VI: investigations of a Romano-Libyan Farm, part 1. *Libyan Studies,* **15**, 1–44.

Barker, G. W. W., Gilbertson, D. D., Griffin, C. M., Hayes, P. P. and Jones, D. A. (1983). The UNESCO Libyan Valleys Survey V: sedimentological properties of Holocene wadi floor and plateau Deposits in Tripolitania, North West Libya. *Libyan Studies,* **12**, 69–85.

Barker G. W. W., Gilbertson D. D., Jones, G. D. B. and Welsby, D. A. (1991). The UNESCO Libyan Valleys Survey XXIII: the 1989 season. *Libyan Studies*, **22**, 31–60.

Bonvallot, J. (1986). Tabias et Jessours du Sud Tunisien. Africulture dans les zones marginales et parade à l'érosion. *Cahiers ORSTOM, Séries Pédologie*, **XXII (19846)**, 163–71.

Brogan, O. and Smith D. J. (1984). Ghirza: a Libyan settlement of the Roman period. Department of Antiquities, Tripoli.

Bruins, H. J., Evenari, M. and Nessler, U. (1986). Rainwater-harvesting for food production in arid lands: the challenge of the African famine. *Applied Geography*, **6**, 13–32.

Bryan, K. (1929). Floodwater farming. *Geographical Review*, **19**, 444–56.

Clark, G. (1986). The UNESCO Libyan Valleys Survey XIV: archaezoological evidence for stock-raising and stock-management in the pre-desert. *Libyan Studies*, **17**, 49–64.

Connor, P. (1984). Can Colonel Gadaffi's desert bloom like Nero's? *The Sunday Times*, 8 July 1984.

Coude-Gaussen, G. and Rognon, P. (1988). The upper Pleistocene loess of southern Tunisia: a statement. *Earth Surface Processes and Landforms*, **13**, 137–51.

Dore, J. N. and Van der Veen, M. (1986). The UNESCO Libyan Valleys Survey XV: radiocarbon dates from the Libyan Valleys survey. *Libyan Studies*, **17**, 65–8.

Dorsett, J. E., Gilbertson, D. D., Hunt, C. O. and Barker, G. W. W. (1984). The UNESCO Libyan Valleys Survey IX: image analysis of Landsat data and its application to environmental and archaeological Surveys. *Libyan Studies*, **15**, 71–80.

El-Amani, S. (1983). Traditional technologies and the development of African environments. Utilization of runoff waters: the 'meskats' and other techiques in Tunisia. *African Environment*, **3**, 106–20.

Evenari, M., Shanan, L. and Tadmor, N. (1982). *The Negev: The Challenge of a Desert*, Harvard University Press, Cambridge, Mass.

Fieller, N. R. J., Gilbertson, D. D. and Olbricht, W. (1984). A new method for the environmental analysis of particle size data from shoreline environments. *Nature*, **311**, 648–51.

Fieller, N. R. J., Flenley, E. C., Gilbertson, D. D. and Hunt, C. O. (1990a). The description and classification of grain size data from ancient and modern shoreline sands at Lepcis Magna using log skew Laplace distributions. *Libyan Studies*, **21**, 49–59.

Fieller, N. R. J., Flenley, E. C., Gilbertson, D. D. and Thomas, D. S. G. (1990b). 'Dumb-bells': a plotting convention for mixed grain size populations. *Sedimentary Geology*, **69**, 7–12.

Fish, S. K. (1985). Prehistoric disturbance floras of the Lower Sonoran Desert and their implications. *American Association of Stratigraphic Palynologists, Contributions Series*, **16**, 77–88.

Flenley, E. C., Fieller, N. R. J. and Gilbertson, D. D. (1987). The statistical modelling of 'mixed' grain size distributions from aeolian sands in the Libyan Pre-desert using log skew Laplace distributions. In Reid, I. and Frostick, L. E. (eds.), *Desert Sediments: Ancient and Modern*, Special Publication of the Geological Society of London, pp. 271–80.

Gale, S. J. and Hunt, C. O. (1986). The hydrological characteristics of a floodwater farming system. *Applied Geography*, **6**, 33–42.

Gale, S. J., Hunt, C. O. and Gilbertson, D. D. (1986). The infill sequence and water carrying capacity of an ancient irrigation channel: Wadi Gobbeen, Tripolitania. *Libyan Studies*, **17**, 1–5.

Gale, S. J., Gilbertson, D. D., Hoare, P. G., Hunt, C. O., Jenkinson, R. D. S., Lamble, A. P., O'Toole, C., van der Veen, M. and Yates, G. (1993). Late Holocene environmental change in the Libyan pre-desert. *Journal of Arid Environments*, **24**, 1–19.

Gilbertson, D. D. (ed.) (1986). Runoff farming in rural arid lands. In *Applied Geography*, Theme Volume 6(1 & 2), 122 pp.

Gilbertson, D. D. and Hunt, C. O. (1988). The UNESCO Libyan Valleys Survey XIX: the Cenozoic geomorphology of the Wadi Merdum, Beni Ulid, in the Libyan Pre-desert. *Libyan Studies*, **19**, 95–121.

Gilbertson, D. D. and Hunt, C. O. (1990). The UNESCO Libyan Valleys Survey XXI: geomorphological studies of the Romano-Libyan Farm at Site Lm4, at the confluence of the Wadi el Amud and the Wadi Umm el Bagul in the Libyan Pre-desert. *Libyan Studies*, **21**, 25–42.

Gilbertson, D. D., Hayes, P. P., Hunt, C. O. and Barker, G. W. W. (1984). The UNESCO Libyan Valleys Survey VII: a classification and functional analysis of ancient irrigation and wall systems in the Libyan Pre-desert. *Libyan Studies*, **15**, 45–70.

Gilbertson, D. D., Hunt, C. O., Briggs, D. J., Coles, G. M. and Thew, N. M. (1987). The UNESCO

Libyan Valleys Survey XVIII: the Quaternary geomorphology and calcretes of the area around Gasr Banat in the Pre-desert of Tripolitania. *Libyan Studies,* **18**, 15–28.

Gilbertson, D. D., Hunt, C. O. and Fieller, N. R. J. (1993). ULVSXXVI: Sedimentological and palynological studies of Holocene Environmental Changes from a plateau basin infill sequence at Grerat D'nar Salem, near Beni Ulid, in the Tripolitanian pre-desert. *Libyan Studies,* **24**, 1–18.

Horowitz, A. (1979). *The Quaternary of Israel,* Academic Press, New York.

Hughes, J. D. and Thirgood, J. V. (1982). Deforestation in ancient Greece and Rome: a cause of collapse. *Journal of Forest History,* **26**, 60–75.

Hunt, C. O. (1985). Recent advances in pollen extraction techniques: a brief review. In Fieller, N. R. J., Gilbertson, D. D. and Ralph, N. G. A. (eds.), *Palaeobiological Investigations: Research Design, Methods, and Data Analysis,* British Archaeological Reports International Series 266, Oxford, pp. 181–8.

Hunt, C. O., Gale, S. and Gilbertson, D. D. (1985). The UNESCO Libyan Valleys Survey VIII: pseudo-karst tectonics and landforms in the Libyan Pre-desert. *Libyan Studies,* **16**, 1–13.

Hunt, C. O., Mattingley, D. J., Gilbertson, D. D., *et al.* (1986). The UNESCO Libyan Valleys Survey XIII: interdisciplinary approaches to ancient farming in the Wadi Mansur, Tripolitania. *Libyan Studies,* **17**, 7–47.

Hunt, C. O., Gilbertson, D. D., Van de Veen, M., Jenkinson, R. D. S., Yates, G. and Buckland, P. C. (1987). The UNESCO Libyan Valleys Survey XVII: the palaeoecology and agriculture of the abandonment Phase at Gasr MM10, Wadi Mimoun in the Tripolitanian Pre-desert. *Libyan Studies,* **17**, 1–14.

Laronde, A. (1986). Roman agricultural development in Libya. *Libya antiqua,* **11**, 13–22.

Laronde, A. (1989). La vie agricole en Libye jusqu'a l'arrivée des Arabes. *Libyan Studies,* **20**, 127–34.

Maley, J. (1977). Paleoclimates of Central Sahara during the Early Holocene. *Nature,* **269**, 573–7.

Mattingly, D. J. (1988). The olive oil boom. Oil surplus, wealth and power in Roman Tripolitania. *Libyan Studies,* **19**, 21–41.

Mattingly, D. J. (1989). Farmers and frontiers. Exploiting and defending the countryside of Roman Tripolitania. *Libyan Studies,* **20**, 135–153.

Nabhan, G. P. (1986a). Papago Indian desert agriculture and water control in the Sonoran Desert, 1697–1934. *Applied Geography,* **6**, 43–42.

Nabhan, G. P. (1986b). Ak-ciñ 'arroyo-mouth' and the environmental setting of Papago Indian fields in the Sonoran Desert. *Applied Geography,* **6**, 61–75.

Pacey, A. with Cullis, A. (1986). *Rainwater Harvesting: The Collection of Rainfall and Run-off in Rural Areas,* Intermediate Technology Publications, London.

Pachur, H.-J. (1980). Climatic history in the Late Quaternary in southern Libya and the western Libyan desert. In Salem, M. J. and Busrewil, M. T. (eds.), *The Geology of Libya,* Volume III, Academic Press, London, pp. 781–8.

Petit Maire, N., Burollet, P. F., Ballais, J.-L., Fontugne, M., Rosso, J.-C. and Lazzar, A. (1991). Paleoclimats holocenes du Sahara septentrional. Depôts lacustres et terrasses alluviales on bordure du Grand Erg Oriental à l'extrême-Sud de la Tunisie. *Comptes Rendues à l'Académie des Sciences, Paris,* 1661–6.

Reij, C, Mulder, P. and Begemann, L. (1988). Water harvesting for plant production. Technical Paper 91, The World Bank, Washington D.C.

Rognon, P. (1980). Comparison between the Late Quaternary terraces around Atakor and Tibisti. In Salem, M. J. and Busrewil, M. T. (eds.), *Geology of Libya,* Volume III, Academic Press, London, pp. 815–35.

Rossignol-Strick, M. and Duzer, D. (1979). West African vegetation and climate since 22,500 B. P. from deep sea core palynology. *Pollen et Spores,* **21**, 105–34.

Sarnthein, M. (1978). Sand deserts during glacial maximum and climatic optimum. *Nature,* **272**, 43–6.

Schulz, E. (1980). An investigation of current geologic processes and an interpretation of Saharan palaeoenvironments. In Salem, M. J. and Busrewil, M. T. (eds.), *The Geology of Libya,* Volume III, Academic Press, London, pp. 789–96.

Van der Veen, M. (1981). The Ghirza plant remains. *Libyan Studies,* **12**, 45–8.

Van der Veen, M. (1985). The UNESCO Libyan Valleys Survey X: botanic evidence of ancient farming in the pre-desert. *Libyan Studies,* **16**, 5–28.

13 Evolutionary Trends in the Wheat Group in Relation to Environment, Quaternary Climatic Change and Human Impacts

M. A. BLUMLER
Department of Geography, State University of New York–Binghamton, USA

ABSTRACT

Evolution in the wheat group (*Triticum/Aegilops*) illustrates salient features of Near Eastern vegetation history and interactions with environment and people. The wheat group is entirely composed of annuals that underwent adaptive radiation in seed and fruit characteristics in relation to edaphics and climate. Wild emmer, apparently the first species to give rise to a domesticate, is a late, polyploid derivative. Many of the features traditionally thought to be the result of domestication were already present in wild emmer and other members of the wheat group; physiological and other differences between wild and cultivated wheats are surprisingly trivial. The popular view that 9000 years of agropastoralism have induced a major coevolutionary response (i.e. the evolution of weeds and domesticates) ascribed to human features that had already evolved prior to the Neolithic, as an adaptive response to environment and climatic change. Nonetheless, introgression and agropastoralism apparently have produced subtle genetic changes in wild wheats, so that their precise original nature remains a detective story worth unravelling. Near Eastern plants may have had more impact on humans than vice versa.

INTRODUCTION

The Near East, and especially the Neolithic hearth of the Fertile Crescent, is universally assumed to have been subjected to gradual environmental devastation over the past 9000 years or so of agropastoralist influence (see D. Zohary & Hopf, 1988, for the timing and location of agricultural origins). Recently, however, I presented evidence suggesting that human impacts on Near Eastern vegetation and soils have been more complicated but less severe than previously believed (Blumler, 1993). Here, in a similar vein, I argue that the same is true of human influences on plant evolution. Because agropastoralist systems have dominated the Near East for so many millennia, most scholars assume that the region has seen considerable extinction and also a great deal of co-evolution, in particular of weeds, domesticated crops and so-called antipastoral (spiny or toxic) plants (e.g. M. Zohary, 1950, 1962, 1973, 1983).

In the course of research in Israel on the relationship between seed weight and environment (Blumler, 1992b), I became aware of the extraordinarily high species diversity of Near

Eastern vegetation, at both local (Naveh & Whittaker, 1979; Blumler, 1992b; Blumler *et al.*, 1992, in preparation) and regional (Davis *et al.*, 1971; M. Zohary & Feinbrun-Dothan, 1966–88) scales; given this, it is unlikely that significant plant extinction has occurred in the recent past. I also noticed to my astonishment that most if not all native 'weeds' occur frequently in natural habitats. This is true even of 'obligate synanthropes' — plants that supposedly only exist in association with humans (M. Zohary, 1962), such as the milk thistle, *Silybum marianum*. This nitrophilic annual sports an elaiosome (Danin & Yom-Tov, 1990), a fat body attached to the achene that attracts harvester ants (*Messor* spp.) which collect the seeds, carry them back to their nest and discard them there after removing the elaisome. The ant nests are unusually fertile; milk thistle plants are concentrated around the nests and also at gazelle defecation spots (Danin & Yom-Tov, 1990). Thus, it appears that the milk thistle evolved originally in adaptation to naturally nitrophilic habitats, and later spread to occupy similar anthropogenic habitats such as sheep bedding grounds and dumps (Danin & Yom-Tov, 1990). Finally, because my research entailed the examination of seeds of several hundred species, I became aware that the Near Eastern flora has undergone a spectacular radiation in seed and fruit types, one that is completely unmatched elsewhere on Earth. Surprisingly, this amazing evolutionary development has gone almost completely unnoticed (Ehrendorfer, 1971; Ellner & Shmida, 1981; Blumler, 1992b). This radiation is concentrated in annual species, which are generally regarded as of comparatively recent origin (Raven, 1973; Stebbins, 1974); consequently, it appears that humans or recent environmental change must have been responsible. Here, I discuss recent revolutionary changes in the Near Eastern flora, evaluating the comparative importance of environment and climatic change, vis-à-vis human impacts.

I examine the question by focusing on one important taxonomic group: *Triticum/Aegilops*, the wheat group. This group is composed of 25–30 wild species and several domesticates, at several ploidy levels (Table 13.1), it is well known because of its importance to temperate agriculture; it exhibits several evolutionary trends characteristic of the flora as a whole; and it was crucially important in agricultural origins and hence in subsequent history.

My thesis is that the major evolutionary trends are due to adaptive radiation with respect to climate and edaphics, resulting especially from recent climatic change to increased seasonality of the rainfall and temperature regime (cf. Byrne, 1987). On the other hand, although humans clearly have caused changes in distribution, evolutionary response to human impacts has been diffuse, and concentrated at the sub-specific level.

THE CAST OF CHARACTERS

The wheat group is composed of seven or eight genomes at the diploid level (Table 13.1), each corresponding to a characteristic diaspore or seed dispersal type (*Aegilops comosa*, genome M, and *Ae. uniaristata*, genome Un, have similar diaspores and their genomes may not be truly distinct; see D. Zohary, 1965, for photographs and descriptions of diaspores). In addition, several allopolyploid combinations have arisen, involving especially U (from *Aegilops umbellulata*), D (from *Ae. squarrosa*) and A (from wild einkorn, *Triticum boeoticum*, or *T. urartu*). All members of the wheat group are annuals. It is generally believed that many of the polyploids, at least, have evolved since the dawn of agriculture under selection from human impacts (Kihara, 1954; Stebbins, 1956; D. Zohary, 1965).

Figure 13.1 illustrates a diaspore of wild emmer (*Triticum dicoccoides*), the wild ancestor

Evolutionary Trends in the Wheat Group

Table 13.1 The members of the wheat group (from D. Zohary, 1965; Chapman, 1985; Miller, 1987)

Species	Genome formula	Ploidy level[b]
Wild		
Aegilops mutica Boiss.	Mt	2×
Ae. speltoides. Tausch	S	2×
Ae. longissima Schweinf. & Muschl.	S¹	2×
Ae. sharonensis Eig	S¹	2×
Ae. searsii Feld. & Kis.	Sˢ	2×
Ae. bicornis (Forsk.) Jaub. & Spach.	Sᵇ	2×
Ae. squarrosa L.	D	2×
Ae. caudata L.	C	2×
Ae. comosa Sibth. & Sm.	M	2×
Ae. uniaristata Vis.	Un	2×
Ae. umbellulata Zhuk.	U	2×
Triticum boeoticum Boiss.	A	2×
T. urartu Tum.	A	2×
Ae. crassa Boiss. 4×	DM^cr	4×
Ae. crassa Boiss. 6×	DD²M^cr	6×
Ae. vavilovii (Zhuk.) Chenn.	DM^cr S^p	6×
Ae. juvenalis (Thell). Eig	DN^j U	6×
Ae. ventricosa Tausch	DUn	4×
Ae. cylindrica Host	CD	4×
Ae. triuncialis L.	UC	4×
Ae. columnaris Zhuk.	UM^c	4×
Ae. biuncialis Vis.	UM^b	4×
Ae. triaristata Willd. 4×	UM^t	4×
Ae. triaristata Willd. 6×	UM^t Un	6×
Ae. geniculata Roth.	UM^o	4×
Ae. peregrina (Hack.) Maire & Weill.	US^v	4×
Ae. kotschyi Boiss.	US^v	4×
T. dicoccoides (Körn.) Schoen.	AB	4×
T. araraticum Jakubz.	AG	4×
Domesticated		
T. monococcum L.	A	2×
T. timopheevi Zhuk.	AG	4×
T. turgidum L.[a]	AB	4×
T. aestivum (L.) Thell.	ABD	6×
T. zhukovskyi Men. & Er.	AAG	6×

[a] Includes emmer, durum, turgidum and other tetraploid wheats.
[b] Diploids (2×) have two sets of chromosomes, as in humans (MtMt in the case of *Aegilops mutica*); polyploids (4×, 6×, etc.) have additional chromosome sets.

of emmer and most other domesticated wheats (see Figure 13.2). On present archaeological evidence this is the most likely candidate for the first species domesticated in the Near East or, indeed, on the entire globe (Blumler & Byrne, 1991). In comparison with other grasses, the diaspore is an evolutionary monstrosity. It is enormous, contains two very large caryopses (seeds) and has a peculiar morphology: that of an arrow-shaped (D. Zohary, 1965), heavily barbed structure which is able upon dispersal to insert itself into the soil (given

Figure 13.1 A diaspore of wild emmer, *Triticum diccoccoides* (x½)

appropriate microsite conditions), and thus be protected from granivores such as harvester ants, birds and rodents. It is generally stated in the literature that this peculiar diaspore structure is an adaption for dispersal (Cook, 1913; D. Zohary, 1965), but this is somewhat misleading as it is only true at a microscale. While the recurved barbs of many Near Eastern grass diaspores enable them to be dispersed in animal fur or human clothing (especially socks) (cf. Shmida & Ellner, 1984), wild emmer is too large to disperse effectively in this manner and typically has only short-distance dispersal (Golenberg, 1986). Hence, Michael Zohary (1937) would have characterised the wild emmer diaspore as 'anti-telechoric', i.e. adapted for *preventing* dispersal (cf. Ellner & Shmida, 1981). On the other hand, the diaspore is well-suited for microdispersal into select microsites, such as soil cracks, thick litter accumulations and adjacent to rocks (Cook, 1913; Blumler, 1991). My research and that of D. Zohary (1969) show that these microsites are largely edaphically controlled; consequently, wild emmer is restricted to hard limestone and basaltic terrains. In essence, wild emmer disaspores differ from those of A genome diploids only in size.

```
    Wild einkorn                    Goatgrass
  Triticum boeoticum     ×     Aegilops speltoides
         AA                            BB
           \                           |
            \                          v
             \                    Wild emmer
              \                Triticum dicoccoides
               \                     AABB
                \                      |
   +-------------\---------------------+----------------------+
   |              \                    v                      |
   |                              Emmer wheat                 |
   |                            Triticum dicoccum   ×  Ae. squarrosa
   |         Einkorn wheat            AABB                  DD
   |       Triticum monococcum          |                     |
   |              AA                    |                     |
   |               \                    |                    /
   |                \                   v                   /
   |                 \             Bread wheat             /
   |              Durum wheat    Triticum aestivum        /
   |            Triticum durum  <---   AABBDD            /
   |                 AABB                               /
   +---------------------------------------------------+
```

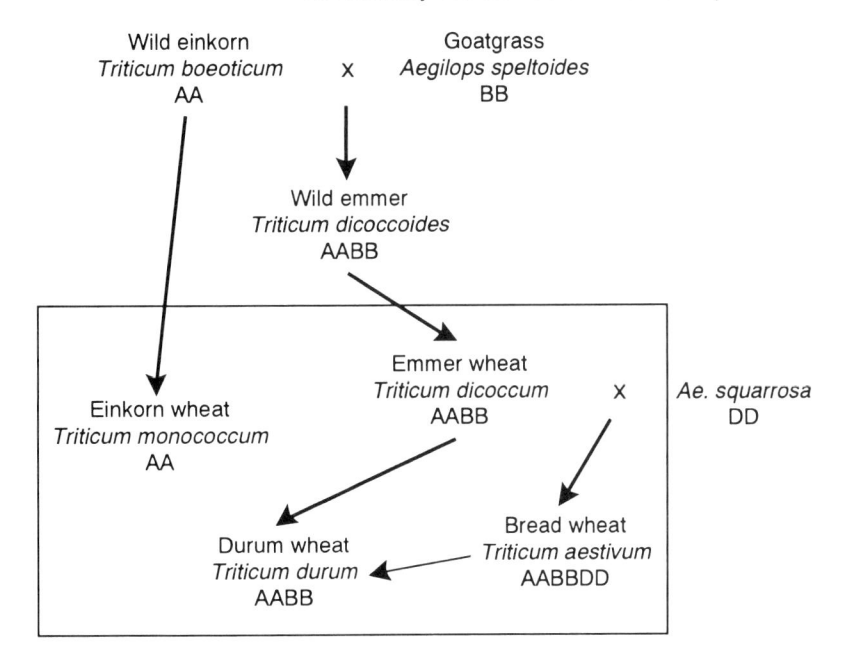

Figure 13.2 A possibly oversimplified outline of the evolution of wheat. Species within the box are domesticates; those outside are wild. Genomes (AA, etc.) as in Table 13.1. Additional wild diploids may have been involved in the origins of wild emmer. My suggestion that durum arose from emmer as a result of introgression from bread wheat is a minority viewpoint

Other wheat group diaspores are analogous. For the most part, they are large-grained, complex, unusual, contain more than one caryopsis/diaspore (synaptospermy), with the caryopses being polymorphic for size and dormancy (heterocarpy), and are adapted for self-implantation in the immediate vicinity of the mother plant. They appear to have only limited dispersal capabilities, although genome U diaspores are 'trample burs', i.e. they may be carried some distance by ungulate hooves, which also serve to grind them into the soil. In general, variation in diaspore morphology among the diploids is related to edaphics and grazing. The edaphic correlation is clearly seen, for instance, in the *Sitopsis* group: *Ae. speltoides* occurs on heavy soils, *Ae. longissima* on coastal sandy loams, *Ae. sharonensis* on consolidated dune sands, *Ae. searsii* on interior sandy soils and *Ae. bicornis* on desert sandy loess (D. Zohary, 1983). Similarly, of the three most common polyploid *Aegilops* spp. in Israel, *Ae. biuncialis* is most frequently found on marly soils because its sharp rachis tip is best able to penetrate the relatively resistant substrate. The other two polyploids, *Ae. peregrina* and *Ae. geniculata,* are often found together but separate somewhat in that the former (with larger grains) is more tolerant of shade or high litter levels. Diaspore diversification with respect to edaphics can be seen also among the annuals of many other Mediterranean and Near Eastern genera, such as wild oats *(Avena*; Ladizinsky, 1971), bur clover *(Medicago*; Lesins & Lesins, 1979), bedstraw *(Galium)* and related taxa (Ehrendorfer, 1971), and filaree *(Erodium)*. For instance, it appears that diaspores of species growing on heavy soils are often comparatively massive, bulky and large-seeded, those on substrates with a tendency to surface crusting are sharp-pointed and those on sands often either slender and pointed, or flattened.

D. Zohary (1965) assumed that the polyploid members of the group evolved and/or spread with humans, and admittedly there is no denying that some species have expanded their ranges enormously since the Neolithic. However, no new diaspore types are represented among the polyploids, and wild emmer, at least, clearly preceded agriculture. Furthermore, although polyploid *Aegilops* spp. are common along roadsides, trampled ground and field edges, they also occur frequently where there is no human disturbance (e.g. D. Zohary & Feldman, 1962). This is especially true in summer-dry regions such as Israel. The important polyploids in Israel are derived from *Ae. umbellulata,* which does not occur there; thus, it appears that either the polyploids migrated into the area or they replaced the diploid in response to a change in environment. The latter seems more likely since the 'umbrella' diaspore type that is characteristic of *Ae. umbellulata* and its polyploid derivates is very successful and one would expect that some member of the group would always have been present. The only clear morphological difference between wheat group diploids and polyploids is that the latter have larger seeds (Fritsch *et al.,* 1977; Blumler, 1992b), but increased seed size is usually *not* associated with weediness, since adaptation to temporarily available disturbed habitats generally requires good dispersal and the production of large numbers of (consequently small) propagules (Harper *et al.,*1970). On the other hand, increased seed size appears to be a common evolutionary response to increasing seasonality of precipitation (Blumler, 1992b).

As indicated in Table 13.2, the trends characteristic of the wheat group are also characteristic of the Near Eastern flora as a whole. Spininess is not found in the wheat group, although the pronounced tendency towards strong awns is perhaps analogous; in any case, spininess is particularly well developed in perennial species and probably therefore represents relatively ancient adaptation; Blumler (1992b) has suggested that it evolves preferentially where there is abundant availabile calcium. Especially striking is the remarkable radiation that has taken place in seed and fruit types, and particularly among annual species (which themselves are far more diverse than elsewhere on Earth; Blumler, 1984). Ehrendorfer (1971) noted that Turkish annuals of the Rubioideae display a greater variety of seed

Table 13.2 Some evolutionary trends in the Near Eastern flora

Annual habit
Heterocarpy (especially dimorphism)
  with respect to size
    dispersal
    dormancy
  amphicarpy
Synaptospermy
Protective fruits
Strange fruits (and seeds)
Antitelechory?
One-seeded fruits?
Seed gigantism

Spininess
Cushion plant growth form

and fruit morphologies than is found in the tribe worldwide. The same could be said of several important herbaceous plant families such as the Gramineae, Cruciferae, Umbelliferae and Papilionaceae. These trends are paralleled in other semi-arid and arid regions (Burtt, 1977; Venable & Lawlor, 1980; Blumler, 1992b), especially in evolutionarily advanced groups such as the tarweeds (Madiinae) of California, suggesting that seasonal or arid climate and the annual habit combine to encourage diversification of seed/diaspore structures. However, by comparison with the Near East these trends are only weakly developed in the analogue regions (Blumler, 1992b).

CAUSES AND CONSEQUENCES OF DOMESTICATION

The archaeological record is inherently incomplete, but the available evidence suggests that agriculture began in the so-called hilly flanks of the Fertile Crescent—the foothill zone extending in an arc from Palestine through south-eastern Turkey to south-western Iran— or in the immediately adjacent arid margin (D. Zohary & Hopf, 1988). The hill zone of the Fertile Crescent is also the centre of distribution of the progenitors of the eight or nine primary (i.e. initial) Near Eastern domesticates (D. Zohary & Hopf, 1988; Blumler, 1992a), including wild emmer and wild einkorn. This region receives plentiful moisture in winter and experiences severe heat and drought in summer, i.e. it has an extreme Mediterranean-type climate. In fact, no other region on Earth has such a pronounced, winter-wet, summer-dry climate (Blumler, 1992b).

As I have discussed elsewhere (Blumler, 1984, 1992b, 1993; cf. Byrne, 1987), seasonality of the precipitation and temperature regime favours the annual habit. This is reflected in the distribution of the annual species of the Triticeae, the tribe to which the wheat group belongs. As West *et al.* (1988) have demonstrated, annual Triticeae are concentrated in the Near East, and specifically in the general vicinity of the Fertile Crescent. Notably, annual Triticeae are *not* common in desert regions of the Near East, in accord with Blumler's (1984, 1992b) conclusion that seasonality of drought is more important than aridity as a determinant of annual plant success (cf. Mandaville, 1986; Danin & Orshan, 1990).

Table 13.3 lists the world's largest-seeded grasses, so far as I have been able to determine them (Blumler, 1992b). The wheat group is particularly dominant. Many progenitors of agricultural crops are on the list, and the progenitors of primary domesticates (the first species to be domesticated in any given region) tend to be nearer to the top of the list than the progenitors of later (secondary) domesticates. Surprisingly, large-seeded annual grasses are more numerous than perennials. Normally, annuals produce relatively small seeds, presumably because they have only a short period of time in which to accumulate photosynthate and ripen seeds, and because often they are adapted to disturbed habitats, which sets a premium on the production of large numbers of easily dispersed propagules. The large-seeded grasses are generally associated with seasonal precipitation regimes, especially in the Near East (a few also are characteristic of wetlands with fluctuating water levels). All this suggests that agriculture began where there were abundant annuals, with large seeds, and that the largest-seeded annuals were cultivated first (Blumler, 1984, 1992b; Blumler & Byrne, 1991; Byrne, 1987). This may be because small-seeded species, though easy to gather in quantity from the wild (Blumler, 1992b), would have been relatively difficult to cultivate with a primitive technology. Seedlings from small seeds are poor competitors, easily uprooted by marauding birds, and perhaps most important, emerge poorly from a

Table 13.3 The world's heaviest-seeded wild grasses[a] (excluding the bamboos) (from Blumler, 1992b)

Species	Common name	Life from[b]	Grain weight[c]	Native distribution	Derived crop
Coix gigantea	Job's tears	P?	>40	SE Asia	
Coix lacryma-jobi	Job's tears	A/P	40	SE Asia	Job's tears
Avena magna	Wild oats	A	35	S Spain	
Hordeum spontaneum	Wild barley	A	35	Near East (NE)	Barley
Zea luxurians	Teosinte	A	35	Guatemala	
Zea mays	Teosinte	A	35	Mex/Guat	Maize
Zea perennis	Teosinte	P	35	C. Mexico	
Zizania aquatica	Wildrice	A	35	E. USA	
Zizania palustris	Wildrice	A	35	Canada, E. USA	
Avena murphyi	Wild oats	A	30	Morocco	
Lygeum spartum	Esparto	P	30	Mediterranean (Med.)	
Tripsacum dactyloides	Gamagrass	P	30	New World	
Triticum dicoccoides	Wild emmer	A	25	Fertile Crescent	Most wheats
Zea diploperennis	Teosinte	P	25	C. Mexico	
Triticum araraticum	Wild emmer	A	22	Near East	*T. timopheevii*
Aegilops triaristata	Goatgrass	A	20	Med. NE	
Oryza barthii	Rice	A	20	West Africa	African rice
Stipa spartea	Porcupine grass	P	20	Great Plains	
Aegilops peregrina	Goatgrass	A	19	E. Med. NE	
Triticum urartu	Wild einkorn	A	17	Fertile Crescent	
Oryza stapfii	Rice	A	16	West Africa	
Aegilops columnaris	Goatgrass	A	15	Asia Minor	
Aegilops crassa	Goatgrass	A	15	Near East	
Avena fatua	Wild oats	A	15	N. Med. C. Asia	Oats
Avena sterilis	Wild oats	A	15	Med. NE	Oats
Stipa pennata		P	15	Med. NE/Balkans	
Stipa lagascae		P	15	Med. NE	
Triticum boeoticum	Wild einkorn	A	15	Near East	Einkorn wheat
Aegilops comosa	Goatgrass	A	14	Aegean	
Oryza spontanea	Rice	A	14	SE Asia	
Aegilops geniculata	Goatgrass	A	13	Med. NE	
Aegilops juvenalis	Goatgrass	A	13	Near East	
Avena longiglumis	Wild oats	A	13	S. Med.	
Bromus diandrus	Ripgut brome	A	13	Med. NE	
Sorghum intrans		A	13	N. Australia	
Sorghum stipoideum		A	13	N. Australia	
Aegilops cylindrica	Jointed goatgrass	A	12	Near East	
Oryza nivara	Rice	A	12	SE Asia	Rice
Aegilops biuncialis	Goatgrass	A	11	Med. NE	
Aegilops kotschyii	Goatgrass	A	11	Near East	
Aegilops triuncialis	Barb goatgrass	A	11	Med. NE	
Bromus stamineus	Chilean brome	A/P	11	C. Chile	*B. mango?*
Secale montanum	Rye	P	11	Med. NE	Rye
Spartina anglica	Cordgrass	P	11	England	

[a]The following taxa, for which data are not available, may also produce large caryopses: *Coix* spp., *Oryza* spp., *Tripsacum* spp., perennial *Zizania* spp. and a few *Stipa* spp.
[b]P = perennial; A = annual.
[c]Estimated approximate mean kernel (caryopsis) weight (mg) when growing under favourable conditions.

rough or litter strewn substrate; therefore, careful preparation of the seedbed using specialised tools may be necessary for their incorporation into an agricultural system (Blumler, 1992b).

Since their derivatives were involved in the origins of agriculture, it appears that wild einkorn, *Ae.speltoides,* and wild emmer must have originated beforehand (Figure 13.2). One or two additional diploids may also have been involved in the origins of wild emmer (Blumler & Byrne, 1991). In any case, since wild emmer is a polyploid it is likely that all or almost all diploid species, and possibly the other polyploids as well, had evolved by this time. Since large seed size seems to have been a prerequisite for agriculture and large seed size is associated with seasonality of precipitation, polyploidy in the wheat group may be related primarily to this factor and not to human disturbance.

Evolution under Domestication

By definition, domestication entails genetic change (Harlan *et al.,* 1973; Blumler & Byrne,1991); however, as I have discussed elsewhere (Blumler, 1992b), the magnitude of the differences between wild progenitor and domesticated derivative is usually greatly exaggerated. Table 13.4 lists some of the evolutionary changes generally posited to have occurred as a result of domestication. Humans have taken over the protection, competition and dispersal functions. Consequently, photosynthate that the wild plant must reserve for these purposes (e.g. in the construction of wild emmer's elaborate diaspore) can go to seed-filling instead; i.e. a number of changes associated with increasing harvest index (percentage of total crop biomass actually harvested for food) and ease of handling (uniformity of maturation and germination) have unquestionably taken place. In addition, a few wild perennials such as cotton have become annual plants under cultivation, and many species have undergone an increase in seed size though to a far lesser extent than generally believed. The primary progenitors, in particular, were *already* large-seeded. Furthermore, it is doubtful whether this increase was adaptive, as often suggested (e.g. Harlan *et al.,* 1973), or simply a consequence of the release of constraints on seed size imposed by tight-fitting husks or

Table 13.4 Evolutionary trends under domestication

Genuine
    Indehiscence (loss of seed dispersal)
    Uniform germination
    Uniform maturity
    Increased harvest index
    Loss of competitive ability (tillering)
    Loss of antiherbivore defence

Greatly exaggerated
    Perennial to annual habit
    Increased seed size

Claimed but without supporting evidence
    Increased adaptation to fertile soils
    Increased seedling vigour
    Consistent physiological changes

other structures in the wild species. Finally, there is no evidence that domesticates are better adapted to fertile soils, have more vigorous seedlings or exhibit consistent physiological changes from the wild (Blumler, 1992b). In fact, wild seedlings may be more vigorous than those of their domesticated descendents (Chapin *et al.*, 1989). Only one new species (bread wheat, *T. aestivum*) is known to have evolved under domestication, aside from recent artificial derivatives such as triticale; domesticates are often given distinct specific epithets for convenience and because of morphological differentiation from their progenitors, but they remain firmly connected genetically to their ancestors (Harlan, 1975). Thus, almost all evolution has occurred at the intra-specific level.

WEEDINESS

'Weed' is a slipshod term (Baker, 1965, 1974; Harlan, 1975). For our purposes it is perhaps best to use Blumler & Byrne's (1991, p. 24) definition: '...an uncultivated plant taxon that benefits from human impacts...'. Table 13.5 lists some traits commonly associated with weediness. Of course, these traits are not found in *all* weeds; rather, in modern parlance they are 'vicious stereotypes' that biologists associate with such plants. Generally speaking, then, one might expect human disturbance to select for these traits in the weed flora.

The progenitors of primary domesticates are often surprisingly non-weedy, wild emmer being the classic example (Aaronsohn, 1910; D. Zohary, 1969; Blumler & Byrne, 1991). For instance, Blumler (1992b, 1993; Blumler *et al.*, in preparation) found that on fertile soils in Israel, wild cereals (wild emmer, wild barley—*Hordeum spontaneum*—and wild oats *Avena sterilis*) decrease in importance as grazing pressure increases, while perennials are unaffected or even perhaps favoured (cf. Litav *et al.*, 1963; D. Zohary, 1969; Noy Moir, 1990). Table 13.6 summarises data of a Japanese collecting expedition through the Taurus

Table 13.5 Some traits that are popularly associated with the evolution of weediness

Annual habit
Rapid growth under fertile conditions
Phenotypic plasticity
Colonizing ability
Excellent seed dispersal
Polymorphic germination (heterocarpy)

Table 13.6 Habitat of specimens of the wheat group collected in the Taurus and Zagros Mountain regions (percentages calculated from data of Sakamoto, 1982)

Habitat	Wild Tetraploids	Wild einkorn	*Aegilops speltoides*
Field edge, roadside	0	22	30
Grassland	22	25	38
Oak park-forest	78	53	32

and Zagros Mountains (Sakamoto, 1982). Wild emmer and the morphologically indistinguishable *T. araraticum* were found only in essentially undisturbed habitats; their diploid ancestor, wild einkorn, was found primarily but not entirely in such habitats; and the other presumed diploid ancestor, *Ae. speltoides,* was found quite often in disturbed situations. Interestingly, this last species is one of the most primitive members of the entire wheat group, and certainly of the section *Sitopsis,* as evidenced for instance by its proclivity for cross-pollination. In this case, the more primitive, more generalised species is weedier than the derived, more specialised forms. Similarly, the one-seeded form of wild einkorn is dominant, and strictly weedy, in the Balkans and Crimea, where the precipitation regime is sub-seasonal or aseasonal; on the other hand, the more peculiar (in an evolutionary sense) two-seeded diaspore form is dominant in the winter-wet, summer-dry Fertile Crescent, where the species is abundant in natural habitats and much less weedy (D. Zohary et al., 1969). Thus, it appears that the diverse annual flora of the Near East did not necessarily evolve as a response to agropastoralism; in fact, human impacts may have shifted the competitive balance towards more primitive, generalised types.

Consider now the extent to which weeds may have evolved in the Near East since agriculture began (using Table 13.5 as a reference). In the wheat group, all species are annuals; there does not seem to be any association, therefore, between human disturbance and evolution of the annual habit. Similarly, rapid growth under fertile conditions may be a prerequisite to success in the face of some kinds of human disturbance, but wild emmer and other non-weedy species were already adapted to fertile conditions prior to the Neolithic (cf. the example of the milk thistle discussed above). Phenotypic plasticity—great adjustments in plant size and in number of fruits, flowers, etc., produced depending on environmental conditions—is characteristic of *all* annuals, which essentially kill themselves manufacturing as much seed as possible. However, large-seeded annuals such as the members of the wheat group have inherently less flexibility in this regard (Kutiel, 1985; Blumler, 1992b). Defining 'colonizing ability' is even more problematic than defining a 'weed', since all species must colonise new areas at times or else face extinction. In practice, the term is applied to any species that expands its range or density and thereby poses a management problem— whether or not the species is truly weedy. Excellent seed dispersal is *not* characteristic of the members of the wheat group, nor does it tend to be characteristic of many other annuals with complex diaspores (M. Zohary, 1937). Heterocarpy as expressed in the wheat group primarily involves dimorphism, which Harper (1977) has suggested to be an adaption to disturbance. However, both in the wheat group and in the Near Eastern flora, dimorphic species are as likely to be non-weedy as weedy. Successful ruderals commonly have a more diffuse polymorphic dormancy, such as is also found in many desert annuals, rather than a simple dimorphism. In short, the more peculiar and specialised morphologies, such as those listed in Table 13.2, appear to have evolved in relation to environment prior to agriculture. Furthermore, with the rise of farming, it may be that more primitive or generalised annuals received a new lease on life rather than that new species of weeds evolved.

ENVIRONMENT AND EVOLUTION

If one examines species distribution maps (e.g. Harlan & Zohary, 1966; D. Zohary, 1969; D. Zohary & Hopf, 1988), it is readily apparent that wild emmer and *T. araraticum* are primarily restricted to the most seasonally wet-and-dry zone of the Fertile Crescent, while wild einkorn (with smaller seeds) is much less constrained and dominates natural habitats with a somewhat longer rainy season. I believe this is a general rule: among Near Eastern annuals, high polyploids and taxa with larger seeds are generally found where conditions are comparatively seasonally dry; i.e. primary adaptation is to increasing seasonality of climate. Admittedly, however, information on the habitats of *Aegilops* spp. is scanty, and it is not known whether they fit this model.

In addition, it appears that much of the diversification in seed and fruit types in the wheat group and indeed the Near Eastern flora in general is related to edaphics. Many groups such as the *Sitopsis* section are clearly separated according to soil type. On limestone, with its irregular solution weathering, there is often an extremely high diversity of microsites, which in turn are occupied by an equally high number of species. This may have been the original selective impetus towards diaspore diversity in annuals, which would have had at least a minor role to play even under less seasonal regimes, given that some limestone microsites would have been difficult for perennials to occupy (cf. Ratcliffe, 1961). More important to the evolution of large-seeded diaspores, perhaps, calcium gives aggregation to heavy soils and thereby enhances the possibilities for self-burial (cf. Peart & Clifford, 1987; Blumler, 1991). For example, wild emmer is found only on soils formed over hard limestone, dolomite and basalt, and is usually associated with crevices adjacent to rocks. Consequently, it may need a large seed to provide sufficient reserves to allow the seedling to reach the light before becoming etiolated (wild emmer appears to be considerably more shade-tolerant than its wild cereal associates, wild barley and wild oats; Blumler, 1991).

Calcium may also affect the distribution of available water in the soil by increasing the amount of soil cracking, so that burial may become a more appropriate strategy than on soils of similar texture without calcium. Since calcareous dust may be ubiquitous in the Near East (Yaalon & Ganor, 1973, 1975; Blumler, 1992b), even soils derived from non-calcareous parent material may have sufficient calcium to exhibit these characteristics.

SOME COMPLICATIONS

Despite all this, some interesting genetic changes appear to have occurred in wild emmer, and perhaps other members of the wheat group, as a result of agropastoralism. In particular, the Upper Jordan Valley 'race' of wild emmer (Harlan & Zohary, 1966; D. Zohary, 1969) appears to be a hybrid swarm resulting from introgression from durum. Table 13.7 demonstrates that the Upper Jordan Valley wild emmer is morphologically intermediate between 'typical' wild emmer and domesticated wheat, and that it resembles durum more than emmer. If it were the progenitor race, i.e. the form originally taken under cultivation (as Harlan & Zohary, 1966, suggested), one would expect it to be more similar to the primitive emmer than to durum. In addition, geographical analysis of electrophoretic data suggests introgression from durum (Blumler, in preparation), and the existence of wild emmer × durum hybrids in the Valley has been demonstrated (D. Zohary & Brick, 1961). This introgression has occurred despite the fact that wild emmer is entirely non-weedy.

The most logical explanation is that durum cultivation in the Upper Jordan Valley so completely surrounded small, relict populations of wild emmer that the latter were swamped, over many centuries, by gene flow from their cultivated derivative. Recently, with near elimination of wheat cultivation in the region and relaxation of the grazing pressure that had kept local wild emmer populations in a state of near-extinction, the introgressed wild

Table 13.7 Characteristic morphology of two races of wild emmer, durum and emmer (from Blumler, in preparation). Respects in which the Upper Jordan Valley (UJV) race of wild emmer resembles durum more than emmer are indicated in **bold**

Character	Wild emmer Typical	UJV	Palestinian durum	Domesticated emmer
Grain				
Size	Small-medium	Medium–large	Large	Large
Shape	Elongate	**Elongate to rounded**	Oval to rounded	Elongate to oblong
Brush	Long	Short	Short	Short
Dimorphism	Pronounced	Variable	Slight	More than durum
Dormancy	Pronounced	Variable	None	None?
Glutenin A1-1	Present	**Generally absent**	Absent	Present
Glutenin A1-2	Absent	Sometimes present	Absent	Absent
Glume (outer Husk)				
Angle	Nearly parallel	Somewhat divergent	Divergent	Somewhat divergent
Shape	Lanceolate	**Lanceolate to ovate**	Lanceolate to ovate	Lanceolate
First tooth	Medium, straight	**Medium to long, straight to curved**	Often long, often curved	Short to medium, straight or beaked
Second tooth	Present	Less developed	Poorly developed	Poorly developed
Colour	Black	**Yellow, black**	Yellow	Yellow, red
Wax	None	Rare	Frequent	Glaucous
Pubescence	Glabrous	**Often pubescent**	Often pubescent	Glabrous
Spikelet width	Narrow	**Broad to very broad**	Very broad	Broad
Vegetative				
Habit	Prostrate	Semi-erect, erect	Erect	Semi-erect, erect
Dimensions	Slender	Robust	Robust	Usually robust
Anthocyanin	Present	Absent or reduced	Absent	Often present
Photosynthetic	High efficiency	Moderate	Low	Low
Flowering	Late	**Early**	Early	Often early

emmer has spread and become abundant. Interestingly, under the new conditions of low grazing pressure, some durum genes may actually be proving adaptive in the wild. In particular, plant productivity is extremely high in the Upper Jordan Valley, so that very thick layers of grass litter remain at the end of the growing season under ungrazed or lightly grazed conditions. This may select for large seed size (characteristic of durum), since the seedling will need to achieve considerable vertical extension growth prior to beginning photosynthesis. Also, there are a few local areas of extremely karstic Eocene limestone where emerging seedlings are shaded by tall rocks and may need large seeds to emerge into the light. This sort of habitat occurs on Transect 'C' at Amiad (Anikster, 1988), where putative durum genes are especially common (Anikster, 1986, 1988; Blumler, in preparation).

In addition, Palestinian durum is usually erect (spring habit), while wild emmer tends to be prostrate in early growth—perhaps in itself an adaptation to heavy livestock grazing; genes for a more upright habit may be adaptive under ungrazed conditions, since competition for light is likely to be intense early in the growing season. Finally, wild emmer often has high levels of anthocyanin, which may deter herbivory (Golenberg, 1986; Blumler, in preparation), while durum lacks anthocyanin; introgressed individuals may grow more rapidly than other wild emmer plants because they do not need to invest in anthocyanin production and, therefore, out-compete the latter if there is no grazing.

If this interpretation is correct, the evolutionary pattern in wild emmer since domestication may have been complicated. On the one hand, heavy grazing had reduced populations to near-extinction at times and perhaps selected for prostrate growth and anthocyanin production (Blumler, in preparation). On the other hand, introgression has occurred around cultivated areas, i.e. in a somewhat haphazard spatial pattern. While much of this introgression is just genetic noise, a few alleles may be improving wild emmer fitness; some, such as for erect growth habit, may actually represent reintroductions into the wild gene pool of alleles formerly present but eliminated by grazing. It appears, then, that the widespread assumption that the Upper Jordan Valley race is the ancestor of domesticated emmer is in need of revision. The Upper Jordan Valley race resembles emmer because of introgression from durum, which it resembles more. To determine the precise genotype(s) and geographical region of domestication will require a complicated analysis to identify introgressed alleles and attempt to determine what additional alleles may have been eliminated from wild populations by agropastoralism (see Blumler, 1992a, for a review of the kinds of analyses that are presently being carried out on other crop progenitors). Whether similar complications apply to polyploid *Aegilops* spp. is unknown. However, D. Zohary's (1965) argument that these species function as genetic sponges, rapidly picking up variation from related diploids and other polyploids, and that they consequently are able to undergo rapid evolutionary change in response to altered levels of human impacts is quite plausible (D. Zohary & Feldman, 1962).

CONCLUSIONS

In sum, it appears that the major evolutionary radiation within the wheat group and among Near Eastern annuals in general occurred in relation to climate and soils. Some evolution must have occurred in response to agropastoralism, but this may have been generally opposed to the trends that had developed as a result of edaphic specialisation and in response to increasing seasonality of climate. Moreover, while climatic change since the Miocene, at least, has been unidirectional—towards increasing seasonality—and, consequently, evolution also has responded in a directional manner, human impacts may have produced conflicting selection regimes and therefore a less clear-cut effect. Agropastoralism may favour generalised species where seasonality is severe (i.e. where advanced types are most abundant naturally), but probably allowed some of the latter, such as several polyploid *Aegilops* spp., to spread into sub-seasonal regions—because plough and cow substitute for severe seasonal drought as mechanisms for eliminating perennials. Finally, given the critical role that large seed size and the annual habit may have played in fostering the Neolithic Revolution, it may not be injudicious to suggest that Near Eastern plants have had more impact on humans than vice versa.

ACKNOWLEDGEMENTS

The author owes a great deal to Roger Byrne for his enthusiastic and willingness to allow the author to follow his research nose, so to speak. Discussions with Aaron Liston improved understanding of the Near Eastern flora immeasurably. Finally, no analysis of members of the wheat group can be undertaken without consultation with Daniel Zohary, to whom the author is most grateful.

REFERENCES

Aaronsohn, A. (1910). Agricultural and botanical explorations in Palestine. *U.S. Department of Agriculture Bureau of Plant Industry Bulletin,* **180**, 1–66.

Anikster, Y. (1986). The biological structure of native populations of wild emmer wheat (*Triticum turgidum* var. *dicoccoides*) in Israel, 2nd Annual Report, June 1985–May 1986, U.S. Department of Agriculture, Agricultural Research Service, Corvallis, Oregon.

Anikster, Y. (1988). The biological structure of native populations of wild emmer wheat (*Triticum turgidum* var. *dicoccoides*) in Israel, Final Report, May 1984–May 1987, U.S. Department of Agriculture, Agricultural Research Service, Corvallis, Oregon.

Baker, H. G. (1965). Characteristics and modes of origin of weeds. In Baker, H. G. and Stebbins, G. L. (eds.), *The Genetics of Colonizing Species*, Academic Press, New York, pp. 147–72.

Baker, H. G. (1974). The evolution of weeds. *Annual Review of Ecology and Systematics,* **5**, 1–24.

Blumler, M. A. (1984). Climate and the annual habit. Unpublished MA thesis, University of California, Berkeley.

Blumler, M. A. (1991). Fire and agricultural origins: preliminary investigations. In Nodvin, S. C. and Waldrop, T. A. (eds.), *Fire and Environment: Ecological and Cultural Perspectives, Proceedings of an International Symposium, March 20–24, 1990, Knoxville,* U.S. Department of Agriculture, Southeastern Forest Experiment Station, General Technical Report SE-69, Ashville, NC, pp. 351–8.

Blumler, M. A. (1992a). Independent inventionism and recent genetic evidence on plant domestication. *Economic Botany,* **46**, 98–111.

Blumler, M. A. (1992b). Seed weight and environment in mediterranean-type grasslands in California and Israel. Unpublished PhD thesis, University of California, Berkeley.

Blumler, M. A. (1993). Successional pattern and landscape sensitivity in the Mediterranean and Near East. In Thomas, D. S. G. and Allison, R. J. (eds.), *Landscape Sensitivity*, Wiley, Chichester, pp. 287–305.

Blumler, M. A. (in preparation). A preliminary appraisal of intraspecific variation in wild emmer, and introgression from durum.

Blumler, M. A. and Byrne, R. (1991). The ecological genetics of domestication and the origins of agriculture. *Current Anthropology,* **32**, 23–54.

Blumler, M. A., Olsvig-Whittaker, L. and Naveh, Z. (1992). Microsite/disturbance heterogeneity and species diversity in a Mediterranean oakpark, Association of American Geographers 88th Annual Meeting, Abstracts, San Diego, p. 21.

Blumler, M. A., Olsvig-Whittaker, L. and Naveh, Z. (in preparation). Between a rock and a hard place: succession, vegetation resilience, and microsite heterogeneity on calcareous substrate in Israel.

Burtt, B. L. (1977). Aspects of diversification in the capitulum. In Heywood, V. H., Harborne, J. B. and Turner, B. L. (eds.), *The Biology and Chemistry of the Compositae,* Volume 1, Academic Press, New York, pp. 41–59.

Byrne, R. (1987). Climatic change and the origins of agriculture. In Manzanilla, L. (ed.), *Studies in the Neolithic and Urban Revolutions. The V. Gordon Childe Colloquium, Mexico, 1986,* British Archaeological Reports, Oxford, pp. 21–34.

Chapin, F. S., Groves, R. H. and Evans, L. T. (1989). Physiological determinants of growth rate in response to phosphate supply in wild and cultivated *Hordeum* species. *Oecologia,* **79**, 96–105.

Chapman, C. G. D. (1985). *The Genetic Resources of Wheat: A Survey and Strategies for Collecting,* International Board of Plant Genetic Resources Secretariat, Rome.

Cook, O. F. (1913). Wild wheat in Palestine. *US Department of Agriculture. Bureau of Plant Industry Bulletin,* **274**, 1–56.

Danin, A. and Orshan, G. (1990). The distribution of Raunkiaer life forms in Israel in relation to the environment. *Journal of Vegetation Science,* **1**, 41–8.

Danin, A. and Yom-Tov, Y. (1990). Ant nests as primary habitats of *Silybum marianum* (Compositae). *Plant Systematics and Evolution,* **169**, 209–17.

Davis, P. H., Harper, P. C. and Hedge, I. C. (eds.) (1971). *Plant Life of South West Asia,* Royal Botanical Garden, Edinburgh.

Ehrendorfer, F. (1971). Evolution and eco-geographical differentiation in some south-west Asiatic Rubiaceae. In Davis, P. H., Harper, P. C. and Hedge, I. C. (eds.), *Plant Life in South West Asia,* Royal Botanical Garden, Edinburgh, pp. 195–213.

Ellner, S. P. and Shmida, A. (1981). Why are adaptations for long-range dispersal rare in desert plants? *Oecologia,* **51**, 133–44.

Fritsch, R., Kruse, J., Ohle, H. and Schäfer, H. I. (1977). Vergleichend-anatomische untersuchungen im verwandtschaftskreis von *Triticum* L. und *Aegilops* L. (Gramineae). *Kulturpflanze,* **25** (Suppl.), 155–265.

Golenberg, E. M. (1986). Multilocus structures in plant populations: population and genetic dynamics of *Triticum dicoccoides.* Unpublished PhD thesis, State University of New York, Stony Brook.

Harlan, J. R. (1975). *Crops and Man,* American Society of Agronomy, Madison.

Harlan, J. R. and Zohary, D. (1966). Distribution of wild wheats and barley. *Science,* **153**, 1074–80.

Harlan, J. R., de Wet, J. M. J. and Price, E. G. (1973). Comparative evolution of cereals. *Evolution,* **27**, 311–25.

Harper, J. L. (1977). *Population Biology of Plants,* Academic Press, London.

Harper, J. L., Lovell, P. H. and Moore, K. G. (1970). The shapes and sizes of seeds. *Annual Review of Ecology and Systematics,* **1**, 327-56.

Kihara, H. (1954). Considerations on the evolution and distribution of *Aegilops* species based on the analyser-method. *Cytologia,* **19**, 336–57.

Kutiel, P. (1985). Competition between annual plants along a soil depth gradient. Unpublished PhD thesis, Hebrew University, Jerusalem.

Ladizinsky, G. (1971). Biological flora of Israel. 2 *Avena* L. *Israel Journal of Botany,* **20**, 133–51.

Lesins, K. A. and Lesins, I. (1979). *Genus Medicago (Leguminosae): a Taxogenetic Study.* Dr W. Junk, The Hague.

Litav, M., Kupernik, G. and Orshan, G. (1963). The role of competition as a factor determining the distribution of dwarf shrub communities in the Mediterranean territory of Israel. *Journal of Ecology,* **51**, 467–80.

Mandaville, J. P. (1986). Plant life in the Rub'al-Khali (the Empty Quarter), south-central Arabia. *Proceedings of the Royal Society of Edinburgh,* **89B**, 147–57.

Miller, T. E. (1987). Systematics and evolution. In Lupton, F. G. H. (ed.), *Wheat Breeding,* Chapman and Hall, London, pp. 1–30.

Naveh, Z. and Whittaker, R. H. (1979). Structural and floristic diversity of shrublands and woodlands in northern Israel and other mediterranean countries. *Vegetatio,* **41**, 171–90.

Noy-Meir, I. (1990). The effect of grazing on the abundance of wild wheat, barley and oats. *Biological Conservation,* **51**, 299–310.

Peart, M. H. and Clifford, H. T. (1987). The influence of diaspore morphology and soil-surface properties on the distribution of grasses. *Journal of Ecology,* **75**, 569–76.

Ratcliffe, D. (1961). Adaptation to habitat in a group of annual plants. *Journal of Ecology,* **49**, 187–203.

Raven, P. H. (1973). The evolution of mediterranean floras. In di Castri, F. and Mooney, H. A. (eds.), *Mediterranean-Type Ecosystems: Origin and Structure,* Springer-Verlag, Berlin, pp. 213–24.

Sakamoto, S. (1982). The Middle East as a cradle for crops and weeds. In Holzner, W. and Numata, M. (eds.), *Biology and Ecology of Weeds,* Dr W. Junk, The Hague, pp. 97–109.

Shmida, A. and Ellner, S. (1984). Seed dispersal on pastoral grazers in open Mediterranean chaparral, Israel. *Israel Journal of Botany,* **32**, 147–59.

Stebbins, G. L. (1956). Cytogenetics and evolution in the grass family. *American Journal of Botany,* **43**, 890–905.

Stebbins, G. L. (1974). *Flowering Plants: Evolution above the Species Level,* Belknap Press of the Harvard University Press, Cambridge, Mass.

Venable, D. L. and Lawlor, L. (1980). Delayed germination and dispersal in desert annuals: escape in space and time. *Oecologia,* **46**, 262–72.

Yaalon, D. H. and Ganor, E. (1973). The influence of dust on soils during the Quarternary. *Soil Science,* **116**, 146–55.

Yaalon, D. H. and Ganor, E. (1975). Rates of aeolian dust accretion in the Mediterranean and desert fringe environments of Israel. *Congress of Sedimentology,* **2**, 169–74.

Zohary, D. (1965). Colonizer species in the wheat group. In Baker, H. G. and Stebbins, G. L. (eds.), *The Genetics of Colonizing Species,* Academic Press, New York, pp. 403–19.

Zohary, D. (1969). The progenitors of wheat and barley in relation to domestication and agricultural dispersal in the Old World. In Ucko, P. J. and Dimbleby, G. W. (eds.), *The Domestication and Exploitation of Plants and Animals,* Duckworth, London, pp. 47–66.

Zohary, D. (1983). Wild genetic resources of crops in Israel. *Israel Journal of Botany,* **32**, 97–127.

Zohary, D. and Brick, Z. (1961). Triticum dicoccoides in Israel: notes on its distribution, ecology and natural hybridization. *Wheat Information Service,* **13**, 6–8.

Zohary, D. and Feldman, M. (1962). Hybridization between amphidiploids and the evolution of polyploids in the wheat *(Aegilops–Triticum)* group. *Evolution,* **16**, 44–61.

Zohary, D. and Hopf, M. (1988). *Domestication of Plants in the Old World,* Oxford University Press, Oxford.

Zohary, D., Harlan, J. R. and Vardi, A. (1969). The wild diploid progenitors of wheat and their breeding value. *Euphytica,* **18**, 58–65.

Zohary, M. (1937). Die verbreitungsekologischen verhältnisse der flora Palaestinas. I. Die anti-telechorischen erscheinungen. *Beihefte zum Botanischen Zentralblatt,* **56A**, 1–155.

Zohary, M. (1950). The segetal plant communities of Palestine. *Vegetatio,* **2**, 387–411.

Zohary, M. (1962). *Plant Life of Palestine,* Ronald Press, New York.

Zohary, M. (1973). *Geobotanical Foundations of the Middle East,* 2 volumes, Gustav Fischer, Stuttgart.

Zohary, M. (1983). Man and vegetation in the Middle East. In Holzner, W., Werger, M. J. A. and Kusima, I. (eds.), *Man's Impact on Vegetation,* Dr W. Junk, The Hague, pp. 287–96.

Zohary, M. and Feinbrun-Dothan, N. (1966–88). *Flora Palaestina,* 4 volumes, Israel Academy of Sciences and Humanities, Jerusalem.

14 Post-European Changes in Creeks of Semi-Arid Rangelands, 'Polpah Station', New South Wales

J. PICKARD
Graduate School of the Environment, Macquarie University, NSW, Australia

ABSTRACT

Grazing of sheep and cattle is the largest Holocene environmental change in semi-arid Australia. The semi-arid rangelands have been grazed for at least 150 years, and domestic and feral herbivores are now the dominant agents of geomorphological change. Consequences include extensive soil erosion and substantial sedimentation in reservoirs. Most research has concentrated on erosion but has overlooked deposition on floodplains. On 'Polpah Station' channels of ephemeral creeks have migrated substantially over the last century, often associated with attempts to dam or divert the creeks. Typical rates of migration are 1–2 m yr^{-1} since 1883; rates of downcutting in the late nineteenth century were approximately 0.06 m yr^{-1}. Up to 1 m of post–European sedimentation occurs over 10 km of floodplain. The bulk of the erosion and deposition probably occurred before the 1950s. Although increased rainfall and better land management are the most likely reasons for the reduced rates after 1950, their relative contributions cannot be determined. Increased rainfall predicted for the Greenhouse Effect is unlikely to increase erosion, sedimentation or creek channel change beyond the rates operating at present.

INTRODUCTION

The geological record is a story of both cataclysmic and gradual changes. Entire land masses have appeared and disappeared, climates have changed, ice sheets waxed and waned, orders of plants and animals have appeared and then become extinct. Although a cliche, there is no doubt that change is the only constant. If this is the case, what is so different about the last 200 years of environmental history of semi-arid Australia? The difference is the rate at which the changes have occurred since the dry 70% of Australia was first invaded and settled by European pastoralists.

Water and wind have removed enormous quantities of soil exposed by loss of protective vegetation and loosened by trampling of sheep and cattle (Noble & Tongway, 1986). Although the scars of this erosion are obvious, the fate of the material is not. Not all the soil blew to New Zealand in dust storms between 1900 and the 1940s (McTainsh *et al.*, 1990); a large proportion was deposited locally.

Wasson & Galloway (1986) compared sedimentation in Umberumberka Reservoir (Figure 14.1) and on a nearby Holocene alluvial fan. They found mean sediment yield in the early Holocene (3000–6000 yr BP) of 0.8–1.0 m^3 ha^{-1} yr^{-1}. From 3000 yr BP to 1850 AD (the arrival of European settlers) the rate fell to 0.04 m^3 ha^{-1} yr^{-1}. From 1915 to 1941 sediment

Environmental Change in Drylands: Biogeographical and Geomorphological Perspectives.
Edited by A. C. Millington and K. Pye. © 1994 John Wiley & Sons Ltd

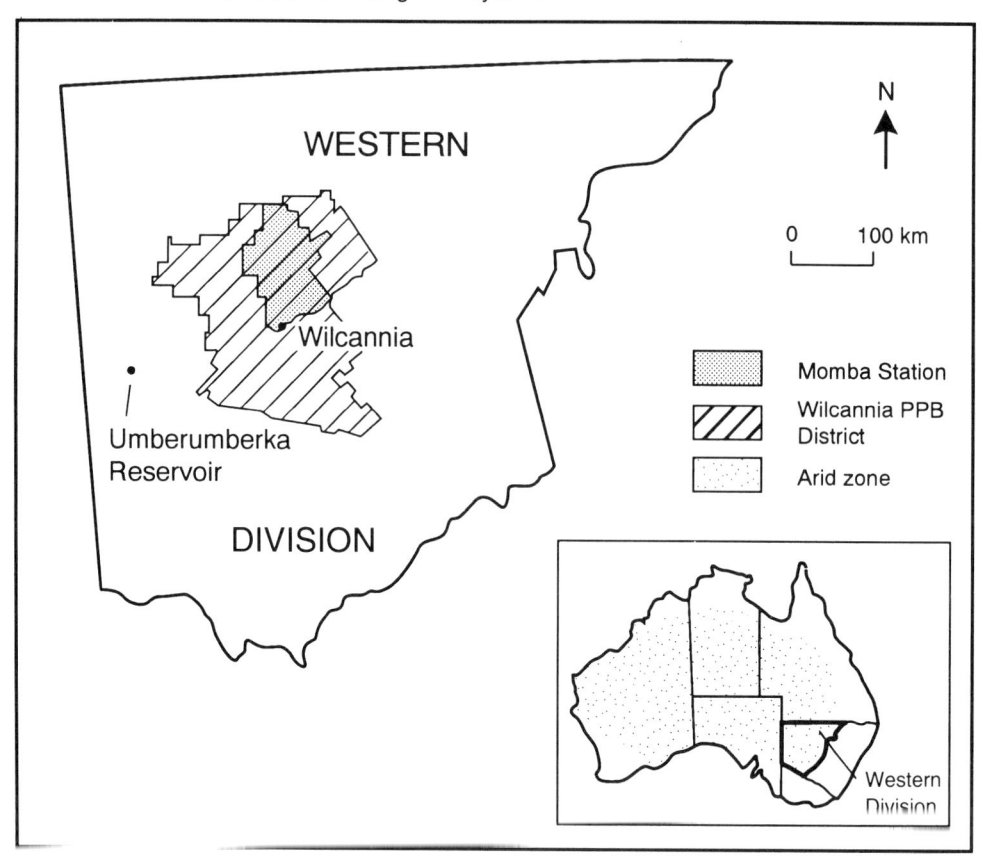

Figure 14.1 Map of the Western Division of New South Wales showing the location of 'Momba Station', Wilcannia Pastures Protection Board District, and Umberumberka Reservoir. Inset map of Australia shows location of the Western Division in the arid zone

was delivered to the reservoir at the rate of 3.1 m^3 ha^{-1} yr^{-1}, but fell in the period 1941–82 to 1.3 m^3 ha^{-1} yr^{-1}. Overall, the post-European sediment yields are 50–90 times higher than the Holocene rates.

The initial very high rates after European settlement are ascribed to stock management. However, the reasons for the decline in sediment yield after 1941 are more obscure. An increase in rainfall (Nicholls & Lavery, 1992) over south-eastern Australia played a part, and Wasson & Galloway (1986) conclude that '...the reduced sediment yields since the 1940s are no more than partially related to [grazing] management and the reduction of rabbits'. This is relevant to the present study as one of the objectives is to partition the cause for changes.

Wasson & Galloway's study is one of several that document post-European geomorphological changes, but one of the very few that is qualitative. Most others concentrate on erosion, e.g. Holmes (1938) and Beadle (1948). While the general story of changes is well known and well documented, it is only poorly known in detail. There are few answers to key questions:

1. What was the sequence of events in a single locality?
2. What is the relationship between stocking rates and erosion over different seasons?
3. Did the drought in the 1940s cause major erosion?

These questions are important in their own right, but gain extra significance as input to a more general question: why is the land in its present condition? Currently we have very little idea of the exact nature of the changes that occurred after European occupation of these semi-arid rangelands. I am currently studying an area exceeding 0.8 Mha near Wilcannia in western New South Wales (NSW) that was originally a single property 'Momba Station' (Figure 14.1). In 1884, 'Momba' was the largest property to ever exist in NSW and after a series of subdivisions (Pickard, 1990) is now occupied by some 27 individual properties. The objective of the MOMBA Program is to provide an integrated and coherent explanation of how the land got into its present condition. A key question is the relative importance of climate and grazing management. Dramatic changes in creek channels and sediments which occur along many creeks in the area may help answer these questions. Early survey plans accurately show channels, and comparison with modern air photos indicates substantial changes. Further evidence is provided by silted-up dams and buried fences built almost 100 years ago. In this paper, I describe work in progress and provide preliminary interpretations of the results to date.

A CENTURY OF ENVIRONMENTAL CHANGES IN SEMI-ARID NEW SOUTH WALES: RAINFALL AND HERBIVORES

In south-eastern Australia, there have been three distinct periods of rainfall in the last 100 years (Nicholls & Lavery, 1992). The periods before 1910 and after 1947 were wetter than the intervening period 1910 to 1947. Droughts (years with <10% of long-term mean annual rainfall) occur at irregular but frequent intervals (Figure 14.2).

In the 1850s, Europeans introduced domestic stock into this semi-arid environment of irregular and highly variable precipitation. Settlement initially centred on the (almost) perennial rivers and a few waterholes. These were quickly augmented by ground tanks and wells. For a short time, stock numbers rose very steeply (Figure 14.3) until graziers were rudely awakened to the harsh realities of semi-arid ecosystems. The large biomass of perennial grasses, herbs and small shrubs belied the very low annual productivity. Some watering points carried over 15 000 sheep (Pickard, 1991a) which denuded circular areas up to 10 km across. Palatable shrubs were lopped to provide feed for starving stock. Other trees and large shrubs were felled to fire boilers which powered steam pumps on the wells, and for fences (Pickard, in press).

By the 1890s, the lethal combination of drought and too many stock had effectively removed all the feed. Up to 50% of the sheep died from starvation. Subsequently, stock numbers have never recovered to such levels (Figure 14.3). Although there is a general relationship between rainfall and stock numbers over the last 100 years (cf. Figures 14.2 and 14.3), the correlation coefficient is only 0.23. Regressions of sheep numbers against rainfall, lagged rainfall and similar linear predictors are all weak (maximum $r^2 = 15.5\%$). Rainfall of the current and previous years are not the sole driving forces; commodity prices play a critical role.

At the time of European settlement, the major agents of geomorphological change were irregular rainfall and wind. Subsequently, these have been supplanted by introduced dom-

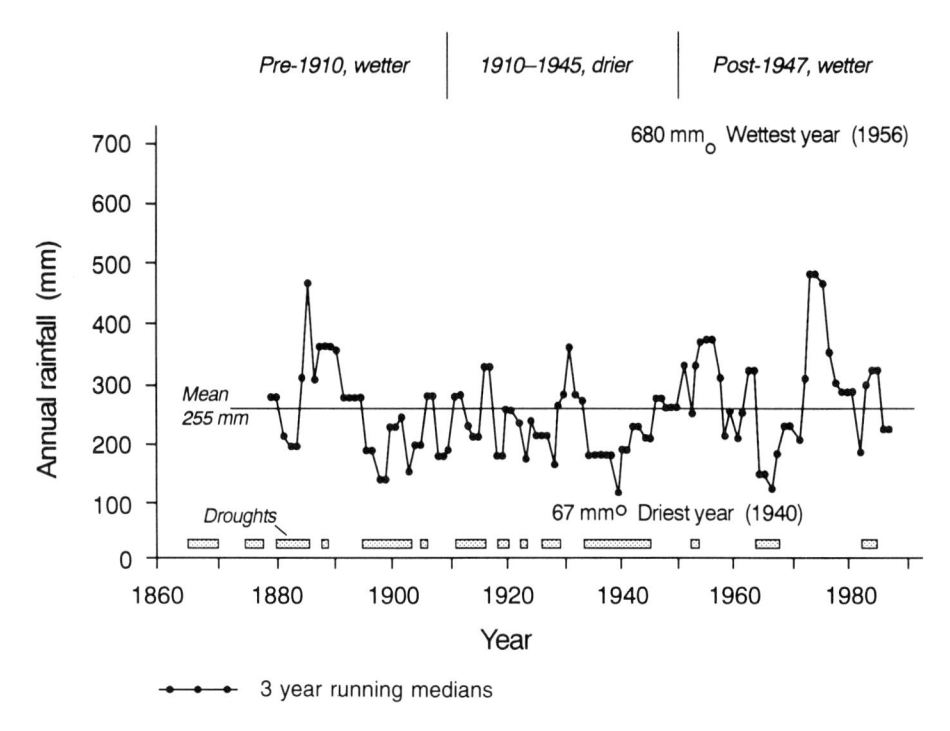

Figure 14.2 Annual rainfall records at Wilcannia from the 1880s to the present. Wetter and drier sequences for south-eastern Australia are also shown (from Bureau of Meteorology records)

estic and feral herbivores (sheep, cattle, rabbits). There are few parts of southern semi-arid Australia that do not show both the past and present impact of their numbers: eating available plant material and trampling the soil surface. The soil is then considerably more exposed to erosion by wind and water. Once mobilised, the soil has impacts on creek channels and sedimentation far beyond its place of origin.

CHANNEL CHANGES IN WANNARA AND FAULKANHAGAN CREEKS, 'POLPAH STATION'

Wannara Creek is the major creek draining 'Polpah Station' (Figure 14.4). It runs south-easterly across the property before reaching the Paroo River some 20 km to the east. Faulkanhagan Creek is the major tributary of Wannara Creek (Figure 14.5). Both are ephemeral — the few large waterholes last a few months after heavy falls of rain. Both creeks exhibit evidence of major channel changes over the last century, and longer. Stream traces are readily visible on the air photos as lines of the rheophyte tree River Red Gum (*Eucalyptus camaldulensis*). Older channels are marked by decaying stumps of *E. camaldulensis*, more recently abandoned channels by lines of dying trees. Some of these changes to channels are linked to European activity, such as dam or diversion construction, others are probably just a continuation of longer-term channel migrations that pre-date European settlement.

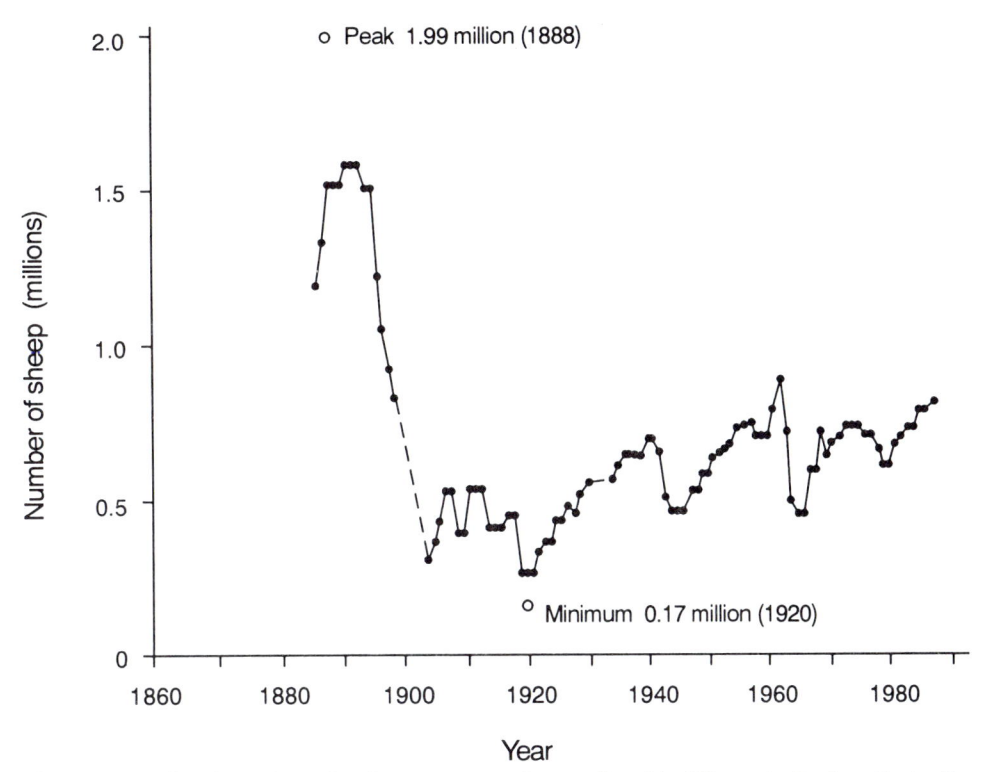

Figure 14.3 Stock numbers (as three-year running medians) in Wilcannia Pasture Protection Board District from the 1880s to the present (from Annual Reports of the Western Lands Commission)

Here I describe in semi-quantitative terms some of the changes. Research in progress is aimed at determining the ages of palaeochannels, and the rates of channel formation and migration, and the processes involved.

Two dams and one diversion have been constructed on 'Polpah' (Figure 14.5). Wannara Dam is an earth dam running approximately west–east and was built before October 1883. It is rectangular, 70 m long × 3 m wide and 1.5 m high (above the plain) with rounded ends protected or reinforced with vertically placed logs up to 20 cm diameter, and boulders to 0.2–0.3 m diameter. On the ground, there is now little sign of a channel above or below the dam. By 9 July 1885, the creek was already flowing around the dam (Figure 14.6) in a 'byewash' (sic). The dam was probably totally silted up and worthless by 1905 because it is omitted from an otherwise detailed appraisal of the area (d'Apice, 1905, 1907 Plan). By the late 1890s, waterholes and dams all across 'Momba' were silting up, so the events at Wannara Dam are by no means unique. There was considerable erosion regionally (Royal Commission, 1901) and, by implication, deposition.

Although the surveys in 1883 and 1885 are very restricted, several changes can be inferred by comparison with the 1980 air photos (Figure 14.6). The present channel separates from the 1883 channel approximately 1 km above the dam and then bifurcates above the dam, leaving the dam isolated on an island some 1 km long × 400 m wide. A nest of channel traces occur below the dam, and the 1885 'byewash' trace is clearly visible. The 1883

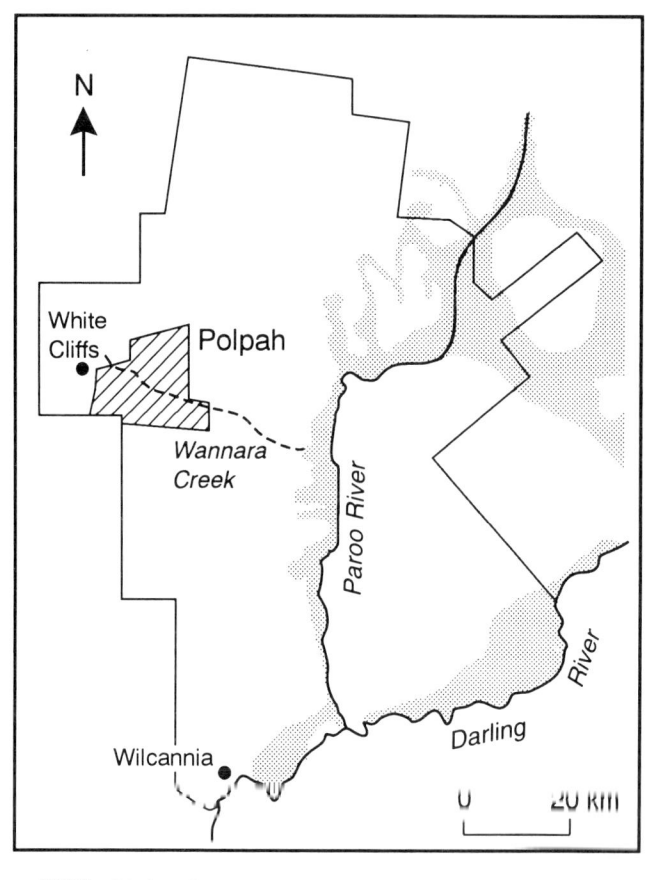

Figure 14.4 Map of 'Momba Station' at its maximum size showing the locations of 'Polpah Station' and Wannara Creek

survey ran north from the dam for almost 1.5 km. The 1885 survey extended south from the dam and up on to the stony slope which adjoins the floodplain. Neither plan shows the present main channel some 40 m wide and 2.5 m deep, which is 250 m south of the dam, or the smaller secondary channel, some 20 m wide and 2.5 m deep running 200 m north of the dam. Although such negative evidence may not be reliable, I suggest that the plans indicate major changes in the channels at Wannara Dam. Basically, the channels have split and moved an average of 225 m from their 1883 position. The overall rate of channel migration at the dam is approximately 2 m yr^{-1} over the past 100 years. While doing so, the local direction of flow has altered by almost 90°.

The date of construction of Lower Wannara Dam (Figure 14.5) is unknown. The wall is approximately 60 m long, 5 m wide and 1 m high (above the plain), with boulders and vertical posts armouring the northern end. Immediately upstream of the dam and ending at the wall is the infilled trace of a channel that still supports living *E. camaldulensis*. The nearest present channel is some 100 m north of the dam, indicating channel migration of this amount over perhaps a century at an average rate of 1 m yr^{-1}.

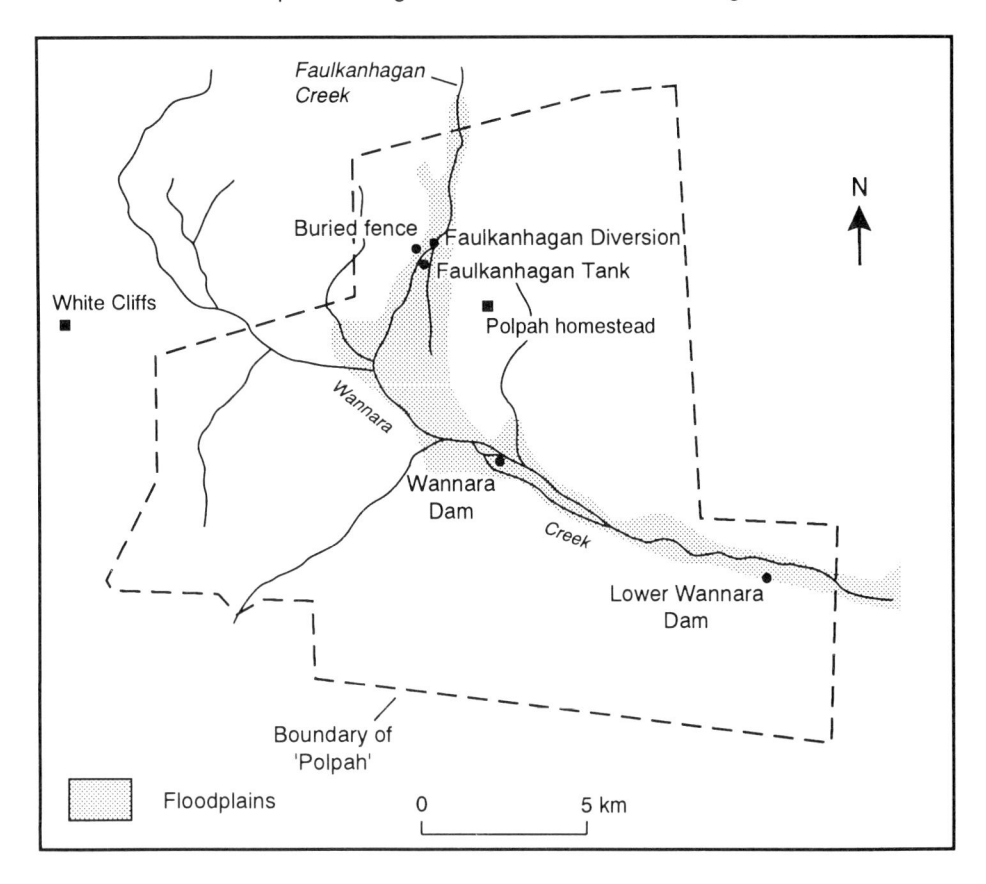

Figure 14.5 Map of 'Polpah' showing the locations of sites described in the text, and floodplains of part of catchments of Wannara and Faulkanhagan Creeks

The consequences of Faulkanhagan Diversion (Figure 14.5) have been more complex and more dramatic. The diversion wall was at least 45 m long × 4 m wide, and 1 m above the bank of the creek. Today, remnants stand on the north and south banks of the creek, immediately downstream of the diverted channel. The top of the wall is approximately 2.2 m above the creek bed. The wall was constructed across Faulkanhagan Creek to divert water into Faulkanhagan Tank (Figure 14.5). The tank was constructed in the early 1880s (Plan Runs 592.3067; Plan 54.2188). By this stage graziers may have learnt that dams across creeks quickly silted up, so they built the tank on the flood plain and hoped to divert water into it. Many other tanks in the Momba study area dating from this period show similar planning.

The major creek is now 15 m wide with 1.2 m high eroded vertical walls at the diversion bank. The side channel is typically 1.5 m deep between the remains of the diversion wall and Faulkanhagan Tank. While the diversion wall may have been successful for a few years, it did not last long and was breached. Probably prior to this, the flow down the diversion was sufficient to deepen any pre-existing channel to below the intake to Faulkanhagan Tank. Today, the creek is approximately 1.5 m lower than the inlet to the tank. In 1905 (d'Apice 1907 Plan) the tank was described as silted, but I believe that the problem was more one of loss of catchment after the diversion channel overdeepened. Assuming that the diversion

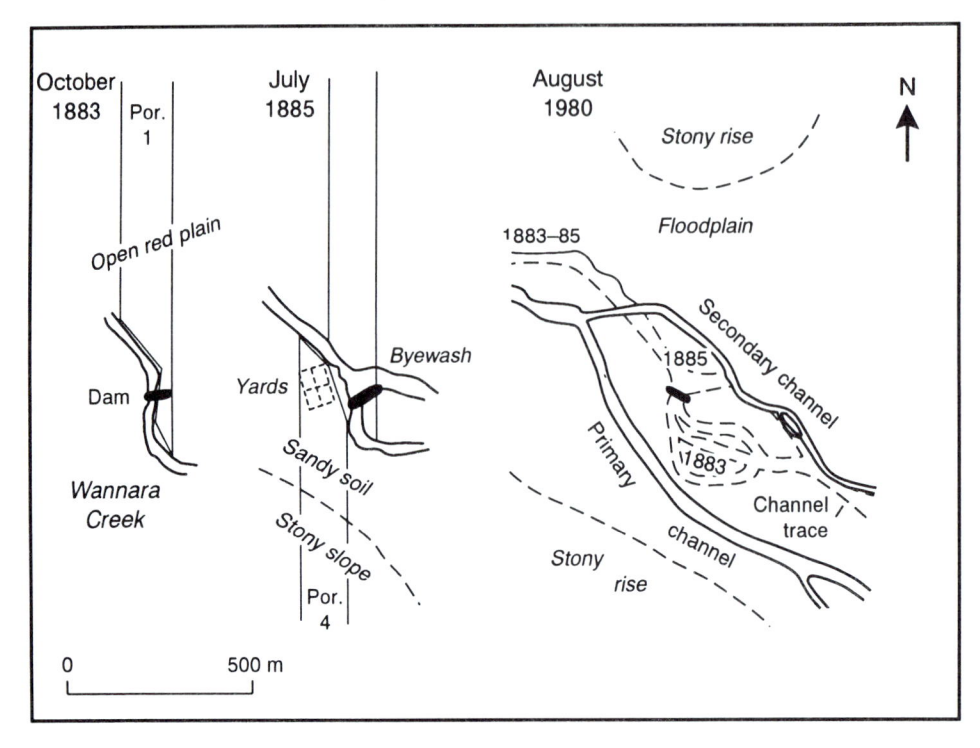

Figure 14.6 Changes in channels of Wannara Creek near Wannara Dam, 1883–1980, compiled from survey plans and air photos. Thin straight lines in 1883 and 1885 are surveyed lines defining boundaries of cadastral portions 1 and 4. Orientation of plans is based on north–south surveys along the marked lines. The 1980 diagram is oriented by comparison with the channels in 1883 and 1885. Thus changes in the orientation of the dam wall are an artefact of how the surveyors depicted the dam. Note the nested channel traces below the dam in 1980. These are tentatively dated on the basis of their orientation (from Plans 20.2188, 1883, and 55.2188, 1885; air photos, 1980; and field observations)

was built in 1880 and that the channel reached its present depth by 1905, this indicates a rate of downcutting of 0.06 m yr^{-1} over the 25 years.

The diverted flow exploited an older abandoned stream trace running all the way down the flood plain to Wannara Creek. *Eucalyptus camaldulensis* has colonised almost the full length of the channel except for the middle section between the tank and Wannara Creek. Here the incised channel disappears in a maze of disconnected elongate shallow ditches generally aligned north–south. The southern end of the incised channel is ill-defined as the channel is being progressively infilled by sediment lodged behind dams of accumulated plant debris.

FENCE POSTS AS INTEGRATED INDICATORS OF EROSION OR DEPOSITION ON WANNARA AND FAULKANHAGAN CREEKS, 'POLPAH STATION'

Channels are not the only elements of creeks that have changed on 'Polpah'. Buried fences indicate that considerable sedimentation on the floodplains of both Faulkanhagan and Wan-

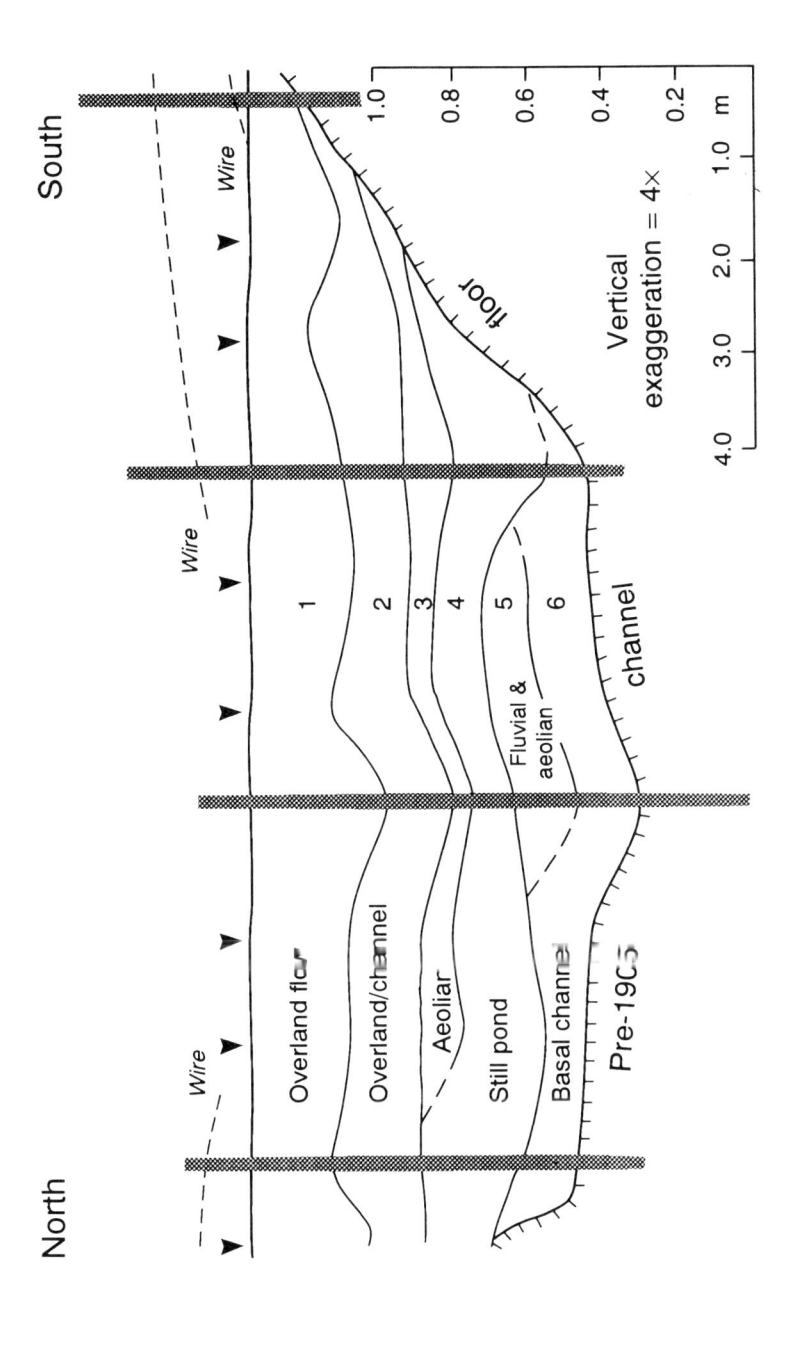

Figure 14.7 Measured cross-section of buried fence on floodplain of Faulkanhagan Creek on 'Polpah'. Section measured in excavated trench at each fence post (solid black) and at vertical arrows. Buried strands of fence wire omitted for clarity. Generalised interpretations of conditions of deposition are shown for each of the six stratigraphic components of the infill (modified from Pickard, 1991b)

nara Creeks. North of Faulkanhagan Tank (Figure 14.5) old fence posts on an apparently level floodplain became shorter and shorter and then increased in height again for no apparent reason (Figure 14.7). An irregular line of dead and decaying stumps of *E. camaldulensis* crossing the fence line suggested that the shorter fence posts had been placed in a now-infilled channel of the creek. Excavation along the fence revealed a sequence of fluvial and aeolian sediments, confirming the existence of a channel that had been totally filled by sediment after the fence was constructed before 1905 (Figure 14.7) (Pickard, 1991b).

There is a consistent layer of modern sediment 0.8–1.0 m deep over a distance of >10 km along the floodplains of both Faulkanhagan and Wannara Creeks. These deposits are difficult to recognise, and local graziers were completely unaware of them. At each site, the key evidence is provided by the very old buried fences. Following Wasson & Galloway's (1986) experience and the general history of the area, I believe that the bulk of deposition ceased about 1940. Dating with ^{210}Pb was unsuccessful, as the sediments contain no detectable levels of the isotope, probably because they are derived from sub-soils. If this is the case, then there are important implications for the processes operating. Sub-soils would only be exposed by scalding (aeolian removal of sandy surfaces leaving a relatively impermeable clay sub-soil) or by lateral expansion of gullies by sidewall erosion. These possibilities have not yet been investigated to determine the more likely source of sediment.

INTEGRATION: CHANGES IN RAINFALL AND STOCK, CHANGES IN CREEKS AND SEDIMENTS

What are the primary causes of the channel changes in Wannara and Faulkanhagan Creeks and of the massive post-1905 sedimentation? Rainfall was lower through the critical period 1910–45 (Figure 14.2), but this alone is probably insufficient to initiate or maintain either effect. In fact, under generally lower rainfall the creeks are more likely to silt up. Occasional extreme floods could reverse this, and one of the characteristics of semi-arid streams is rare extreme floods. There are no data on the magnitude and frequency of such events. Sheep numbers have gradually increased since 1900 following the crash of the 1890s (Figure 14.3). Together, decreased rain and increasing sheep numbers could have both initiated and maintained the geomorphological changes. Sheep grazing alone would probably be sufficient.

Despite the wide extent of land degradation in semi-arid Australia, the causes are still acrimoniously debated. In the most general terms, the fundamental cause is too many herbivores trying to eat too little plant biomass on land of inherently low productivity (Christie, 1984; Pickard, 1993). This explains why degradation continues today although many graziers prefer to argue that the cause is either droughts ('God did it') or overgrazing at the turn of the century ('Grand-dad did it') (Palmer, 1991). Neither explanation is at all convincing, but the results from the present project are relevant to the debate.

There is little doubt that the rates of changes have increased since European occupation (Wasson & Galloway, 1986), but can we say the same of the magnitude? Stream traces on airphotos of Wannara Creek and buried channels being exposed by the current creek channel indicate that channel changes pre-date European occupation of the land. For example, extensive charcoal from logs burnt *in situ* occurs below gravel lenses of an old channel exposed in the bank of Wannara Creek between Wannara Dam and Lower Wannara Dam. This is dated at 21 000 ± 1000 ^{14}C yr BP (LHRL 216–2). The buried channel here suggests that the changes are comparable with those of today. Perhaps the magnitude is controlled by intrinsic properties of the creek system. Current changes are merely the most recent expression of continuing morphological development of these semi-arid creeks adjusting to different combinations of flow and sediment. The major geomorphological agents in these semi-arid lands are now introduced herbivores (sheep, cattle and rabbits). Combined, these animals exert more impact on processes than do rainfall, wind, sediment type and others.

Even if the bulk of the recent sedimentation on the floodplains occurred during the drier period of 1910–45 we have no independent evidence that early graziers and their sheep were the cause. The late 1940s are a key period as they mark the end of a 40 year sequence of drier years. New technology was introduced after World War II: numbers of watering points increased substantially (Pickard, 1990), polythene pipe was used to extend permanent waters into previously dry areas and stock management changed with better road transport networks. However, as both rainfall and technology changed simultaneously, we are currently unable to resolve the cause. Perhaps the only real test will come during the next 40 year period of dry years. Even this test will be confounded because technology and economic conditions are never static. Most likely we will never find an unambiguous answer as we cannot carry out the requisite unambiguous full-scale experiment over a sufficient time period.

The most extensive relevant literature exists for the 'arroyo problem' of the arid southwest of the United States of America (e.g. Graf, 1983, 1988). There is no available evidence (photographic, anecdotal or Aboriginal oral tradition) to indicate that the creeks on 'Momba' or 'Polpah' have incised pre-existing meandering channels. The single known exception is the channel diverted from Faulkanhagan Creek described above. Despite this, the 'arroyo problem' literature is relevant as it canvasses three main causes: grazing, climatic changes and catastrophic events. Graf (1983, p. 293) summarised the position as 'a widely accepted, well-integrated theory for arroyos has not yet been developed because of three major problems: limited testing, lack of sediment data, and lack of long-term historical records.' As I suggest for 'Polpah', we may never be able to apportion cause.

CHANGES IN THE FUTURE

Forecasting changes is hazardous in the highly variable and poorly understood semi-arid zone of Australia. Graetz et al.(1988) have modelled the likely course of arid and semi-arid Australia over the next 50 years under the best available scenario predicted for the Greenhouse Effect. For south-eastern Australia generally, rainfall will increase in the warm seasons, thus continuing the trends of the last several decades described earlier (Nicholls & Lavery, 1992). Sheep grazing will remain the predominant land use.

The projected increase in warm-season rain will increase the distribution and abundance of grasses as ground cover in all arid zone vegetation. With the projected higher temperatures, Graetz et al. (1988) expect that this grassy vegetation will be more prone to wild fires. The consequences of these secondary changes are hard to predict. Even with more frequent and widespread wild fires, erosion (and hence sedimentation) may not increase as conditions are conducive to rapid regrowth of the grasses. Indeed, it may well be the case that the bulk of easily eroded material was liberated in the first few decades of European settlement and the consequent lack of erodible material, not changes in stock numbers or rainfall, is what has limited erosion. In any event, I believe that the impact of stock management will swamp all but the most extreme rainfall or wildfire events.

ACKNOWLEDGEMENTS

The author's research in semi-arid New South Wales is supported by grants from the Australian Institute of Nuclear Science and Engineering, Australian Research Council, Macquarie University Research Grants and NSW Department of Planning National Estate Grants. Graziers at White Cliffs have helped in many ways; the author wishes to thank them all, especially Annette and Barry Turner of 'Polpah' for their country hospitality and friendship. Damian Gore helped with the field work and his alternative interpretations considerably improved the author's arguments.

REFERENCES

Beadle, N. C. W. (1948). *The Vegetation and Pastures of Western New South Wales with Special Reference to Soil Erosion,* New South Wales Government Printer, Sydney.
Christie, E. K. (1984). Production and stability of semi-arid grassland. In Parkes, D. (ed.), *Northern Australia. The Arenas of Life and Ecosystems on Half a Continent,* Academic Press, Sydney, pp. 157–71.
d'Apice, L. (1905). Recommending area to be offered under Part VII of the Western Lands Act within the Momba Resumed Area No. 55, Counties of Fitzgerald, Killara, Yungnulgra and Young, Western Division. Unpublished report 05/64 of 28 August 1905 on File WLC 51–4060, New South Wales Government Records Repository #258124.
Graetz, R. D., Walker, B. H. and Walker, P. A. (1988). The consequences of climatic change for seventy percent of Australia. In Pearman, G. I. (ed.), *Greenhouse. Planning for Climatic Change,* CSIRO, Melbourne, pp. 199–420.
Graf, W. L. (1983). The arroyo problem — palaeohydrology and palaeohydraulics in the short term. In Gregory, K. J. (ed.), *Background to palaeohydrology,* Wiley, London, pp. 279–302.
Graf, W. L. (1988). *Fluvial Processes in Dryland Rivers,* Springer-Verlag, Berlin.
Holmes, M. (1938). The erosion-pastoral problem of the Western Division of New South Wales. *University of Sydney Publications in Geography,* 2, 1–51.
McTainsh, G. H., Lynch, A. W. and Burgess, R. C. (1990). Wind erosion in eastern Australia. *Australian Journal of Soil Research,* 28, 323–39.
Nicholls, N. and Lavery, B. (1992). Australian rainfall trends during the twentieth century. *International Journal of Climatology,* 12, 153–63.
Noble, J. C. and Tongway, D. J. (1986). Pastoral settlement in arid and semi-arid rangelands. In Russell, J. S. and Isbell, R. F. (eds.), *Australian Soils: The Human Impact,* University of Queensland Press, Brisbane, pp. 217–42.
Palmer, R. H. (1991). The Western Division of New South Wales: a miracle of recovery. *Australian Journal of Soil and Water Conservation,* 4(1), 4–8.
Pickard, J. (1990). Analysis of stocking records from 1884 to 1988 during the subdivision of *Momba,* the largest property in semi-arid New South Wales. *Proceedings of the Ecological Society of Australia,* 16, 245–53.
Pickard, J. (1991a). Land management in semi-arid New South Wales. *Vegetatio,* 91, 191–208.
Pickard, J. (1991b). Quaternary studies and land rehabilitation in semi-arid New South Wales; or when is a scald not a scald? In Brierley, G. (ed.), *Quaternary Applications in Australia,* Department of Biogeography and Geomorphology, Australian National University, Canberra, pp. 109–17.
Pickard, J. (1993). Land degradation, land rehabilitation and conservation in the semi-arid zone of Australia: grazing is the problem. . .and the cure, In Moritz, C. (ed.), *Conservation Biology in Australia and Oceania,* Surrey Beatty, Chipping Norton, pp. 131–7.
Pickard, J. (in press). Do old survey plans help us discover what happened to western New South Wales when Europeans arrived? *Australian Zoologist.*

Royal Commission (1901) *Royal Commission to Inquire into the Condition of the Crown Tenants of the Western Division of New South Wales*, 2 volumes, Legislative Assembly of New South Wales, Sydney.

Wasson, R. J. and Galloway, R. W. (1986). Sediment yield in the Barrier Range before and after European settlement. *Australian Rangeland Journal,* **8**, 79–90.

UNPUBLISHED SURVEY PLANS, MAPS AND AIR PHOTOS AVAILABLE FROM GOVERNMENT AGENCIES

Archives Office of New South Wales, Sydney, NSW

d'Apice (1907) Plan: Plan showing areas offered for lease under Part VII of the Western Lands Act, Counties of Fitzgerald, Killara, Yungnulgra and Young, Western Division. Scale 2 miles to an inch (Archives Office of New South Wales file 10/43901, PL 55 'Momba').

Plan Room, Department of Conservation and Land Management, Sydney, NSW

Plan Runs 592.3067: *Plan showing part of the boundaries of Yungnulgra Plains N., Yungnulgra Plains and Yungnulgra Plains S. Runs, Albert District.* Survey date unknown, but late 1870s; scale 1: 31 680.

Western Lands Commission, Dubbo, NSW

Plan 20.2188: Plan of Portion 1, Parish of Rosstrevor, County of Yungnulgra. Surveyed 5 October 1883; scale 1:15 840.

Plan 54.2188: Plan of Portion 4, Parish of Kirk, County of Yungnulgra. Surveyed 11 July 1885; scale 1:15 840.

Plan 55.2188: Plan of Portion 4, Parish of Rosstrevor, County of Yungnulgra. Surveyed 9 July 1885; scale 1:15 840.

New South Wales Land Information Centre, Bathurst NSW

1980 air photos: White Cliffs, roll NSW 2880, negatives 101–103, 13 August 1980, approximate contact scale 1:60 000.

15 Anthropogenic Factors in the Degradation of Semi-arid Regions: A Prehistoric Case Study in Southern France

J. WAINWRIGHT*

Department of Geography, University of Southampton, UK

ABSTRACT

Degradation of the Mediterranean landscape has occurred intermittently since the Early Neolithic in south-west France and north-east Spain, as shown by palaeobotanical and sedimentological data. A case study is made of intensified erosion in the Early Bronze Age (ca. 2000–1600 BC) in Languedoc. The use of computer simulation demonstrates that the pattern of erosion seen today could only have occurred if the joint conditions of extreme rainfall events and total clearance of the land were present. This clearance is probably related to the successive exploitation of increasingly marginal land, upon which contemporary communities were increasingly reliant. This erosive phase appears to be one reason for the collapse in the Early Bronze Age of Languedoc, and subsequent poor fertility of vast areas of this landscape. Comparisons are made with other prehistoric and historic examples in the area to show the persistence of this phenomenon, and the importance of using (pre)historic data in understanding the present and future landscapes of the region.

INTRODUCTION

The present Mediterranean landscape is the result of a wide range of processes occurring at different times in the past and present. From the later prehistoric periods (since the Neolithic), human influences have been increasingly important in their effects on the landscape (cf. Butzer, 1974). Understanding the present landscape is therefore a question of studying the processes at work in terms of their anthropogenic and natural components, and their interlinkages. This allows a better knowledge not only of the past conditions for human occupation of the landscape but also its potential under present conditions, and as the result of further changes.

To demonstrate the potential of this approach, a case study from the Early Bronze Age of south-west France is analysed in detail. The background to the degradation in this area from the Neolithic period is drawn from existing palaeobotanical, sedimentological and archaeological data. By simulating the erosion of a hillslope under various scenarios, the relationship between an archaeological site and the landscape will be shown in dynamic

---

*Present address: Department of Geography, King's College London, UK.

Environmental Change in Drylands: Biogeographical and Geomorphological Perspectives.
Edited by A. C. Millington and K. Pye. © 1994 John Wiley & Sons Ltd

terms. The cause of the degradation to the present-day garrigue is elucidated by the elimination of various hypotheses. From this, implications are drawn for the development of garrigue associations elsewhere in the south of France. Comparisons with other erosion studies in the Mediterranean further show the importance of looking at anthropogenic factors, and the relevance of additional extensions of this approach.

THE EARLY BRONZE AGE CRISIS IN LANGUEDOC

The Early Bronze Age in the south-west of France is represented by a distinct break in the archaeological record. Whereas material is known from a small number of stratified cave sites, the continuing use of dolmens and from isolated finds, open air settlements are virtually unrepresented. Guilaine (1972) describes only three sites (Condamine, Ladern [Aude]; Foun d'en Peyre II, Conilhac-Corbières [Aude]; Les Companelles, le Soler [Pyrénées-Orientales]) that can be assigned to the period, on the basis of pottery decorated with impressed cordons. All of these sites contain earlier material of Chasséen and Bell Beaker styles (cf. Table 15.1). By 1990, only two further sites had been discovered in western Languedoc (Guilaine et al., 1989) (Station de Rosier, Ventenac-Cabardès [Aude]; Laval de la Bretonne, Monze [Aude]). In eastern Languedoc, on the limestone plateaux north of Montpellier, evidence for Chalcolithic settlement is abundant in the form of the Fontboüisse culture, with continuity from the Mid-Neolithic Chasséen settlement (Delano Smith, 1972; Gascó, 1980). Early Bronze Age material is again only present in very rare cases, e.g. at Le Lébous, where it is associated with a destruction horizon (Arnal et al., 1963).

Several explanations may be put forth to explain this absence:

1. The cultural sequences used to define the Early Bronze Age in Languedoc are incorrect, and sites presently defined as belonging either to earlier or later periods actually correspond to this part of the chronological sequence.

Table 15.1 Chronology of later prehistoric periods in Southern France

Archaeological period	Date ^{14}C calibrated BC
Early Neolithic	
Cardial	7000–4500
Epicardial	5000–4000
Middle Neolithic	
Chasséen	4000–2900
Final Neolithic	
Véraza	2900–2200
Chalcolithic	
Fontboüisse	2500–2000
Bell Beaker	
Early Bronze Age	2000–1600
Middle Bronze Age	1600–1250

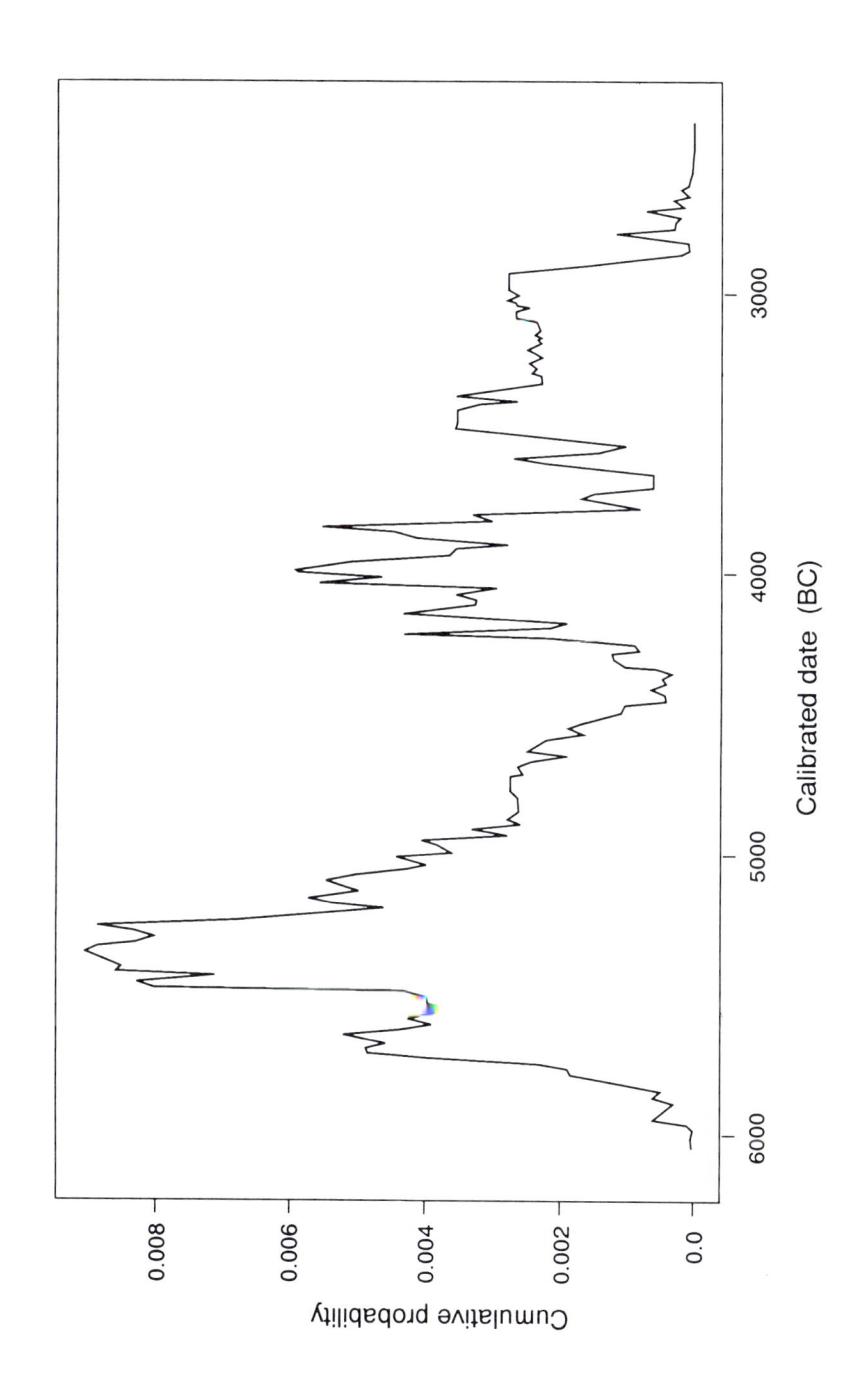

Figure 15.1 Combined ¹⁴C dates for sites showing clearance in Languedoc. The curve is defined by the cumulative probabilities of the calibrated ¹⁴C dates in the text, and gives a probabilistic representation for the dating of clearance phases

2. The Early Bronze Age was a period of economic collapse, in which populations became relatively sparse and occupied only marginal sites.
3. Sites belonging to the Early Bronze Age have been exposed to selective erosion, and have thus been removed from the available archaeological record.

Since there is now available a consistent, calibrated ^{14}C-date sequence covering the relevant periods (Guilaine & Gascó, 1989), the first argument cannot be said to hold, so that we are able to accept the conventional culture material–chronological associations. It will be proposed here that the true pattern is in fact due to a combination of the second and third explanations.

LAND USE IN PREHISTORIC LANGUEDOC

The history of degradation of the landscape in Languedoc is closely related to the settlement of Neolithic populations, with the introduction of new technologies and social structures. Table 15.2 outlines the relevant data. The earliest dated evidence for this comes from the cave site of Camprafaud (Hérault), with a ^{14}C date for the lowest Cardial level falling in a calibrated 1σ range of 7050–6560 BC (all dates have been calibrated from ^{14}C to solar years following the methods of Pearson *et al.*, 1986; Pearson & Stuiver, 1986; Stuiver & Reimer, 1986). More securely dated evidence for the change to pastoral economies comes from Gazel (Aude), with a calibrated 1σ range of 5960–5720 BC. At this stage, populations are still small, and evidence for degradation, if any, remains slight and localized.

The first evidence for more widespread clearance is associated with evidence for more sedentary occupations on the coastal plain. A degraded association of holm oak (*Quercus ilex*), aleppo pine (*Pinus halepensis*), juniper (*Juniperus* spp.) and *Phillyrea* spp. is found at the drowned coastal site of Ile Corrège-Leucate (1σ range, 5740–5580 BC); cereal cultivation is attested from the settlement of Peiro Signado (Hérault), dated to 5490–5240 BC. In the second half of the sixth millennium BC, pollen and charcoal data attesting anthropogenic interference in the landscape are found at Augery and Les Frignants on the Rhône delta (Pons *et al.*, 1979), Baume de Montclus (Gard), Abri II de Puechmargues (Aveyron), Camprafaud (Hérault) and St. Pierre de la Fage (Hérault). At the same time there is pollen evidence from a peat core on the Donezan (Aude: altitude 1510 m a.s.l.) for deforestation, together with the presence of increased grass and other non-arboreal pollen (Ruisseau du Fournas: 1σ range, 5310–4940 BC; Jalut, 1977).

Dated ecosystem change continues through the Epicardial and Chasséen (Figure 15.1). During the Chasséen, there is a massive expansion throughout Languedoc of sedentary village communities (Vaquer, 1990), with a corresponding increase in the exploitation of the landscape. At the rock shelter of Font Juvénal (Aude), Brochier (1984) has shown direct evidence for hillslope erosion occurring contemporaneously with anthropically induced vegetation changes (levels: C6 [1σ range 3495–3105 BC] and C5 [3350–3035 BC]).

At this time, the typical garrigue area surrounding Montpellier became relatively densely populated (cf. Delano Smith, 1972). This density increases rapidly in the succeeding Chalcolithic. Gascó (1980) has argued from locational data, which show a marked preference for the edges of cultivable basins, that this shows the first development of the 'Ager, Saltus, Sylva' system of exploitation (cf. Bloch, 1931). There is a further implicit assumption in this, in that for the basins to have been more attractive for cultivation, there must already

Table 15.2 Sites showing evidence of early degradation in Languedoc and Eastern Spain

Site	1σ date BC	Indicators of degradation	Reference
Leucate	5740–5580	Holm oak, aleppo pine, juniper, phillyrea	Vernet (1980)
Jean Cros	5550–5245	Holm oak	Vernet (1972)
Abri II de Puechmargues	5500–5320	Holm oak	Vernet (1980)
Peiro Signado	5490–5240	Cereals	Freises & Montjardin (1981)
Augéry	5480–5245	Holm oak, juniper, plantain	Pons et al. (1979)
Baume de Montclus	5370–5085	Holm oak	Vernet (1972)
Ruisseau du Fournas	5310–4945	Increasing grass pollen	Jalut (1977)
Châteauneuf-lez-Martigues	5215–5000	Aleppo pine, juniper	Vernet (1972)
La Borde	4370–4040	Increasing NAP, cereals	Jalut (1977)
Camprafaud	4230–3820	Holm oak	Vernet (1972)
Les Frignants	4210–3790	Holm oak	Pons et al. (1979)
Freychinède	4040–3790	Decreasing beech, increasing grasses, plantain, heather, bracken, cereals	Jalut et al. (1982)
Font-Juvénal	3780–3530	Box, holm oak	Brochier (1984)
Marsillargues	3100–2915	Holm oak	Planchais & Parra Vergara (1984)
Salpêtière	2865–2480	Box, holm oak, juniper, phillyrea	Bazile-Robert (1980)
Cova de l'Or (Valencia)	5980–5245	Holm oak, aleppo pine, cereals	Vernet et al. (1984)
Torreblanca (Castellón)	5330–5210	Increasing grass pollen	Menéndez Amor & Florschütz (1961)
Cova del Toll (Catalunya)	4995–4685	Box, holm oak, juniper	Vernet et al. (1984)
Cova de les Cendres (Valencia)	4895–4530	Holm oak, increasing grass pollen, pistacia, plantain	Badal et al. (1989)

have been a certain amount of soil erosion on the surrounding slopes. Contemporary erosion on a large scale may be seen at St-Guilhem-le-Désert, where colluvial deposits of up to five metres in thickness blocked the basin and led to the formation of a lake. Early Bronze Age material is found in the basal deposits of the latter (Ambert & Gascó, 1989). A similar hillwash sequence, incorporating Late Neolithic artefacts, is seen at Tournemire (Ambert, 1986). This hypothesis is also strengthened stratigraphically, in that the dry-stone walls of the settlement of Boussargues are placed on an eroded red-soil horizon (Colomer et al., 1990). Data from the Montagne d'Alaric, an uplifted dome of Cretaceous and Tertiary limestones and marls 15 km to the east of Carcassonne, however, produce a slightly different picture. Although Vérazian occupation (partly syncronous with the Fontbuxian of the Montpellier garrigue) is attested here, the erosion of the plateau does not seem to occur until at least the Early or Middle Bronze Age. The evidence for this is again stratigraphic, in the form of the deep burial of the Early Bronze Age site of Laval de la Bretonne.

STATEMENT OF THE PROBLEM

The data outlined above suggest an asynchronous and localized scale of change in the development of the garrigues. They further imply that the cause is unrelated to a period of changed climate. This can be confirmed by observation of bioclimatic data throughout the Western Mediterranean for this and subsequent periods.

The aim here is a preliminary analysis of the Laval de la Bretonne sub-catchment of the Montagne d'Alaric, in an attempt to provide comparative relative and absolute dating evidence for the erosive processes. In the classic study of the development of the Montpellier Garrigue, Dugrand (1964) outlines its biogeography (Figures 15.2 to 15.4). This suggests that if deep soils are still present after cultivation, the holm oak association can re-establish itself after 70–80 years. However, there is a highly sensitive period of 5–10 years where the stability of the ecosystem is dependent on the localized growth of erosion-resistant species. It is in this phase that the forest soils may also become rapidly truncated. Once the upper humic horizons are removed or mineralized, residual red soils are left on which plant colonisation is much less rapid, forming a positive feedback for the erosion. As a result various types of regosol are formed, with slope position the principal controlling factor. The method followed here is to concentrate on this sensitive period in order to demonstrate the conditions in which degradation following agricultural abandonment may take place.

METHOD

The sub-catchment of the Montagne d'Alaric containing the Early Bronze Age site of Laval de la Bretonne was digitized and divided into cells of 74.8 m in the downslope direction and 18.3 m in the across-slope direction. This digital terrain model (Figure 15.5) was used as the input to a finite difference model which was used to simulate the erosion on the slope. The model contains a hydrological component, based on a simple storage-type analogy (cf. Gilman & Thornes, 1985), a sediment transport component, and simulates the interactions between these and the growth of vegetation. The equations dealing with these are shown in Table 15.3.

The model is in fact a simplified version of one comprising a stochastic rainfall generator and stochastic sediment transport model (Wainwright & Thornes, 1991). This, however, proved too slow in producing useful simulations, but was used to parameterize the Musgrave formula (equation 15.1 in Table 15.3), giving results within the range of published data. The assumption of Band (1985) and Pearce (1976) of the relationship between average annual rainfall and net erosion has been used in the actual simulations. This is also supported by a series of comparative model runs (Wainwright, 1991). The model is fully distributed. Hydrological parameters were derived from field experiments on uneroded soils on similar

	Forest climax →	Initial clearance →	Continuing clearance →	Complete clearance of woody species →	After continued grazing
Limestone	Quercus ilex and Q. pubescens with understorey	Broom Juniperus oxycedrus Cistaceae	Q. coccifera (south) Box (east and high) Cistaceae	Brachypodium ramosum	Euphorbia spp. Asphodelus spp. (plus Thymus spp., Lavandula spp., wild rose, pear, plum in east)
Marl	Quercus ilex and Q. pubescens with understorey	Broom Juniperus oxycedrus Cistaceae	If little erosion Q. coccifera Otherwise: Heather Rosmarinus officinalis Lithospermum fruticosum Schoenus nigricans	Aphyllanthes monspeliensis	In basins: Deschampsia media
Siliceous	Quercus ilex and Q. pubescens with understorey	Broom Juniperus oxycedrus Cistaceae	Arbutus maquis	Heather, ivy, Brachypodium phoenicoides	

Figure 15.2 Biogeography of the Garrigue regressive phase (after Dugrand, 1964)

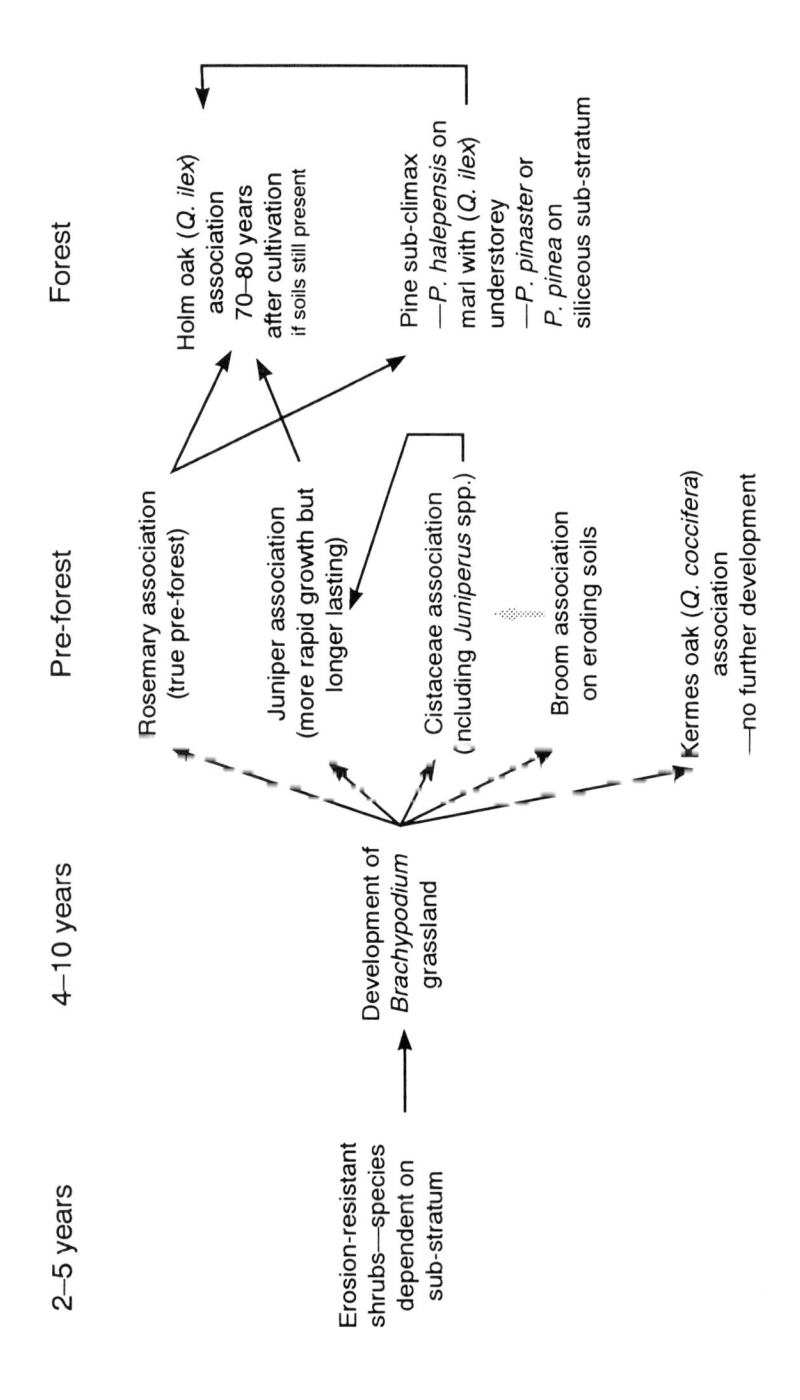

Figure 15.3 Biogeography of the Garrigue progressive phase (after Dugrand, 1964)

Massive limestone and siliceous bedrock

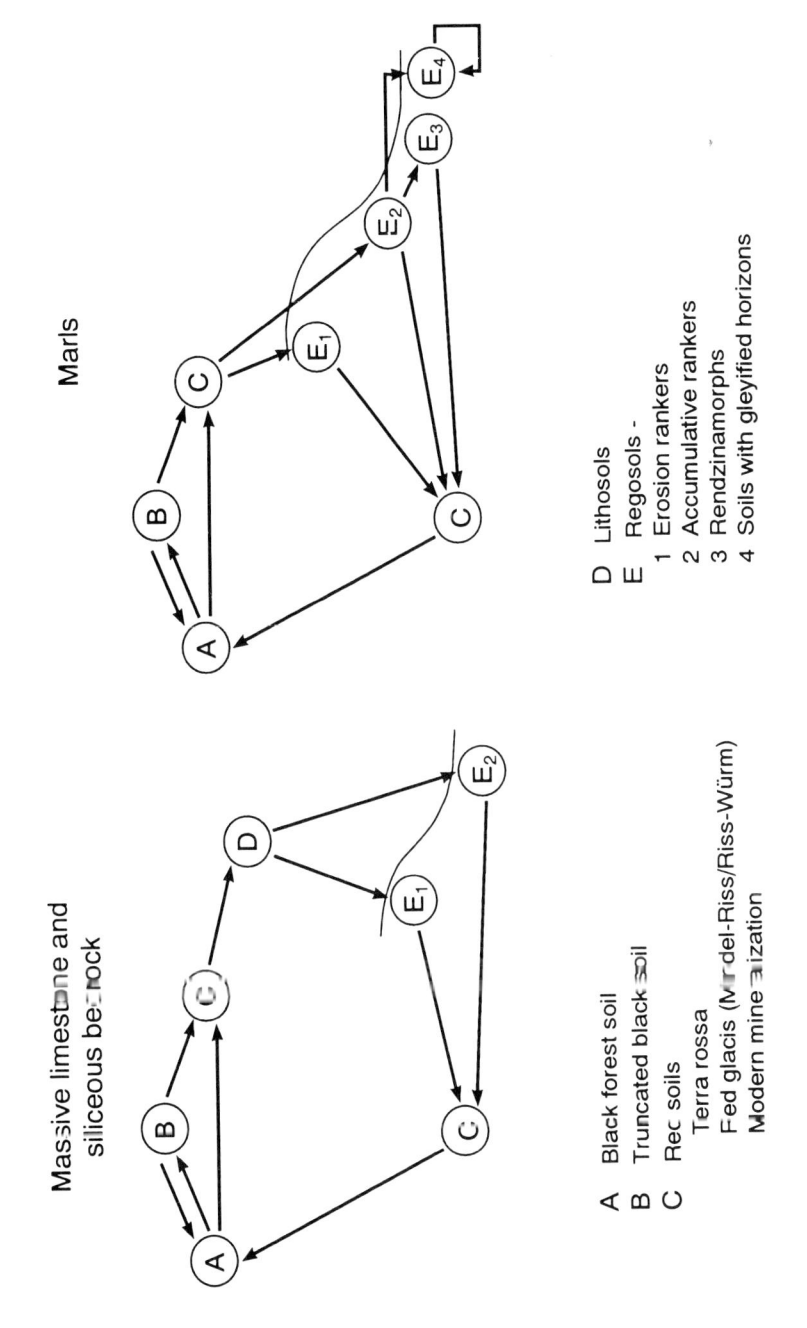

Marls

A Black forest soil
B Truncated black soil
C Rec soils
  Terra rossa
  Fed glacis (Mindel-Riss/Riss-Würm)
  Modern mineralization

D Lithosols
E Regosols -
  1 Erosion rankers
  2 Accumulative rankers
  3 Rendzinamorphs
  4 Soils with gleyified horizons

Figure 15.4 Biogeography of the Garrigue pedogenesis (after Dugrand, 1964)

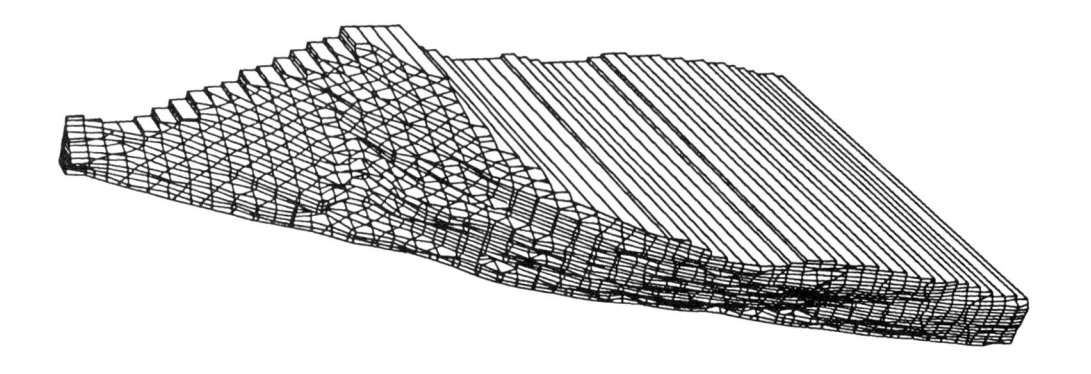

Figure 15.5 Digital terrain model of Laval de la Bretonne sub-catchment of the Montagne d'Alaric

bedrock in another section of the Montagne d'Alaric. Simulations were carried out using the conditions given in Table 15.4.

For all events V_0 was taken as 10 g m^{-2}. Additional simulations were made for the extreme event with V_0 as 25 and 1000 g m^{-2} (equivalent to 10 and 50 years growth respectively). The extreme event is the largest recorded falling on the catchment since 1959, and occurred on 7 November 1962.

RESULTS

The results show a consistent pattern of erosion over the catchment. The spatial pattern for the average storm event is shown in Figure 15.6 after one and 50 iterations. The upper part of the catchment is characterised by shallow erosion; this is to an average depth of 0.18 mm after one iteration, and reaches a depth of 3.9 mm after 50 years. The lower part sees a very irregular pattern, with the greater amounts of erosion, but also deposition. The deepest part of this is precisely in the part of the slope where the Laval de la Bretonne site is to

Table 15.3 Principal equations used in the erosion model

Erosion:
$$q = k_7 \, of^2 \, s^{1.67} \tag{15.1}$$

where q = volume of erosion (m³ s⁻¹)
  k_7 = coefficient (dimensionless)
  of = overland flow discharge (m³ s⁻¹)
  s = slope (m m⁻¹)

Vegetation growth:
$$V_t = V_{t-1} + g \, V_{t-1} \, (1-V_{t-1}/V_{max}) - k_1 \, V_{t-1} \, z_- \tag{15.2}$$

where V = vegetation biomass (at time $t-1, t, \ldots$) (g m⁻²)
  g = initial growth coefficient (dimensionless)
  V_{max} = maximum supportable biomass (g m⁻²)
  k_1 = coefficient reflecting the removal of biomass (m⁻³)
  z_- = amount of soil lost (m³)

Vegetation cover:
$$v_c = 1.66 \, (V/62.5)^{0.5} \tag{15.3}$$

where v_c = ground cover afforded by a given biomass (dimensionless)

Effect on erosion:
$$q' = q \, e^{0.7 v_c} \tag{15.4}$$

Table 15.4 Rainfall conditions used in simulations

	Rainfall (mm)	Duration (h)
Average event	11.936	6.904
	40.0	5.0
	80.0	20.64
Extreme event	156.0	22.0

be found, showing that its position was ideally suited for burial and preservation. Parts of this slope area were buried to a depth of 2.5 mm after 1 year, and 14.6 cm after 50 years. If the successive amounts of erosion are plotted (Figure 15.7, top), then a clear drop is observed through time. This shows a marked linear relationship ($r^2 = 0.998$, $n = 50$, t-ratios on a and $b = 106.5$ and -84.86, both significant at >0.999) which intersects the x axis after approximately 50 years. At this point, the vegetation is able to reduce the erosion effectively to zero.

For the 40 mm event this picture is repeated very closely (Figure 15.7, bottom). Indeed, in this range, the catchment appears highly insensitive to erosion, with the average erosion only rising to 4.4 mm after 50 years, and maximum deposition to 16.3 cm. Vegetation growth again prevents erosion after 51 years ($r^2 = 0.996$, $n = 50$, t-ratios $= 88.86$ and -71.26, both significant at >0.999).

296

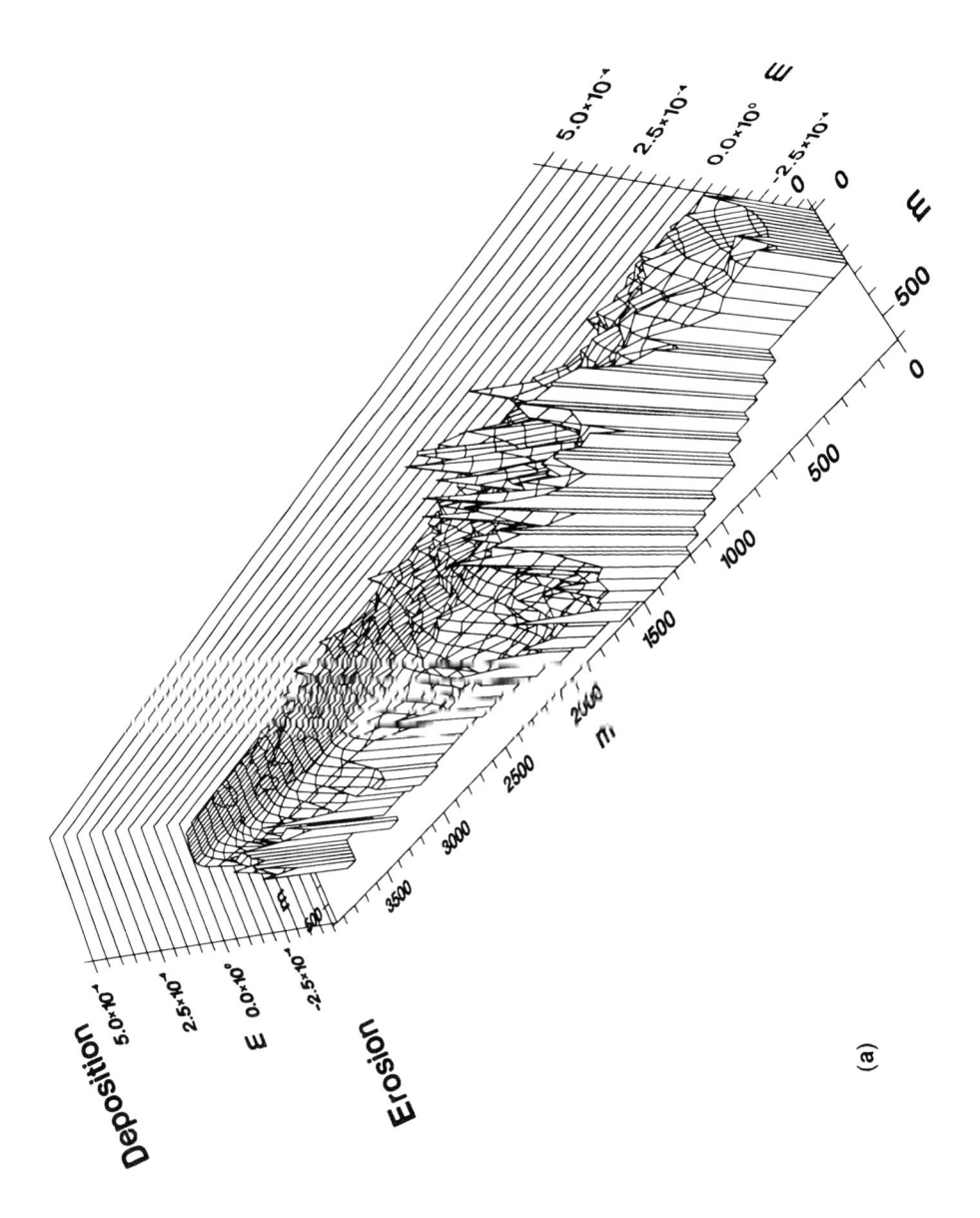

(a)

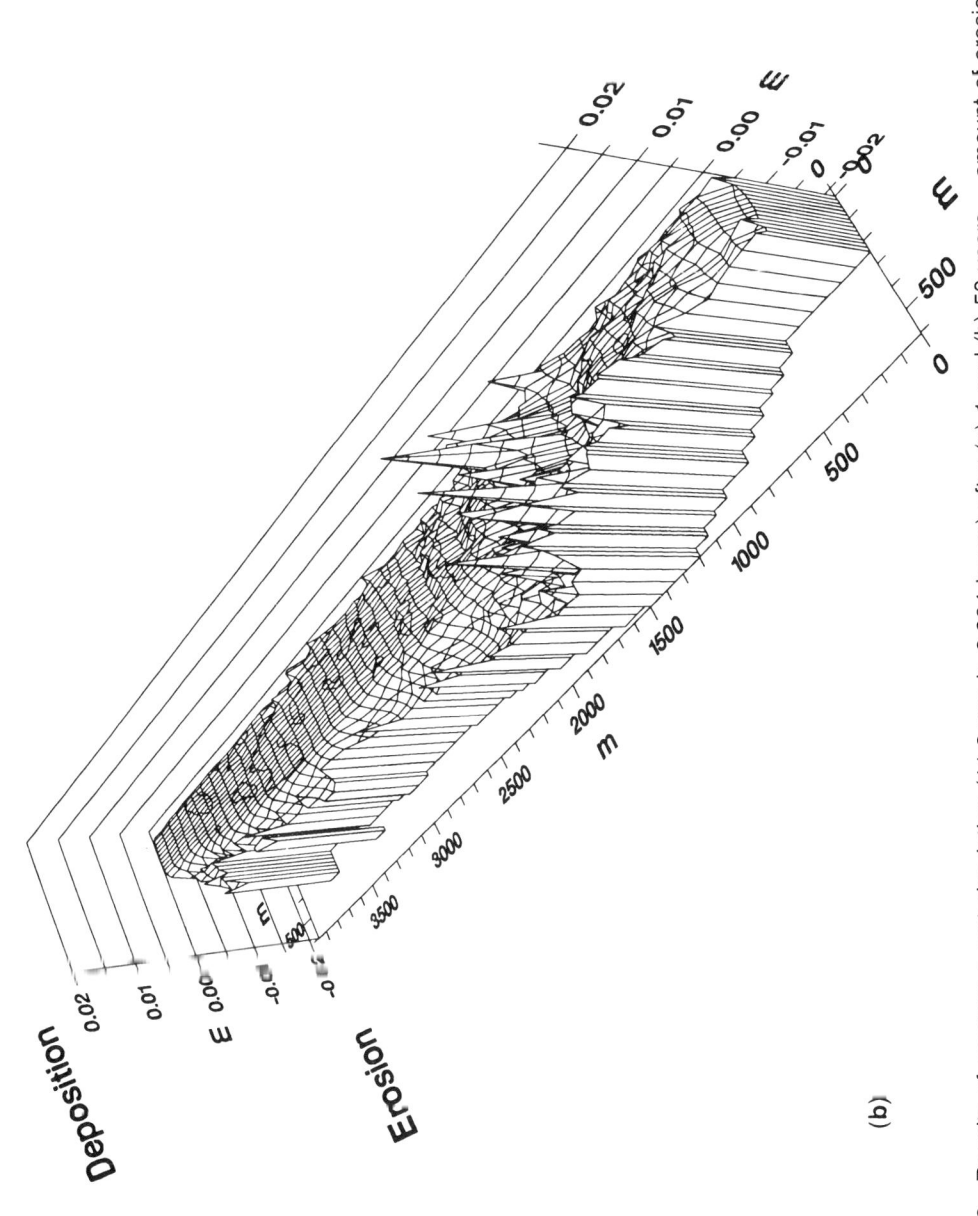

Figure 15.6 Results of average event simulation (11.9 mm in 6.904 hours) after (a) 1 and (b) 50 years — amount of erosion

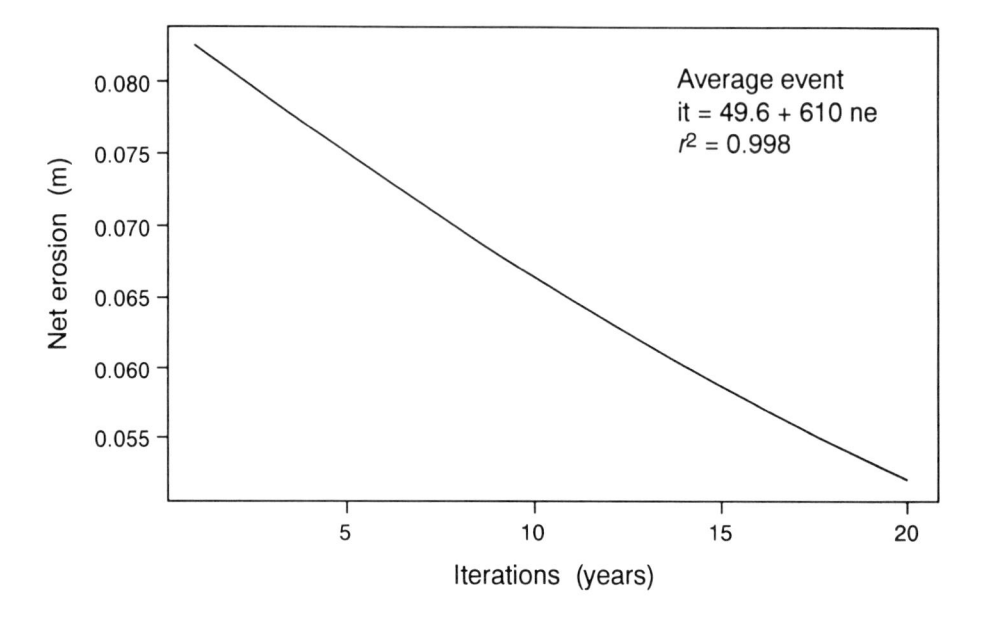

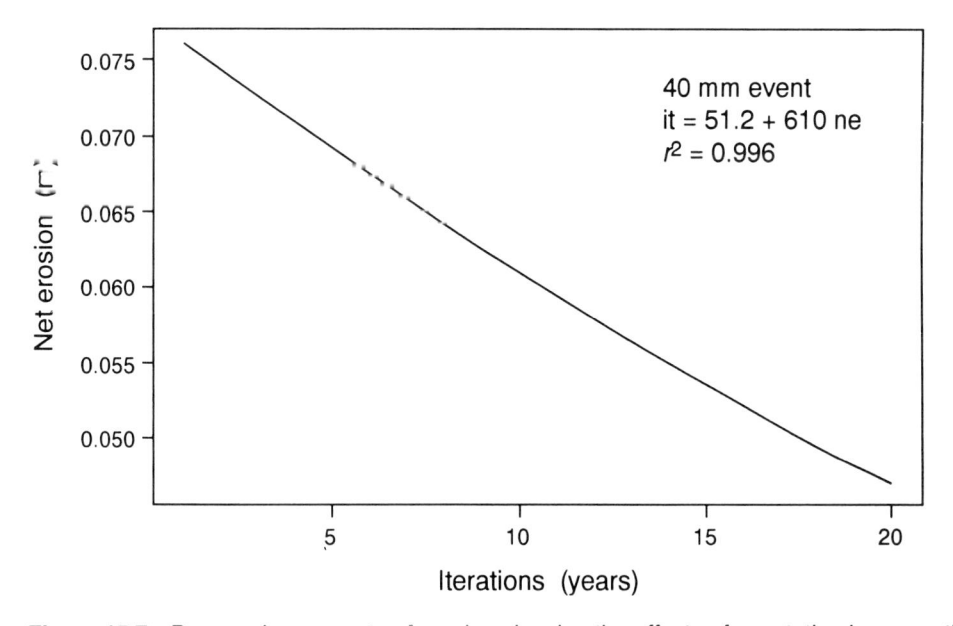

Figure 15.7 Progressive amounts of erosion showing the effects of vegetation in preventing erosion. Top: average event, it = 49.6+610ne, r^2 = 0.998. Bottom: 40 mm event, it = 51.2+640ne, r^2 = 0.996

However, these conditions appear unable to reproduce what is observed in the field, in that most of the soil is removed from the upper slopes, and that the Laval de la Bretonne site is buried by up to 1.5 m of soil. Thus it was decided to simulate the effects of the extreme rainfall event, the result of which is shown in Figure 15.8. In this case, parts of

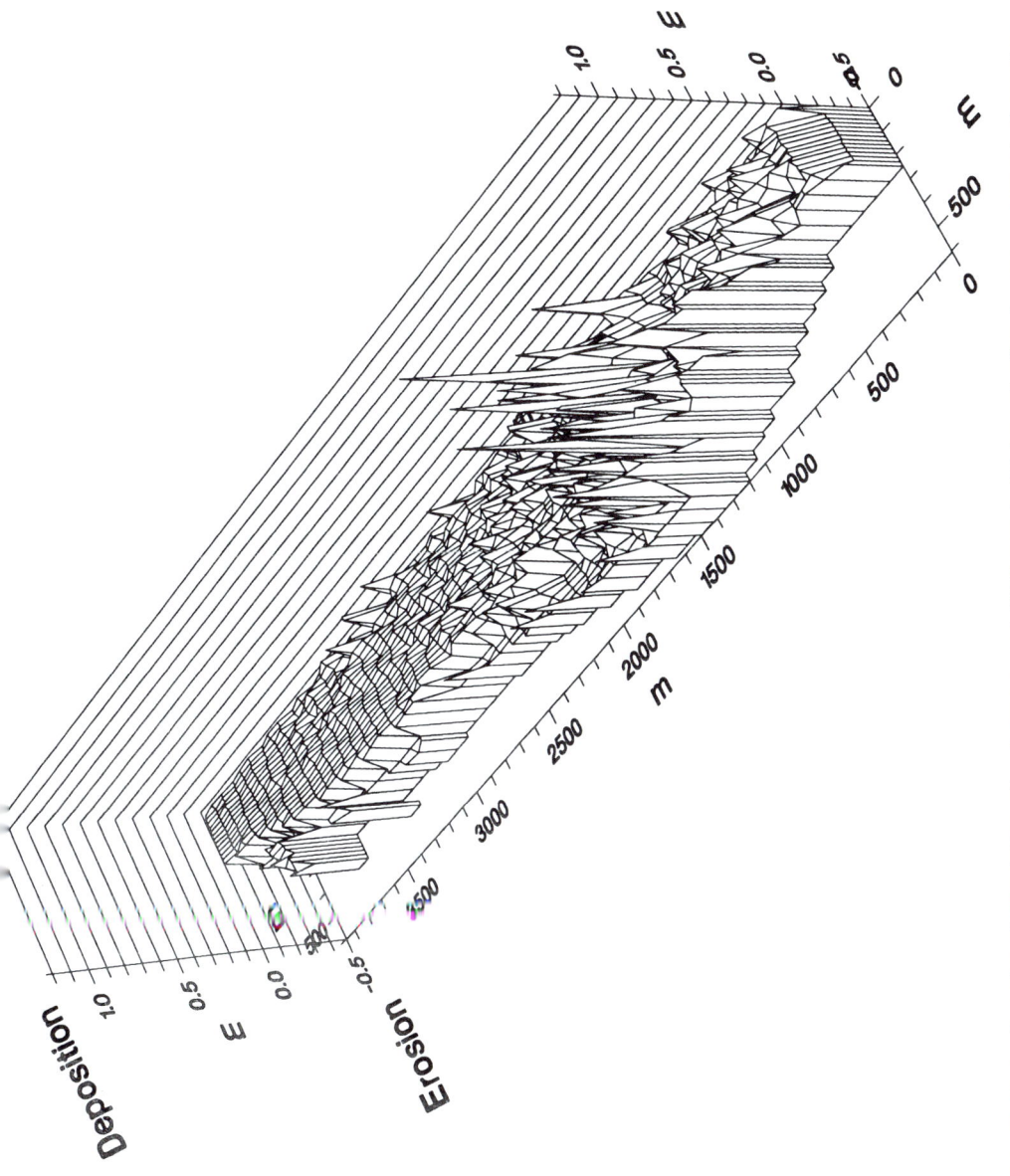

Figure 15.8 Results of simulation of extreme event (156 mm in 22 hours) after 1 year—amount of erosion

the catchment lose all soil after a single event, with an average erosion of 9 cm. The lower slope sees deposition to 1.24 m. Following erosion to this depth, plant growth is significantly reduced due to the removal of the thin humic layer, modelled by the soil loss term in the logistic growth curve. In fact, if a second extreme event is simulated, the system becomes entirely unstable.

To predict the effects of vegetation and therefore the effects of anthropogenic disturbance on the extreme event, two further simulations were carried out, the first with an equivalent of 10 years of regrowth and the second with 50 years (Table 15.5). In the first, although in some areas almost all of the soil is removed, the mean erosion drops by a half to 4.5 cm, as does the maximum amount of deposition (to 62.3 cm). The latter sees mean erosion drop to 0.9 mm, with deposition peaking at 1.5 cm.

From these results we may conclude that the catchment is relatively stable to erosion, even with the removal of vegetation. For the observed patterns of erosion to have occurred, it appears necessary for there to have been an extreme rainfall event, or series of events. This must have been associated with widespread clearance of the catchment, as even small amounts of vegetation are able to restabilize the system in extreme conditions. The counter-argument that repeated small events could have caused the erosion in the subsequent period is untenable, both stratigraphically and due to the fact that such intense, repeated activity in the area would be expected to leave at least some archaeological trace.

IMPLICATIONS

From these initial considerations it seems that in the latter part of the Chalcolithic and the Early Bronze Age in Languedoc, a diachronous change occurred in land use related to intense degradation of upland areas. This degradation was apparently caused by overexploitation of the environment, coupled with extreme storm events. The localized nature of the exploitation and the infrequency of extreme events are probable causes of the diachroneity. The fact that apparently intact forest soils remain on steep slopes on another part of the catchment partly supports this. An initial hypothesis is for a progressive degradation as settlement moved from degraded to undegraded areas. Further comparative work will be required to test this generalization from the small areas observed here. From the predicted insensitivity of the catchment in the lower rainfall range, it seems unlikely that climatic change may be attributed as a major factor.

Table 15.5 Summary of simulation results

Event (mm)	Period (years)	Mean erosion (mm)	Maximum erosion (mm)	Maximum burial (mm)
11.936	1	0.18	2.1	2.5
	50	3.90	139	146
40.0	1	0.19	2.3	2.7
	50	4.37	155	163
80.0	1	16.2	262	304
156.0	1	90.5	500	1236
(10 years' vegetation)	1	45.5	499	623
(50 years' vegetation)	1	8.78	13.3	15.4

The result of these palaeoenvironmental reconstructions is that in this period, large quantities of sediment were being transported from upland to lowland areas. This massive reduction in cultivable land may have led to extreme social stress and collapse. There is almost no evidence for inter-regional exchange, a greatly different picture from the preceding periods (cf. Petrequin *et al.,* 1988). Populations may have been much reduced and resettled on the plains. This hypothesised move, coupled with the movement of sediment, would cause the deep burial of many of the archaeological remains, as seen with the Laval de la Bretonne site. Under present conditions, these sites are archaeologically invisible. This decline was only reversed towards the start of the Middle Bronze Age (*ca.* 1600 BC), when wider contacts were re-established, shown, for example, by the appearance of Baltic amber at the Grotte du Collier (Lastours: Guilaine *et al.,* 1989).

If the degradation hypothesis is correct, then it is possible to say that short-term implications for human populations lasted for up to half a millennium (cf. Table 15.1). However, it would also be fair to say that the ramifications extend into the Mediaeval and modern periods. Although partly related to political factors (Braudel, 1986, p. 72, says Languedoc was 'almost a colony' of northern France), the infertility of vast areas of Mediterranean France cannot be ignored in the creation of this imbalance. The economic factors must be considered as prime movers.

COMPARATIVE STUDIES

In their study of erosion around archaeological sites in south-east Spain, Gilman & Thornes (1985) and Wise *et al.* (1982) were able to show that accelerated erosion tends to be very localised and not due to large-scale climatic factors (cf. Huntley & Prentice, 1988). Their view that few archaeological sites have subsequently been eroded is subject to some bias, in that it fails to take account of sites that have been completely destroyed by erosion. The sites that have been eroded — Cerro del Gallo (Wise *et al.,* 1982), Zapata (Gilman & Thornes, 1985), Terrera Ventura (Gilman & Thornes, 1985) and the ravines at Los Millares (Gilman & Thornes, 1985) — again show that erosion is sporadic and localised. The first two examples show that erosion post-dates the Argaric (Early Bronze Age), the third post-dates the Late Neolithic and the latter occurs before the Chalcolithic.

Stevenson & Moore (1988) believe that large-scale erosion due to deforestation led to major changes in the regional hydrology in Huelva (south-west Spain) in the fifth millennium BC. This in turn caused the formation of large marshes. Further destruction in this area appears to have been prevented by a more successful management of the environment in prehistoric and historic times (Stevenson & Harrison, 1992).

In north-eastern Spain, Burillo Mozota *et al.* (1986) show that two erosive phases are important. The first, thickest sequence consists of hillwash and debris flows, and is linked with Iberian (Iron Age) material. A subsequent, less-severe phase appears to be associated with Mediaeval occupation.

Balista & Leonardi (1985) present a similar scenario in the Veneto region of Italy, with widespread expansion in the Middle to Late Bronze Age being followed by erosion and postulated population decline in the Final Bronze and Early Iron Ages. In the Adige Valley to the north, degradation apparently occurs from the Late Neolithic/Chalcolithic, reaching a peak intensity in the first half of the first millennium BC. Further, intermittent degradation has continued from *ca.* 600 AD to the present day (Coltorti & Dal Ri, 1985).

Additional support for the intermittent erosion hypothesis can be found in the eastern Mediterranean. Van Andel *et al.* (1986) have published a sequence for the southern Argolid with four erosion cycles interbedded with soil horizons, from the Early Helladic (*ca.* 3200–2000 BC) to the present day, each of which can be correlated with apparent episodes of population expansion. In a subsequent paper, Van Andel *et al.* (1990) summarise data from the whole of Greece. Here they suggest that erosion occurs from the Early to Middle Neolithic in Macedonia and Thessaly, becoming increasingly prevalent from the first millennium BC. The work of Wagstaff (1981) on alluvial sequences reinforces this.

For Turkey, Roberts (1990) has also hypothesised the occurrence of clearance–degradation–regeneration cycles. The principal effects of these are seen at the present time and in the first millennium BC, although smaller-scale effects are again on-going from the Neolithic.

A similar model of erosion in sub-Mediterranean conditions to the case study presented above has been historically documented in the Pyrenees by Antoine (1988, 1989) and Métaillie (1986). They show that torrential debris flows tend to occur following extreme rainfall events, principally towards the end of the last century, at the maximum extent of Pyrenean settlement. The absence of a 'folk memory' relating to catastrophic events demonstrates the importance of combined geomorphological, environmental and (pre)historical approaches in this field.

CONCLUSIONS

This paper has shown, by the modelling of a catchment in a garrigue régime, the approximate timing and rapidity of such degradation. The use of various scenarios indicates that, under conditions of natural vegetation, the catchment in question is extremely stable, even in severe climatic conditions. However, when clearance occurs, most probably by human use of the landscape, the erosional pattern seen today can be generated after one or two extreme rainfall events. Similar patterns can be seen elsewhere in the Mediterranean region. By combining these data with archaeological and historical data, it is possible to show the effects on a human scale. A clear picture of long-lasting implications resulting from short-term effects is demonstrated. These should provide sufficient warning when considering the processes at work in the present.

ACKNOWLEDGEMENTS

The author wishes to thank Jean Gascó and Laurent Carozza for help in the field and guidance concerning the Bronze Age of Languedoc. Local meteorological data were kindly supplied by M. Bourjade of Monze. This paper has also benefited from comments by John Thornes, Richard Harrison and two anonymous reviewers. Its remaining shortcomings are, as tradition has it, the author's own responsibility.

REFERENCES

Ambert, P. (1986). Les tufs Holocènes du Plateau du Larzac: données actuelles. *Méditerranée,* **57** (1–2), 61–70.

Ambert, P. and Gascó, J. (1989). Les tufs de Saint-Guilhem-le-Désert. Évolution Holocène et pression anthropique sur le milieu karstique (Languedoc, France). *Bulletin du Musée d'Anthropologie Préhistorique de Monaco,* **32**, 63–85.

Antoine, J.-M. (1988). Un torrent oublié mais catastrophique en Haute-Ariège. *Revue Géographique des Pyrénées et du Sud-Ouest,* **59**(1), 73–88.

Antoine, J.-M. (1989). Torrentialité en Val d'Ariège: des catastrophes passées aux risques présents. *Revue Géographique des Pyrénées et du Sud-Ouest,* **60**(4), 521–34.

Arnal, J., Martin Granel, H. and Sangmeister, E. (1964). Lébous. *Antiquity,* **38**, 191–200.

Badal, E., Bernabeu, J., Fumanal, M. P. and Dupré, M. (1989). Secuencia cultural y paleoambiente en el yacimiento neolítico de la cova de les cendres (Moraira-Teulada, Alicante). *Communicación presentada a la 2ª Reunión sobre Quaternario Iberico,* Madrid, Septembre 1989.

Balista, C. and Leonardi, G. (1985). Hill slope evolution: pre and protohistoric occupation in the Veneto. In Malone, C. and Stoddart, S. (eds.), *Papers in Italian Archaeology,* Volume IV, Part i, *The Human Landscape,* BAR International Series 243, Oxford, pp. 135–52.

Band, L. E. (1985). Field parameterization of an empirical sheetwash transport equation. *Catena,* **12**, 281–90.

Bazile-Robert, E. (1980). Macrorestes végétaux carbonisés d'Âge Subboréal (Chalcolithique, Fontbouïsse) de la Salpêtrière (Remoulins, Gard). In Guilaine, J. (ed.), *Le Groupe de Véraza et la Fin des Temps Néolithiques dans le Sud de la France et la Catalogne,* CNRS, Toulouse, pp. 252–6.

Bloch, M. L. B. (1931). *Caractères Originaux de l'Histoire Rurale Française,* Oslo, Paris.

Braudel, F. (1986). *L'Identité de la France I: Espace et Histoire,* Artaud-Flammarion, Paris.

Brochier, J. E. (1984) Chênes à feuillage Caduc, chênes verts et stabilité des versants. In *Influences Méridionales dans l'Est et le Centre-Est de la France au Néolithique: Le Rôle du Massif Central. Actes du 8ᵉ Colloque Interrégional sur le Néolithique,* Le Puy, 1981, CREPA, Clermont Ferrand, pp. 321–7.

Burillo Mozota, F., Gutiérrez Elorza, M., Peña Monné, J. L. and Marcén, C. S. (1986). Geomorphological processes as indicators of climatic changes during the Holocene in North-East Spain. In *Quaternary Climate in the Western Mediterranean. Proceedings of the Symposium on Climatic Fluctuations during the Quaternary in the Western Mediterranean Regions,* Madrid, 1986.

Butzer, K. W. (1974). Accelerated soil erosion: a problem of man–land relationships. In Manners, I. R. and Mikesell, M. W. (eds.), *Perspectives on Environment,* Association of American Geographers, Washington D.C., pp. 57–77.

Colomer, A., Coularou, J. and Gutherz, X. (1990). *Boussargues (Argelliers, Hérault). Un habitat ceinturé chalcolithique: les fouilles du secteur ouest,* DAF 24, Paris.

Coltori, M. and Dal Ri, L. (1985). The human impact on the landscape: some examples from the Adige Valley. In Malone, C. and Stoddart, S. (eds.), *Papers in Italian Archaeology,* Volume IV, Part i, *The Human Landscape,* BAR International Series 243, Oxford, pp. 105–34.

Delano Smith, C. (1972). Late Neolithic settlement, land-use and garrigue in the Montpellier Region, France. *Man,* **7**(3), 397–407.

Dugrand, R. (1964). *La Garrigue Montpelliéraine,* Presses Universitaires de la France, Paris.

Freises, A. and Montjardin, R. (1981). Le Néolithique ancien côtier du Midi de la France. In *Actes du Colloque International de Préhistoire, Montpellier, 1981 — Le Néolithique Ancien Méditerranéen,* La Féderation Archéologique de l'Hérault, Montpellier, pp. 201–28.

Gascó, J. (1980). L'habitat paysan de la communauté de Fontbouïsse. In Guilaine, J. (ed.), *Le Groupe de Véraza et la Fin des Temps Néolithiques dans le Sud de la France et la Catalogne,* CNRS, Toulouse, pp. 241–6.

Gilman, A. and Thornes, J. B. (1985). *Land-Use and Prehistory in South-East Spain,* The London Research Series in Geography & Allen and Unwin, London.

Guilaine, J. (1972). *L'Âge du Bronze en Languedoc Occidental, Rousillon, Ariège,* Klincksieck, Paris.

Guilaine, J. and Gascó, J. (1989). Médor et la chronologie de la fin de l'Age du Bronze en Languedoc. In Guilaine, J., Vaquer, J., Coularou, J. and Treinen-Claustre, F. (eds.), *Ornaisons-Médor. Archéologie et Écologie d'un Site de l'Age du Cuivre, de l'Age du Bronze final et de l'Antiquité Tardive,* CASR, Toulouse, pp. 217–23.

Guilaine, J., Vaquer, J. and Rancoule, G. (1989). *Carsac et les Origines de Carcassonne,* Musée des Beaux-Arts, Carcassonne.

Huntley, B. and Prentice, I. C. (1988). July temperatures in Europe from pollen data, 6000 years before present. *Science,* **241**, 687–90.

Jalut, G. (1977). *Végétation et Climat des Pyrénées Méditerranéennes depuis Quinze Mille Ans,* Archives d'Écologie Préhistorique 2, EHESS, Toulouse.

Jalut, G., Delibrias, G., Dagnac, J., Mardones, M. and Bouhours, M. (1982). A palaeoecological approach to the last 21 000 years in the Pyrenees: the peat bog of Freychinède (alt. 1350 m, Ariège, S. France). *Palaeogeography, Palaeoclimatology, Palaeoecology,* **40**, 321–59.

Menéndez Amor, J. and Florschütz, F. (1961). La concordancia entre la composición de la vegetación durante la Segunda Mitad del Holoceno en la Costa Levante (Castellón de la Plana) y en la Costa W. de Mallorca. *Boletin del Real Sociedad Español de la Historia Natural (G),* **59**, 97–100.

Métaillie, J.-P. (1986). Photographie et histoire du paysage: un exemple dans les Pyrénées Luchonnaises. *Revue Géographique des Pyrénées et du Sud-Ouest,* **57**(2), 179–208.

Pearce, A. J. (1976). Magnitude and frequency of erosion by Hortonion overland flow. *Journal of Geology,* **84**, 65–80.

Pearson, G. W. and Stuiver, M. (1986). High-precision calibration of the radiocarbon timescale, 500–2500 BC. *Radiocarbon,* **28**(2B), 839–62.

Pearson, G. W., Pilcher, J. R., Baillie, M. G. L., Corbett, D. M. and Qua, F. (1986). High precision ^{14}C measurements of Irish oaks to show the natural ^{14}C variations from AD 1840–5210 BC. *Radiocarbon,* **28**(2B), 911–34.

Petrequin, P., Chastel, J., Giligny, F., Petrequin, A.-M. and Saintot, S. (1989). Réinterprétation de la civilisation Saône-Rhône. Une approche de tendances culturelles du Néolithique Final. *Gallia Préhistoire,* **30**, 1–89.

Planchais, N. and Parra Vergara, L. (1984). Analyses polliniques de sédiments Lagunaires et Côtiers en Languedoc, en Rousillon et dans la Province de Castellón (Espagne); bioclimatologie. *Bulletin de la Société Botanique Française,* **131**, *Actualités Botaniques,* **1984**(2/3/4), 97–105.

Pons, A., Toni, Cl. and Triat, H. (1979). Édification de la Camargue et histoire Holocène de sa végétation. *Terre Vie, Revue Écologique, Supplément,* **2**, 13–30.

Roberts, N. (1990). Human-induced landscape change in south and southwest Turkey during the Later Holocene. In Bottema, S., Entjes-Nieborg, G. and van Ziest, W. (eds.), *Man's Role in the Shaping of the Eastern Mediterranean Landscape,* Balkema, Rotterdam, pp. 53–67.

Stevenson, A. C. and Harrison, R. J. (1992). Ancient forests in Spain: a model for land-use and dry forest management in south-west Spain from 4000 BC to 1900 AD. *Proceedings of the Prehistoric Society,* **58**, 227–47.

Stevenson, A. C. and Moore, P. D. (1988). Studies in the vegetational history of South-West Spain IV Palynological investigations of a valley mire at El Acebrón, Huelva. *Journal of Biogeography,* **15**(2), 339–61.

Stuiver, M. and Reimer, P. (1986). A computer program for radiocarbon age calibration. *Radiocarbon,* **28**(2B), 1022–30.

Van Andel, T. H., Runnels, C. H. and Pope, K. O. (1986). Five thousand years of land use and abuse in the Southern Argolid, Greece. *Hesperia,* **55**, 103–28.

Van Andel, T. H., Zangger, E. and Demitrack, A. (1990). Land use and soil erosion in prehistoric and historical Greece. *Journal of Field Archaeology,* **17**, 379–96.

Vaquer, J. (1990). *Le Néolithique en Languedoc Occidental,* Éditions du CNRS, Paris.

Vernet, J.-L. (1972). Contribution à l'histoire de la végétation du Sud-Est de la France au Quaternaire. Étude de macroflores, de Charbon de Bois Principalement. Thèse de Docteur ès Sciences Naturelles, Université de Montpellier.

Vernet, J.-L. (1980). La végétation du Bassin de l'Aude, entre Pyrénées et Massif Central, au Tardiglaciaire et au Postglaciaire d'après l'analyse anthracologique. *Review of Palaeobotany and Palynology,* **30**, 33–55.

Vernet, J.-L., Badal Garcia, E., Grau Almero, E. and Ros Mora, T. (1984). Charcoal analysis and the Western Mediterranean prehistoric flora. In Waldren, W. H., Chapman, R., Lewthwaite, J. and Kennard, R. C. (eds.), *The Deya Conference of Prehistory,* Volume 1, BAR International Series 229 (i), Oxford, pp. 165–77.

Wagstaff, J. M. (1981). Buried assumptions: some problems in the interpretation of the 'younger fill' raised by recent data from Greece. *Journal of Archaeolgical Science,* **8**, 247–64.

Wainwright, J. (1991). Erosion of semi-arid archaeological sites: a study in natural formation processes. Unpublished PhD thesis, University of Bristol.

Wainwright, J. and Thornes, J. B. (1991). Computer and hardware modelling of archaeological sediment transport on hillslopes. In Lockyear, K. and Rahtz, S. (eds.), *Computer Applications and Quantitative Methods in Archaeology, 1990,* BAR International Series 565, Oxford, pp. 183–94.

Wise, S. M., Thornes, J. B. and Gilman, A. (1982). How old are the badlands? A case study from South-East Spain. In Bryan, R. and Yair, A. (eds.), *Badland Geomorphology and Piping,* Geo Books, Norwich, pp. 259–77.

16 Erosion-Vegetation Competition in a Stochastic Environment Undergoing Climatic Change

J. B. THORNES and J. BRANDT
Department of Geography, King's College London, UK

ABSTRACT

This paper extends earlier modelling work to consider the sensitivity of a simple evolving ecological system to stochastic perturbations in the environment. The system is assumed to consist of deep- and shallow-rooting species evolving from low biomass states in the presence of erosion. Growth is limited by a regional annual precipitation-determined biomass capacity, constrained by soil thickness and a soil moisture water-use function. Growth rate is determined by soil moisture (species A) and overland flow (species B). Erosion is assumed to be a function of soil thickness, soil organic matter profile and vegetation cover, the first two operating through the partition between infiltration and overland flow. Erosion is reduced exponentially by increasing vegetation cover. The system, though not fully parameterised, is based on conditions prevailing in south-east Spain.

The effects of trending and stochastic rainfall variability under favourable and less favourable soil conditions are examined in terms of response by extinction, competitive success and replacement, and by soil erosion. The results indicate the extent to which sensitivity to rainfall trends depends not only on the magnitude of the stochastic component but also on the initial conditions, the soil characteristics and the particular sequence of rainfall events.

INTRODUCTION

The effects of vegetation on erosional response have always been of great interest to geomorphologists under natural or anthropogenic conditions, on geological, Quaternary and Holocene timescales and through speculative, empirical and mathematical modelling. The task has assumed greater urgency as attempts are made to evaluate the possible impacts of the climatic changes projected for the coming decades. A number of approaches can be adopted depending on the level of complexity. At one extreme, simple forecasts on the basis of empirical equations such as the USLE or the Langbein and Schumm (1958) curve might be adequate. At the other end are the complex parameter-rich mathematical simulation models developed from the late 1960s onwards, such as ANSWERS, SPUR and WEPP.

This paper lies about halfway along that spectrum, since it involves physical principles, yet is not too demanding in field parameters. Like the larger models it is capable of progressively more complex development at the expense of utility. The question of interest is the extent to which the system response to a changing mean annual rainfall is likely to be affected by inter-annual rainfall variability when these inputs operate through a non-linear system coupling the amount of erosion to vegetation cover with two different competing species, behaving under different rules. Attention is focused on the inter-specific and plant-erosion interactions as they might appear in semi-arid environments. A simple hydrological budget and rain-dependent plant growth are assumed and evapotranspiration, nutrient and temperature controls on plant growth are not considered. Other models are available in which these aspects are more extensively investigated (e.g. Eagleson, 1978; Kirkby & Neale, 1986) and some of these have been used to predict climatic change effects.

The paper first outlines the basic model of Thornes (1990) and extends it to two species whose growth parameters have different moisture dependencies. The third section considers the situations investigated theoretically by May (1973) and Roughgarden (1975), in which trends and stochastic variations in the inputs and parameters determine survival susceptibility in a single species. The fourth section evaluates response in terms of replacement, extinction and soil erosion for the two-species system subjected to deterministic and stochastic forcing by generated and actual series from Murcia, south-east Spain. The final section discusses the general lessons to be learned from these experiments and draws some conclusions.

BASIC COMPETITION MODEL

Assume that a species grows logistically, i.e.

$$dV_1/dt = r_1 V_1 (1-V_1/V_r) - d_1 V_1 - e_1 Z \tag{16.1}$$

where, in the first term on the right-hand side, r_1 is the growth parameter, V_1 is the species biomass (g m^{-2}) and V_r is the maximum vegetation capacity. V_r is the regional precipitation-determined vegetation capacity, V_{max}, conditioned by soil moisture and soil thickness by

$$V_r = V_{max}/\{1+P_1 \exp(-P_2 \Theta (Z_1-Z))\} \tag{16.2}$$

where Θ is soil water per unit depth, Z_1 the soil profile depth and Z the depth of erosion (positive downwards). P_1 and P_2 are coefficients. This function (Biot, 1988) is illustrated in Figure 16.1 for different soil depths and water contents using parameters typical of south-east Spain (Table 16.1). For a fixed value of r_1 the impact of rainfall operates through the soil moisture to fix the upper bound V_r to which vegetation can grow. In this sense growth is resource-constrained in the ecological sense. With two species the combined biomasses may not exceed V_r.

The annually available soil moisture is determined for bare soil to be the lesser of the mean storm rainfall rate and the mean storm infiltration rate multiplied by the number of storms, and is assumed to be distributed across the entire soil profile depth (Z_1-Z). In addition, in profiles truncated by an amount Z, the bare soil infiltration rate (F_0) is assumed to decrease as an exponential function of Z and of the soil stone content or bulk density (R). On vegetated soils the bare soil infiltration rate is enhanced by the impact of surface litter, assumed (through a coefficient A) to be a linear function of vegetation biomass. This litter effect decays exponentially with depth as a function of an organic matter decay rate, b, so that for vegetated soils the infiltration rate is

$$F_v = F_0 \exp(-RZ) + AV_1 \exp(-bV_1) \tag{16.3}$$

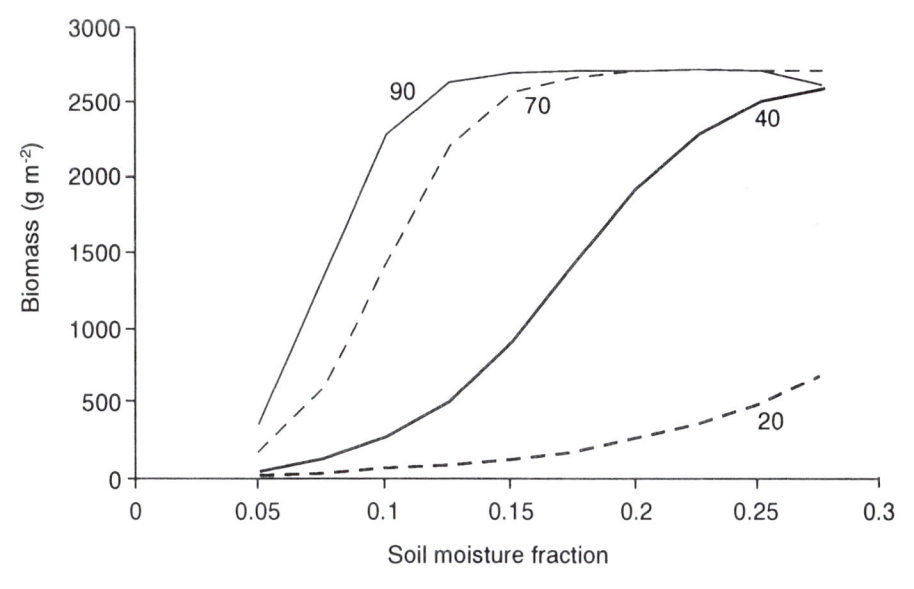

Figure 16.1 The response in biomass (V_r) as a function of soil moisture and depth of soil

Table 16.1 The table shows sensitivity of vegetation cover (k-veg, r-veg) and soil erosion (dz/dt) in terms of mean values and (in brackets) coefficients of variability for three generated rainfall series (Stc 1, 2 and 3) and two historical series (Mu8 and Mu10). There are no climatic changes and the conditions favourable, medium and tough are defined in the text and represent combinations of soil infiltrability and storm intensity. The final column (r^2) shows the correlation of mean annual rainfall and mean annual soil erosion according to the model runs under these conditions

	Condition	k-veg	r-veg	dz/dt	r^2 dz/dt
Stc1	Favourable	993 [1.19]	44 [1.91]	0.029 [6.38]	56.5
Stc2	Medium	224 [0.75]	645 [2.01]	0.164 [3.06]	83.6
Stc3	Tough	32.3 [1.57]	1139 [1.25]	0.249 [2.62]	86.7
Mu8	Favourable	1060 [0.86]	260 [2.74]	0.113 [3.31]	70.8
Mu10	Tough	60 [0.81]	1244 [1.35]	0.387 [2.20]	84.7

When vegetation is absent the second term falls to zero. Otherwise the infiltration is determined as the sum of the infiltration rates for the two cover conditions (vegetation and bare soil) proportional to their relative areas. If there is more than one species, then the surface vegetation cover is the sum of the biomasses of both species. Any excess in the mean storm rainfall over the mean infiltration rate is assumed to be annual overland flow per unit area.

The second term on the right-hand side of equation (16.1) is a death rate, and on integration is exponential with a rate coefficient of d_1. It is based on the assumption that there is a maintenance cost for plant support which is proportional to biomass. If there is no fresh growth, then the plant will consume mass to maintain itself and progressively die off at a rate dependent on d_1.

The third term in equation (16.1) reduces plant growth in relation to the amount of soil truncation, and this again will accelerate the collapse of the system in the absence of growth, as well as reducing the rate of growth overall. It is based on the assumption that erosion reduces the growth capacity by reduction of nutrient stocks and organic matter, and there is a large literature on erosion effects on productivity. In this paper the growth rate is assumed to be affected linearly, though in practice, given work on erosion rate and productivity, an exponential rate might be more appropriate. In this literature there is some confusion between the effects of organic matter concentration and other effects such as moisture availability, runoff and infiltration, so for the moment we have retained the linear expression. Some effects of various fixed values of Z on equilibrium and the trajectories towards it are illustrated in Figure 16.2.

The erosion rate is determined by a simple power-type equation reduced exponentially by vegetation cover:

$$dZ/dt = q_{tot}ms^n\{1-\exp(-0.7V_1)\} \qquad (16.4)$$

where q_{tot} is the sum of the overland flow from the vegetated and unvegetated parts of the surface, s is the ground slope and m and n are parameters. The exponential reduction in erosion due to vegetation cover follows Elwell and Stocking (1976).

For this model with one species, the pair of differential equations, (16.1) and (16.4), can be solved for a variety of parameter values and, more interestingly, for the stability of the equilibria and the flow vectors of change for various starting conditions. The model can be easily extended to more elaborate hydrology, e.g. with evapotranspiration or with a more elaborate set of soil erosion processes, and can be solved for different parameter values of

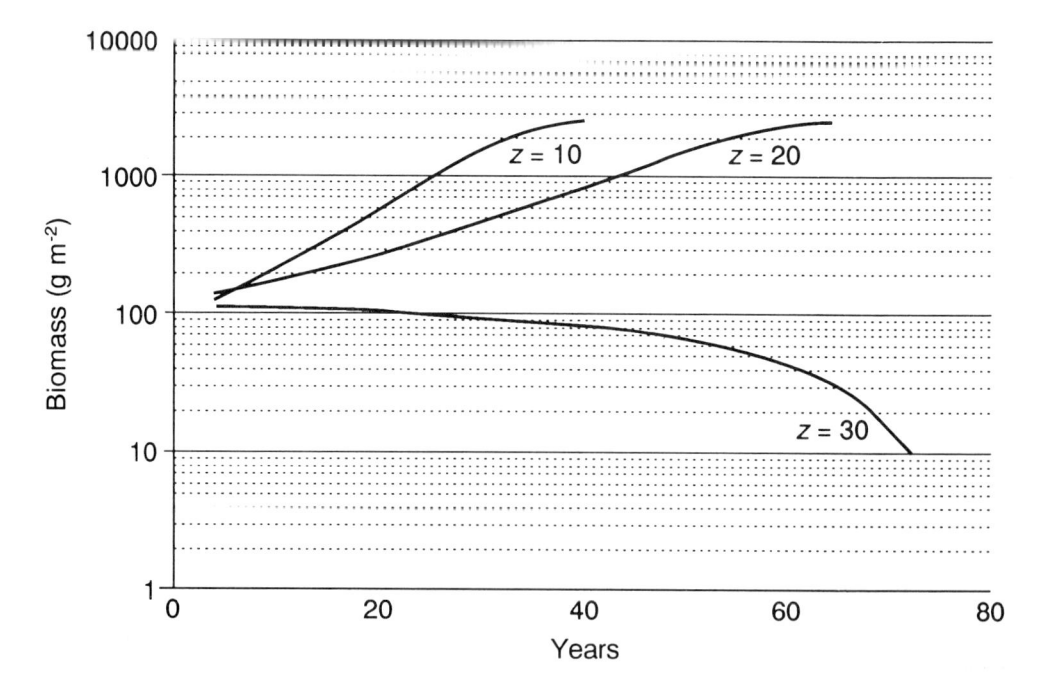

Figure 16.2 The time-based growth curves for vegetation with different amounts of surface loss of soil (in cm)

the water-use function to derive separate spatial domains for different species (Thornes, 1990).

Here we wish to extend the model to consider: (i) species with different water-use characteristics and (ii) dynamic rather than fixed-growth coefficients to provide r-dependency in the ecological sense. This means that we separate the behaviour of the invasive opportunistic r-type strategy species, dependent on surface runoff from individual storm events, from inertial persistent k-type species which can tap deeper, annually available and more reliable water resources. In this paper we assume that the r-type strategists are represented by annual grasses and herbs, the k-type strategists by bushes.

For the existing species (species A), with biomass V_1, growth is controlled by the soil moisture. The simple assumption is made that the moisture is uniformly distributed throughout the profile and that all the moisture is available for plant growth. This is crude but, insofar as growth is strongly related in Mediterranean environments to spring-available soil moisture, it is a reasonable working proposition. It is likely that this species will be deep rooting, have a low death rate coefficient (d_1) and considerable resistance to drought. It can be thought of as a bush species such as *Thymus* or *Anthyllis* (though the latter also has special nutrient utilisation strategies and exhibits long-term stability in the face of perturbations, even though establishment may be slow). By contrast, consider that a second species (species B) with biomass V_2 which is very shallow rooting, responds quickly to rainfall but dies off when no water is available, such as a grass. This is an r-strategist in the ecological sense. Its growth rate is defined as follows:

$$dV_2/dt = r_2V_2\{1-V_2/(V_{max}-V_1)\} - d_2V_2 - e_2Z \qquad (16.5)$$

This is similar in form to equation (16.1). We make r_2 depend on the amount of overland flow, so that species B can be thought of as having a high r-type dependency and is an r-strategist in ecological terms. Although growth is still logistic, the upper limit to growth (the k-dependency) is now constrained not only by soil moisture but also 'spare' vegetation capacity, as determined by the maximum climatic potential biomass (V_{max}) minus the amount of biomass of the other species (V_1). The maximum climatic potential biomass is assumed to be determined by a climatically controlled function (e.g. Whittaker & Marks, 1975). Hence competition between the plant species comes from the limit to total resource available after the deep-rooting species have satisfied their need. Since the infiltration rates are a function of total surface biomass, increase in *both* species reduces the propensity to overland flow and hence the growth rate for species B. Finally, if d_2 is made large then the half-life of species B is made small. For annual species it might be assumed that at the beginning of each year V_2 is zero. In practice there is always some potential growth (seedcorn) for both species left over, even after the complete collapse as a result of seed dispersal, seed banks or root stock, and in the digital simulations below this is accommodated by allowing a small but fixed amount of vegetation at the commencement of every iteration. Finally we need r_1 also to depend on available soil moisture to bring more realism to the model. This is the r-strategy component in its growth.

In this developed model the interactions are now a good deal more complex, not least because we now have three competitors, since soil erosion rates are now reduced by both V_1 and V_2. It is still possible to obtain analytical solutions, although, when dealing with stochastic inputs, the problem is far from trivial. By contrast it is much easier to simulate various scenarios. We suppose for subsequent runs that the rainfall is 300 mm, but that the infiltration rates and the number of events per year can vary. A tough environment for species A is then one that has low infiltration rates (average 0.5 cm per storm) and fewer events (more rainfall per event). These conditions favour both species B and erosion. Tough environments are also those with skeletal soil profiles ($Z > 0$), which favour overland flow and erosion but reduce growth rates for both species through erosional competition.

An example of interaction and evolution at these two extremes over time from a nearly bare soil is given in Figure 16.3. In the upper diagram the conditions are favourable (infiltration rate 2 cm per event, 10 effective events per year), species A grows rapidly and

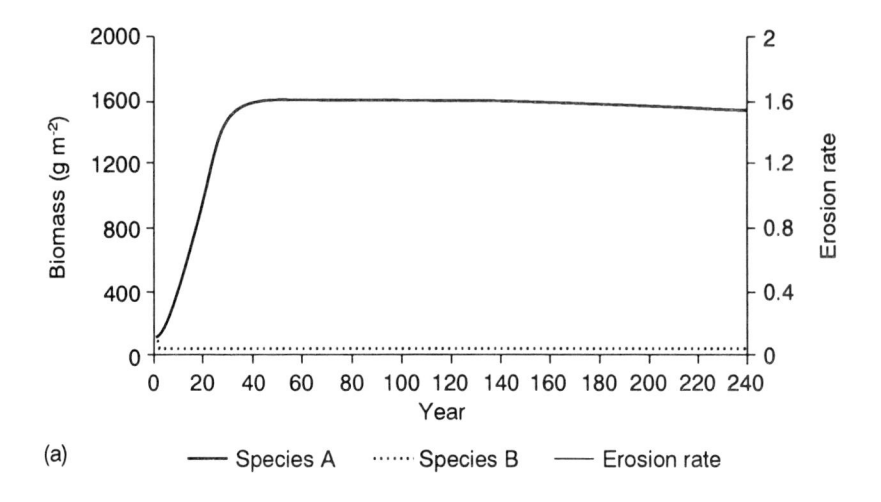

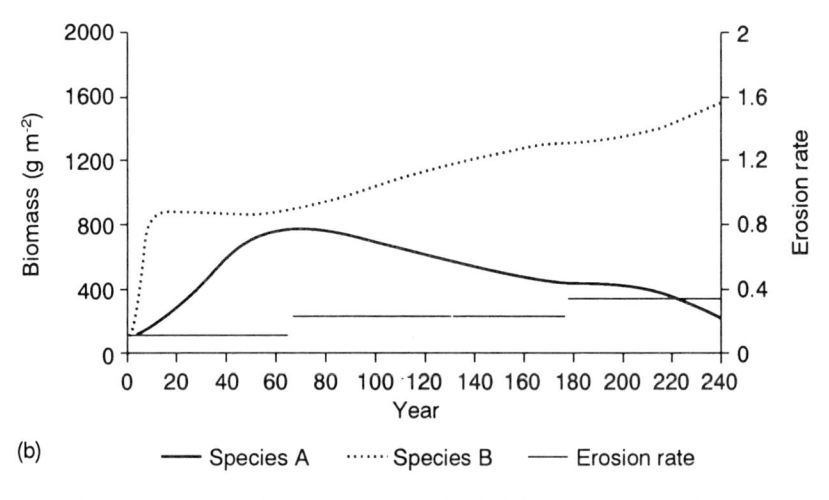

Figure 16.3 Deterministic erosion competition with (a) favourable conditions and (b) intermediate conditions of infiltrability and storm intensity. In both cases species A has maintenance costs in the form of biomass required for respiration. Rainfall conditions are assumed constant

logistically, and species B and erosion are negligible. In the lower graph conditions are tough, infiltration rates are low (0.5 cm per event), rainfall is more intense (5 effective events per year) and as a consequence the initial growth of both species moves nearly to equilibrium after about 80 iterations. However, the progressive erosion gives advantage to species B which continues to dominate until the complete extinction of species A after about 240 iterations.

In Figure 16.4(a), with tough conditions, rainfall declines at a rate of 3 mm yr^{-1} (a figure comparable to the mean rainfall between 1890 and 1934 in Murcia), so that after 100 years the system must collapse (note that it does not quite go to zero — there is still the seedcorn referred to above). The evolution over this time shows a rapid rise in species B, but gradually, despite the decreasing rainfall, it is overtaken by species A, although by year 70 this too falls rapidly after growth has ceased due to lack of sufficient rainfall. Figure 16.4(b), for favourable starting conditions, illustrates the effects of a step change to higher intensities with constant rainfall at year 50. Here the increase in overland flow promotes species B which is able to grow at the expense of species A and eventually overtake it, while the rate of erosion progressively increases.

STOCHASTIC ENVIRONMENTS — SINGLE SPECIES

We could continue to experiment with the deterministic system, as defined, to better understand its behaviour using an erosional model with plant growth by inputting steps or ramped climatic changes. We are, however, particularly interested in climatic change in the context of the strong inter-annual variations such as occur in semi-arid environments and the effects they could have on the response. There are several ways of approaching the problem. In plant ecology analytical solutions to the one-dimensional diffusion equation have been used to evaluate the effect of perturbations on deterministic systems on the assumption that the noise element is uncorrelated (May, 1973). Where this latter assumption does not hold, simpler growth models have also provided analytical solutions for autocorrelated inputs (Roughgarden, 1975). Most recently Pease et al. (1989) have obtained analytical solutions to a competition model in which the species are spatially arranged in an environment across which the climatic domains shift in response to climatic change.

Stochastic Generation of Rainfall Series

Before simulations can proceed, a characteristic climatic regime has to be provided from which departures can be considered. While the actual record at a place is a single realisation of an underlying stochastic process, what is needed is a generator that can provide many realisations of the process, having essentially the same characteristics but different specific realisations. We have used the data from 30 years of daily rainfall at 28 stations in the Murcia region of south-east Spain.

The stochastic generator for daily rainfall series consists of two parts. The first is a Markov chain which determines the probability of rain occurring and the second determines the magnitude of rainfall from a distribution function. The Markov transition matrix for dry or wet conditions on successive days is determined from empirical data and three matrices are used for three seasons as follows:

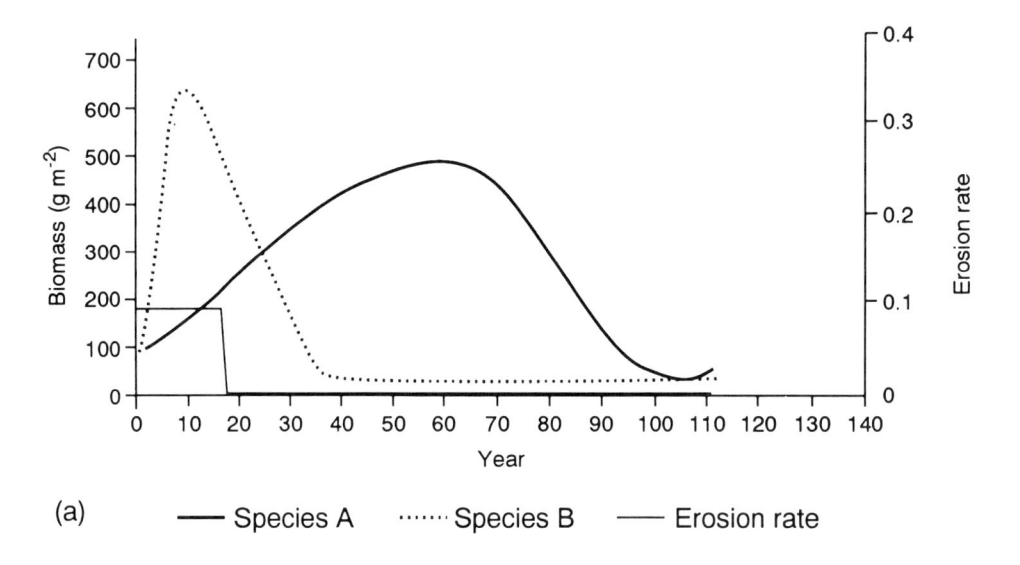

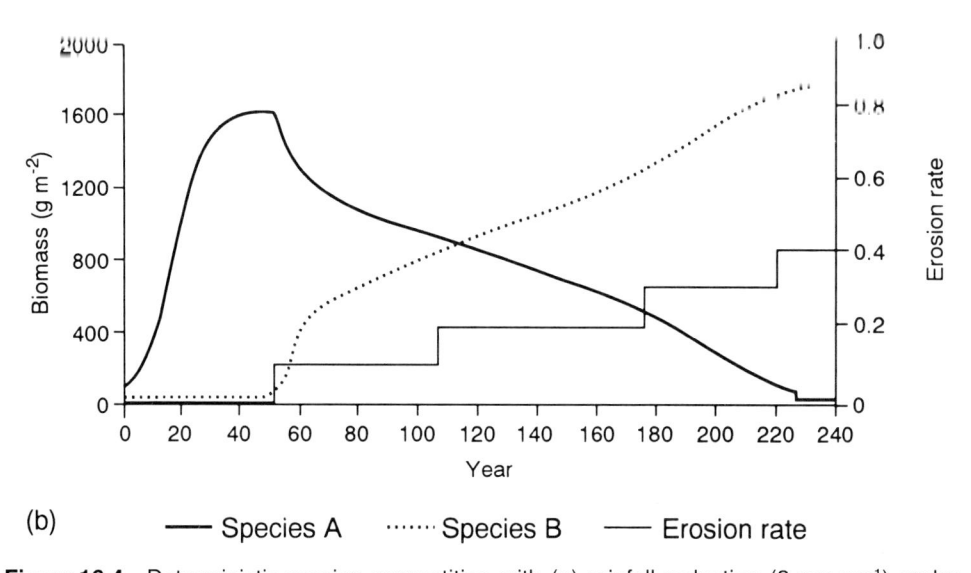

Figure 16.4 Deterministic erosion competition with (a) rainfall reduction (3 mm yr^{-1}) under tough conditions and (b) step change in rainfall intensity after 50 years with favourable conditions

Season 1: September–January
Season 2: February–May
Season 3: June–August

Daily rainfall magnitude values approximate a two-parameter gamma distribution after square root transformation. Alpha and beta values were calculated for seasons as defined above except for seasons with less than 5 rain days.

The stochastic generation assumes stationarity and independence between seasons. The wet–wet and dry–dry transition probabilities are chosen from two-parameter gamma distributions for the season and then used to determine the sequence of wet and dry days. For wet days the magnitude of rainfall is obtained by generating alpha values for the distribution of magnitudes, calculating the associated beta values from regression and then generating the magnitude from the distribution. For annual rainfalls the daily simulated sequences are summed. Deterministic signals can be added to the generated series and/or the parameters of the generators can be changed to simulate climatic change. Further details can be found in Brandt and Thornes (1991).

Single Species, Single Realisations

By way of illustration, consider first a single-species model with fixed amounts of erosion ($dZ/dt=0$) and the species A type of growth (strong k-strategy). The annual rainfall provided has the stochastic properties for the Province of Murcia as outlined above. Consider the two outcomes of the same single realisation of annual rainfall amount for a hundred years, as shown in Figure 16.5. In Figure 16.5(a) the environment is favourable and the evolution is one of rapid growth to equilbrium, despite the rainfall variability. Only year 77 was able to markedly perturb the equilibrium, which was restored after about 25 years. In Figure 16.5(b), the same annual rainfall series has been used with favourable infiltration conditions but with fewer events per year and a rainfall decrease of 3 mm yr^{-1} superimposed. Near equilibrium is attained much later, at a lower level, and the equilibrium is unstable. After an earlier collapse it was not able to recover. The point is that the stochastic fluctuations were able to bring the system down beyond the recovery point in the rather tougher environment.

Single Species, Multiple Realisations

Single realisations are informative, but the average performance of the system is more important, especially in a predictive context. To illustrate this, consider the results of 100 simulations of 100 years of rainfall with a rainfall decrease at two rates and with one rate under three different soil erosion conditions (Figure 16.6). Here survival refers to the continuance of biomass above the seedcorn level. In the first case one set uses the actual Murcia stochastic generator data, the other two have superimposed trends of decreasing annual rainfall. The distributions for 0 (actual generated data) and 1 mm yr^{-1} decrease are similar, with about 34% of the realisations lasting beyond 100 years. By contrast there is a 0% probability of survival beyond that time with a 3 mm yr^{-1} reduction. In the second case, with a 3 mm decrease under 0, 10 and 15 cm erosion losses, the figure shows the probability of survival for different amounts of profile truncation. The interesting result

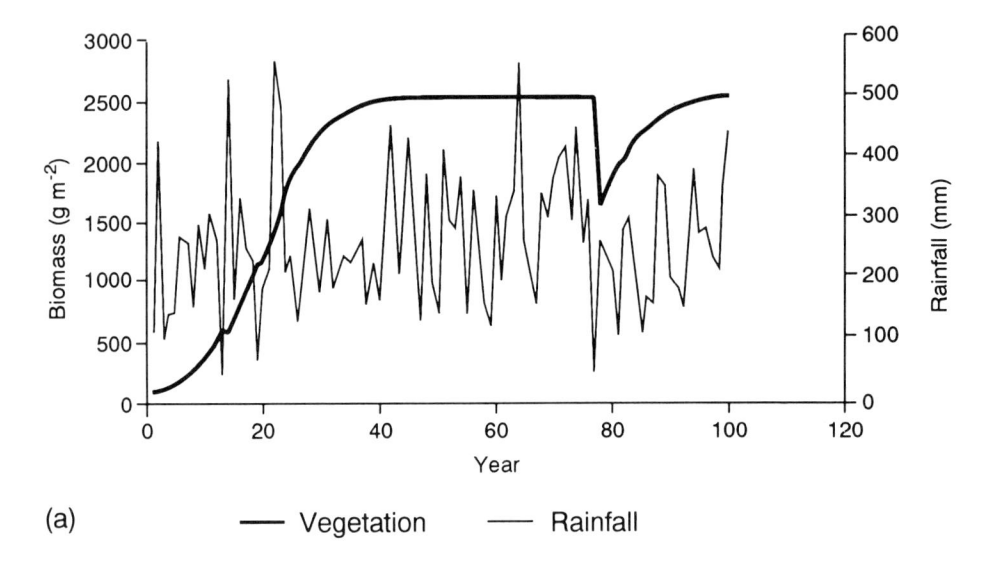

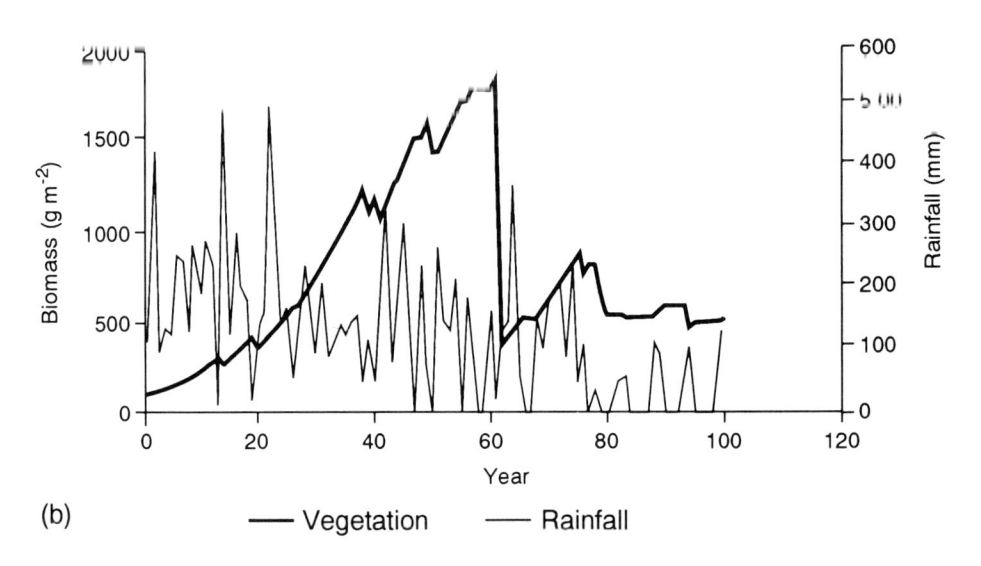

Figure 16.5 Single species evolution with stochastic rainfall generated by the rainfall generator for Murcia under (a) favourable conditions but 10 cm soil already lost and (b) with decreasing rainfall (3 mm yr^{-1}) and tough conditions

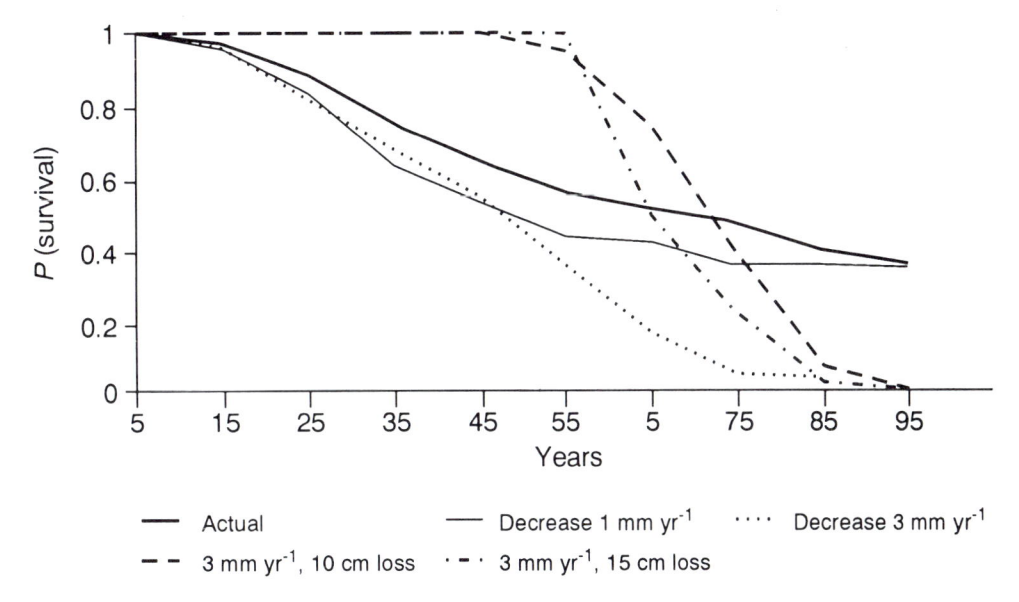

Figure 16.6 Probability of survival at various intervals for different rainfall and soil loss conditions

here is that, with no truncation, the probability of collapse is higher. This is because here the relatively favoured case tends to overproduce in wet years and collapse in subsequent very dry years. The probability of survival is almost normally distributed. In the more severe systems the vegetation cover is smaller, oscillations in growth rate are smaller and the system is better able to survive in the earlier years.

STOCHASTIC ENVIRONMENTS — TWO SPECIES

Generated Series

Consider now the case of the competing species with different behavioural characteristics. This is first illustrated with the Murcian stochastic generator in Figure 16.7. In Figure 16.7(a) the species are in an environment favourable to species A, with a high infiltration rate and constant mean annual rainfall through time. Here the dominant k-dependent species (A) reaches equilibrium and despite the perturbation at 77 years restores itself to equilibrium again, albeit more slowly, through time. By contrast species B, which is r-dependent and relies on overland flow, remains more or less at seedcorn level because of the paucity of overland flow. In Figure 16.7(b), on lithology with a lower infiltration rate, overland flow dominates. Species A has insufficient moisture to assert itself, peaking weakly at 76 years, and species B fluctuates according to the level of overland flow but sustains a very low cover at about 500 g m^2. Soil erosion is storm dominated and there are some large peaks at the early stage of cover development. However, these are suppressed somewhat by the cover generated by both species. In Figure 16.8(a) we illustrate the effects of a sharp fall in rainfall under the same conditions as in Figure 16.7(b). All cover is eventually supressed (except for seedcorn) after about 80 years and never manages to recover because the rainfall

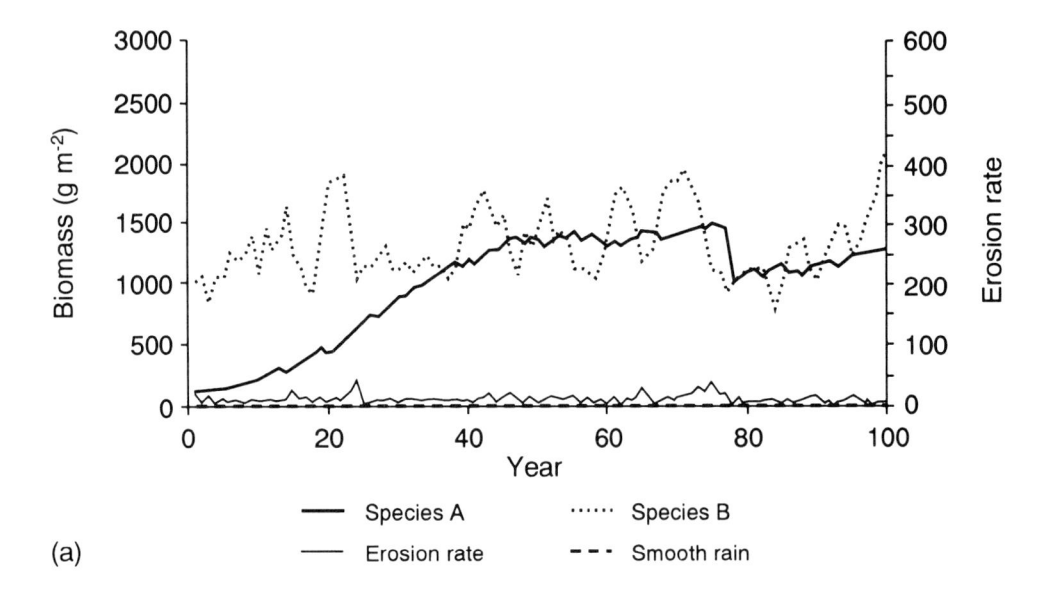

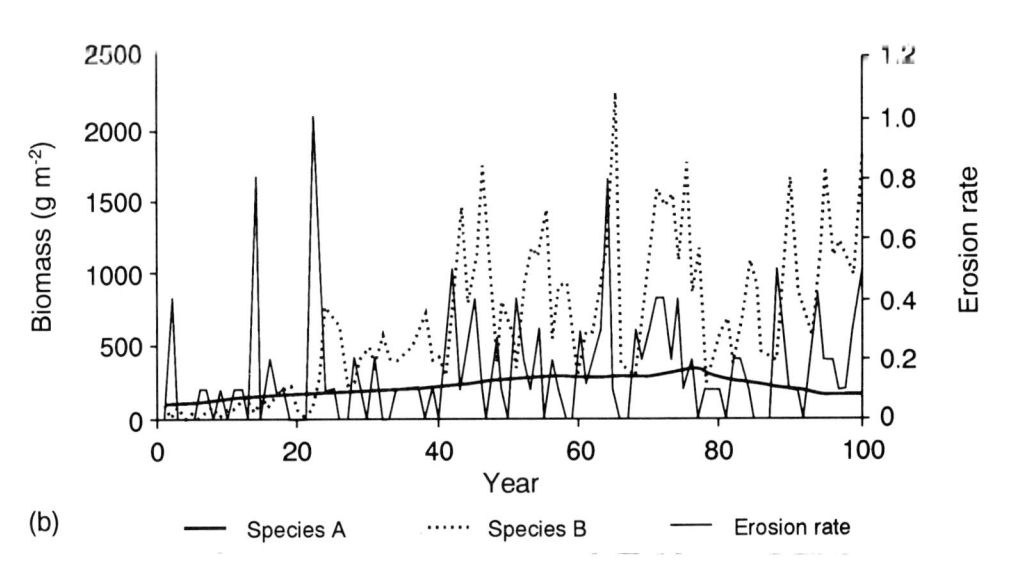

Figure 16.7 Two-species simulations with stochastic inputs under (a) favourable and (b) less favourable conditions

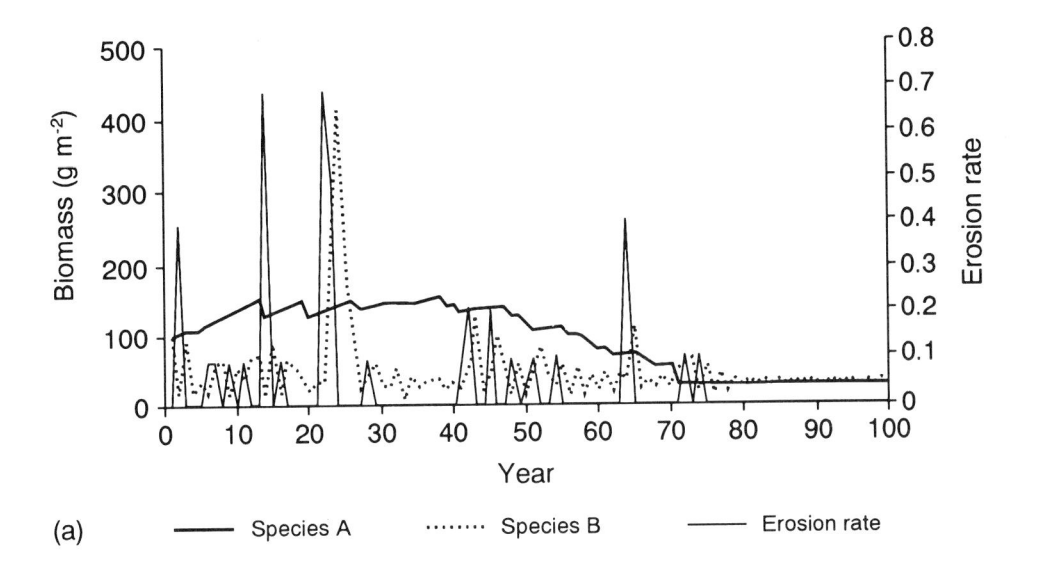

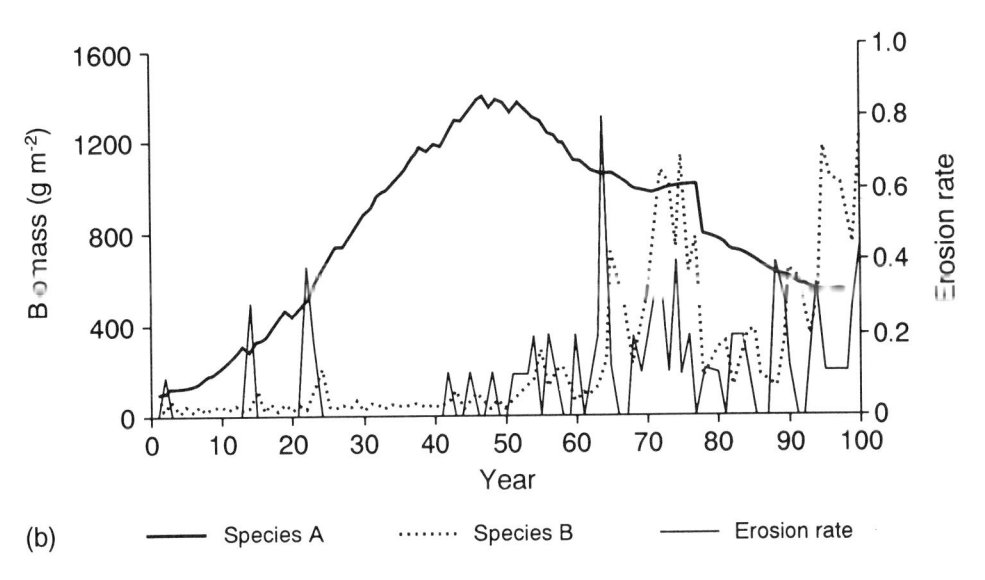

Figure 16.8 Two-species competition (a) under less favourable conditions with stochastic inputs of decreasing rainfall and (b) under favourable conditions with a stepped increased in rainfall intensity after 50 years

318   Environmental Change in Drylands

per event is lower than that required for overland flow production and insufficient to sustain species A towards the end of the period. Finally we note the effects of an increase in intensity after year 50 *without* reduction in overall rainfall with relatively high infiltration rates, as in Figure 16.8(b).

HISTORICAL RECORD

We now consider an actual rainfall record. This is the long series of annual data for the city of Murcia. We would like to know how the model output responds to a natural series on assumptions about how the rainfall is partitioned by storms. Figure 16.9 shows the two species over the same period of time as they respond in a tough environment (where tough has the specific meaning described earlier), with low infiltration rates and the number of storms per year set at 10. Both species are assumed to be starting on bare ground in 1864. In this case no maintenance costs are imposed on the bush species A. The result shows that species A was able to grow logistically until about 1915 when there was a crash resulting from the progressively worsening drought. This was coupled with the overall fall in annual rainfall totals during the 1890–1934 period and it was not until the late 1940s that species A began to recover. By the late 1970s it nearly reached values comparable to those at the end of the last century, only to die off dramatically in the 1980s. Species B shows a rapid growth on bare soil in the 1880s but at first was unable to compete with species B. From then on both species B and erosion remained supressed until the crash of 1915, after which it was able to compete on a nearly equal footing and overtook species A in the drought of the 1980s.

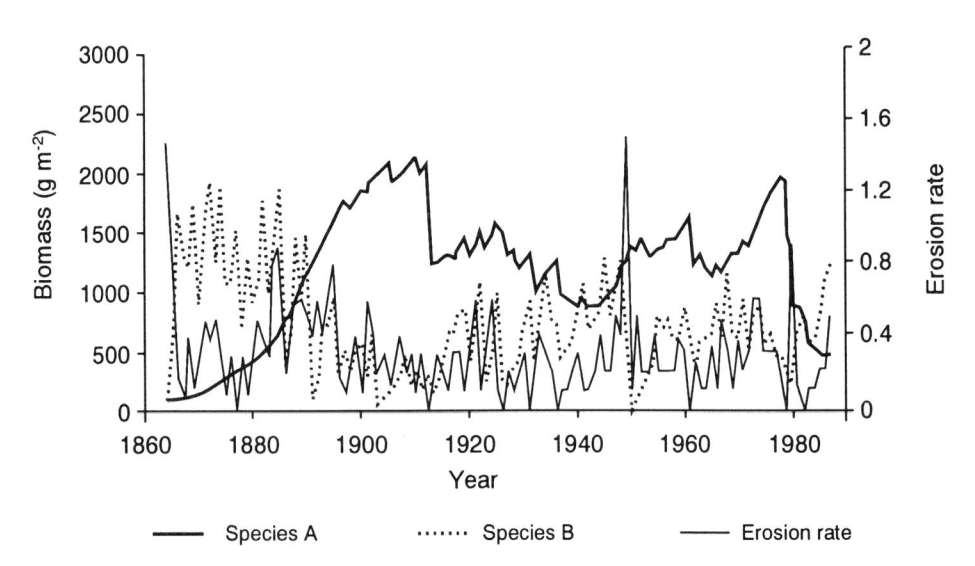

Figure 16.9 Simulation of erosion and vegetation competition for two competing species with different strategies in a tough environment (low infiltration capacity soils) using actual rainfall data for Murcia, 1864–1985

DISCUSSION

In the previous sections we have illustrated the impact of stochastic variations on a simple two-species model of plant response to runoff, soil moisture and erosion. The examples illustrate the complexity of response in terms of changes in the deterministic and stochastic inputs. They show that with the inertial k-type dependency quite severe deterministic and stochastic shocks to the system can be buffered, at least in relatively favourable environments. As the environments become tougher, e.g. under spatial variations of soil type, the ability to withstand both types of shock is significantly diminished and the r-type species are better able to resist climatic and erosional impacts. These results are consistent with expectations on a theoretical basis for deterministic competing systems subject to random shocks, at least for the single-species case. The second feature of interest is the delay in either the impact of individual shocks or of a long-term decrease of rainfall in the model system. In favourable environments the impact of change may be appreciably delayed or undetectable for significant periods of time or recovery may be almost immediate. By contrast, in tougher environments quite small perturbations may result in a rapid response.

These ideas may be summarised in terms of sensitivity to change of the two species and erosion relative to the variability of the input. This is shown in Table 16.1 where the ratio of the coefficient of variation of the output to the coefficient of variation of the stochastic rainfall input is given. The first three lines show the sensitivity of three cases, in progressively more difficult environments, using the Murcia stochastic generator data. In any single row the sensitivity of soil erosion is seen to be much greater than that of the plant species and, as expected and modelled, the variability in the r-dependent species is significantly greater than that of species A except in the tough environment. This is thought to reflect the greater instability of erosion when in competition with species A in favourable and 'medium' environments. Overall rates of erosion are smaller in the favourable environments and the few extreme events are more marked when they do occur. As the environment gets tougher, more erosion occurs, the average rate increases and there are relatively fewer extreme events. The k-dependent species is most sensitive to rainfall in the toughest environment whereas the r-dependent species is less sensitive.

In the lower two lines we illustrate the cases using the actual Murcia rainfall series. Most interesting here again is the sharp reduction in the sensitivity (and hence greater resilience) of species B and soil erosion reaction in the tougher environment, indicating again its higher level of adaptation to these conditions. In the last column are given the correlations between the rainfall inputs and soil erosion outputs. In general as the environment is more difficult the correlation is higher, which is in agreement with general observations and the expectations of the model.

We may conclude that the response by erosion to changes in climate is not only vegetation-dependent but also depends on the ecological nature of the species, the severity of the environmental conditions, the pre-existing extent of erosion and the *particular* sequence as well as the general character of the stochastic component of the rainfall involved.

CONCLUSIONS

We have illustrated from simple models the complexities that can arise in biomass and erosional response with different plant behaviour, different environmental conditions and with some simple types of climatic change. The results suggest that quite different responses can be observed over short distances due to differences in edaphic (especially erosional) conditions. The results indicate substantial variations in sensitivity of erosional response to rainfall as mediated through simply defined plant covers. Further investigations of the actual plant historical record over this critical period are required and this might be achieved through dendrochronological and pollen-rain studies.

The development of models of any type is a progressive process of successive approximations (Thornes, 1990). In this paper we have attempted to take a different approach to that normally adopted. The model should therefore be regarded as very primitive, lacking as it does any reference to other limiting factors in vegetation growth, such as nutrients, and failing to accommodate genetic adaptation, migration and invasion strategies and more complex erosional processes. Some of these are now being attempted (e.g. Baird *et al.*, 1992).

ACKNOWLEDGEMENTS

This work was initiated under a grant from the Natural Environment Research Council and has continued under a grant from the European Community EPOCH programme (EV4C), both of which sources are gratefully acknowledged.

REFERENCES

Baird, A., Thornes, J. B. and Watts, G. (1992). Extending overland-flow models to problems of slope evolution and the representation of complex slope-surface topographies. In Parsons, A. J. and Abrahams, D. (eds.), *Overland Flow*, University College Press, London, pp. 199–224.

Biot, Y. (1988). Forecasting productivity losses by sheet and rill erosion in semi-arid rangeland. Unpublished PhD thesis, University of East Anglia, Norwich.

Brandt, J. and Thornes, J. B. (1991). Third Annual Report on Project EV4C/European Community. Available from authors.

Eagleson, P. S. (1978). Climate, vegetation and soils. *Water Resources Research*, 14, 705–76.

Elwell, H. A. and Stocking, M. A. (1976). Vegetation cover to estimate soil erosion hazard in Rhodesia. *Geoderma*, 15, 61–70.

Kirkby, M. J. and Neale, R. H. (1986). A soil erosion model for incorporating seasonal factors. In Gardiner, V. (ed.), *International Geomorphology 1986*, Part II, Wiley, Chichester, pp. 189–210.

Langbein, W. B. and Schumm, S. A. (1958). Yield of sediment in relation to mean annual precipitation. *Transactions American Geophysical Union*, 39, 1076–84.

May, R. M. (1973). *Stability and Complexity in Model Ecosystems*. Princeton University Press, Princeton, 263 pp.

Pease, C. M., Land, R. and Bull, J. J. (1989). A model of population growth, dispersal and evolution in a changing environment. *Ecology*, 70, 1657–64.

Roughgarden, J. (1975). A simple model for population dynamics in a stochastic environment. *American Naturalist*, 109, 713–36.

Thornes, J. B. (1990). The interaction of erosional and vegetational dynamics in land degradation: spatial outcomes: In Thornes, J. B. (ed.), *Vegetation and Geomorphology*, Wiley, Chichester, pp. 41–55.

Thornes, J. B. (1991). Geomorphology and grass roots models. In Macmillan, W. (ed.), *Remodelling Geography*, Blackwell, Oxford, pp. 3–21.

Whittaker, R. H. and Marks, P. L. (1975). Methods of assessing terrestrial productivity. In Leith, H. L. and Whittaker, R. H. (eds.), *Primary Productivity of the Biosphere*, 55–118, Springer Verlag, New York, pp. 55–118.

17 Environmental Change, Disturbance and Regeneration in Semi-Arid Floodplain Forests

F. M. R. HUGHES
Department of Geography, University of Cambridge, UK

ABSTRACT

Research in both tropical and temperate semi-arid regions shows that while vegetation patterns within floodplain forests indicate a low tolerance to prolonged flooding, minimum-size floods are necessary for successful regeneration and establishment of seedlings. During the Holocene, floodplain vegetation has adjusted to changing disturbance regimes associated with hydrological and sedimentological changes. The implications of continuous adjustment by floodplain vegetation are discussed in the context of the downstream impacts of dams. This paper emphasizes the need to assess the significance of different magnitude and frequency floods for floodplain forest regeneration in the context of environmental change over a range of timescales.

INTRODUCTION

Floodplains in semi-arid regions are natural resources of far greater importance than their total area would suggest because of the 'oasis-like' role that they play within both the natural and human environment (Scoones, 1991). Characteristically, floodplains in all climatic regions are composed of quite well-defined assemblages of trees, shrubs and herbaceous species. These are frequently associated with particular geomorphological features in the floodplain (Hupp & Osterkamp, 1985) and are responsive to the hydroperiods (length, duration and seasonality of flooding) they experience (Conner *et al.*, 1981; Lugo *et al.*, 1989). Consequently, location of species assemblages is also closely related to elevation within the floodplain (Hughes, 1990; Titus, 1990) and to a lesser extent substrate (Huffman & Forsythe, 1981; Dunn & Stearns, 1987).

This paper focuses on forested floodplains in semi-arid regions, with particular examples being drawn from Africa and North America. Whereas in humid regions floodplain forests tend to grade gradually into the surrounding upland forests, in semi-arid regions there is generally an abrupt transition from the floodplain vegetation to that on the surrounding drylands. This transition is related to the position of the water table within the floodplain, since floodplain trees can only survive in semi-arid areas where their roots have access to the water table all year round.

Floodplains are physically very active environments, experiencing a number of external disturbances in the form of flooding, erosion, deposition and channel change. These directly and indirectly affect the vegetation of different parts of a floodplain through their influence on regeneration patterns. It has been observed that individual species have quite specific regeneration requirements (Mahoney & Rood, 1991) which tend to be closely adapted to flooding regimes. While the general influence of flooding on vegetation is acknowledged (as, for example, in 'disturbance' and 'recovery' models; Wissmar & Swanson, 1990), a number of studies also show a strong dependence on certain conditions created by floods for providing suitable regeneration sites (Huffman, 1980; Bradley & Smith, 1986; Junk, 1989).

It is important to understand the extent to which floodplain trees are dependent on a particular flood regime because in many rivers these are being deliberately altered by dam construction. The ability of floodplain vegetation to respond to the prevailing disturbance pattern (created by floods) and to changes in the disturbance regime over time will be determined by its regeneration mechanisms and the regeneration niches available. This paper aims to consider regeneration responses by floodplain forest species to changing disturbance regimes in semi-arid areas over a range of timescales from 10^1 to 10^4 years. The implications of these responses for the particular impacts of dams on downstream floodplains in semi-arid areas is then briefly discussed. Examples from humid regions are included where they shed light on general principles.

THE DYNAMICS OF SPATIAL AND TEMPORAL FLOODPLAIN VEGETATION PATTERNS

Prior to the 1960s relatively few studies of floodplain vegetation had been carried out in North America, and most had concentrated on semi-arid areas because of the luxuriant growth of their riparian communities (Johnson & Lowe 1985). Subsequently, floodplains have been studied in many regions, in part because of their economic and environmental importance. Probably the most intensively studied rivers are those of humid south-east United States, where detailed work has been carried out on community patterns, structural characteristics, biomass, productivity and nutrient cycling (see, for example, Brinson, 1989). Early studies in colonial Africa concentrated on detailed floristic descriptions and vegetation mosaic mapping, and related the results to substrate or geomorphology (Bégué, 1937; Trochain, 1940; Keay, 1949; Kemp, 1961).

Some researchers in North America commented at an early stage on the specific needs of some riparian species (e.g. willows and poplars) to regenerate successfully (Moss, 1938; Ware & Penfound, 1949; Buell & Wistendahl, 1955; Weaver, 1960; Everitt, 1968). More recently, successional patterns observed on floodplains have been linked to the flooding history of the river (Johnson et al., 1976; Teversham & Slaymaker, 1976; Nanson & Beach, 1977; Baker, 1990). It has become increasingly apparent that a dynamic approach to studying floodplain forest ecology is necessary, involving a detailed understanding of:

(i) successional trends;
(ii) their relationships to the physical processes of floodplain disturbance; and
(iii) most importantly, the way in which these physical processes provide regeneration sites and affect regeneration potential throughout the floodplain.

In general terms, the distribution of vegetation within a floodplain is closely related to elevation and therefore to its hydroperiod, and to the mosaic of floodplain geomorphological features (e.g. ridges, swales and cut-off channels) and their associated sediment suites. As a result, complex floodplain vegetation mosaics mirror the patterns of floodplain landforms and sediments (e.g. Figure 17.1). The fact that vegetation associations tend to be found in predictable locations within a floodplain indicates that vegetation responds in a constant and sometimes rapid way to changing hydrological and sedimentological conditions as channels change course.

In many floodplains, predictable successional trends caused by changes in drainage and sediment characteristics have been documented (Viereck, 1970; Gill, 1971; Kellerhals & Gill, 1973; Johnson et al., 1976; Pautou, 1984; Amoros et al., 1986; Hughes, 1988). In the Beatton River of British Columbia, Nanson & Beach (1977) studied vegetation succession in relation to elevation and sedimentation in the floodplain over a time period of about 500 years. They concluded that balsam poplar (*Populus balsamifera*) colonizes fresh point-bar sediments and that after 50 years the maximum poplar density has been reached and sedimentation has raised the floodplain surface to above bankful level. This leads to an abrupt decline in sedimentation and white spruce (*Picea glauca*) seedlings establish themselves. As long as further disturbance by fire or floods does not occur, white spruce will continue to regenerate. Succession here would seem to be predictable until a section of floodplain is reworked during a major flooding period. Although this and similar examples do not relate specifically to semi-arid rivers, many of the principles are likely to be relevant in drier climates because of the prime influence of channel processes.

The relationship between the time frame of vegetation succession and that of the floodplain sediments reworking is crucial to understanding the role played by disturbance processes on floodplain vegetation. It might be expected that where floodplain turnover is rapid, a large percentage of the vegetation would tend to be at a pioneer stage on new sediment. Where channel change is slower vegetation has longer to adjust and a larger proportion of the floodplain is likely to be occupied by mature vegetation stands. Within any individual floodplain, the active channel margins will be reworked with much greater frequency and support short-lived pioneer communities. Rivers in semi-arid areas tend to have sandy non-cohesive perimeters and extreme events are also usually more important than low-magnitude–high-frequency events in causing major changes in channel form (Patton and Baker, 1980), although recovery can also occur more rapidly in semi-arid areas (Schumm & Lichty, 1963; Wolman & Gerson, 1978). Turnover time for sediments in semi-arid floodplains therefore tends to be shorter than that of humid floodplains. For example, the turnover time in many rivers in the semi-arid North American prairies is thought to be around 200–300 years (Rood, personal communication; Cordes, personal communication), whereas in the Beatton River it is about 700 years (Nanson & Beach, 1977). In the Tana River (semi-arid south-east Kenya) floodplain turnover may be as frequent as every 150 years, and a similar time has been postulated for the Animas River (Colorado) floodplain (Baker, 1990). In a mapped stretch of the Tana River floodplain, mature forest occupies 63.6% of the floodplain (Figure 17.1) but the most diverse forest type, which only grows on stable sandy levees, occupies only 1.5% of the mapped area. Active point-bars and oxbows occupy approximately 13% of the floodplain.

Kangas (1989) expounds an energy-related approach for the simultaneous examination of the relative time frames of the geomorphological and ecological systems which together form a floodplain. Using the successional time to climax forest (in his example from south-east Michigan, about 500 years) and the length of time for the floodplain landform to develop (about 10 000 years in the area he studied), Kangas obtains 'embodied energy' values which serve to demonstrate that the forest ecosystem turned over 20 times faster than

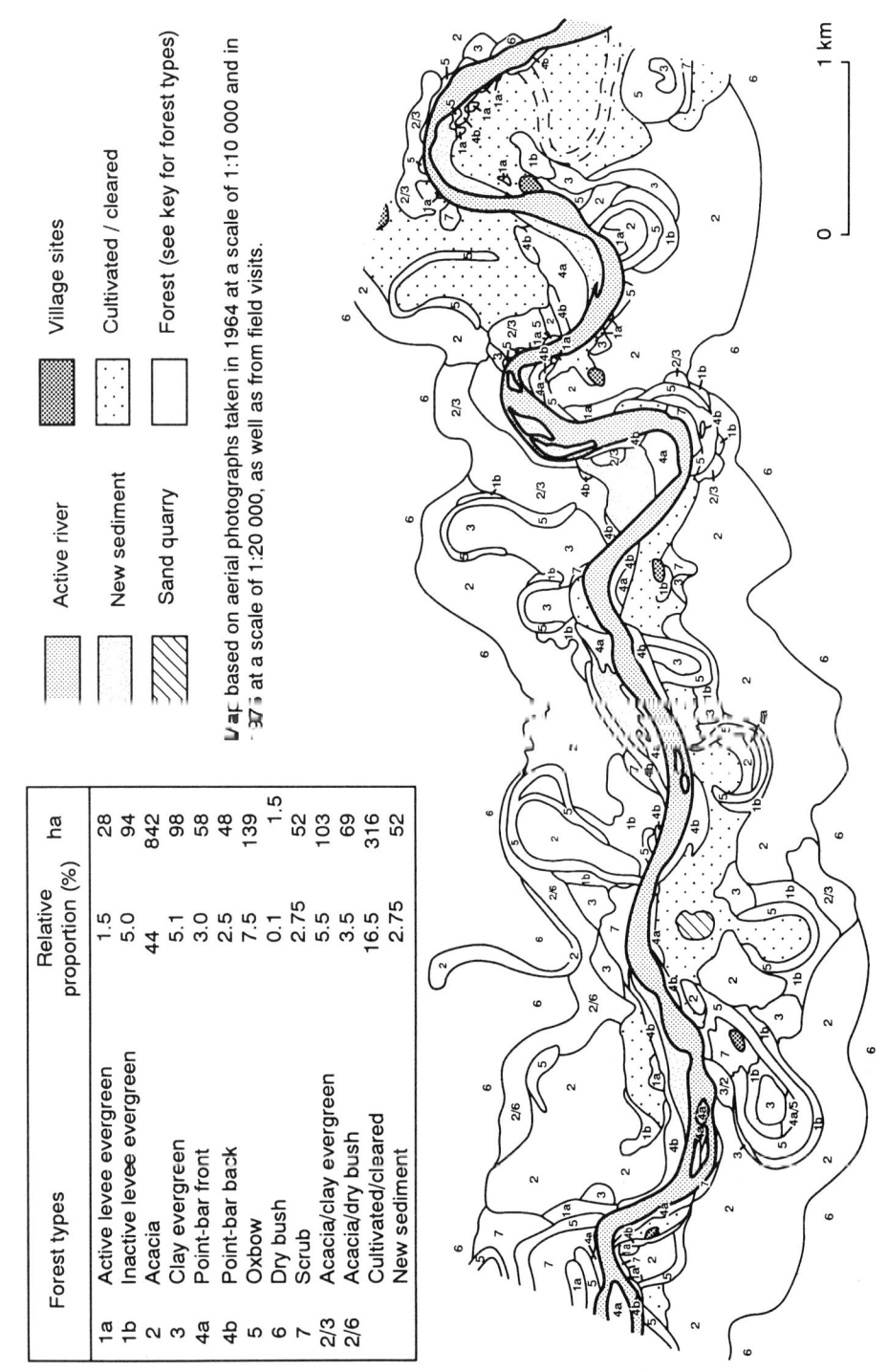

Forest types		Relative proportion (%)	ha
1a	Active levee evergreen	1.5	28
1b	Inactive levee evergreen	5.0	94
2	Acacia	44	842
3	Clay evergreen	5.1	98
4a	Point-bar front	3.0	58
4b	Point-bar back	2.5	48
5	Oxbow	7.5	139
6	Dry bush	0.1	1.5
7	Scrub	2.75	52
2/3	Acacia/clay evergreen	5.5	103
2/6	Acacia/dry bush	3.5	69
	Cultivated/cleared	16.5	316
	New sediment	2.75	52

Figure 17.1 Distribution of forest types near the village of Pumwani in the Tana River floodplain of Kenya. (Reproduced from Hughes, 1990, by permission of British Ecological Society)

the landform. Obviously, the closer the turnover rates of the landform and the ecosystem the more likely it is that a small change in one will result in a larger change in the other. On a long timescale this kind of approach gives a good overview of the floodplain system over time by concentrating on successional development of an ecosystem through its total accumulated energy. However, it gives little impression of the dynamism of a floodplain in terms of its temporary energy losses and gains through erosion, deposition and sediment reworking. It also does not take into account the spatial and temporal variation in distribution of different parts of the ecosystem, which result from periodic disturbance.

The importance of disturbance regimes to ecosystems has long been acknowledged, though not explicitly discussed, being rooted in studies like those of Aubreville (1967) on mosaics and Watt (1947) on heather cycles in the Breckland of England. Whereas early studies tended to view disturbance as an occasional event, interrupting succession, it is now realised that disturbance permeates ecosystems at all spatial and temporal scales (Pickett & White, 1985), and it is particularly well appreciated in tropical rain forest ecosystems. Equilibrium at a local scale, even over long time periods, is rarely reached, and in the context of constant change through time it is difficult to define what is disturbance and what is normal change (Sousa, 1984). An environmental fluctuation causing disturbance and mortality will select positively towards those species most able to tolerate it. Eventually a similar environmental fluctuation may not cause disturbance because the original community will have evolved into a more tolerant one. Sousa (1984) suggests that a change in 'disturbance threshold' occurs through a change in species composition and perhaps also through species evolution to a more tolerant community. Taking this argument further, it is possible to envisage a continuum of community response to disturbance from one of evolving in a way that reduces the impact of stress to one of actually requiring short-term disturbance for long-term survival.

In the context of floodplain forests in semi-arid areas it is necessary to look initially at the regeneration needs and mechanisms of floodplain species to understand the role that disturbance plays in their distribution. On a longer timescale it might be possible to assess how changing flooding and sedimentological regimes have altered the 'disturbance thresholds' of floodplain species to give the highly dependent relationships that floodplain species now have to particular magnitude-frequency floods.

FLOODPLAIN VEGETATION RESPONSES TO DISTURBANCE, THROUGH REGENERATION

The regeneration response of a community will determine its recovery following a disturbance. In some floodplains the scale of geomorphological disturbance each year is enormous compared to the total floodplain area; in others there may be prolonged floods but the amount of geomorphological change may be slight. In the highly mobile floodplains typical of semi-arid areas there will be a high proportion of new sites available for regeneration each year. In less mobile rivers most disturbance will be in the form of overland floods in which new sediment (and often seeds) may be deposited in established stands of trees. In the former, species tolerant of, or even favouring, physical disturbance will tend to colonize and the area of mature forest may be limited. Into this category fall many North American prairie rivers where poplars tend to be the main and sometimes only tree species, requiring new sediment sites for regeneration. In less mobile floodplains most areas (unless they are permanently flooded) will tend to be occupied by mature forests. In these forests, and on stable sites at higher elevations in more mobile floodplains, overland floods and their sediment loads are also critical in any single year. As well as recharging the water table, they affect regeneration potential by creating or destroying regeneration niches (*sensu* Grubb, 1977). In most forest types, availability of resources for new plants decreases after the initial establishment period, but in floodplain forests the floodwaters and their sediments can replenish resources even without the creation of gaps. This may lend competitive advantages to some species in their seed or seedling phases on the canopy floor in readiness for a gap in the more advanced successional stages.

Spatially, therefore, forests on floodplains demonstrate different regeneration strategies in different locations, ranging from regeneration of pioneer species, where active sedimentation occurs, to gap-phase type regeneration in mature forests. In these older stands, past geomorphological history in the form of substrate and topographic variations may be reflected in species diversity. For example, Reid (1974) documents catastrophic floods (1 in 150 years events) in 1970 and 1971 in the Hume River on the west side of the Mackenzie Delta in Canada. These floods toppled many poplar and spruce trees that were not carried away. In 1972 and 1973, abundant suckers of balsam poplar, alder and willow were observed at lower elevations while many spruce seedlings were observed in the white spruce stands. In years to come this event will be reflected in the narrow age classes of trees in those stands and the changed mix of species.

The exact regeneration needs and tolerances of floodplain species vary hugely, not only between parts of a floodplain but also between floodplains in different geographical regions. The importance of regeneration requirements to subsequent development of floodplain communities is implicit in a number of studies, although much of the detailed work on regeneration has been carried out through laboratory studies and fewer studies have looked in detail at regeneration in the field. In semi-arid floodplains there is a marked vegetation zonation associated with elevation and susceptibility to flooding with, generally, forest growth only supported above a particular elevation. Specht (1989) identifies such zonation in *Eucalyptus camaldulensis* forests in semi-arid Australia; other examples have been noted in the Chobe (north-east Botswana) and Pongolo (South Africa) Rivers by Simpson (1975) and Furness & Breen (1980) respectively and in the Turkwell River (Kenya) (Oba, 1991). In the lower Tana River forest growth is only sustained above a level receiving less than 28 days of continuous flooding (Hughes, 1990) (Figure 17.2). The implication of this is that most forest species have a low tolerance to flooding, with the species at lower elevations having the greatest tolerance. An exception to this general observation is demonstrated by many poplars. In the Tana River the endemic poplar, *Populus ilicifolia*, is found at lower elevations, usually on point-bars, demonstrating a need for open well-watered sites (Figure 17.3).

Interestingly, in floodplain forests in humid areas the emphasis is also on the tolerance of forest species to prolonged flooding because the duration of flooding in these forests is usually much more prolonged than in drylands (Junk, 1989; Robertson *et al*., 1978). Many species in the bottomland forests of the south-east United States show special physical adaptations to flooding, enabling them to survive in an anoxic environment (Gill, 1970; Hook & Brown, 1973). The distribution of floodplain species in both humid and semi-arid environments is thus determined by both adult and seedling tolerance to stresses caused by flooding. Generally, tolerance to flooding increases with age and size (Gill, 1970; Kozlowski, 1984) so that it is the successful germination of seeds and establishment of seedlings in relation to a flood that is critical. In humid areas the development and establishment of

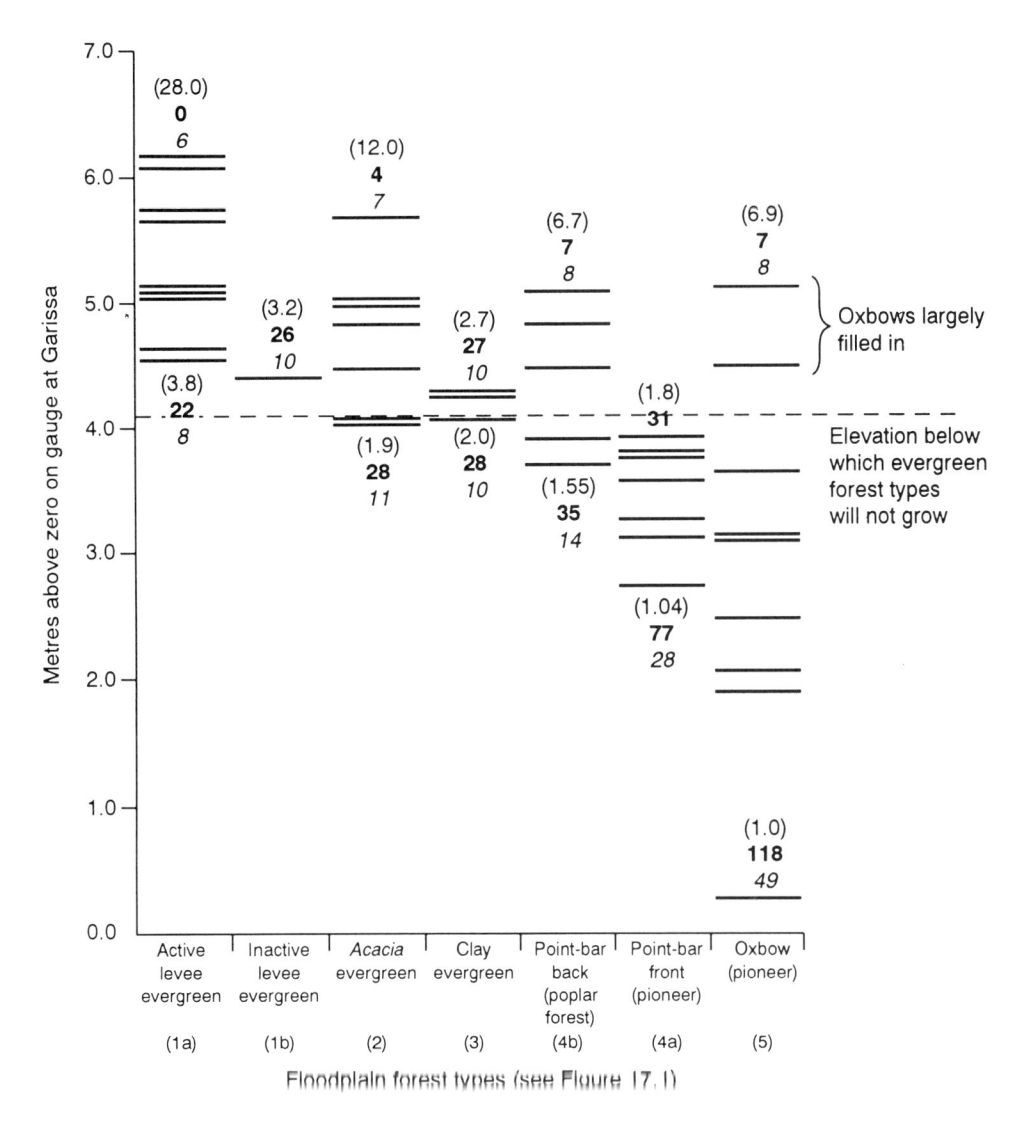

Environmental Change, Disturbance and Regeneration

[Chart showing elevations (metres above zero on gauge at Garissa) from 0.0 to 7.0, for floodplain forest types. Each entry shows (flood return period in years), maximum duration in days, and average duration in days.]

Active levee evergreen (1a):
- (28.0) 0 6
- (3.8) 22 8

Inactive levee evergreen (1b):
- (12.0) 4 7
- (3.2) 26 10

Acacia evergreen (2):
- (1.9) 28 11

Clay evergreen (3):
- (2.7) 27 10
- (2.0) 28 10

Point-bar back (poplar forest) (4b):
- (6.7) 7 8
- (1.55) 35 14

Point-bar front (pioneer) (4a):
- (1.8) 31
- (1.04) 77 28
- (1.0) 118 49

Oxbow (pioneer) (5):
- (6.9) 7 8

Oxbows largely filled in

Elevation below which evergreen forest types will not grow

Floodplain forest types (see Figure 17.1)

(12.0) Flood return period in years (based on a flood return graph produced by the Kenya Ministry of Water Development in 1978)
4 **The maximum duration in days of any flood during 1934–81 (excluding 1961)**
7 *The average duration in days of the longest flood in the years in which flooding occurred during 1934–81*

Figure 17.2 Elevations (corrected for river surface slope) relative to the Garissa gauge (on the lower Tana) and vegetation type of sample plots in the Tana floodplain forests, including the lowest and highest plot of each type. In calculating flood return periods, it was assumed that all plots at equal elevation had an equal chance of being flooded, regardless of distance from the river. (Reproduced from Hughes, 1990, by permission of British Ecological Society)

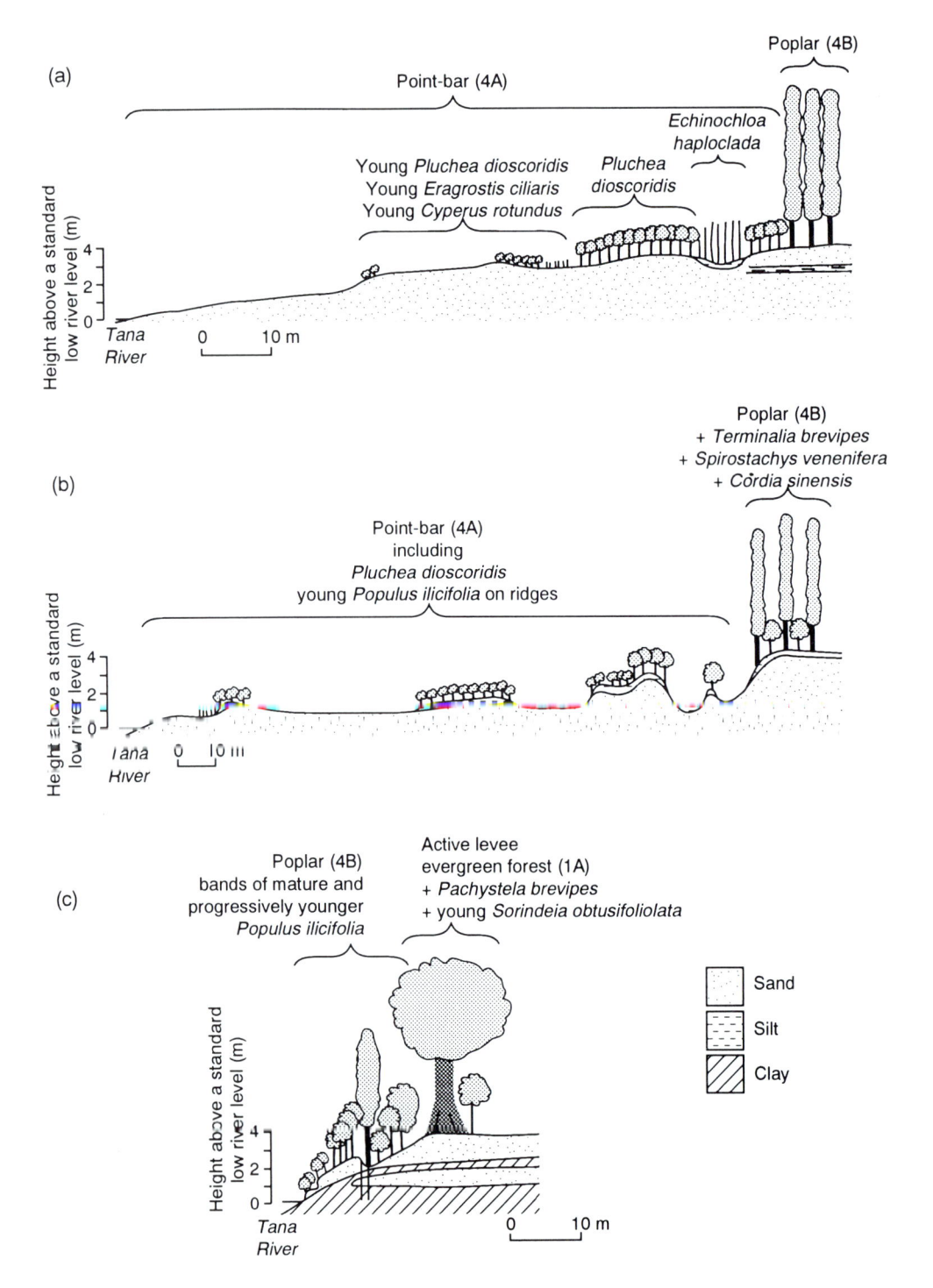

Figure 17.3 Surveyed point—bar transects at various stages of development in the Tana River floodplain, Kenya (after Hughes, 1985)

floodplain forests depend on the coincident availability of viable seeds with low water levels during the germination and seedling establishment stages (McKnight et al., 1981; Sharitz & Lee, 1985; Streng et al., 1985; Junk, 1989; Alvarez-Lopez, 1989; Bacon, 1989). Thus soil moisture conditions act selectively in determining seedling success during the growing season (Sigafoos, 1964; Burgess et al., 1973; Larsen et al., 1981).

In North American semi-arid floodplains, the literature concentrates on poplar ecology and some general models of poplar regeneration have been developed. They place emphasis less on the tolerance to flooding than on the need for flooding to provide suitably moist sites for regeneration, and on the need for active sedimentation and the maintenance of water tables for seedling survival.

Two models have been developed for poplar replenishment mechanisms in semi-arid southern Alberta (Bradley et al., 1991). The first, the 'incremental replenishment model', describes forest replenishment as an ongoing process on the tips of meander lobes (Bradley & Smith, 1986) and is based on studies of *Populus deltoides* seedlings along the Milk River (see Figure 17.4). They found that seedlings established successfully in years when a stage of at least the two-year return flood was coincident with successful seed production, high sediment loads and active channel migration. This combination occurred at approximately five-year intervals, as reflected in the age structures of the floodplain poplars. Their model was refined by Reid (1991) who defined the most successful establishment years as those with late June/early July maximum daily flows between 60 and 100 m^3 s^{-1}. This emphasises a need for sustained, high water tables during the early growing season a need also noted by Hughes (1990) in the Tana River.

The 'incremental replenishment model' is certainly relevant to floodplains with shallow gradients and fine sediments, but on the slightly steeper and more gravelly floodplains approaching the foothills of southern Alberta a 'general replenishment model' appears more applicable (Virginillo et al., 1991). Here, major overbank events are more important stimulii for poplar regeneration. Such events might occur every 30 or 50 years (Bradley et al., 1991) or even less frequently, and appear to allow widespread regeneration across the floodplain. Successful poplar recruitment in this model is related to high spring floods, again followed by maintained water tables within the floodplain sediments through the ensuing summer, possibly over a series of consecutive years. A similar situation has been described for a gravel bed reach of the Animas River (Baker, 1990).

It is possible that willow and poplar seeds can germinate underwater, allowing them to establish rapidly on sandbars and other fresh deposits (Hosner, 1958), although it has been suggested that willows in the Mackenzie River usually release seeds after floods recede and then colonize open sites (Lees, 1964). As well as seeding phenology, digestion of seeds by animals can be critical in germination as found by Oba (1991) in the Turkwel floodplain (Kenya). Although a number of the poplars in North American prairie rivers produce prolific numbers of seeds, their viability is usually only two to four weeks (Fenner et al., 1984; Moss, 1938; Noble, 1979; Bradley et al., 1991). Successful regeneration in all these semi-arid environments, however, tends to occur not at the lowest most mesic locations but at slightly higher sites because of the stress of physical erosion at lower elevations. In semi-arid rivers experiencing a winter freeze and a spring snow-melt regime (e.g. western North America), ice-scouring is the major source of erosion at these lower elevations.

Mahoney and Rood (1991) have carried out experimental studies using 'rhizopods' which permit controlled lowering of water levels within tubes of soil to study the influence of declining water levels on poplar growth and survival. They used a natural poplar hybrid

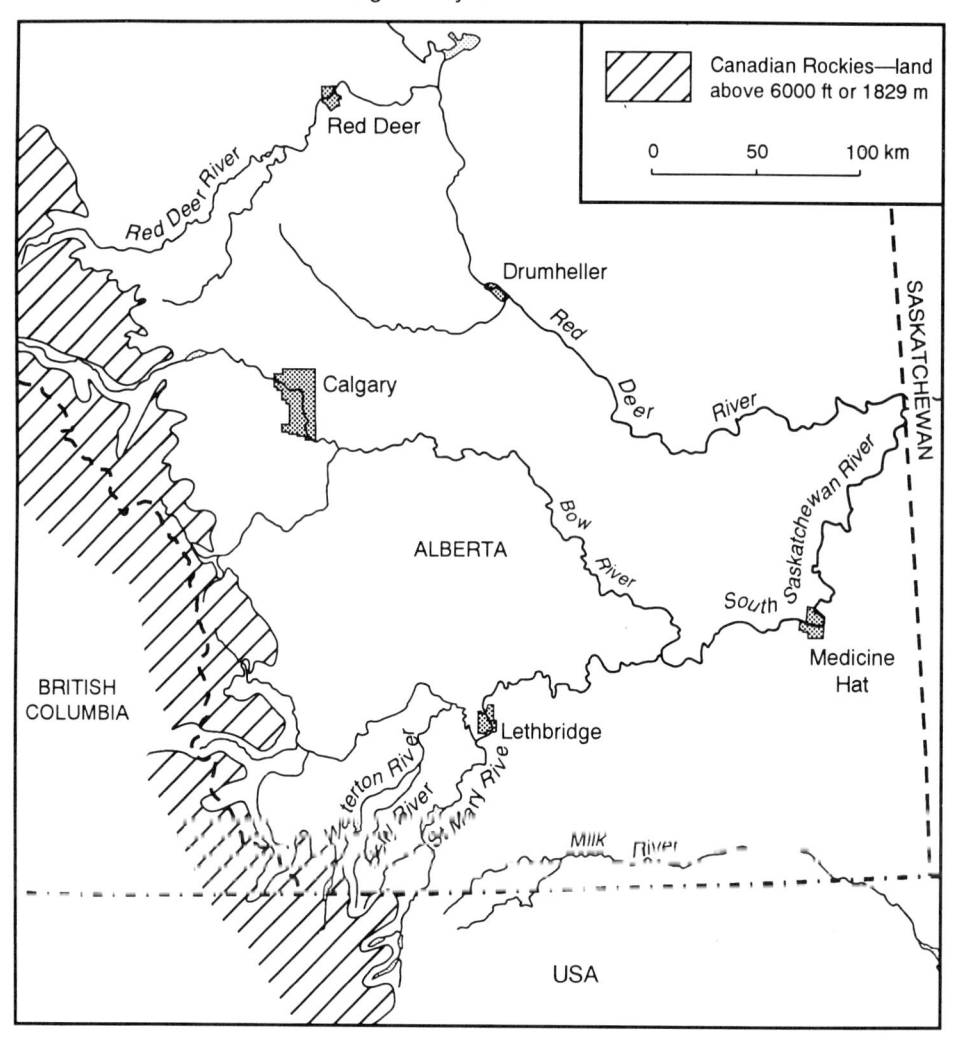

Figure 17.4 Map of Southern Alberta, Canada, showing the locations of principal rivers whose floodplain vegetation has been studied

from southern Alberta (*P. deltoides* × *P. balsamifera*) and found that seedling survival was over 90% when the water table declined by 2 cm day^{-1} or less, but was reduced to about 40 and 25% in rhizopods in which the water table declined by 4 and 8 cm day^{-1} respectively. The implications are that, for poplar seedlings at least, the shape of the flood attenuation hydrograph is critical in determining whether or not seedling establishment will take place. Thus, for successful regeneration to take place it is necessary for late spring and summer flows to be maintained at a minimum height and for quite specific rates and timings of flows to occur following a spring flood. Such general requirements are noted by a great number of authors working in semi-arid parts of North America (Moss, 1938; Read, 1958; Farmer & Bonner, 1967; Noble, 1979; Crouch, 1979; Groenveld & Griepentrog, 1985; Baker, 1990; Bradley *et al.*, 1991; Stobbs *et al.*, 1991).

The implications of all these studies are that minimum-size floods, or perhaps series of floods, are critical for regeneration and successful establishment of seedlings in semi-arid floodplains. The frequency of these events may be quite variable, even in rivers flowing from the same mountain range and experiencing similar flow regimes. Thus, in the Red Deer River (southern Alberta) most trees appear to have regenerated between 40 and 80 years ago and some are 135–140 years old (Marken 1991), compared to the five-year frequency hypothesized for the nearby Milk River (Bradley & Smith, 1986) (Figure 17.4).

The well-marked age structure of most floodplain forests is a good indication that conditions necessary for regeneration are quite specific and not always frequent, and that in general some form of fluvial disturbance process is necessary for successful regeneration. The vegetation of semi-arid floodplains seems to have evolved to a point where it cannot survive without certain levels of disturbance and it could be envisaged that over time the disturbance threshold, as defined by Sousa (1984), has changed. Thus, although a particular flood event might cause immediate destruction, if regeneration is to occur at all that flood is also necessary because the floodplain vegetation has evolved a long-term dependence on it through its regeneration requirements.

Much of this discussion has focused on the floristically simple floodplains of semi-arid North America. In most humid floodplains and in many tropical semi-arid floodplains (e.g. the Tana River), the situation is more complicated. In the Tana, point-bar regeneration similar to that described in Alberta occurs, with *Populus ilicifolia* being the dominant tree, although it is often associated with the shrub *Pluchea dioscoridis* (Figure 17.3). Tree-dating has not proved possible in the Tana, but the clear size classes of poplars indicates periodic regeneration. Other tree species in the Tana only grow above a certain elevation, implying a low tolerance to inundation. Generally, these species are arranged in well-defined vegetation types associated with elevation, and particular landforms and substrates (Hughes, 1988, 1990). There is, however, inevitable overlap of species in different parts of the floodplain associated with both their own variable environmental tolerances, the lag-time between changed river position and associated site conditions, and completion of plant life cycles. Initial evidence from seedling studies in the Tana River implies that even in stable forests at higher elevations, periodic minimum floods are necessary to recharge water tables and permit the successful establishment of these trees (Hughes, 1985).

As a river moves within its floodplain the susceptibility to flooding of any particular site will change, but in the stable areas that rarely receive floods, the influence of elevation and hydroperiod become less important and competition, soil texture, soil fertility and amount of sunlight become more important factors in seedling survival. Thus autogenic processes become more important than the allogenic processes associated with the fluvial regime. Variability in these older forests reflects past depositional and other geomorphological processes that are no longer discernable (Van Beek *et al.*, 1979; McKnight *et al.*, 1981) but which acted selectively at seedling establishment stage. It seems clear that all sections of a floodplain have evolved with disturbance of some kind and that evidence of past disturbance events is provided by species patterns in the older vegetation stands. It is also clear that these disturbances are necessary for regeneration in all parts of a floodplain and that the magnitude–frequency relationships of floods necessary is very variable.

HOLOCENE VEGETATION CHANGES IN FLOODPLAINS

Whilst studies of age structures and regeneration patterns of present-day floodplain forests give us some idea of which magnitude–frequency floods are important to their continued regeneration and survival, it is also important to try to understand the longer-term patterns of channel change that have led to the contemporary floodplain environments.

Although there is considerable scope for detailed palaeohydrological studies in the semi-arid tropics (Baker, 1991), in temperate zones it is suggested that the principal morphological changes to fluvial systems in the last 15 000 years have been changes in the length of drainage networks, change from single thread to multithread channels and changes in the size and shape of channel cross-sections (Gregory & Maizels, 1991). To understand their influence on floodplain vegetation development these morphological changes need to be discussed in the context of their relationships to changes in sediment availability and transport, and associated flood magnitudes and frequencies.

It is clear that the relationships between sediment deposition and palaeohydrology are extremely complex and cannot readily be extrapolated from our understanding of present-day magnitude–frequency relationships of floods and sediment transport. Patton & Baker (1980) demonstrate the difficulty in establishing palaeoflood frequency from available hydrological records and alluvial chronology in the Pecos River of semi-arid Texas. In particular, it is difficult to establish detailed changes in a fluvial system and to correlate these with climatic change. In some temperate rivers preliminary correlation between palaeodischarges and palaeoenvironmental factors are now available (e.g. Rotnicki, 1991; Bohncke & Vandenberghe, 1991; Starkel, 1991a), often based on deposits in stable slackwater areas that retain a sediment record and permit ^{14}C dating and palaeoecological analysis. It is possible that if, as is suggested in the previous section, particular magnitude–frequency floods are necessary in semi-arid areas for floodplain forest regeneration, evidence of particular species assemblages in palaeosediments in semi-arid floodplains might contribute to reconstruction of palaeodischarges. For example, in the British palaeoecological record, it is apparent that alder (*Alnus glutinosa*), a floodplain tree, established intermittently in space and time throughout the Holocene, not only when suitable habitats became available but also when particular 'weather events provided opportunities for establishment, regeneration and expansion' (Bennett & Birks, 1990, p. 127).

Starkel (1991b) suggests that generally the sequences of alluvial fills and abandoned palaeochannels dating from the Holocene indicate 'longer phases of stability (1200–1500 years) alternating with shorter phases of higher flood activity (300–700 years)' (Starkel, 1991b, p. 478). In semi-arid temperate areas gravelly, meandering rivers dominated during the Holocene (Starkel, 1991b), although there were clearly many local variations (Schumm, 1968). In the semi-arid western United States extreme events have also had important influences on the Holocene fluvial landscape. Catastrophic floods caused by the rapid outflow from great palaeolakes (e.g. Lake Bonneville around 14 000 yr BP) have been well documented (Baker, 1983) and have left large-scale deposits and channel forms.

In semi-arid Africa numerous studies of Holocene climatic change have been made, many using data from closed-basin lake-level studies or large-scale palaeohydrological evidence to demonstrate the direction and magnitude of change. There is, however, relatively little detailed palaeohydrological information from individual river valleys. From about 20 000 to 12 000 yr BP arid conditions prevailed in many parts of Africa (broken by short moist periods; Gasse *et al.*, 1990) and there is evidence that the lower Senegal River ceased to flow because it was obstructed by dunes (Grove & Warren, 1968; Grove, 1985). From around 12 000 yr BP moister conditions than the present prevailed in west and east Africa. Very large-scale fluvial features were associated with outflows from the palaeolake Mega-Chad, which was about ten times as large as Lake Bonneville and reached its greatest extent from about 10 000–6000 yr BP (Grove, 1985). There is evidence that many rivers excavated deep channels at the start of this period when discharges were high but sea levels still low (Grove, 1985). For example, there are buried channels at depths of about 40 m in the sediments of the lower Benue and Senegal Rivers. Massive alluviation subsequently occurred as sea levels rose to their present levels, and it is within the upper layers of these recent Holocene alluvial sediments that present-day rivers have created their floodplains.

Grove (1985) suggests that conditions were too dry for trees to grow in and around river channels during the early arid period, but the moister conditions from 12 000 yr BP produced datable woody material in river sediments (Thomas & Thorpe, 1980). Palynological evidence from west Africa (Lezine & Casanova, 1989) and from the eastern Sahara (Ritchie & Haynes, 1987) suggests that, generally, by 9500 yr BP forests had expanded. The extent of forest fluctuated, but by 2000 yr BP the main Holocene pluvial period had ended and the forest had largely given way to semi-arid vegetation.

A fluctuating climate and continuously changing base levels in many river basins require constant adjustment by all fluvial parameters (Nilsson et al., 1991). This is well demonstrated by the presence of river terraces: for example, in the Hoh River valley in Washington State, USA, four discrete vegetation communities have been documented by Fonda (1974) on surfaces varying in age from 80–100 > 750 years. The vegetation communities on them represent a successional progression linked to channel downcutting and associated changes in water tables and flooding influences as well as to autogenic changes in microenvironments. In this example only the lowest terrace is directly influenced by current fluvial processes so that there is a time frame of only about 100 years for understanding the impacts of these processes on current regeneration patterns.

As well as the use of paleohydrological techniques, in many rivers the use of historical documentation and dendrochronology can help establish their hydrological history. A number of studies have looked at historical changes in alluvial areas in some detail (e.g. Amoros et al., 1986; Decamps et al., 1989; Roux et al., 1989). In Europe, where deliberate manipulation of river systems can be traced to Roman times (Petts, 1989), the effects of climatic change and human impacts are difficult to separate. In North America, where documented human impacts only extend through the last two centuries, river systems can show us the natural patterns of fluvial and vegetation adjustment related to climatic change until much more recent times.

It is a matter of some conjecture when and where present-day floodplain forests in semi-arid parts of Africa had their origins. It is likely that some are isolated relicts of a more extensive forest which have adjusted their species composition through several disturbance thresholds to suit the prevailing hydrological conditions over the last 2000 years. Most show floristic and structural links between savanna woodland and tropical moist rainforest (White, 1983; Hamilton, 1974). In some areas they are an extension of the tropical moist forest, e.g. along valleys cut into the miombo woodlands of the Angolan Plateau (White, 1983), and in others are the only place where certain true savanna species that are not found in moist forests can grow. Early on, Keay (1959), working in Nigeria, distinguished between floodplain forests that were outliers from lowland rainforests and those in the drier Sudanian and Sahelian zones which had a quite different composition and structure. For example, species like *Diospyros mespiliformis* and *Tamarindus indica* have a pan-African distribution and can grow in moist areas away from floodplains, as well as on floodplains, while *Trichilia emetica* and *Ficus sycomorus* have a strictly riparian distribution.

In the evergreen floodplain forests of the Tana River species of both affinities have been found and it has been suggested that the two endemic monkey species found in the Tana forests are evidence that these floodplain forests were once part of a more extensive evergreen forest (Andrews et al., 1975; Homewood, 1976). It must be assumed that subsequently, during the last several thousand years of a predominantly semi-arid climate, the species composition of the lower Tana forests has changed in response to lowered discharges. However, these adjustments in species composition have not been so great that the forests cannot still support in their most diverse sections the two endemic monkey species which depend on a particular range of forest trees for a year-round food source and their long-term survival (Homewood, 1978; Marsh, 1986; Medley, 1992).

There is written and recorded evidence that between the sixteenth and eighteenth centuries wetter conditions than at present prevailed in east Africa, followed by a rainfall decline in the early nineteenth century (Nicholson, 1978). In the late nineteenth century, records from a number of lakes and rivers in East Africa indicate another period of higher rainfall. These climate changes must have affected discharges in the Tana basin. A comparison of the levels of Lake Naivasha and Lake Victoria with Tana River discharges (Figure 17.5) show some similarities in trend for the period since 1933. Since these two lakes share catchment areas with the Tana, it could be inferred that the Tana would also have had higher discharges towards the end of the nineteenth century. It has been suggested that many of the mature forest trees now present on the Tana floodplain might have become established during this period (Hughes, 1984). From 1933, records show that hydrological regimes and sediment inputs in the Tana floodplain have fluctuated, most likely in response to climatic change and increased cultivation in the upper catchment (Dunne & Ongweny, 1976). The records also include a very large flood event in 1961 which was considered to have an 80-year recurrence interval (Dunne & Leopold, 1978). These changes might partially explain the low regeneration levels of many forest species recorded by Hughes (1985).

This discussion of the Tana floodplain forests suggests that changes in Holocene climate and forest extension help to explain their species composition and also the presence of both an endemic poplar and two endemic primate species. However, historical and current changes in flooding regime associated with climatic change and complicated by human use are more likely to explain present-day forest dynamics and regeneration patterns.

It seems likely that in many river basins the present-day floodplain forests originated under the influence of fluvial regimes that have subsequently changed. It is therefore also likely that the regeneration conditions that permitted development of these mature forests will not reoccur within the lifespan of the current tree assemblages. For example, in the Bow River (Alberta) wetter conditions prior to and around the turn of the century coincided with an increased suspended sediment load created by logging activities in the foothills of the Rockies. It is suggested that these conditions may have promoted widespread regeneration of *Populus deltoides* in lower reaches which had previously been less forested and are again today showing few signs of significant regeneration (Cordes, 1991). Similarly, Baker (1990) describes a post-mid-1800s increase in *Populus angustifolia* in the Animas River floodplain. This was associated with a cooler and wetter post-Little Ice Age climate which generated more frequent years of flooding conditions suitable for regeneration than have been experienced since. There is also evidence for naturally occurring drought stress on floodplain vegetation during the 1930s in parts of the northern and central American prairies when reduced stream flows led to widespread poplar mortality (Weaver & Albertson, 1936; Ellison & Woolfolk, 1937; Albertson & Weaver, 1945).

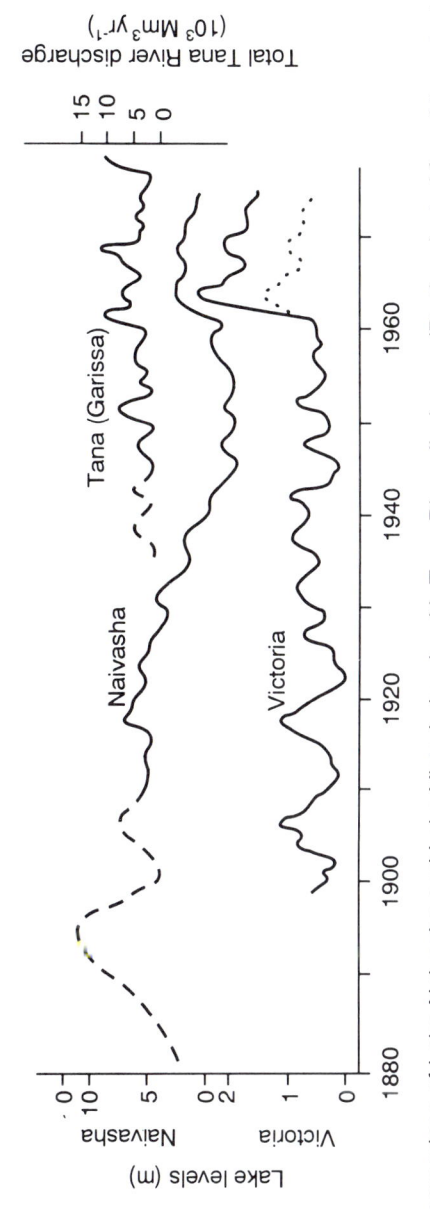

Figure 17.5 A comparison of Lake Naivasha and Lake Victoria levels with Tana River discharge. (Partly adapted from Vincent et al., 1979. Reproduced by permission of Kluwer Academic Publishers.) See text for a fuller discussion

For the study of contemporary floodplain forest ecology it is helpful to put some form of time frame on the development of present-day fluvial regimes. However, it has become clear that the concept of contemporary fluvial processes is difficult to define since there has been constant adjustment of fluvial regimes in response to changing climate, changing base levels and changing availability of post-glacial sediment sources. Floodplain forests have adjusted to this situation of constant change in disturbance regimes through their regeneration responses over time. Inevitably there is a lag period between regeneration and tree maturation at a site, so that present-day forest extent and diversity may not be an indication of present-day fluvial processes. The maintenance of some form of dynamic stability in the fluvial regime can occur, though this obviously varies enormously in its time frame from river to river. Some rivers such as the Beatton River (Nanson & Beach, 1977) have maintained predictable successions for up to 500 years, others for far less time. The implications of continuous changes in the fluvial regime for the assessment of the impact of river manipulation, and particularly of dam construction, on downstream floodplain ecosystems are considerable.

SOME IMPLICATIONS OF DAM IMPACTS ON SEMI-ARID FLOODPLAIN FORESTS

Dams have considerable and usually adverse impacts on downstream riparian ecosystems. In doing so they provide experimental manipulation on a massive scale which allows the exploration of natural change at an accelerated rate.

Downstream of dams there is usually a reduction in the frequency of overbank floods by reduced flood peaks. They also trap sediment, leading to clearer downstream water and increased erosion. The downstream changes to a river channel are complex and variable, depending on channel characteristics (e.g. whether meandering or braided, or they have gravel or sand beds). The amount of sediment reduction and changed discharge pattern are particularly crucial for downstream vegetation. In semi-arid rivers where vegetation regeneration depends on specific conditions provided by relatively infrequent floods, the downstream impacts are likely to reduce regeneration potential. Again there is great variability between rivers with different types and proportions of bed and suspended sediment load.

The effects of dams are classified by Richards (1982) as medium-term adjustments that create a temporary disequilibrium which then passes through 'transient states' while reaching a new equilibrium. The critical question in considering the floodplain vegetation is whether the new equilibrium reached generates a similar disturbance regime to the previous state, or whether a new 'disturbance threshold' will gradually be attained through adjustment in species composition. It is probably impossible to predict exact channel morphological changes due to dam building because so many variables are involved, but where river regimes in semi-arid areas have predictable flow patterns it seems clear that if flows that normally allow floodplain vegetation regeneration to occur are reduced then reduction in downstream regeneration is likely to occur.

Rood & Mahoney (1990, 1991) summarize the literature from the western prairies and conclude that dams contribute to the failure in regeneration of floodplain poplars by reducing downstream flows, altering flow patterns to attenuate spring flooding and stabilizing summer flows. Reduced river meandering caused by reduced spring floods and sediment loads have been cited as the prime causes of reduced floodplain regeneration in sandy floodplains. For example, Johnson *et al.* (1976) and Reily & Johnson (1982) note the reduction in river meandering below the Garrison Dam (Missouri River, North Dakota) and suggest that it has led to the loss of habitat creation and regeneration potential in the floodplain downstream. Many other authors cite similar declines in poplar species downstream of dams in semi-arid North America (Crouch, 1979; Bradley & Smith, 1986; Rood & Heinz-Milne, 1989; Rood & Mahoney, 1991; Fenner *et al.*, 1985).

Rood and Heinz-Milne (1989) study the influence of river damming on the riparian poplar forests on the parallel St Mary's, Waterton and Belly Rivers (southern Alberta) (Figure 17.4). The first two were dammed in 1951 and 1964 respectively whereas the Belly River is undammed. Comparisons of sites upstream and downstream of the dams and on the Belly River using aerial photographs from 1961 and 1981 showed a 48% reduction in poplar abundance downstream of the dam on the St Mary's River and a 23% decline downstream of the dam on the Waterton River over the 20-year study period. The Belly River showed less than 1% decline over the same period and forest declines were only 5 and 6% respectively upstream of the dams.

An intensive programme of dam building on the upper Tana River in Kenya since the 1960s will eventually have eleven dams in place, five of which have already been built. It is possible that control of river flows exercised by these dams will prevent regeneration events, even if higher discharges like those experienced at the end of the last century were to be repeated. There are already signs from aerial photography that meander sinuosity is reduced and that the river may no longer be as dynamic as it used to be (Hughes, 1990) (Figure 17.6, and also Figure 17.1). The sediments in the Tana floodplain range from sandy to clayey and there are perched water tables within the floodplain where clay bands are located. In such floodplains, and in floodplains where there may be a high percentage of clay lining the wetted perimeter of the channel, overbank floods are needed to recharge the floodplain water table since lateral movement of water from the channel into the floodplain is not straightforward. In the Tana, seeds of some forest species did germinate but there was a notable absence of established seedlings and saplings (Hughes, 1985). Rainfall or temperature variations may well have triggered germination but it is possible that, without the replenishment of floodplain water tables and their subsequent maintenance, successful establishment is unlikely to occur.

Where vegetation replenishment events are naturally infrequent there should be particular concern over the long-term impacts of dam construction. This is particularly the case where floodplains are dominated by only one tree species. In such rivers, the frequency of the floods which permit regeneration to occur may be variable as long as they are more frequent than the average life expectancy of the species involved to ensure continuity of a species at a site. Where forests are more varied, adjustments towards a new disturbance threshold by changing species composition can be expected, as demonstrated by Stevens & Waring (1985) in the Colorado River downstream of the Glen Canyon Dam.

Where poplars have declined downstream of dams in Alberta, recommendations have been made to simulate natural floods by planned water releases, and in particular to slow the decline in flow following spring peak floods in order to promote poplar regeneration. However, Bradley & Smith (1986) note on the Milk River that reduced geomorphological activity and sediment load are more detrimental to regeneration than reduced summer flows because sites for regeneration are not created. In reaches that follow the incremental replenishment model, therefore, simply manipulating flows may not have the desired effect of

338 Environmental Change in Drylands

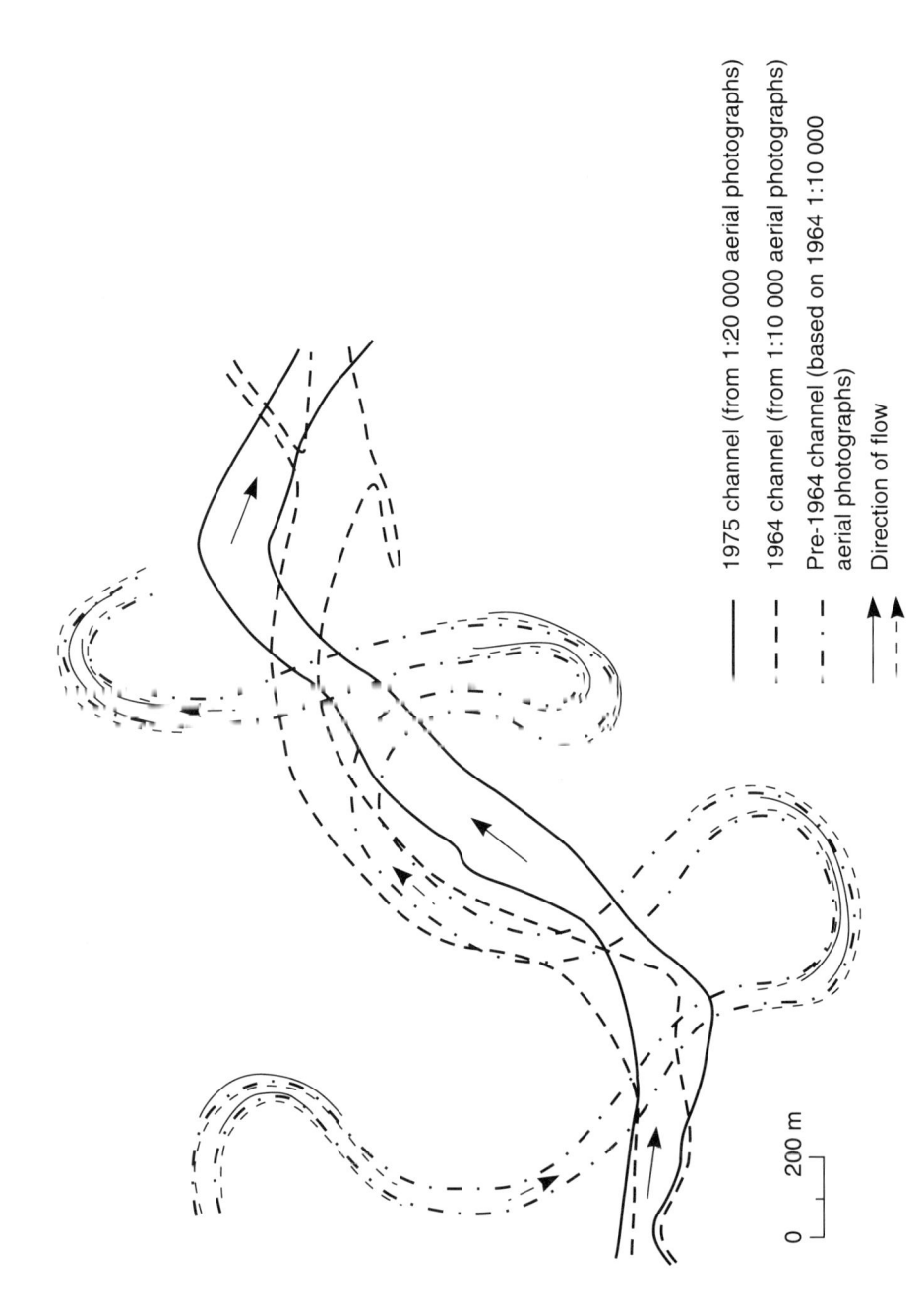

Figure 17.6 Meander movement near Pumwani, on the lower Tana River, Kenya

promoting poplar regeneration. The indication is that a holistic approach to water management is necessary, including as full an understanding as possible of the relationships between dynamic geomorphology, hydrology and vegetation. Furthermore, such studies need to be set within as long a time frame as possible since natural and man-made alterations to river regimes may already be altering them in a way detrimental to the regeneration potential of the floodplain. In these cases dam construction could aggravate to a critical extent an already declining situation for vegetation establishment.

Initially, a new reduced disturbance regime resulting from dam construction could lead to a temporary increase in mature stands as high disturbance events are reduced and with them the frequency of destruction of mature forests. However, long-term losses in species diversity through reduction in the creation of regeneration niches can be expected and would reflect the importance attached to regeneration niches by Grubb (1977).

CONCLUSIONS

Studies of the magnitude–frequency relationships of floods in semi-arid fluvial systems emphasize the importance of different-sized floods and their inter-arrival times in determining channel form. Although flood events with return periods from <1 to several years can be the most significant for sediment movement (Wolman & Miller, 1960) in semi-arid rivers about 40% of sediment transport is by <10 year floods (Richards, 1982). However, the recovery of channel form following very high magnitude–low-frequency floods depends not only on subsequent flow regimes but also on the ability of vegetation to regenerate (Wolman & Gerson, 1978; Schumm & Lichty, 1963). In this paper, emphasis has been placed on the importance of particular magnitude and frequency floods in creating the conditions necessary for regeneration of floodplain trees. This is a different emphasis from one which looks at the role of vegetation in influencing the recovery of channel form following high magnitude events. Thus, floods with a recurrence interval between 2 and 50 years play an important role in shaping floodplain communities in both temperate and tropical semi-arid areas. However, it is also clear that when viewed in the context of long-term environmental change, present-day floodplain vegetation communities are transient and constantly evolving in response to changing disturbance thresholds. The significance of particular magnitude–frequency floods for floodplain vegetation is therefore constantly changing as it adjusts to new disturbance thresholds, making measurement of that significance very difficult.

Concern today over the future of floodplains following dam construction has to be addressed through furthering the understanding of which magnitude–frequency floods optimize regeneration potential for particular species at the present time. This is important because in semi-arid areas people's livelihoods often depend on the current status of floodplain vegetation. However, in order to predict the impacts of changed flow regimes, short-term impacts have to be compared with long-term environmental change. In some cases, these comparisons indicate that the present regime would not have been maintained even in the absence of human impact. If any meaningful attempt is to be made to achieve long-term sustainability of ecosystems it is very important that more research is carried out on detailed regeneration requirements of floodplain species and on environmental change in floodplains in order to provide a reference frame within which to set today's ecosystem dynamics. It is already clear, however, that there are no straightforward answers to questions on the long-term future of floodplains in semi-arid regions.

ACKNOWLEDGEMENTS

This paper was conceived while the author was a visitor in the Department of Geography, University of Calgary, during 1991. The visit was partly funded by a study visit grant from the Royal Society and NSERC of Canada. The author would like to thank Larry Cordes, Sandy Marken, Derald Smith and Nigel Waters of that department, Cheryl Bradley of W.E.S.T., and Stewart Rood and John Mahoney of the Department of Biology, University of Lethbridge, for many useful discussions during that time. The author would also like to thank Bill Adams, Dick Chorley, Andrew Millington, Keith Richards and an anonymous reviewer for helpful comments on various drafts.

REFERENCES

Albertson, F. and Weaver, J. E. (1945). Injury and death or recovery of trees in prairie climate. *Ecological Monographs,* **15**, 395–433.

Alvarez-Lopez, M. (1989). Ecology of *Pterocarpus officianalis* forested wetlands in Puerto Rico. In Lugo, A., Brinson, M. and Brown, S. (eds.), *Forested Wetlands*, Elsevier, New York, pp. 251–65.

Amoros, C., Bravard, J. P., Castella, C., et al. (1986). Recherches interdisciplinaires sur les ecosystemes de la basse-plaine de l'Ain (France): potentialités évolutives et gestion. Documents de Cartographie Ecologique 29, Université de Grenoble, 166 pp.

Andrews, P., Groves, C. P. and Horne, J. F. M. (1975). Ecology of the Lower Tana River floodplain (Kenya). *Journal of East African Natural History Society and National Museum,* **151**, 1–31.

Aubreville, A. (1967). The unusual forest-savanna mosaics of the upper part of the loop of the Ogooue River. *Gabon. Adansonia,* **7**, 13–22.

Bacon, P. R. (1989). Ecology and management of swamp forests in the Guianas and Caribbean Region. In Lugo, A., Brinson, M. and Brown, S. (eds.), *Forested Wetlands*, Elsevier, New York, pp. 213–50.

Baker, V. R. (1983). Large-scale fluvial palaeohydrology. In Gregory, K. J. (ed.), *Background to Palaeohydrology*, Wiley, Chichester, pp. 453–78.

Baker, V. R. (1991). A bright future for old flows. In Starkel, L., Gregory, K. J. and Thornes, J. B. (eds.), *Temperate Palaeohydrology*, Wiley, Chichester, pp. 497–514.

Baker, W. L. (1990). Climatic and hydrologic effects on the regeneration of *Populus angustifolia* James along the Animas River, Colorado. *Journal of Biogeography,* **17**, 59–73.

Bégué, L. (1937). *Contribution a l'Étude de la Végétation Forestière de la Haute-Côte d'Ivoire*, Publications du Comite d'Etudes Historiques et Scientifiques de l'Afrique Occidentale Francaise, Librairie Larose, Paris.

Bennett, K. D. and Birks H. J. B. (1990). Post-glacial history of alder (*Alnus glutinosa* (L.) Gaertn.) in the British Isles. *Journal of Quaternary Studies,* **5**, 123–33.

Bohncke, S. J. P. and Vandenberghe, J. (1991). Palaeohydrological development in the southern Netherlands during the last 15,000 years. In Starkel, L., Gregory, K. J. and Thornes, J. B. (eds.), *Temperature Palaeohydrology*, Wiley, Chichester, pp. 253–81.

Bradley, C. E. and Smith, D. G. (1986). Plains cottonwood recruitment and survival on a prairie meandering river floodplain, Milk River, southern Alberta and nothern Montana. *Canadian Journal of Botany,* **64**, 1433–1442.

Bradley, C., Reintjes, F. and Mahoney, J. (eds.) (1991). *The Biology and Status of Riparian Poplars in Southern Alberta*, World Wildlife Fund Canada and Alberta Fish and Wildlife Division, Forestry, Lands and Wildlife, Government of Alberta, Canada, 85 pp.

Brinson, M. M. (1989). Riverine forests. In Lugo, A., Brinson, M. and Brown, S. (eds.), *Forested Wetlands*, Elsevier, New York, pp. 87–141.

Buell, M. F. and Wistendahl, W. A. (1955). Floodplain forests of the Raritan River. *Bulletin of the Torrey Botanical Club,* **82**, 463–72.

Burgess, R. L., Johnson, W. C. and Keammerer, W. R. (1973). Vegetation of the Missouri River floodplain in North Dakota. Report to the Office of the Water Resources Research, U.S. Department of the Interior, Washington D.C., 161 pp.

Conner, W. H., Gosselink, J. G. and Parrondo, R. T. (1981). Comparisons of the vegetation of three Louisiana swamp sites with different flooding regimes. *American Journal of Botany*, **68**, 320–31.

Cordes, L. D. (1991). The distribution and age structure of cottonwood stands along the lower Bow River. In Rood, S. B. and Mahoney, J. M. (eds.), *The Biology and Management of Southern Alberta's Cottonwoods*, University of Lethbridge, Alberta, pp. 13–23.

Crouch, G. (1979). Changes in the vegetation complex of a cottonwood ecosystem on the South Platte River. *Great Plains Agricultural Council Publication*, **91**, 19–22.

Decamps, H., Fortune, M. and Gazelle, F. (1989). Historical changes of the Garonne River, southern France. In Petts, G. E., Moller, H. and Roux, A. L. (eds.), *Historical Changes of Large Alluvial Rivers: Western Europe*, Wiley, Chichester, pp. 249–69.

Dunn, C. P. and Stearns, F. (1987). Relationship of vegetation layers to soils in southeastern Wisconsin forested wetlands. *American Midland Naturalist*, **118**, 366–74.

Dunne, T. and Leopold, L. B. (1978). *Water in Environmental Planning*, Freeman, San Francisco.

Dunne, T. and Ongweny, G. S. O. (1976). A new estimate of sedimentation rates on the upper Tana River. *The Kenyan Geographer*, **2**, 109–26.

Ellison, L. and Woolfolk, E. J. (1937). Effects of drought on vegetation near Miles City, Montana. *Ecology*, **18**, 329–36.

Everitt, B. (1968). Use of cottonwoods in an investigation of the recent history of a floodplain. *American Journal of Science*, **266**, 417–39.

Farmer, R. E. and Bonner, F. T. (1967). Germination and initial growth of eastern cottonwood as influenced by moisture stress, temperature and storage. *Botanical Gazette*, **128**, 211–5.

Fenner, P., Brady, W. and Patton, D. (1984). Observations on seeds and seedlings of Fremont cottonwood. *Journal of Desert Plants*, **6**, 55–8.

Fenner, P., Brady, W. and Patton, D. (1985). Effects of regulated water flows on regeneration of Fremont cottonwoods. *Journal of Range Management*, **38**, 135–8.

Fonda, R. W. (1974). Forest succession in relation to river terrace development in Olympic National Park, Washington. *Ecology*, **55**, 927–42.

Furness, H. D. and Breen, C. M. (1980). The vegetation of seasonally flooded areas of the Pongolo River floodplain. *Bothalia*, **13**, 217–30.

Gasse, F., Tehet, R., Durand, A., Gilbert, E. and Fontes, J.-C. (1990). The arid-humid transition in the Sahara and the Sahel during the last glaciation. *Nature*, **346**, 141–6.

Gill, C. J. (1970). The flooding tolerance of woody species — a review. *Forestry Abstracts*, **31**, 671–88.

Gill, D. (1971). Vegetation and environment in the Mackenzie River Delta, Northwest Territories, a study in subarctic ecology. Unpublished PhD thesis, University of British Columbia.

Gregory, K. J. and Maizels, J. K. (1991). Morphology and sediments: typological characteristics of fluvial forms and deposits. In Starkel, L., Gregory, K. J. and Thornes, J. B. (eds.), *Temperate Palaeohydrology*, Wiley, Chichester, pp. 31–59.

Groenveld, D. P. and Griepentrog, T. E. (1985). Interdependence of groundwater, riparian vegetation and streambank stability: a case study. In *Riparian Ecosystems and Their Management — Reconciling Conflicting Uses*, USDA Forest Service, General Technical Report RM-120, pp. 44–8.

Grove, A. T. (1985). The physical evolution of the river basins. In Grove, A. T. (ed.), *The Niger and Its Neighbours: Environmental History and Hydrobiology, Human Use and Health Hazards of the Major West African Rivers*, Balkema, Rotterdam and Boston, pp. 21–60.

Grove, A. T. and Warren, A. (1968). Quaternary landforms and climate on the south side of the Sahara. *Geographical Journal*, **134**, 194–208.

Grubb, P. J. (1977). The maintenance of species-richness in plant communities: the importance of the regeneration niche. *Biological Review of the Cambridge Philosophical Society*, **52**, 107–45.

Hamilton, A. C. (1974). The history of vegetation. In Lind, E. M. and Morrison, M. E. S. (eds.), *East African Vegetation*, Longman, London, pp. 188–209.

Homewood, K. M. (1976). Ecology and behaviour of the Tana Mangabey (*Cercocebus galeritus galeritus*). Unpublished PhD thesis, University of London.

Homewood, K. M. (1978). Feeding strategy of the Tana Mangabey (*Cercocebus galeritus galeritus*) (Mammalia: primates). *Journal of Zoology*, **186**, 375–91.

Hook, D. D. and Brown, C. L. (1973). Root adaptations and relative flood tolerance of five hardwood species. *Forestry Science*, **19**, 225–9.

Hosner, J. F. (1958). The effects of complete inundation upon seedlings of six bottomland trees species. *Ecology*, **39**, 371–3.

Huffman, R. T. (1980). The relation of flood timing and duration to variation in bottomland hardwood community structure in the Ouachita Basin of southeastern Arkansas. US Army Engineering Waterways Experiment Station, Miscellaneous Paper E-80-4, Vicksburg.

Huffman, R. T. and Forsythe, S. W. (1981). Bottomland forest communities and their relation to anaerobic soil conditions. In Clark, J. R. and Benforado, J. (eds.) *Wetlands of Bottomland Hardwood Forests*, Elsevier, New York, p. 187–95.

Hughes, F. M. R. (1984). A comment of the impact of development schemes on the floodplain forests of the Tana River of Kenya. *Geographical Journal*, **150**, 230–45.

Hughes, F. M. R. (1985). The Tana River floodplain, forest, Kenya: ecology and the impact of development. Unpublished PhD thesis, University of Cambridge.

Hughes, F. M. R. (1988). The ecology of African floodplain forests in semi-arid zones: a review. *Journal of Biogeography*, **15**, 127–40.

Hughes, F. M. R. (1990). The influence of flooding regimes on forest distribution and composition in the Tana River floodplain, Kenya. *Journal of Applied Ecology*, **27**, 475–91.

Hupp, C. R. and Osterkamp, W. R. (1985). Bottomland vegetation distribution along Passage Creek, Virginia, in relation to fluvial landforms. *Ecology*, **56**, 670–81.

Johnson, R. R. and Lowe, C. H. (1985). On the development of riparian ecology. In *Riparian Ecosystems and Their Management: Reconciling Conflicting Uses*, USDA Forest Service, General Technical Report RM-120, pp. 112–5.

Johnson, W. C., Burgess, R. L. and Keammerer, W. R. (1976). Forest overstorey vegetation and environment on the Missouri River floodplain in North Dakota. *Ecological Monographs*, **26**, 59–84.

Junk, W. J. (1989). Flood tolerance and tree distribution on central Amazonian floodplains. In Holm-Nielsen, L. D., Nielson, I. C. and Balslev, H. (eds.), *Tropical Forests: Botanical Dynamics, Speciation and Diversity*, Academic Press, New York, pp. 47–64.

Kangas, P. C. (1989). Long-term development of forested wetlands. In Lugo, A. E., Brinson, M. M. and Brown, S. (eds.), *Forested Wetlands*, Elsevier, New York, pp. 25–51.

Keay, R. W. J. (1949). An example of Sudan zone vegetation in Nigeria. *J. Ecology*, **37**(2), 335–364.

Keay, R. W. J. (1959). *An Outline of Nigerian vegetation*, Federal Government Printer, Lagos.

Kellerhals, R. and Gill, D. (1973). *Observed and Potential Downstream Effects of Large Storage Projects in Northern Canada*, Commission Internationale des Grands Barrages, Madrid, Q.40-R-46, pp. 731–54.

Kemp, R. H. (1961). Growth and regeneration of Khaya grandiflora in 'Kurame' in southern Zaria Province. Technical Note 10, Department of Forestry Research, Nigeria.

Kozlowski, T. T. (1984). Response of woody plants to flooding. In Kozlowski, T. T. (ed.), *Flooding and Plant Growth*, Academic Press, New York, pp. 129–63.

Larson, J. S., Bedinger, M. S., Bryan, C. F., *et al.* (1981). Transition from wetlands to uplands in southeastern bottomland hardwood forests. In Clark, J. R. and Benforado, J. (eds.), *Wetlands of Bottomland Hardwood Forests*, Elsevier, New York, pp. 225–73.

Lees, J. C. (1964). Tolerance of white spruce seedlings to flooding. *Forestry Chronicle*, **40**, 221–5.

Lezine, A.-M. and Casanova, J. (1989). Pollen and hydrological evidence for the interpretation of past climates of tropical West Africa during the Holocene. *Quaternary Science Reviews*, **8**, 45–55.

Lugo, A. E., Brown, S. and Brinson, M. M. (1989). Concepts in wetland ecology. In Lugo, A. E., Brinson, M. M. and Brown, S. (eds.), *Forested Wetlands*, Elsevier, New York, pp. 53–85.

McKnight, J. S., Hook, D. D., Langdon, O. G. and Johnson, R. L. (1981). Flood tolerance and related characteristics of trees of the bottomland forests of the southern United States. In Clark, J. R. and Benforado, J. (eds.), *Wetlands of Bottomland Hardwood Forests*, Elsevier, New York, pp. 29–61.

Mahoney, J. M. and Rood, S. B. (1991). A device for studying the influence of declining water table on popular growth and survival. *Tree Physiology*, **8**, 305–14.

Marken, S. (1991). Plains cottonwoods and riparian vegetation on the lower Red Deer River. In Rood,

S. B. and Mahoney, J. M. (eds.), *The Biology and Management of Southern Alberta's Cottonwoods*, University of Lethbridge, Alberta, pp. 11–13.

Marsh, C. M. (1986). A resurvey of the Tana River primates and their forest habitat. *Primate Conservation*, **7**, 72–81.

Medley, K. E. (1992). Patterns of forest diversity along the Tana River in Kenya. *J. Tropical Ecology*, **8**(4), 353–373.

Moss, E. H. (1938). Longevity of seed and establishment of seedlings in species of *Populus*. *Botanical Gazette*, **99**, 529–47.

Nanson, G. C. and Beach, H. F. (1977). Forest succession and sedimentation on a meandering river flooplain, northeast British Columbia, Canada. *Journal of Biogeography*, **4**, 229–51.

Nicholson, S. E. (1978). Climatic variations in the Sahel and other African regions during the past five centuries. *Journal of Arid Environments*, **1**, 3–24.

Nilsson, C., Grelsson, G., Dynesius, M., Johansson, M. E. and Sperens, U. (1991). Small rivers behave like large rivers: effects of postglacial history on plant species richness along riverbanks. *Journal of Biogeography*, **18**, 533–41.

Noble, M. G. (1979). The origin of *Populus deltoides* and *Salix interior* zones on point bars along the Minnesota River. *American Midland Naturalist*, **102**, 69–102.

Oba, G. (1991). The ecology of the floodplain woodlands of the Turkwel River, Turkana, Kenya. UNESCO–TREMU Technical Report No. D-2, pp. 31–105.

Patton, P. C. and Baker, V. R. (1980). Geomorphic responses of central Texas stream channels to catastrophic rainfall and runoff. In Doehring, D. O. (ed.), *Geomorphology in Arid Regions*, Allen and Unwin, London, p. 189–217.

Pautou, G. (1984). L'organisation des forets alluviales dans l'axe Rhodanien entre Geneve et Lyon comparison avec d'autres systemes fluviaux. *Documents de Cartographic Ecologique, Grenoble*, **27**, 43–64.

Petts, G. E. (1987). Timescales for ecological change in regulated rivers. In Craig, J. and Kemper, J. (eds.), *Regulated Streams: Advances in Ecology*, Plenum Press, New York, pp. 257–66.

Pickett, S. T. A. and White, P. S. (1985). *The Ecology of Natural Disturbance and Patch Dynamics*, Academic Press, Orlando.

Read, M. A. (1958). Silvical characteristics of plains cottonwood. USDA Forest Service, Rocky Mountain Forest and Range Experimental Station, Paper 33.

Reid, D. E. (1974). *Vegetation of the Mackenzie Valley — Part 1*, Volume 3, North Engineering Services for Canadian Arctic Gas Study Limited Biological Report Series.

Reid, D. E. (1991). Cottonwoods of the Milk River. In Rood, S. B. and Mahoney, J. M. (eds.), *The Biology and Management of Southern Alberta's Cottonwoods*, University of Lethbridge, Alberta, pp. 35–42.

Reily, P. and Johnson, W. (1982). The effects of altered hydrologic regime on tree growth along the Missouri River in North Dakota. *Canadian Journal of Botany*, **60**, 2410–23.

Richards, K. (1982). *Rivers: Form and Process in Alluvial Channels*, Methuen, London and New York.

Ritchie, J. C. and Haynes, C. V. (1987). Holocene vegetation zonation in the eastern Sahara. *Nature*, **330**, 645–7.

Robertson, P. A., Weaver, G. T. and Cavanaugh, J. S. (1978). Vegetation and tree species patterns near the northern terminus of the southern floodplain forest. *Ecological Monographs*, **48**, 249–67.

Rood, S. B. and Heinz-Milne, S. (1989). Abrupt riparian forest decline following river damming in southern Alberta. *Canadian J. Botany*, **67**, 1744–1749.

Rood, S. B. and Mahoney, J. M. (1990). The collapse of river valley forests downstream from dams in the western prairies: probable causes and prospects for mitigation. *Environmental Management*, **14**, 451–64.

Rood, S. B. and Mahoney, J. M. (1991). The importance and extent of cottonwood forests decline downstream from dams. In Rood, S. B. and Mahoney, J. M. (eds.), *The Biology and Management of Southern Alberta's Cottonwoods*, University of Lethbridge, Alberta, pp. 1–9.

Rotnicki, K. (1991). Retention of meandering and sinuous alluvial rivers and its palaeohydrological implications. In Starkel, L., Gregory, K. J. and Thornes, J. B. (eds.), *Temperate Palaeohydrology*, Wiley, Chichester, pp. 431–71.

Roux, A. L., Bravard, J.-P., Amoros, C. and Pautou, G. (1989). Ecological changes of the French upper Rhone River since 1750. In Petts, G. E., Moller, H. and Roux, A. L. (eds.), *Historical Change of Large Alluvial Rivers*, Wiley, Chichester, pp. 323–50.

Schumm, S. A. (1968). River adjustment to altered hydrological regimen–Murrumbidgee River and palaeochannels, Australia. USGS Professional Paper 598.

Schumm, S. A. and Lichty, R. W. (1963). Channel widening and floodplain construction along Cimarron River in south-western Kansas. USGS Professional Paper 352, pp. 71–88.

Scoones, I. (1991). Wetlands in drylands: key resources for agricultural and pastoral production in Africa. *Ambio*, **20**, 366–71.

Sharitz, R. R. and Lee, L. C. (1985). Limits on regeneration processes in southeastern riverine wetlands. In *Riparian Ecosystems and Their Management: Reconciling Conflicting Uses*, US Department of Agriculture, Forest Service, General Technical Report RM-20, pp. 139–43.

Sigafoos, R. S. (1964). Botanical evidence of floods and floodplain deposition. USGS Professional Paper 485A.

Simpson, C. D. (1975). A detailed vegetation study on the Chobe River in N. E. Botswana. *Kirkia*, **10**, 85–227.

Sousa, W. P. (1984). The role of disturbance in natural communities. *Annual Review of Ecological Systematics*, **15**, 353–91.

Specht, R. L. (1989). Forested wetlands in Australia. In Lugo, A. E., Brinson, M. M. and Brown, S. (eds.), *Forested Wetlands*, Elsevier, New York, pp. 387–406.

Starkel, L. (1991a). The Vistula Valley: a case study for central Europe. In Starkel, L., Gregory, K. J. and Thornes, J. B. (eds.), *Temperate Palaeohydrology*, Wiley, Chichester, pp. 171–88.

Starkel, L. (1991b). Long-distance correlation of fluvial events in the temperate zone. In Starkel, L., Gregory, K. J. and Thornes, J. B. (eds.), *Temperate Palaeohydrology*, Wiley, Chichester, pp. 473–95.

Starkel, L., Gregory, K. J. and Thornes, J. B. (eds.) (1991). *Temperate Palaeohydrology: Fluvial Processes in the Temperature Zone during the last 15,000 years*, Wiley, Chichester.

Stevens, L. E. and Waring, G. L. (1985). The effects of prolonged flooding on the riparian plant community in Grand Canyon. In *Riparian Ecosystems and Their Management: Reconciling Conflicting Uses*, USDA Forest Service General Technical Report RM 120, pp. 81–6.

Stobbs, K., Corbiere, A., Mahoney, J. M. and Rood, S. B. (1991). The influence of rate of water table decline on establishment and survival of hybrid poplar seedlings. In Rood, S. B. and Mahoney, J. M. (eds.), *The Biology and Management of Southern Alberta's Cottonwoods*, University of Lethbridge, Alberta, pp. 47–55.

Streng, D. R., Glitzenstein, J. S. and Harcombe, P. A. (1989). Woody seedling dynamics in an East Texas floodplain forest. *Ecological Monographs*, **59**, 177–204.

Teversham, J. M. and Slaymaker, O. (1976). Vegetation composition in relation to flood frequency in Lillooet Valley, British Columbia. *Catena*, **3**, 191–201.

Thomas, M. F. and Thorpe, M. B. (1980). Some aspects of the geomorphological interpretation of Quaternary alluvial sediments in Sierra Leone. *Zeitschrift Geomorphologie für Supplementband*, **36**, 140–61.

Titus, J. H. (1990). Microtopography and woody plant regeneration in a hardwood floodplain swamp in Florida. *Bulletin of the Torrey Botanical Club*, **117**, 429–37.

Trochain, J. (1940). Contribution a l'étude de la végétation du Sénégal. Unpublished PhD thesis, Faculté des Sciences de l'Université de Paris.

Van Beek, J. L., Haymon, A. L., Wax, C. L. and Wicker, K. M. (1979). Operations of the Old River Control Project, Atchalfaya Basin: an evaluation from a multiuse management standpoint. US Environmental Protection Agency, EPA-600/4-79-073.

Viereck, L. A. (1970). Forest succession and soil development adjacent to the Chena River in interior Alaska. *Arctic and Alpine Research*, **2**, 1–26.

Vincent, C. E., Davies, T. D. and Beresford, A. K. C. (1979). Recent changes in the level of Lake Naivasha, Kenya as an indicator of equatorial westerlies over East Africa. *Climatic Change*, **2**, 175–89.

Virginillo, M., Mahoney, J. M. and Rood, S. B. (1991). Establishment and survival of poplar seedlings along the Oldman River, Southern Alberta. In Rood, S. B. and Mahoney, J. M. (eds.), *The Biology and Management of Southern Alberta's Cottonwoods*, University of Lethbridge, Alberta, pp. 55–63.

Ware, G. H. and Penfound, W. T. (1949). The vegetation of the lower levels of the floodplain of the South Canadian River in central Oklahoma. *Ecology*, **30**, 478–84.
Watt, A. S. (1947). Pattern and process in the plant community. *Journal of Ecology*, **35**, 1–22.
Weaver, J. E. (1960). Floodplain vegetation of the central Missouri Valley and contacts of woodland with prairie. *Ecological Monographs*, **30**, 37–64.
Weaver, J. E. and Albertson, F. W. (1936). Effects of the Great Drought on the prairies of Iowa, Nebraska and Kansas. *Ecology*, **17**, 567–639.
White, F. (1983). *The Vegetation of Africa*, UNESCO, Paris.
Wissmar, R. C. and Swanson, F. J. (1990). Landscape disturbances and lotic ecotones. In Naiman, R. J. and Decamps, H. (eds.), *The Ecology and Management of Aquatic Terrestrial Ecotones*, UNESCO/MAB Series 4, Parthenon, pp. 65–89.
Wolman, M. G. and Miller, J. P. (1960). Magnitude and frequency of forces in geomorphic processes. *Journal of Geology*, **68**, 64–74.
Wolman, M. G. and Gerson, R. (1978). Relative scales of time and effectiveness of climate in watershed geomorphology. *Earth Surface Processes*, **3**, 189–208.

18 Monitoring the Flooding Ratio of Tunisian Playas Using Advanced Very High Resolution Radiometer (AVHRR) Imagery

N. A. DRAKE
Department of Geography, King's College London, UK

and

R. G. BRYANT*
Department of Geography, Postgraduate Research Institute of Sedimentology, University of Reading, UK

ABSTRACT

Monthly advanced very high resolution radiometer (AVHRR) imagery was used to estimate the flooding ratio of nine Tunisian playas from 1985 to 1987. The flooding ratio is the average period a playa contains water over the time of observation. In Tunisia, this ratio generally decreases with increasing aridity along a north–south climate gradient. Contemporary flooding ratios were compared to limited historical records of lake levels for Chott Djerid and Sebkah Kelbia. Both comparisons suggest that the flooding ratio has decreased during this century. There is evidence that this has been caused by damming of rivers and reduction in the water table due to increased groundwater abstraction, though increasing aridity may also be significant. Flooding ratios can also provide general indications about the rate of playa deflation, relative differences in sedimentation rates and biological diversity. The relationship between the flooding ratio and these variables is discussed and evaluated.

INTRODUCTION

Temporal changes in the area and volume of perennial lakes in arid closed basins can be evaluated to provide palaeoclimatic records (Street-Perrott and Harrison, 1985). However, in many deserts perennial lakes are rare and ephemeral lakes, which exhibit rapid changes in both area and volume, dominate the lacustrine landscape. A different approach has to be taken when analysing the areal and volumetric changes in ephemeral lakes. One method by which such changes can be determined is to estimate the flooding ratio, the average period of time an ephemeral lake contains water during an extended period of observation.

The field-based data required to calculate the flooding ratio are rarely available but it can be estimated from weather satellite data (e.g. AVHRR or METEOSAT images) that have a highly temporal resolution. This is possible because the visible and near-infrared

---

*Present address: Department of Environment, University of Stirling, Stirling, FK9 4LA, UK.

Environmental Change in Drylands: Biogeographical and Geomorphological Perspectives.
Edited by A. C. Millington and K. Pye. © 1994 John Wiley & Sons Ltd

radiation measured by sensors on these satellites indicates the presence of water as water absorbs the majority of light, while the sediments and salts on a dry lake bed are highly reflective. In addition, these satellites also provide synoptic image coverage that allows simultaneous monitoring of lakes at the regional scale.

This paper evaluates the use of AVHRR for estimating the flooding ratio of nine Tunisian playas between 1985 and 1987. Different image processing methods were evaluated to determine the most effective method for monitoring the flooding ratio. The estimated flooding ratios for two ephemeral lakes were then compared to limited historical data to indicate possible changes in the flooding ratio this century.

Geomorphological and biological inferences can also be made about an ephemeral lake once the flooding ratio has been calculated. Motts (1965) classified the different playas according to their flooding ratios (Table 18.1) and stated that these different types of playas have different geomorphological, sedimentological and biological regimes. Therefore once the flooding ratio of a playa is known these factors can be inferred. For example, net aggradation can be inferred from a high flooding ratio as deflation can only occur when the surface is dry (and the flooding ratio is low) (Mabbutt, 1977). Motts (1975) suggested that sedimentation rates increase with the flooding ratio. However, there is evidence that this assumption may not be the case for two reasons. Firstly, the seminal work of Langbein and Schumm (1958) showed that sediment yields along a similar climatic gradient to that studied in Tunisia do in fact vary, being low in arid areas because of the lack of rainfall, and they rise with increasing rainfall until they peak in the semi-arid zone. These observations have subsequently been confirmed by Walling and Kleo (1983). Secondly, the flooding ratio is unlikely to exert much influence on aeolian deposition on playas.

The flooding ratio may also be indictive of biological production and diversity. DeDekker (1988) stated that the longer a playa is submerged the greater its biological diversity will be. As biological production on playas is restricted to wet periods the flooding ratio may also be thought of as an indicator of biological production.

TUNISIAN LAKES AND PLAYAS

The lakes and playas in Tunisia occur along a climatic gradient that extends from the Mediterranean in the north to the Sahara Desert in the south (Figure 18.1). Consequently, they exhibit large variations in hydrology, sedimentology, salinity and biology. Nine lakes and playas were studied (Figure 18.1) and these are introduced below.

Lake Bizerte, the most northerly lake in the country, has been connected to the sea by

Table 18.1 Classification of playa lakes according to their flooding ratios (after Motts, 1975)

Lake type	Flooding radio (%)
Lake	> 0.66
Playa lake	> 0.33 < 0.66
Playa	< 0.33

an artificial channel since the mid 1800s. Since then its water level and chemistry have been strongly controlled by the sea as its catchment area, and therefore the overall freshwater inflow, is small.

Lake Ichkeul is fresh in winter when rivers are in flood, but is salty in summer because of evaporation and inflow of saline water from Lake Bizerte. The lake is 90 km^2 in area and is 1.5 m deep in summer but rises to 2.5 m during some winters (Hollis, 1986). The sedimentation rate at the lake's deepest point was estimated from ^{210}Pb-dated sediments (Hollis, 1986). The historic sedimentation rate accelerated from 0.28 mm yr^{-1} between 1819 and 1890 to 0.97 mm yr^{-1} between 1890 and 1923, a 3.5 times increase. From 1923 to 1966 it increased a further 1.5 times, and since 1966 there has been a further fourfold increase, the sedimentation rate during this latter period being 6 mm yr^{-1}. This acceleration in sedimentation has been attributed to human activity (Hollis, 1986).

Sebkah el Hani and Sebkah Kelbia lie in shallow depressions on the eastern Plain of Kairouan. Sebkah Kelbia has a lake area of 120 km^2 and a catchment area of 15 000 km^2, much of which is mountainous. Low salinity of the ephemeral lake water makes it an important area for overwintering waterfowl. When the lake dries out the local water table falls, suggesting that the playa has a role in local water table recharge. It is fed by three rivers which rise in its mountainous upper catchment. Discharges from these rivers into the sebkah are diminished by infiltration and evaporation on the Plain of Kairouan. Although surface water from Oued Merguellil and Oued Zeroud flow into Sebkah Kelbia, groundwater seeping from the base of these two rivers may flow through the sediments under the Plain of Kairouan and into either Sebkah el Hani or Sebkah Kelbia, although flow into the latter is probably dominant (Hollis and Kallel, 1986). Consequently the groundwater systems of Sebkah Kelbia and Sebkah el Hani are interconnected; they are open hydrological systems. Sebkah Kelbia also has an open surface water hydrology. During large floods, the lake overflows a sill and water flows along Oued Sed to the sea. Such periodic floods flush the salts from the basin, giving rise to the sebkah's low salinity and the lack of salt crust development. By way of contrast, Sebkah el Hani has a closed surface water hydrology and exhibits a well-developed salt crust when dry.

Seven topographic surveys since 1910 have shown that Sebkah Kelbia has a very high sedimentation rate, 82 million m^3 of sediment being deposited between 1933 and 1969, and 67 million m^3 between 1969 and 1979. This gives a sedimentation rate of 5.6 mm yr^{-1} over the entire sebkah between 1969 and 1979, and the catchment sediment yield (259 t km^{-2} yr^{-1}) is comparable to that of the Yangtze (Hollis and Kallel, 1986). Calculations based on data from 1930 to 1955 show that the flooding ratio for Sebkah Kelbia was 68.6% during that time. Hollis and Kallel (1986) indicated that the sebkah has been wet for 89.4% of the last 76 years, and they predicted that in the future the lake will dry out 2.5 times more frequently than in the past because of dam construction on the main rivers that feed the sebkah.

Sebkah Moknine is found at a similar latitude to Sebkah el Hani though the climate is less arid due to the maritime influence (Figure 18.1). To the author's knowledge there have been no detailed studies of the hydrology, sedimentology or geomorphology of Sebkah Moknine. Field reconnaissance and topographic maps show that, like Sebkah el Hani, it has a small subdued catchment and, though it is close to the sea, there is no surface water connection.

Sebkah Noual and Chott el Guettar straddle the 200 mm isohyet and are located on the semi-arid/arid climatic boundary. Their catchments are more mountainous than those further

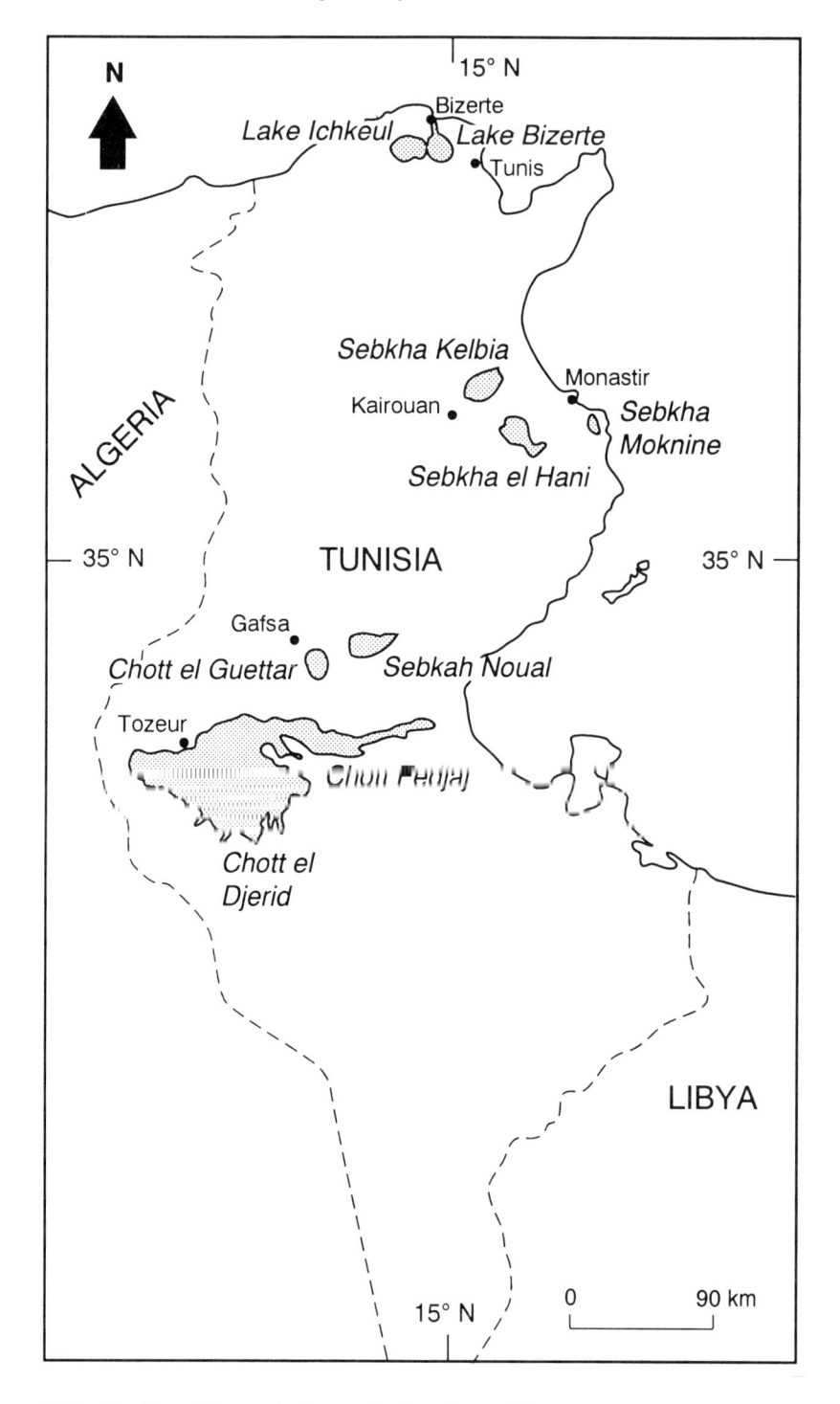

Figure 18.1 Tunisian lakes and playas; the locations of the synoptic climate stations are indicated

north. Millington *et al.* (1987) showed that Chott el Guettar receives surface waters from major channels to the east and west and from smaller channels that drain directly from the surrounding mountains during floods. The western channel forms a large delta complex that is perennially moist due to groundwater seepage. There are no studies of Sebkah Noual, but topographic maps show that it is fed by a number of substantial ephemeral channels and it has a salt crust when dry.

The Djerid–Fedjaj basin has an aerial extent of 10 500 km². Chott el Djerid itself has a surface area of approximately 5360 km², and its elongate northern arm — the Chott el Fedjaj — has an area of 770 km² (Millington *et al.*, 1987). Surface water influxes to the two basins are thought to be subordinate to the groundwater influxes (Coque, 1962; Drake, 1992), though surface water can have significant effects. The Djerid–Fedjaj basin is situated at one of the lowest and most northern extremities of the Bas Sahara artesian basin which covers most of southern Algeria and Tunisia and parts of Morocco and Libya. The basin contains two main aquifers, the Continental Intercalaire and the Complex Terminal. The two chotts derive considerable amounts of water from these aquifers via springs on the basin floor (Mamou, 1976).

The current state of these aquifers has been investigated by Mamou (1976). Groundwater movement in the Continental Intercalaire is from east to west and in the Complex Terminal is from south-east to north-west. There is evidence that the large number of boreholes sunk during this century have had an affect on the local water table. In 1900 the estimated yield from the natural springs in the Nefzaoua area was 660 l s^{-1}. The first artificial borehole, yielding 604 l s^{-1}, was installed in 1907. Further borehole abstraction reduced the yield of the springs to 560 l s^{-1} by 1950 and 250 l s^{-1} by 1976, when borehole abstractions reached 2570 l s^{-1} and some springs had dried up.

Because the Chott Djerid and Chott Fedjaj are playas where the groundwater provinces are considerably larger than their topographic basins, any fluctuations in the water table of the basin will be controlled by the temporary balance between artesian recharge and evaporation (Coque, 1962). The presence or absence of surface water on Chott Djerid was recorded between 1947 and 1958 (Table 18.2). During most years, at least part of the playa was

Table 18.2 Presence or absence of water on the Chott el Djerid (1947–58) (from Coque, 1962. Reproduced by permission of Armand Colin)

	Jan	Feb	Mar	Apr	May	Jun	Jul	Aug	Sep	Oct	Nov	Dec
1947		−								+		
1948		−	−	−	−	−	−	−	−	−	−	−
1949	−	−			−	−	−	−	−	−	−	−
1950				−		−						+
1951	+	+		−	−			−			+	+
1952	+		−		−	−	−	−				
1953			−	−		−						
1954					−	−		−		−	+	+
1955	+		−	−	−			−		−		
1956	+	+	+	−				−		−		
1957		+	+		−	−	−	−	−		+	+
1958	+	+	+		−	−	−	−	−		+	+

Water present = +, water absent = −, blank means no data collected.

covered by shallow water between November and February. From these observations Coque (1962) concluded that in the winter months when there is less evaporation than aquifer recharge the water table rises to cover the surface and that during the summer this balance is reversed.

PREVIOUS STUDIES OF EPHEMERAL LAKES USING SATELLITE IMAGERY

A number of people have utilised weather satellite imagery to monitor ephemeral lakes. Rosema and Fiselier (1990) used METEOSAT data to study the Niger Delta and the Lake Chad and, by interpreting thermal inertia images to determine the area of water, they were able to analyse wet season flooding for different years. These changes were related to climatic changes and human-induced flooding. Prata (1990) used AVHRR data to monitor the 1984 floods on Lake Eyre, Australia. The contraction of the lake was monitored by visual interpretation of the area of standing water. By assuming that exchanges due to groundwater, rainfall and runoff were negligible, this information was used to estimate evaporation rates which compared favourably with field estimates. Kuipers and Menenti (1986) used Landsat MSS data to monitor groundwater-fed lakes in Libya. They delineated the maximum lake area and calculated its mean reflectance for each date. These data were then plotted against time and used as a relative measure of lake area. Fluctuations in the area of these lakes were found to be due to variations in groundwater inputs.

AVHRR Image Processing and the Estimation of Flooding Ratios

Monthly advanced very high resolution radiometer (AVHRR) imagery collected between 1985 and 1987, and compiled by Kennedy (1989), was used in this study. The pertinent characteristics of the AVHRR sensor are given in Table 18.3.

A 512×512 sub-area of LAC data (1.1 km resolution at nadir) covering most of Tunisia was collected monthly on the day with least cloud cover. These data were registered to the 21 June 1985 image because of its near-nadir viewing angle and very low cloud cover using >50 ground control points per image. Nearest neighbour re-sampling was used for image registration and maximum errors were in the order of 1 km. As there are no published UTM maps of Tunisia the data were then geometrically corrected to a Tactical Pilotage Chart (Kennedy, 1989). Channel 1 and 2 data were used and atmospherically corrected using Singh and Cracknell's (1986) algorithm.

The method employed by Kuipers and Menenti (1986) outlined in the previous section was adapted in order to determine the flooding ratio of Tunsian playas. Initially this was done by delimiting the playas which necessitated including the mudflats that border the ephemeral lakes. The mean reflectance of the region delimited for each playa was then calculated from each date. As water has a much lower reflectance in AVHRR channels 1 and 2 than the clastic sediments and salts found on playas, its presence gives a low mean reflectance while a dry playa has a higher overall reflectance. When plotted as a time series, changes in the mean reflectance will provide a relative measure of the change in the area of standing water on a playa. This time series is termed the mean reflectance vector (MRV). Estimation of the flooding ratio from the MRV involved determining the mean reflectance of the playa immediately after the ephemeral lake has evaporated. This value can be used

Table 18.3 Characteristic of the NOAA satellites AVHRR sensors

Coverage cycle	9 days
View angle	± 55.4°
Ground coverage	2700 km
Orbit inclination	98.8°
Orbital height	833 km
Orbital period	102'
Instantaneous field of view	1.39–1.41 mrad
Launch date	
NOAA-7	23/06/1981
NOAA-9	12/12/1984
NOAA-11	24/09/1988
Equator crossing time[a]	
NOAA-7 & NOAA-9	14:30
NOAA-11	13:30
Spectral range	μm
Channel 1	0.580–0.680
Channel 2	0.725–1.100
Channel 3	3.500–3.930
Channel 4	10.300–11.300
Channel 5	11.500–12.500

[a]At time of satellite launch.

as a threshold, below which an ephemeral lake will exist and above which there will be no standing water. The MRV therefore provides a continuous relative time series of lake area that can be used in combination with a threshold to interpolate the time in the month when the standing water totally evaporated. The accuracy of this interpolation depends on the accuracy of the threshold, which is in turn controlled by the assumption that it is only the area of water that controls the reflectance of the playa. This assumption is investigated later.

AVHRR bands 1 and 2 reflectance data are sensitive to water in slightly different ways due to the variations in the absorption coefficient which increases towards longer wavelengths. Band 1 radiation penetrates clear water to a depth of approximatley 2 m, but for band 2 the penetration depth is only 0.4 m (Prata, 1990). Thus band 2 will be more effective than band 1 at identifying standing water, as in shallow water there will be less contribution from the bottom sediments. Therefore, band 2 was used to calculate the MRV in this study.

The second method of estimating the flooding ratio involved the visual inspection of band 2 images to determine whether any standing water was present on a playa. The problem with this method is that there is no simple way of interpolating the time that the ephermeral lake dried up. The method is therefore suited to intensive monitoring using data acquired on a daily or near-daily basis.

Satellite imagery can also be used to estimate the area of standing water on each playa for each month. Such information was used here to determine whether the MRV is a good estimate of the amount of water on a playa, and thus whether the MRV is a reliable estimate of the flooding ratio. The area of water was computed by visual inspection of each playa for each image acquisition date. Visual inspection is preferred over computational techniques (e.g. classification) as the user's knowledge of the drainage basin can be employed to discriminate between standing water and moist areas due to factors such as groundwater seepage and river inflow. Standing water will generally be found in the lowest part of the basins, or downwind of a topographic low in the dominant seasonal wind direction. In this research the area of standing water was estimated by defining a region of interest around the wetted area, counting the number of pixels and converting the pixel count to an areal measurement. The resulting time series of the area of water provides information on the area and length of time the playa was flooded, and was used to calculate the flooding ratio.

The influence of climate on the flooding ratio was analysed by using monthly meteorological data from stations adjacent to the lakes and playas provided by the Institute National de la Meterologie.

COMPARISON OF METHODS OF ESTIMATING THE FLOODING RATIO

The MRV time series for three sebkahs (Kelbia, el Hani and Moknine), along with the appropriate climatic data, are shown in Figures 18.2, 18.3 and 18.4. All of these playas exhibit large seasonal variations in the amount of water covering their surfaces. Though Sebkah Kelbia and Sebkah el Hani are geographically close they have different cycles of wetting and drying. Sebkah el Hani exhibits a rhythmic, seasonal curve whereas that of Sebkah Kelbia appears to be more random and Sebkah Kelbia is wetter for longer periods of time than Sebkah el Hani. These differences can be explained in terms of the differences in their catchment geomorphology. Sebkah Kelbia is fed by three major rivers draining a large, mountainous catchment, whereas Sebkah el Hani has a smaller, less steep catchment with a shorter channel system and lower inflow. Months with high rainfall totals produce marked spikes in the Sebkah Kelbia MRV series, because runoff responds quickly to rainfall However, some spikes in MRV series are not related to monthly rainfall at Kairouan and are presumably related to spatially restricted convective storms elsewhere in the catchment. Consequently there is a low correlation between the MRV for Sebkah Kelbia and the monthly rainfall at Kairouan. For Sebkah el Hani all of the months with high rainfall totals are discernable in the MRV. In 1987, when monthly rainfall totals were relatively low, the MRV series showed a remarkable similarity to the insolation at Kairouan (Figure 18.3). Such a relationship would be expected if evaporation of a reasonably constant water influx (such as groundwater seepage) was occurring. The Sebkah Moknine MRV series (Figure 18.2) exhibits a smooth cyclical pattern similar to that of Sebkah el Hani.

Estimation of the flooding ratio using MRV data necessitated the determination of the mean reflectance of the playa just after an ephemeral lake has evaporated. Figure 18.5 illustrates how this was achieved for Sebkah el Hani. During the three-year monitoring period there were two dates when the lake was at the final point of desiccation and these periods were chosen for calculation of the threshold. However, if the lowest mean value is used (threshold A) the flooding ratio for the first year would be 81%, while the threshold B flooding ratio was 77%. Because of the problems associated with an unstable threshold some measure of stability was introduced by using the mean of all images that satisfy the criteria for choosing a threshold and by also indicating the range of the thresholds. Using this method, Sebkah el Hani has a flooding ratio of $79.5 \pm 4\%$. The flooding ratio for all of the playas analysed is shown later in Figure 18.9.

Figure 18.2 The MRV and climatic data for Sebkah Moknine

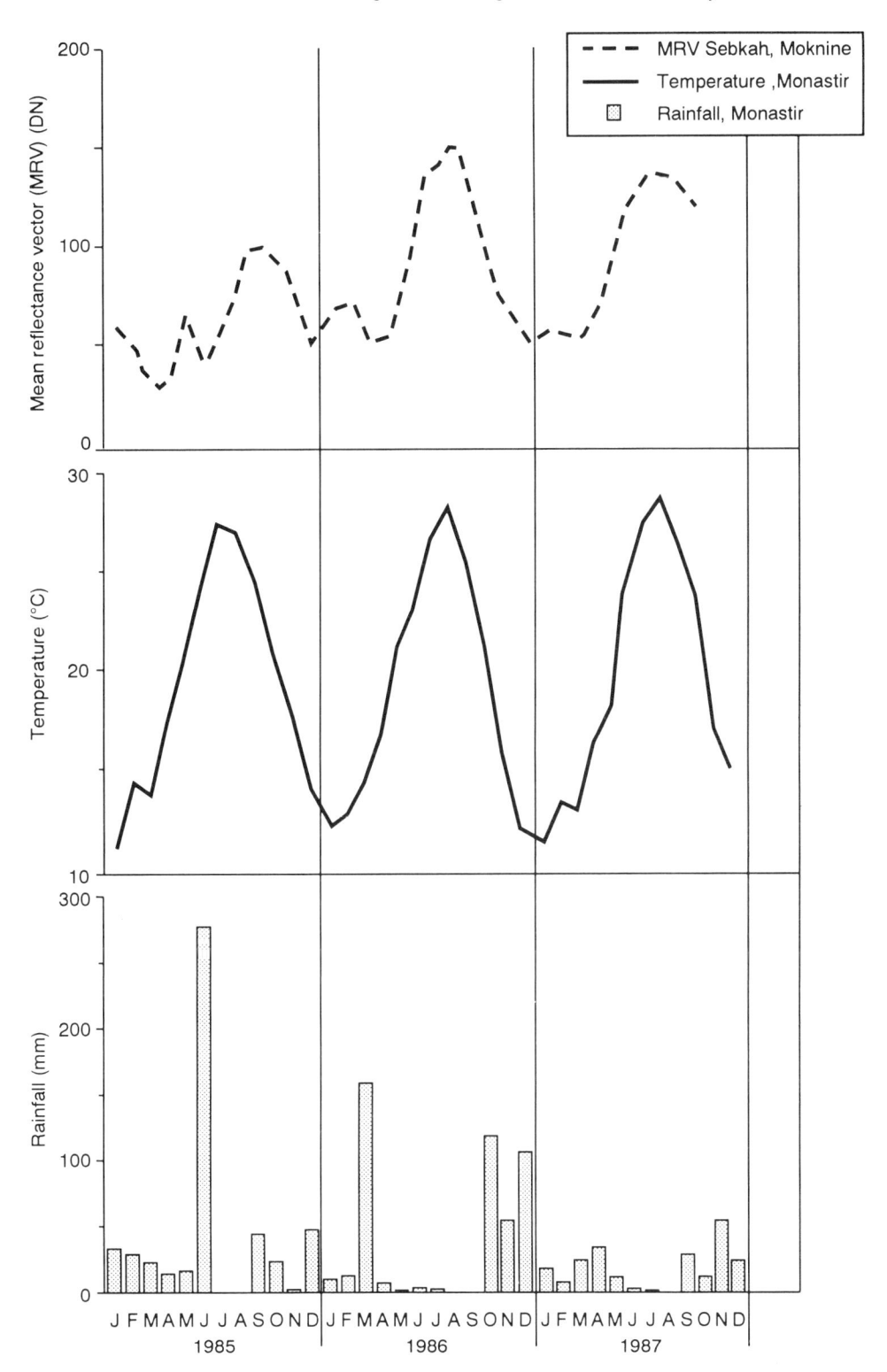

Monitoring the Flooding Ratio of Tunisian Playas   355

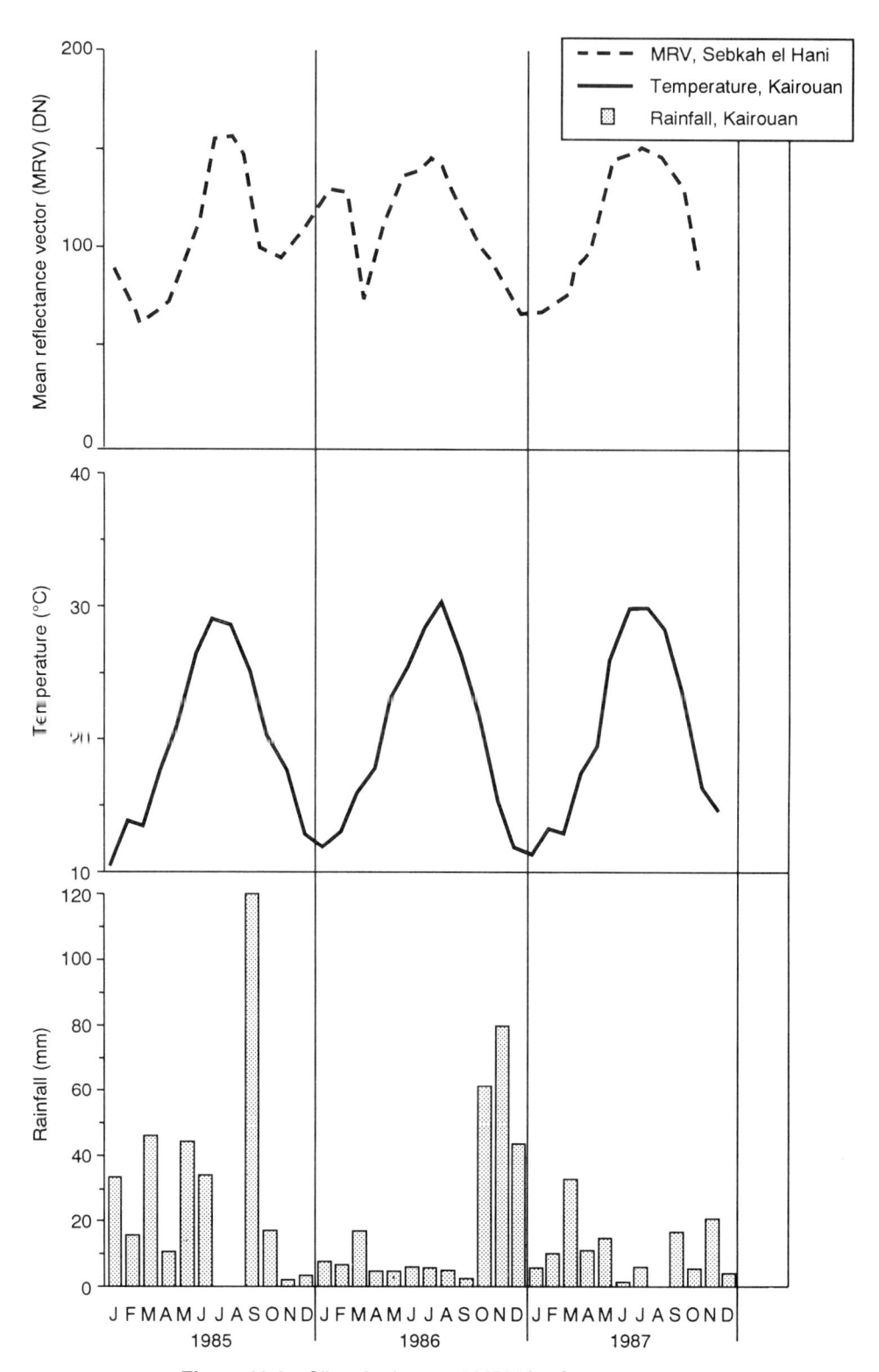

Figure 18.3 Climatic data and MRV for Sebkah al Hani

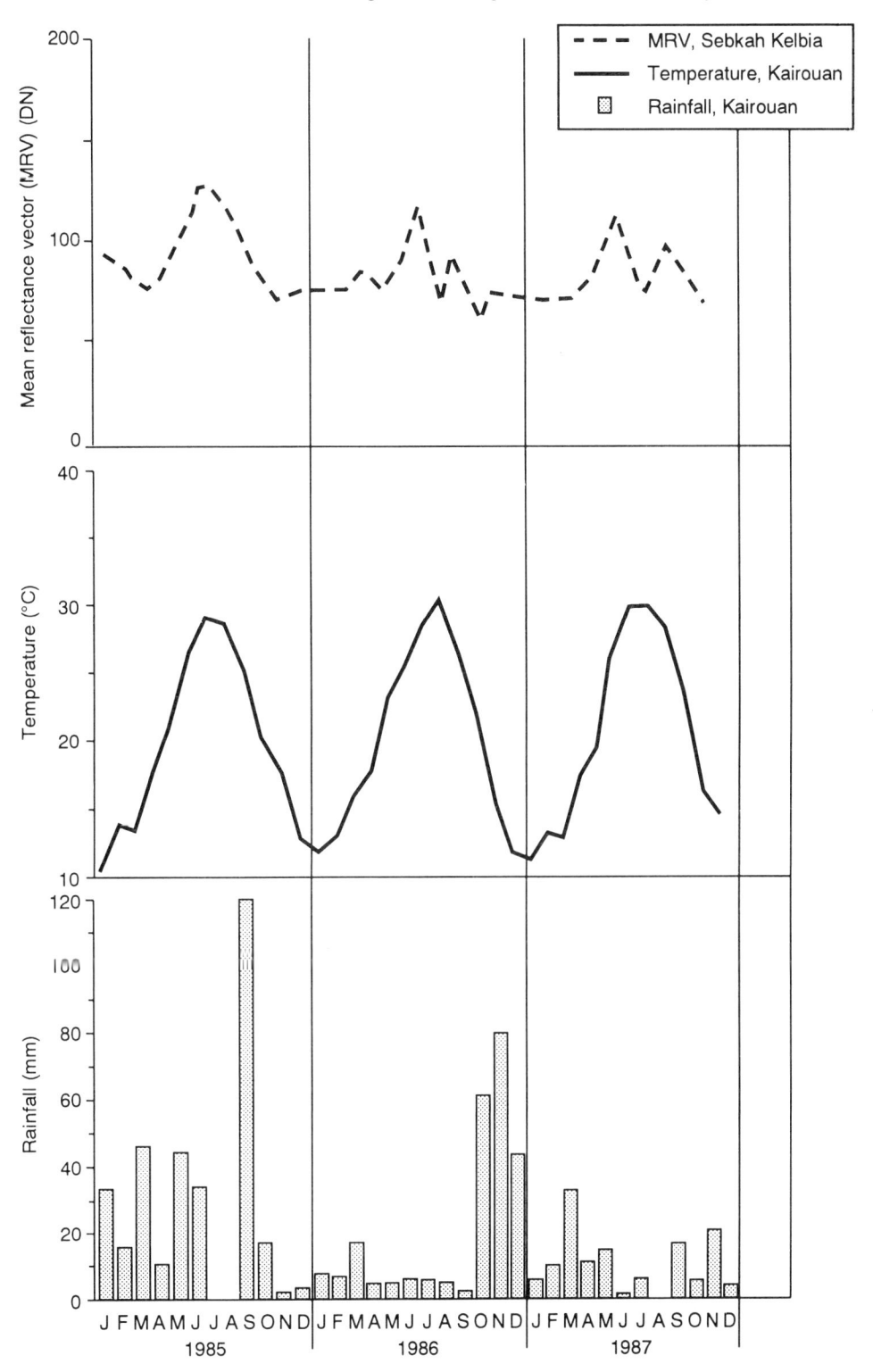

Figure 18.4 Climatic data and MRV for Sebkah Kelbia

358   Environmental Change in Drylands

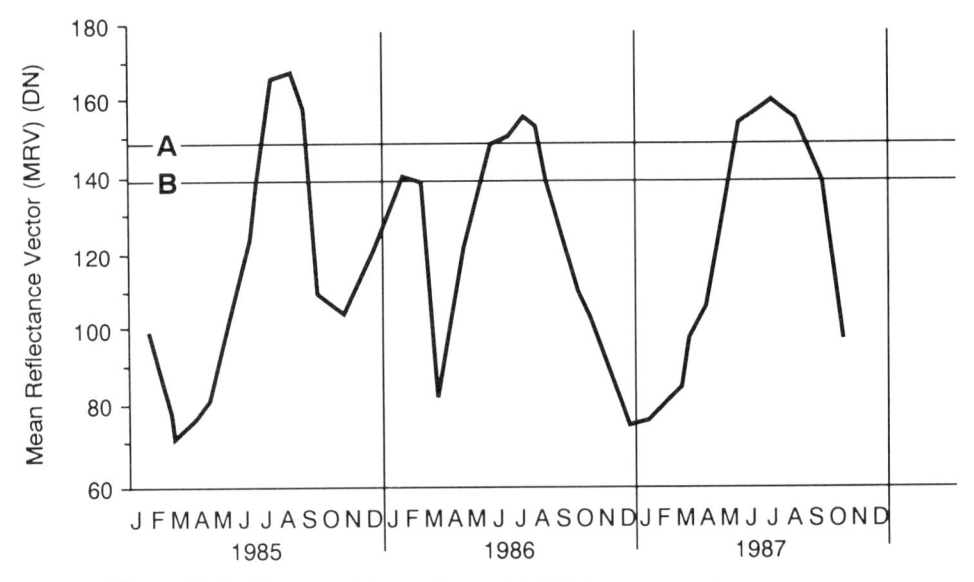

Figure 18.5 Upper and lower threshold MRV values for Sebkah el Hani

The area of standing water was also calculated for each playa. The areas for Sebkahs el Hani, Moknine and Kelbia are shown in Figure 18.6. The water area and the MRV show variable correlation coefficients for these three playas: Sebkah Kelbia 0.75, Sebkah Moknine 0.55 and Sebkah el Hani 0.27. These low correlations are largely caused by changes in the reflectance of the playa upon desiccation. After the standing water has evaporated the reflec

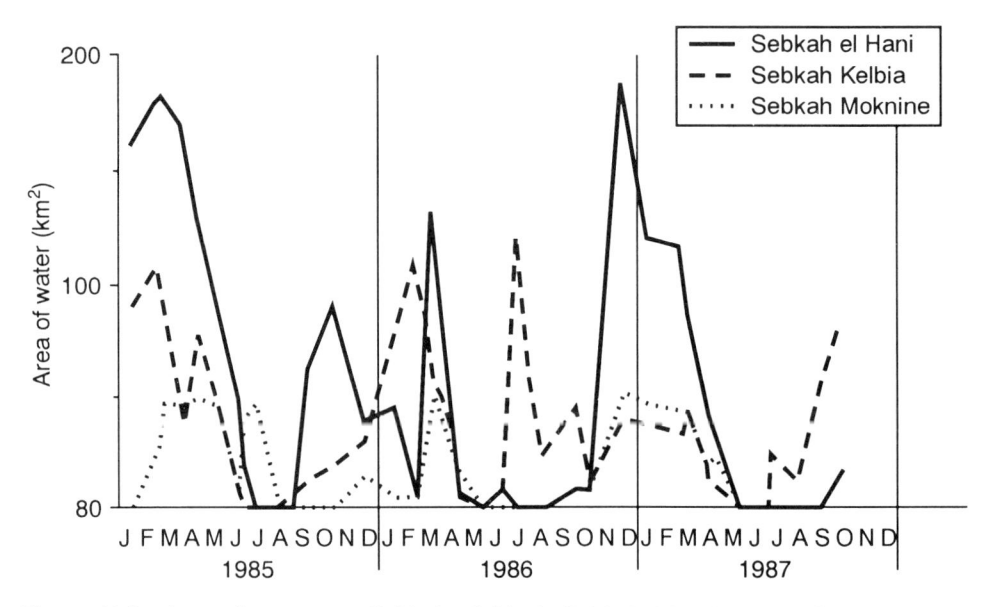

Figure 18.6 Area of water on Sebkah el Hani, Sebkah Moknine and Sebkah Kelbia, 1985–87

tance continues to rise, as moisture is lost from the surface. This indicates that the MRV is sensitive to crust moisture as well as standing water in an ephemeral lake.

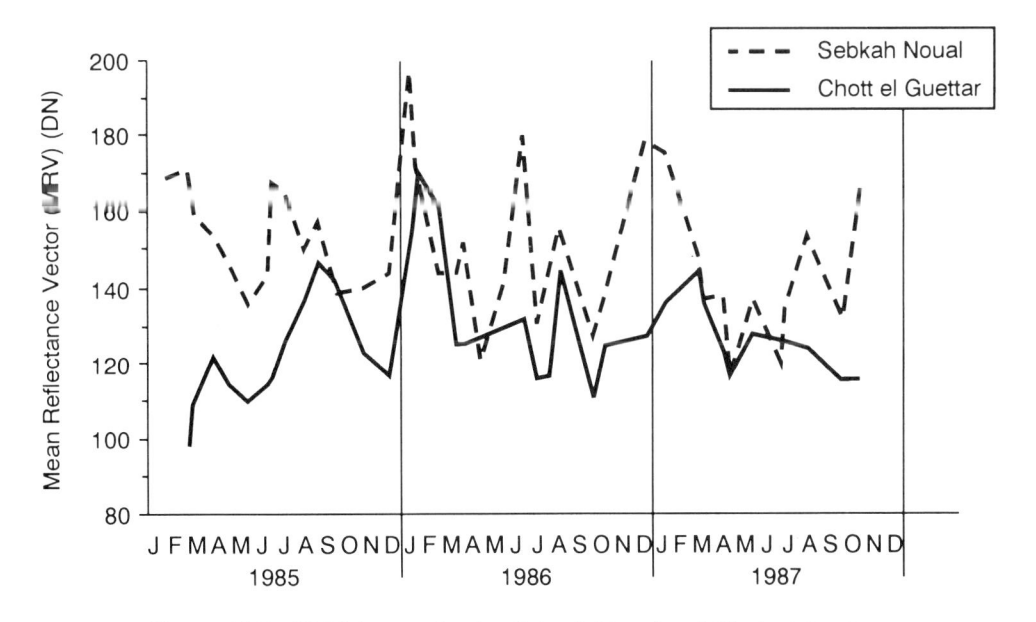

The MRV time series for Sebkah Noual and Chott el Guettar (Figure 18.7) exhibit erratic behaviour. Neither time series shows any correlation with the meterological data from Gafsa. Chott el Guettar was dry throughout the monitoring period and consequently had a flooding ratio of zero. Sebkah Noual also exhibited an erratic MRV time series, although it differs significantly from that for Chott El Guettar. The flooding ratio for Sebkah Noual was 10.2 ± 16.5%. The large range associated with this flooding ratio is due to a complexity of the playa surface. Sebkah Noual has two different surface types when dry — mudflats fringing a central salt pan. The amount of moisture in the surface sediments of the mudflats varies spatially and temporally causing changes in mudflat reflectance which is unrelated to the area of standing water. The areal extent of salt cover on the mudflat also varies, due to periodic inundation with water from storms within the catchment. Such inundations are followed by the establishment of a salt crust by precipitation and efflorescence. Consequently, the MRV time series for Sebkah Noual is controlled by variations in the area of standing water and variations in the amount of moisture in, and salt efflorescene on, the mudflats. These complexities mean that the MRV is unsuitable for estimating the flooding ratio and that it is necessary to determine it by another method.

Chott Fedjaj was dry throughout the monitoring period and, like Chott el Guettar, had a flooding ratio of zero. It appeared dry because water in the Chott Fedjaj catchment feeds into Chott Djerid via the Djerid–Fedjaj channel. Chott Djerid, like Sebkah Noual, shows variations in the MRV time series that are unrelated to the area of standing water. Therefore the areal extents of water for Sebkah Noual and Chott Djerid were determined by visual interpretation of the AVHRR imagery (Figure 18.8). Nonetheless, the correlation between the MRV and the area of surface water for both of these two playas is low (Chott Djerid, $r^2 = 0.02$; Sebkah Noual, $r^2 = 0.11$) due to the factors previously discussed.

Figure 18.7 MRV time series for Sebkah Noual and Chott el Guettar

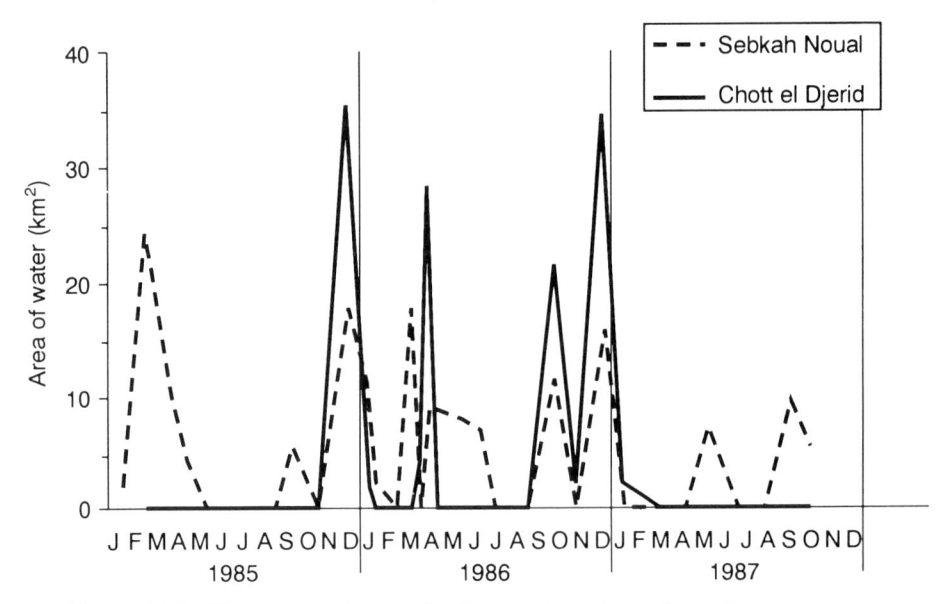

Figure 18.8 The extent of water for Sebkah Noual and Chott Djerid, 1985–87

It appears that the flooding ratio can be estimated with reasonable accuracy by the MRV for simple playas, i.e. those that alternate between an ephemeral lake and a single dry surface type. However, for those playas with more than one type of surface the MRV is inaccurate and the presence or absence of water must be calculated and used to estimate the flooding ratio.

Estimation of the flooding ratio for Sebkah Noual was further complicated by the fact that the high evaporation rates in southern Tunisia mean that small floods will usually evaporate within a month. When the residence time of an ephemeral lake in a playa is shorter than the monitoring period, e.g. Sebkah Noual, it is impossible to infer how long the playa was flooded. Sebkah Noual had six such floods in the monitoring period. If it is assumed that the presence of water means that the sebkah was flooded for the *entire* month, the flooding ratio was 66%. This is much higher than the 10.2 ± 16.5% estimated from the MRV and is probably an overestimate. For the Chott Djerid there were no single-month floods in the monitoring period, and the flooding ratio of 23.5% may be assumed to be reasonably accurate.

Figure 18.9 shows the estimates of the flooding ratios derived from the MRV and the area-of-water methods. The flooding ratio calculated by the area-of-water method falls within the range of the flooding ratio estimated by the MRV for all sebkahs, except Sebkah Noual and Chott Djerid. The accuracy of the flooding ratio estimated using the area-of-water method is clearly dependent on the residence times of the ephemeral lakes and the interval monitoring. The residence time is controlled by (i) the size of the catchment, (ii) the size and duration of the individual rainfall events and (iii) the rate of water loss from the playa through evaporation and, in the case of playas with open groundwater systems, groundwater seepage. Playas with multiple surface types, especially if they are found in small catchments with high evaporation rates, appear to require a monitoring period of less than one month.

RELATIONSHIP BETWEEN FLOODING RATIO AND ENVIRONMENTAL FACTORS

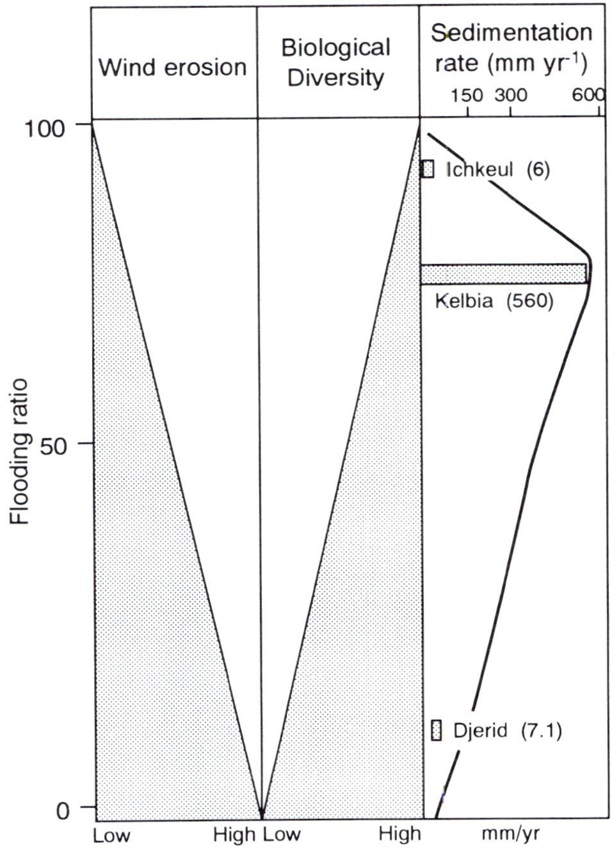

Wind erosion	Biological Diversity	Sedimentation rate (mm yr⁻¹)
100 — Low	Low	150 300 600
50 — High	High	Ichkeul (6), Kelbia (560)
0 — Low	High	Djerid (7.1)

Flooding ratio (y-axis: 0, 50, 100)

TUNISIAN FLOODING RATIOS

Playa / Lake name	Class	Flooding ratio % (NRV)	Flooding ratio % (AOW)	Latitude
Lake Bizerte	Lake	100	100	37.3
Lake Ichkeul	Lake	100	100	37.2
Sebkah Kelbia	Lake	77.7 ± 10.0	87.3	35.9
Sebkah Moknine	Lake	70.7 ± 5.3	75.2	35.6
Sebkah el Hani	Lake	79.5 ± 3.0	76.7	35.6
Sebkah Noual	Playa / Lake	10.2 ± 16.5	66.0	34.5
Chott el Guettar	Playa	0.0	0.0	34.2
Chott el Djerid	Playa	–	23.5	34 - 33.3
Chott el Fedjaj	Playa	0.0	0.0	33.8

Figure 18.9 Flooding ratios, latitude information and the general information on Tunisian playas provided by the flooding ratio classification

DISCUSSION

Long-term records of the flooding ratio of individual playas can provide data for the determination of historical changes in catchment hydrology. AVHRR data could be used to provide a global data set on such changes, though the AVHRR data archive only goes back to 1981. Therefore they have to be extended using historical records for longer-term analyses. Such records may be in the form of either observed or modelled flooding ratios. Hollis and Kallel (1985) modelled the hydrology of Sebkah Kelbia to determine the effect of climate and man-induced changes within its catchment. The model was validated using lake level records between 1942 and 1945. They simulated lake levels, salinity and the flooding ratio from 1900 to 1975, and estimated a flooding ratio of 89.8%. They predicted that dam construction planned for the catchment would reduce this to 75.5% during the 1980s. The flooding ratio calculated by the MRV method for the period 1985–87 was 77.7 ± 10% and that estimated by the area-of-water method was 87%; though there is a 10% difference in the flooding ratio estimates, they are both less than the mean 1900–75 flooding ratio. The reduction may be because the dams built on the tributaries have reduced the flow into Sebkah Kelbia, though it could also be that the three years used to monitor the sebkah were drier than the long-term mean.

For Chott Djerid there is limited historical data on standing water observations between 1947 and 1958 (Coque, 1962) (Table 18.2). Between 1948 and 1958 there were 65 months without observations, 19 months with water on the playa and 59 dry months. If the months without observations are ignored, the mean flooding ratio is 32.2%. The flooding ratio between 1985 and 1987 was 23.5%, possibly implying that Chott Djerid has become slightly drier. The desiccation may be due to either climatic or man-induced causes. For example, there has been a large increase in groundwater extraction during this period although climatic influences cannot be excluded.

Having estimated the flooding ratio of these playas it is possible to classify them according to the criteria used by Motts (1965) (Figure 18.9). In general, the flooding ratios decrease from north to south along the climatic gradient, confirming that climate has a large influence on playa surface hydrology. However, the geomorphology and the geology of the catchments are also significant; Chott el Guettar and Chott Fedjaj were dry throughout the monitoring period but are at similar latitudes to two playas that have flooding ratios between 20 and 30%. In Chott Fedjaj the low ratios are due to the fact that it drains rapidly into Chott Djerid (i.e. it is not a closed basin). In Chott el Guettar the low ratios are due to the reduced inflow from the west, because the water is used for irrigation before it enters the playa, although groundwater seepage of used irrigation water is important. The size of the catchment may also be significant.

The flooding ratio classification can provide generalised information about many aspects of playas (Figure 18.9). There is only very limited information concerning sedimentation rates in Tunisian lakes and playas. In 1990 and 1992 two transects were surveyed at different locations across the causeway that crosses the northern Cott Djerid. In both transects the playa was higher to the north of the causeway — 5 cm in the central salt pan facies in 1990 and 11 cm in the marginal mudflats in 1992. The causeway was constructed in 1983 and this gives maximum sedimentation rates between 7.1 and 12.2 mm yr^{-1}. These are maximum rates because if the causeway had not been built the sediment would have been deposited over a larger area. The sedimentation rate of Sebkah Kelbia was calculated from data provided by Hollis and Kallel (1986). It was assumed that sedimentation rates throughout the sebkah were equal; this gives a mean sedimentation rate of 5.6 cm yr^{-1} which represents a minimum value. The sedimentation rate in the centre of Lake Ichkeul is 6 mm yr^{-1} (Hollis, 1986). Therefore, it appears that the sedimentation rates in Tunisia *might* be related to the flooding ratio and *may* follow Langbein and Schumm's curve (Figure 18.9).

The flooding ratio can also be used to relate the susceptibility of a playa to aeolian deflation, as playas with low flooding ratios will have more frequent and longer periods when the water table is low. At such times the relatively dry surface will be susceptible to wind erosion. Chott el Guettar, Chott Fedjaj and Chott Djerid may be expected to be highly susceptible to wind erosion on this criteria.

The flooding ratio also indicates the levels of biological productivity and diversity. Field observations of the near-surface sediments of Sebkah el Hani and Sebkah Moknine (Millington *et al.*, in press) (and the studies of Hollis, 1986, on Lake Ichkeul) suggest that those with high flooding ratios do contain an appreciable amount of organic matter, whereas on Chott el Guettar and Chott Djerid organic-rich sediment only occurs in the vicinity of springs and seepage zones.

CONCLUSIONS

Monitoring the flooding ratio of playas on a regional, or possibly global, basis is a promising application for satellite imgery. However, accurate estimation of the flooding ratio using monthly AVHRR imagery is problematic. Simple visual estimation of the presence or absence of standing water appears to be more reliable than the MRV, particularly as the MRV is unreliable for playas with multiple surface types. However, the presence and absence of standing water does not allow interpolation between image acquisition dates, as the MRV does, and thus requires more intensive monitoring because lakes with low flooding ratios and small catchments usually dry out quickly after a storm. The length of monitoring period needed for such work is a function of the playa in the region that experiences the shortest residence time for an ephemeral lake after a flood; in this study it was Sebkah Noual. For certain aims the MRV method may be a useful technique; for example, if the aim of a study was to compare a small number of playas in different region, those with simple surfaces could be chosen, and the MRV approach would be cost-effective as considerably less data would be needed.

It was possible to classify Tunisian playas from their flooding ratios calculated from monthly AVHRR imagery. The classification can then be used to rank the lakes in terms of their sedimentation rates, their susceptibility to deflation and biological diversity.

Long-term estimates of the flooding ratio can also provide important information on the changing hydrology of a playa. The data presented here is of too short a timescale to provide a particularly revealing time series information. However, for Sebkah Kelbia and Chott Djerid, where there were additional limited historical records, it can be demonstrated that both playas probably have become drier during this century. Some of this change is certainly due to anthropogenic activity, although there may be a climatic change element as well.

ACKNOWLEDGEMENTS

The authors wish to thank Pam Kennedy for providing the co-registered satellite imagery (which was originally supplied by the University of Dundee) and Andrew Millington for helpful advice. Rob Bryant carried out this work under NERC grant G4/89/GS/101.

REFERENCES

Coque, R. (1962). *La Tunisie Presaharinne. Etude Geomorphologique*. Armand Colin, Paris.
DeDekker, P. (1988). Biological and sedimentary facies of Australian salt lakes. *Palaeogeography, Palaeoclimatology, Paleoecology*, **62**, 237–70.
Drake, N. A. (1992). Mapping and monitoring surficial materials, processes and landforms in southern Tunisia using remote sensing. Unpublished PhD thesis, University of Reading.
Hollis, G. E. (1986). *The Modelling and Management of the Internationally Important Wetland at Garaet el Ichkeul, Tunisia*, International Waterfowl Research Bureau, Special Publication 4.
Hollis, G. E. and Kallel, M. R. (1986). Modelling natural and man-induced changes on Sebkhet Kelbia, Tunisia. *Transactions Institute British Geographers*, **N.S. 11**, 86–104.
Kennedy, P. J. (1989). A study of Tunisian grazing lands using data from the advanced very high resolution radiometer. Unpublished PhD thesis, University of Reading.
Kuipers, H. and Menenti, M. (1986). Groundwater-fed lakes in the Libyan desert: their varying area as observed by means of Landsat-MSS data. *Proceedings ISLSCP Conference*, Rome, Italy, 2–6 December, pp. 467–71.
Langbein, W. B. and Schumm, S. A. (1958). Yield of sediment in relation to mean annual precipitation. *Transactions of the American Geophysical Union*, **39**, 1076–84.
Mabbutt, J. A. (1977). *Desert Landforms*, Australian National University Press, Canberra.
Mamou, A. (1976). Contribution a l'etude hydrogeologique (de la presqu'ile de Kebili). Unpublished PhD thesis, Université Pierre et Marie Curie, Paris.
Millington, A. C., Drake, N. A., White, K. H. and Bryant, R. G. (in press). Salt ramps: wind-induced depositional features on Tunisian playas. *Earth Surface Processes and Landforms*.
Millington, A. C., Jones, A. R., Quarmby, N. and Townshend, J. R. G. (1987). Remote sensing of sediment transfer processes in playa basins. In Frostick, L. E. and Reid, I. (eds.), *Desert Sediments: Ancient and Modern*, Blackwell Scientific Publications, Oxford, pp. 369–83.
Motts, W. S. (1965). Hydrologic types of playas and closed valleys and some relations of hydrology to playa geology. In Neal, J. T. (ed.), *Geology, Mineralogy and Hydrology of U.S. playas*, Airforce Cambridge Research Laboratories, Environmental Research Paper 96, pp. 73–104.
Prata, A. J. (1990). Satellite-derived evaporation from Lake Eyre, South Australia. *International Journal of Remote Sensing*, **11**, 2051–68.
Rosema, A. and Fiselier, J. L. (1990). Meteosat-based evapotranspiration and thermal inertia mapping for monitoring transgression in the Lake Chad region and Niger Delta. *International Journal of Remote Sensing*, **11**, 741–52.
Singh, S. M. and Cracknell, A. P. (1986). The estimation of atmospheric effects for SPOT using AVHRR Channel 1 data. *International Journal of Remote Sensing*, **7**, 361–77.
Street-Perrott, F. A. and Harrison, S. P. (1985). Lake levels and climate reconstruction. In Hecht, A. D. (ed.), *Paleoclimate Analysis and Modelling*, Wiley, New York, pp. 291–340.
Walling, D. E. and Kleo, A. H. A. (1979). Sediment yields of rivers in areas of low precipitation: a global view. In *The Hydrology of Areas of Low Precipitation*, IAHS-AISH Publication 128, Wallingford, pp. 479–493.

19 Waterlogging and Soil Salinity in the Newly Reclaimed Areas of the Western Nile Delta of Egypt

R. GOOSSENS
Laboratory for Regional Geography and Landscape Science, University of Gent, Belgium

T. K. GHABOUR
Remote Sensing Center, Academy of Scientific Research and Technology, Cairo, Egypt

T. ONGENA
Laboratory for Regional Geography and Landscape Science, University of Gent, Belgium

and

A. GAD
Remote Sensing Center, Academy of Scientific Research and Technology, Cairo, Egypt

ABSTRACT

Land reclamation in the desert areas flanking the Nile Delta has led to waterlogging and an increase in soil salinisation. Satellite imagery has been used to map waterlogged and salinised areas, and field survey has shown that soil salinisation is more closely related to drainage conditions than irrigation water quality. Image classification techniques were used to map the extent of waterlogged and saline soils in 1977 and 1989 from Landsat MSS and SPOT XS data respectively. From these data it was shown that the extent of waterlogged and saline soils increased 4.5 times between 1977 and 1989. A model of the growth of the waterlogged and saline areas is used to predict the expansion of waterlogged soils over the next decade.

INTRODUCTION

Salt-affected soils occur under a wide range of environmental conditions, and it has been estimated that about one-third of the cultivated land under irrigation in the world is already salt-affected (Framji, 1974). Such soils are most frequently found in arid and semi-arid regions, where salt-affected soils often occur in fertile alluvial plains. El Gabaly (1959) estimated that about 800 000 ha of land in Egypt were suffering from primary or secondary salinisation, an area representing approximately one-third of all arable land in the country. Since the late 1950s the area of salt-affected soils has increased as a result of perennial

Environmental Change in Drylands: Biogeographical and Geomorphological Perspectives.
Edited by A. C. Millington and K. Pye. © 1994 John Wiley & Sons Ltd

irrigation without adequate drainage. In addition, the demand for irrigation water has increased since desert soils have been reclaimed for agriculture. This is the case in Baheira Governorate where *good quality* irrigation water is partly diverted towards the Newly Reclaimed Areas, while the older cultivated land (in the Delta) receives less *good quality* irrigation water than it previously did. As a consequnece farmers in the Delta are sometimes obliged to irrigate their fields with saline drainage water. Clearly such a practice will ultimately lead to an increase in the extent of the salt-affected soils in the Delta. Although the Newly Reclaimed Areas only receive *good quality* water there are many areas with saline soils. Their presence is due to the way in which land reclamation has been undertaken on the desert fringes. Generally, land reclamation proceeds with the wind-blown sands, which are commonly found in these areas, being levelled and then irrigation and drainage channels constructed.

The mapping and inventory of soil salinisation can be achieved by using remotely sensed data or field survey. In Pakistan (Rafiq, 1975) and India Landsat Multi-Spectral Scanner (MSS) and Thematic Mapper (TM) data have been evaluated for mapping salt-affected soils. At the large scale, Sehgal *et al.* (1988) used Landsat MSS data for mapping salt-affected soils in the framework of a general soil map of India, whilst at the small scale Sharma & Bhargava (1988) and Rao *et al.* (1991) have used Landsat MSS and TM data respectively for more detailed mapping of limited areas. Dwivedi (1992) used Landsat MSS and TM imagery for mapping and monitoring the salt-affected soils associated with poor drainage in the Indo-Gangetic alluvial plain, and detected a decrease in the areal extent of saline soils between 1975 and 1986. All of these studies have a common theme which is that salt-affected and waterlogged or poorly drained soils are mapped through visual interpretation, whilst in this research the focus is on automated mapping of salinisation and waterlogging through supervised image classification.

Prior to this study, saline and waterlogged soils in Egypt have mainly been mapped using a traditional field and-laboratory approach (Abdel Salam *et al.*, 1972, 1973). In this paper a new approach for Egypt, that of mapping and constructing an inventory of waterlogged and saline soils using digital image processing, is evaluated.

THE STUDY AREA

The study area is located south-east of Alexandria in Baheira Governorate and is in the fringes of the western Nile Delta (Figure 19.1). The area is characterised by a BWh climate according to the Köppen classification. The mean annual rainfall is 22 mm and the main annual temperature is 20.8°C. The altitude of the area varies between 3 and 15 m a.s.l. The area consists of sandy to sandy-loam soils due to the fact that the soils have evolved in aeolian deposits which are underlain by alluvial clays. The average thickness of the aeolian sand deposits is 3 m, but the topography is undulating due to the fact that much of the aeolian deposits are in the form of dunes.

Since the early 1970s the area has undergone incremental reclamation to provide agricultural land. Large-scale projects were established, and state-owned companies given concessions to reclaim the land for agriculture; this led to the establishment of large fields of up to 1 km^2. During reclamation the original dune topography is destroyed in the following manner. The tops of the dunes are levelled and the inter-dune depressions are partially filled up with the sands from the dune crests. The levelling is, however, not perfect as remnants

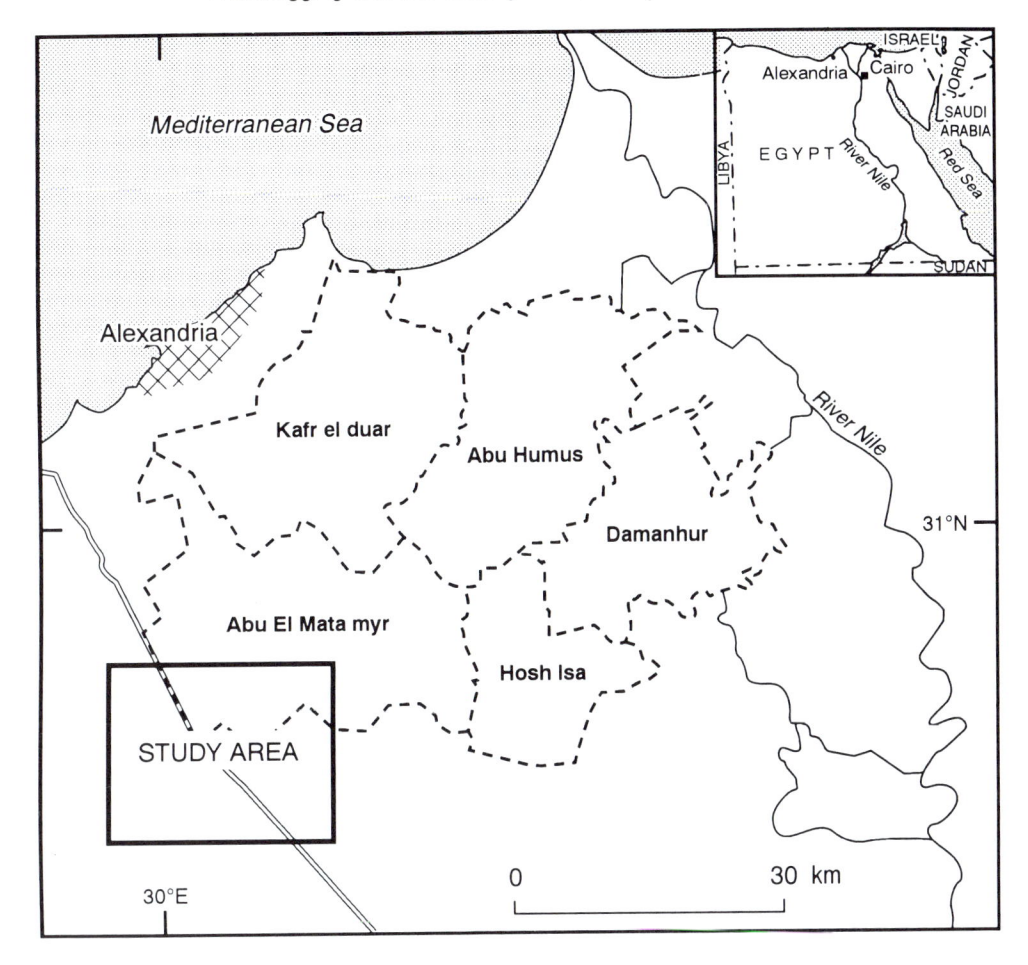

Figure 19.1 Location map

of the original topography can still be seen in some fields. Irrigation and drainage canal systems are constructed on the levelled terrain. The irrigation water for the area is diverted from the Nubaria Canal and is of good quality (the EC varies between 1.2 and 1.4 mS). The land is mainly watered by flood irrigation, although centre pivot systems have also been installed. The main crops grown are wheat, clover and citrus.

FIELD SURVEY

SPOT multispectral (XS) imagery from 1989 was initially interpreted to locate areas of waterlogging and salinisation, and then checked in the field. At this time no quantitative data were available on soil salinity variability throughout the area. The only data available were from some soils profiles described by Abdel Salam *et al.* (1972, 1973). Consequently there was a need for a well-distributed series of soil salinity measurements to enable interpretation and validation of the remotely sensed data. Therefore during the initial period of field work 148 soil samples were taken at a depth of 40 cm, which corresponds to the average rooting depth (Ayers & Westcot, 1976; Loveday, 1984). The electrical conductivity (EC) was measured on the saturated soil paste (United States Salinity Laboratory Staff, 1954). Correlation coefficients between the different parameters measured at the field sample points (Table 19.1) were calculated. The correlations are generally weak, although two are strong correlations. Firstly, there is a strong correlation between ECP (electrical conductivity of saturated soil paste) and ECD (electrical conductivity of drainage water) ($r^2 = 0.905$) which can be explained in two ways:

1. The coarse-textured soils are characterized by low particle surface tension which enables soluble salts to be easily released.
2. The high macroporosity of the coarse-textured soils means that they have a high leaching fraction which results in leaching of the soluble salts over a relatively short time.

Secondly, there is a high correlation between ECI (electrical conductivity of irrigation water) and ECD ($r^2 = 0.854$).

In addition, the relationship between ECP and the depth of the groundwater table was investigated using two parameters, namely the depth of the oxido-reduction zone and the depth of the reduction zone (Figure 19.2). The 'knick-point' on the graph of ECP and the depth of the oxido-reduction zone is found where the depth of the oxido-reduction phenomenon is around 40–50 cm. This observation is in line with that recorded by other workers (e.g. Halvorson & Rhoades, 1974; Rhoades, 1975; Kovda, 1975) who have found that soil salinity is directly related to the depth of the groundwater table. A high groundwater table results in a low leaching capacity, which in turn leads to salt concentration in the upper soil profile. In the area studied the critical depth appears to be about 40–50 cm, and in Figure 19.2 it can be seen that an increase in the groundwater table above about 50 cm results in a strong increase in the soil salinity. This is particularly important since this depth is approximately the same as the mean rooting depth of most of the field crops grown in the area, and any salt buildup at approximately 50 cm will affect crop growth and suppress yields. There is circumstantial evidence to support this as in many parts of Baheira Governorate areas with high groundwater tables can only be used for the cultivation of rice because it is tolerant of wet soils and moderately high salinity levels (Ayers and Westcot, 1976; Loveday, 1984). Compared with other studies the critical depth identified in this work is rather high. Halvorson & Rhoades (1974) and Rhoades (1975) identified critical depths of

Table 19.1 Parameters measured for each soil sample

Symbol	Description
ECP	electrical conductivity of the saturated soil paste
ECI	electrical conductivity of the irrigation water
ECD	electrical conductivity of the drainage water
ECS	electrical conductivity of the soil water
OR	depth of the oxido-reduction zone
R	depth of the reduction zone
pH	pH of the saturated soil paste
CaCO$_3$	amount of CaCO$_3$ (HCl method)
COLOUR	topsoil colour

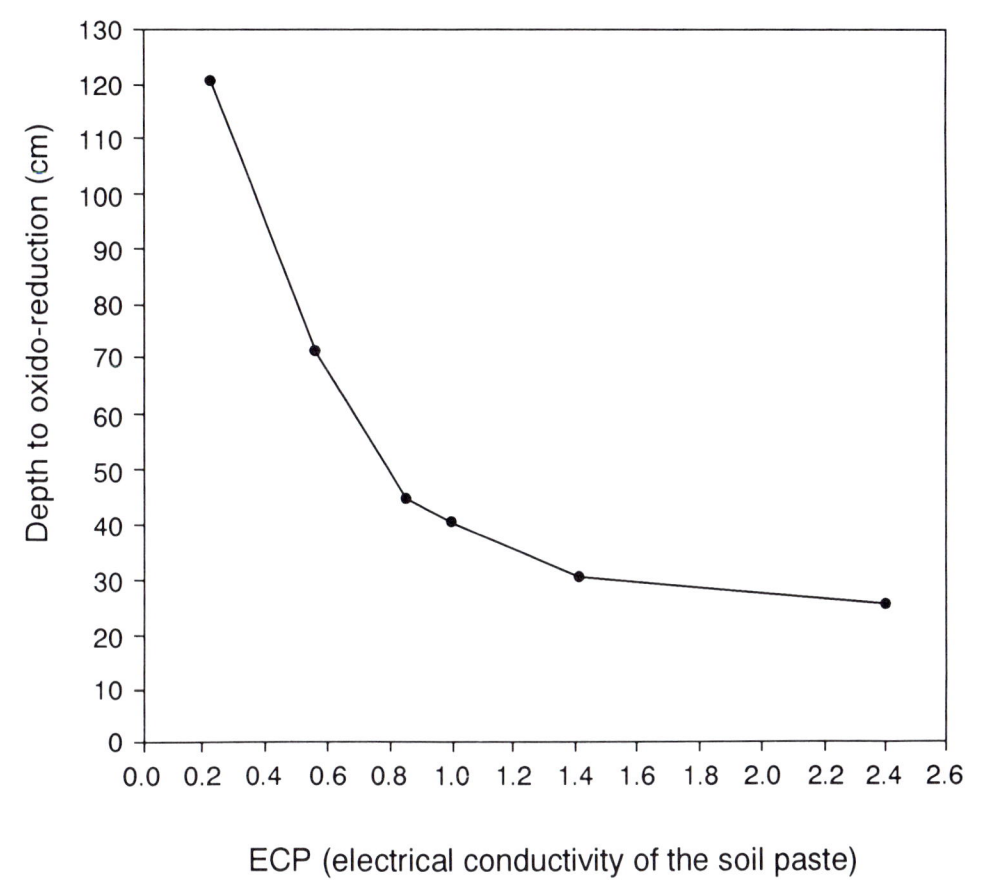

Figure 19.2 Relationship between depth to oxido-reduction and electrical conductivity of soil paste

around 1 metre, and Kovda (1975) measured a critical depth of 5 metres. The reason for these differences is probably that in this study it was not the depth of the groundwater table that was measured but the top of the oxido-reduction zone — in other words, a depth that is influenced by the highest position of the groundwater table (in winter time) and the height of the capillary fringe. In addition, differences in soil texture between these soils and those measured by the other workers, as well as the depth of the soil sampled (40 cm versus 30 cm), contribute to the differences.

It can be concluded that the problems of salinity are directly related to those of soil drainage in this, as in other areas. This is particularly true for the Newly Reclaimed Areas of Baheira Governorate where the quality of the irrigation water is not an issue. As well as poorly drained, irrigated soils there are also areas of naturally occurring waterlogged soils in the Newly Reclaimed Areas. These soils have very high salinity levels (up to 18 mS) but, due to their extreme levels of electrical conductivity and the fact that the groundwater table is above the altitude of land, these soils were not taken into account when establishing the ECP–OR relationship.

IMAGE CLASSIFICATION

As maps with information about the existing areas of waterlogged and saline soils in the Newly Reclaimed Areas were unavailable, an inventory of such areas was made from remotely sensed data. The waterlogging and salinity situation immediately after reclamation was also unknown. Therefore to provide an indication of the past situation an older satellite image was interpreted. Classification was performed on two satellite images: a Landsat MSS image acquired on 29 January 1977 to provide an historical baseline and a SPOT XS image acquired on 20 September 1989.

The two images were acquired in different seasons. The 1977 scene is taken in winter, when the groundwater table was high, while the 1989 scene is from the summer when the groundwater table is low. Furthermore, cropping patterns are different between the winter and summer seasons in this area. The seasonal differences in the water table will lead to differences in the area of waterlogging. These had to be taken into account in the final analysis and consequently the increase in waterlogging quoted represents a minimum increase. The seasonal differences in land cover were easily accounted for during land cover classification, and do not affect the areal estimates of waterlogging. In addition, sun angle corrections were performed on these images to enable effective comparison.

Supervised classification was based on training site data collected from known sites of the following cover types in the Newly Reclaimed Areas: vegetation, two types of bare soils, sands, waterlogged soil, non-irrigated citrus orchards and irrigated citrus orchards (Table 19.2). The difference between the two bare soils was made on the basis of the stage of irrigation. Bare soil 1 was dry at the surface, while bare soil 2 represents recently irrigated soil. The sand class comprised unreclaimed desert soil. The vegetation class consisted mainly of clover and the main harvesting period had finished. The separability of the different classes was examined using commission–omission and divergence matrices derived from the training site data; this resulted in acceptable separabilities.

A maximum likelihood classifier (Figure 19.3) was used and generally the classification of the SPOT XS imagery was acceptable. However, misclassifications occurred: for the SPOT XS imagery these were most frequent between the irrigated citrus orchards and waterlogged soils because in orchards where citrus trees are flood-irrigated there are spectral similarities between the open shallow water of the waterlogged soils class and that under the trees. The confusion between the two classes mainly occurred in the young citrus orchards, where there was a low leaf area index. It is less of a problem in the older plantations because greater proportions of the reflectance derives from the crowns of mature

Table 19.2 Cover types

	1	2	3	4	5	6	7
1. Vegetation	0	100	100	100	100	100	100
2. Bare land (1)		0	100	100	100	98	100
3. Desert sands			0	100	100	100	100
4. Non-irrigated citrus				0	99	100	100
5. Irrigated citrus					0	100	99
6. Bare land (2)						0	100
7. Waterlogged soil							0

Waterlogging and Soil Salinity in the Newly Reclaimed Areas 371

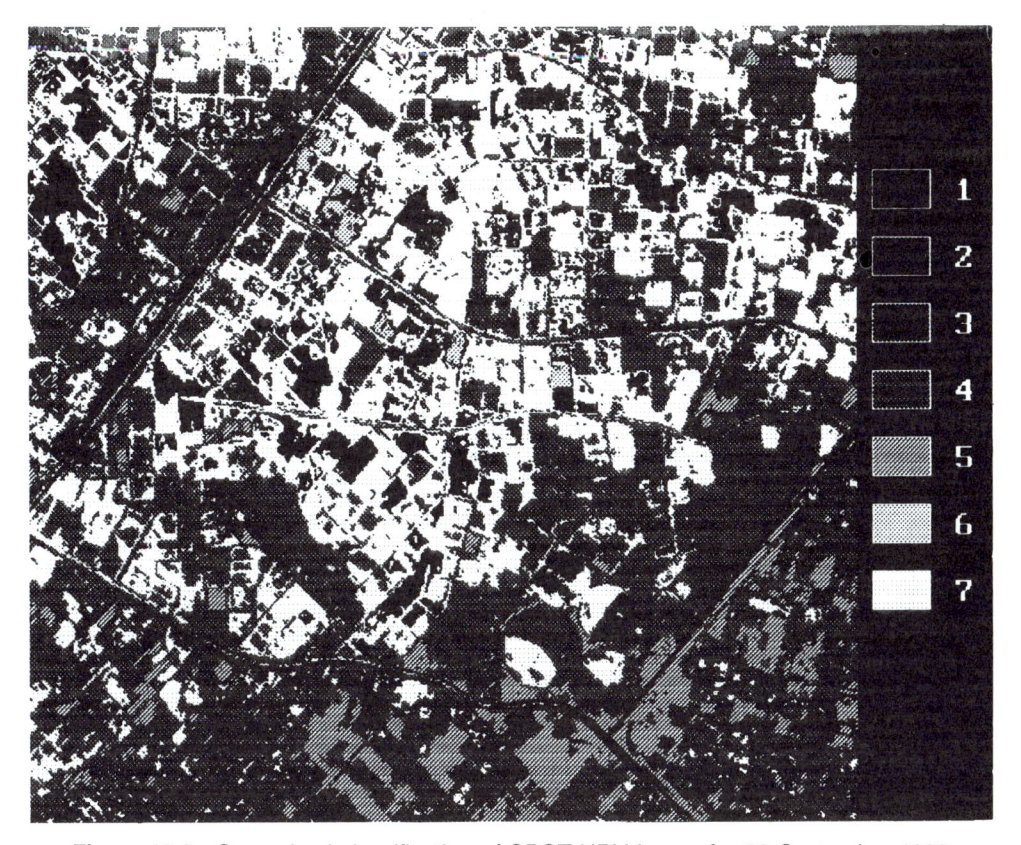

Figure 19.3 Supervised classification of SPOT-HRV image for 20 September 1989

trees. This problem could be solved by the use of multitemporal data sets for each year to examine the seasonal differences in reflectance or spectral unmixing to distinguish the proportions of bare soil water and vegetation.

The Landsat MSS image was also classified by supervised classification (Figure 19.4) with, as far as possible, the same land cover classes used as for the SPOT XS image. However, an irrigated field crop class had to be added, acknowledging the fact that this image was acquired in the winter. The statistical separability of the trained samples was again examined by means of commission–omission and divergences matrices and resulted in acceptable separabilities, although they were not as good as those for SPOT XS imagery. The between-class confusion occurs mainly between the waterlogged soil class and the irrigated field crop and irrigated citrus orchard classes. This spectral overlap is clear in the commission–omission matrix, but is less evident in the divergence matrix. The confusion arises because the soils in all three classes are flooded by shallow water at this time of the year. It is evident that these three classes have more-or-less the same spectral characteristics,

Figure 19.4 Supervised classification of Landsat MSS image for 19 January 1977

and this creates the low separability between classes. The reason that this statistical overlap was not so strongly developed in the SPOT XS data is most probably due to the fact that the citrus trees had smaller crowns in 1977 than in 1989.

MULTITEMPORAL ANALYSIS

The areas classified as waterlogged were extracted from both scenes to enable a comparison in waterlogged areas between 1977 and 1989 (Table 19.3). The two sub-scenes were geo-referenced and warped to each other. The image warping caused some problems since the area in 1977 was mainly desert, while in 1989 an agricultural landscape with fields, roads, and irrigation and drainage channels had evolved. Consequently, a limited number of ground control points — nine — were used, and these points were all located in the north of the study area along the Noubaria (mainly bridges over the canal) and in the south along the

Table 19.3 Comparison between 1977 and 1989

	Number of pixels	ha
Situation in 1977		
1. Waterlogged soil	313	139
2. Vegetation	2 846	1262
3. Bare land	8 597	4258
4. Desert sands	12 114	5375
5. Non-irrigated citrus	74	33
6. Irrigated citrus	1	0.5
Reject	19	
Situation in 1989		
1. Waterlogged soil	1 422	631
2. Vegetation	3 225	1430
3. Bare land	17 824	7909
4. Desert sands	1 669	741
5. Non-irrigated citrus	164	72
6. Irrigated citrus	100	44
Reject	560	

desert road from Cairo to Alexandria (mainly road intersections). The limited number of ground control points was also due to the fact that 80 m Landsat MSS and 20 m SPOT XS spatial resolution data had to be warped together. The ground control points were located on false colour composite images, and the same ground control point file was used to warp the MSS classified image to the SPOT image with a r.m.s. error of 120 m.

To evaluate the expansion of waterlogging and salinity the classified images for 1977 and 1989 were compared. The main results are that:

(i) between 1977 and 1989 most of the desert land had been reclaimed;
(ii) citrus orchards showed the largest expansion of all land cover types between 1977 and 1989; and
(iii) the area of the waterlogged soils had increased approximately 4.5 times (i.e. by almost 500 ha) between 1977 and 1989.

The expansion of the area of waterlogged soils paralleled that of the reclamation of the desert soils, which can be attributed to the fact that reclamation is mainly achieved by flood irrigation.

At the present time an unequivocal explanation cannot be given concerning the creation of the vast areas of waterlogged soils. Nevertheless, it is clear that a number of parameters contribute to create waterlogging. Firstly, all the waterlogged soils are situated in slight depressions. These depressions function as collection sites for surface water and excess drainage water. Secondly, field observations indicate that there is often a layer of heavy, compacted clay at a depth of approximately 3 m in such depressions. This clay horizon is impermeable and results in a perched water table. Thirdly, the desert soils contain high amounts of $CaCO_3$ and during reclamation, when the soils are irrigated, the $CaCO_3$ is leached down the profile. The depth at which it re-precipitates depends on the amount and intensity of the irrigation. The more water that is used, and the more intense the irrigation, the greater the leaching. In waterlogged soils $CaCO_3$ can be found throughout the profile,

374 Environmental Change in Drylands

from the topsoil to the compact subsoil clay horizon, and it impedes drainage (Mohamed et al., 1973; Abd El Raham & Abdel Salam, 1979). Finally, the lack of maintenance of drainage channels is common in waterlogged areas. This leads to the growth of reeds which block the drains and prevent the flow of the drainage water. Stagnating water in the drains seeps into the surrounding fields and the groundwater table rises. All four factors contribute to waterlogging; nonetheless, it is clear that this area has a high potential for waterlogging which is realized during reclamation and exacerbated by mismanagement.

ESTIMATION OF THE FUTURE EXPANSION OF WATERLOGGED SOILS

It has been shown that the area of waterlogged soils expanded by about 500 ha between 1977 and 1989. This equates to an average growth rate of 50 ha yr^{-1} or 6.4%. This rate has been used to make approximate predictions concerning waterlogging in the next decade. The first prediction is based on a linear growth model, which was applied because only two dates had been studied. In addition, it was assumed that the waterlogged soils grow in a radial manner, and that no special drainage works are executed over the period of the prediction. The waterlogged areas mapped from the SPOT XS image (1989) were the basis for the extrapolation (Figure 19.5).

This prediction assumed a radial growth model, but ideally the microtopography should have been taken into account because waterlogging is mainly restricted to depressions. However, this could not be done for two reasons:

1. The only digital elevation model available had a height interval of 5 m, which was too coarse to account for topographic differences required for the spatial modelling of the expansion of waterlogging.

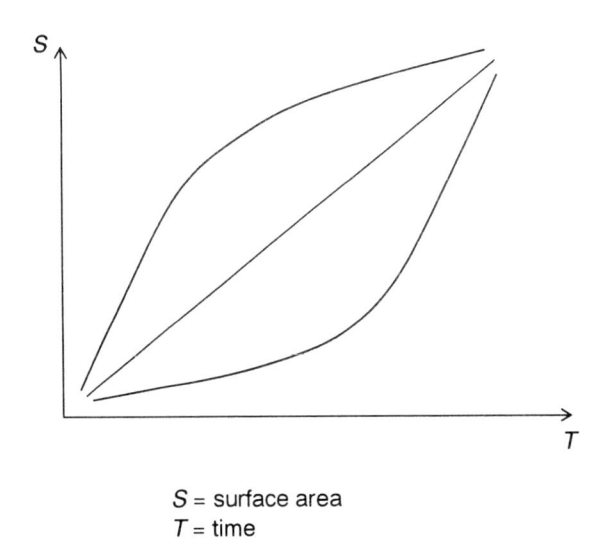

S = surface area
T = time

Figure 19.5 Potential models of the expansion of waterlogged areas

2. The only available topographic map that was digitized dated from 1947 and represented the topography before the reclamation.

The only solution to this problem would be to create a new DEM with a finer height interval based on aerial photographs taken after reclamation. If this could be achieved a far better model and more accurate prediction would result.

It is clear from the image maps showing the growth of waterlogged soils (Figure 19.5) that not only will land be lost due to the expansion of waterlogging but agricultural land will also be lost as it will become completely surrounded by waterlogged soils and access to it restricted. In 1989, 6.5% of the study area was waterlogged. After applying the linear growth model this will increase to 11.1% (1225 ha) in 2000. This indicates that the areas suffering from waterlogging will double in ten years, assuming the linear growth model. However, the linear growth model can be questioned, and in theory three growth models are possible (Figure 19.6):

(i) a logarithmic model,
(ii) a linear model and
(iii) an exponential model.

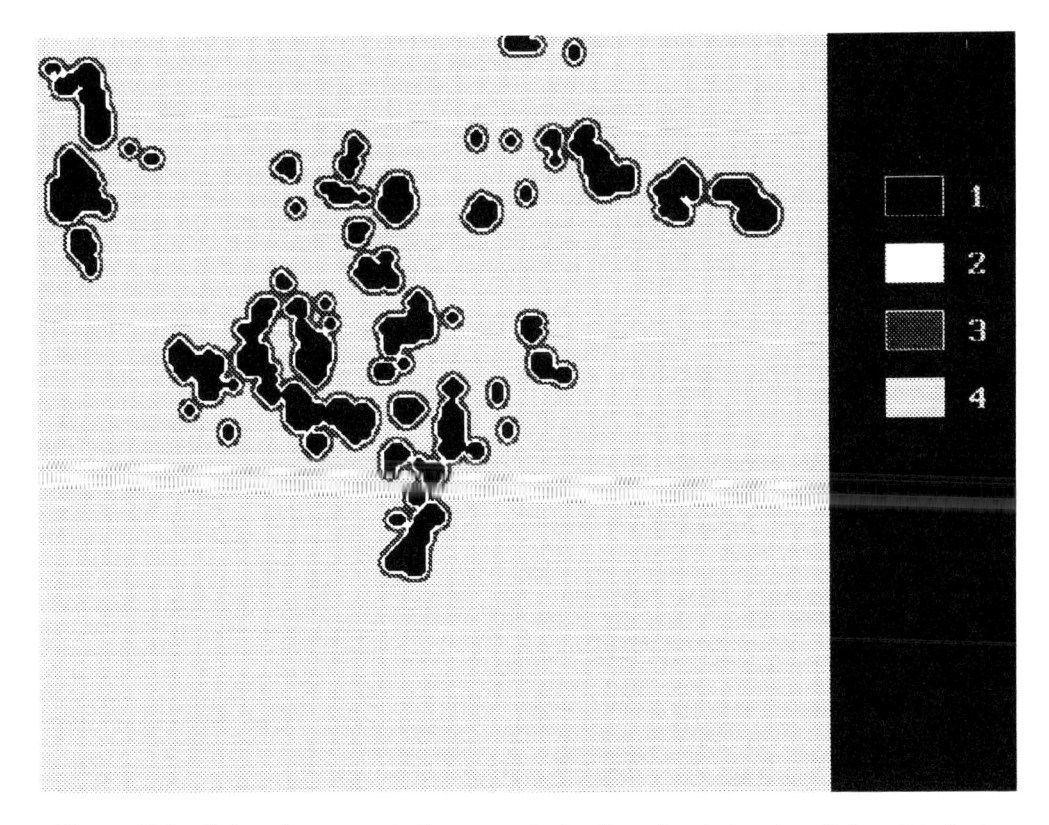

Figure 19.6 Colour 1 represents the present situation of waterlogging. Colour 2 indicates those areas that are immediately in risk of waterlogging. This stage can be reached by 1995. Colour 3 indicates those areas that are at risk of waterlogging by the year 2000. The areas in colour 4 are not immediately endangered

With more data points the validity of the different models could be examined, but that was not possible in this study.

CONCLUSIONS

The waterlogged soils in the western part of the Newly Reclaimed Areas in the Baheira Province in Egypt are created by a clay horizon below the aeolian sands impeding drainage. Their areal extent and spatial pattern is controlled by the microtopography after reclamation. Since the start of the reclamation the excess water applied to the soils has not drained away and has accumulated above the clay horizon until, in many places, the groundwater table has reached the surface.

The use of satellite imagery has proved to be an excellent tool for monitoring of salinisation and waterlogging, particularly since recent topographical maps are not available. By the application of multitemporal image analysis it is possible to estimate the loss of land due to waterlogging and by implementing a spatial growth model it is possible to predict the expansion of waterlogging. Further research is needed to elucidate the growth rates of waterlogging by (i) studying a series of land cover maps and (ii) the use of microtopographic information.

ACKNOWLEDGEMENTS

This research was carried out with the help, advice and organisation provided by Prof. Dr Abdel Hady and Prof. Dr Hussein Younes (Director and Vice-Director), Remote Sensing Center, Academy for Scientific Research and Technology, Cairo; Mr Laurent (Director) and Ms Decadt (Coordinator), Ministry for Scientific Policy, Brussels; and Prof. Dr L. Daels (Director), Laboratory for Regional Geography and Landscape Science, University of Gent. Their efforts are gratefully acknowledged. The authors also wish to thank their colleagues Dr Mohamed Dadawi, Dr Beata De Vlegher and Dr Pascal Brackman for their advice and encouragement.

REFERENCES

Abd El Raham, S. M. H. and Abdel Salam, M. A. (1979). Studies of soils of the north western Nile Delta fringes with special attention to the formation of Calcic Horizon in highly calcareous soils. *Desert Institute Bulletin ARE,* **29**, 287–309.

Abd El Raham, S. M. H., Ismail, H. A. and Elkadi, H. A. (1980). Parent material uniformity in Mariut North Tahrir Area. *Desert Institute Bulletin ARE,* **30**, 181–90.

Abdel Salam, M. A., El Kady, H. A. and Damaty, M. A. (1972). Soil classification of the Amriya–Maryut area, western Mediterranean coast of Egypt. *Desert Institute Bulletin ARE,* **22**, 281–93.

Abdel Salam, M. A., Abdalla, M. M., Zein El Abedine, A. and El Demerdashe, S. (1973). Comparative studies of soils in the Delta region and border fringes. *Desert Institute Bulletin ARE,* **23**, 59–89.

Ayers, R. S. and Westcot, D. W. (1976). Water quality for agriculture. Irrigation and Drainage Paper 29, FAO, Rome.

Dwivedi, R. S. (1992). Monitoring and the study of the effects of image scale on delineation of salt-affected soils in the Indo-Gangetic plains. *International Journal of Remote Sensing,* **13**, 1527–36.

El-Gabaly, M. M. (1959). Improvement of soils, irrigation and drainage in Egypt. In *Research on Crop Water Use, Salt-Affected Soils and Drainage in the Arab Republic of Egypt*, FAO, Cairo.

Framji, K. K. (ed.) (1974). *Irrigation and Salinity: A World Wide Survey*, International Commission on Irrigation and Drainage, New Delhi.

Halvorson, A. D. and Rhoades, J. D. (1974). Assesing soil salinity and identifying potential saline seep areas with field soil resistance measurements. *Soil Science Society of America Proceedings,* **38**, 567–81.

Kovda, V. (1975). Evaluation of soil salinity and waterlogging. In *FAO Soils Bulletin*, No. 31, FAO, Rome.

Loveday, J. (1984). Amendments for reclaiming sodic soils. In Shainberg, I. and Shalhevet, J. (eds.), *Soil Salinity under Irrigation—Processes and Management*, Springer Verlag, Berlin, pp. 220–37.

Mohamed, S. A., El Demerdashe, S. and Abdel Salam, M. A. (1973). Study of some physical and chemical properties of Maryut soils, AR, Egypt. *Desert Institute Bulletin ARE*, **23**, 83–98.

Rafiq, M. (1975) Use of satellite imagery for salinity appraisal in the Indus Plain. In *Prognosis of Salinity and Alkalinity, FAO Soils Bulletin*, No. 31, FAO, Rome, pp. 141–6.

Rao, B. R. M., Dwivedi, R. S., Venkataratnam, L., Ravishankar, T. and Thammappa, S. S. (1991). Mapping the magnitude of sodicity in part the Indo-Gangetic plains Uttar Pradesh, northern India, using Landsat-TM data. *International Journal of Remote Sensing*, **12**, 419–25.

Rhoades, J. D. (1975). Measuring, mapping and monitoring field salinity and water table depths with soil resistance measurements. In *Prognosis of Salinity and Alkalinity, FAO Soils Bulletin*, No. 31, FAO, Rome, pp. 159–86.

Sehgal, J. L., Saxena, R. K. and Verma, K. S. (1988). Soil resource inventory of India using image interpretation technique. *Remote Sensing is a Tool for Soil Scientists, Proceedings of the 5th Symposium of the Working Group on Remote Sensing*, ISSS, Budapest, Hungary, pp. 17–31.

Sharma, R. C. and Bhargava, G. P. (1988). Landsat imagery for mapping saline soils and wet lands in north-west India. *International Journal of Remote Sensing*, **9**, 39–44.

United States Salinity Laboratory Staff (1954). *Diagnosis and Improvement of Saline and Alkali Soils*, USA, Washington, D. C.

20 The Implications of the Altered Water Regime for the Ecology and Sustainable Development of Wadi Allaqi, Egypt

G. DICKINSON
Department of Geography and Topographic Science, University of Glasgow, UK

K. MURPHY
Department of Botany, University of Glasgow, UK

and

I. SPRINGUEL
Department of Botany, Faculty of Science at Aswan, Assiut University, Egypt

ABSTRACT

The object of this study was to examine the environmental changes that have taken place in the Wadi Allaqi area, on the eastern side of Lake Nasser in southern Egypt, as a result of the impoundment of Lake Nasser by the Aswan High Dam. Sub-surface hydrology has been modified fundamentally, and shoreline areas are repeatedly flooded as a result of fluctuating lake levels which are related to seasonal and annual variations in the regime of the lower Nile. The increase in available surface and sub-surface water in the Allaqi area has resulted in colonisation by new vegetation, in particular dense stands of *Tamarix nilotica*. Periodic flooding has also resulted in deposition of lacustrine sediments which has led to modified soil conditions which have aided some of these vegetation changes.

If suitable plants for grazing can be encouraged the possibilities for livestock rearing are considerable, but as the area now has conservation status a general management strategy to cover both sustainable agriculture and resource conservation must be developed.

INTRODUCTION

Wadi Allaqi is a major tributary of the Nile valley, situated 180 km south of Aswan in southern Egypt, on the east side of Lake Nasser (Figure 20.1). The lake, which was created by the construction of the Aswan High Dam, is one of the largest reservoirs in the world (White, 1988). By 1978–9 the lake had attained its maximum level since construction of 179 m a.s.l., just below its planned absolute maximum of 180 m. The lake water level (Figure 20.2) has subsequently fluctuated between 150 and 179 m a.s.l. (Pulford *et al.*, 1992; Ali *et al.*, 1993). Lake Nasser extends for more than 500 km, submerging the Nile Valley

---

Environmental Change in Drylands: Biogeographical and Geomorphological Perspectives.
Edited by A. C. Millington and K. Pye. © 1994 John Wiley & Sons Ltd

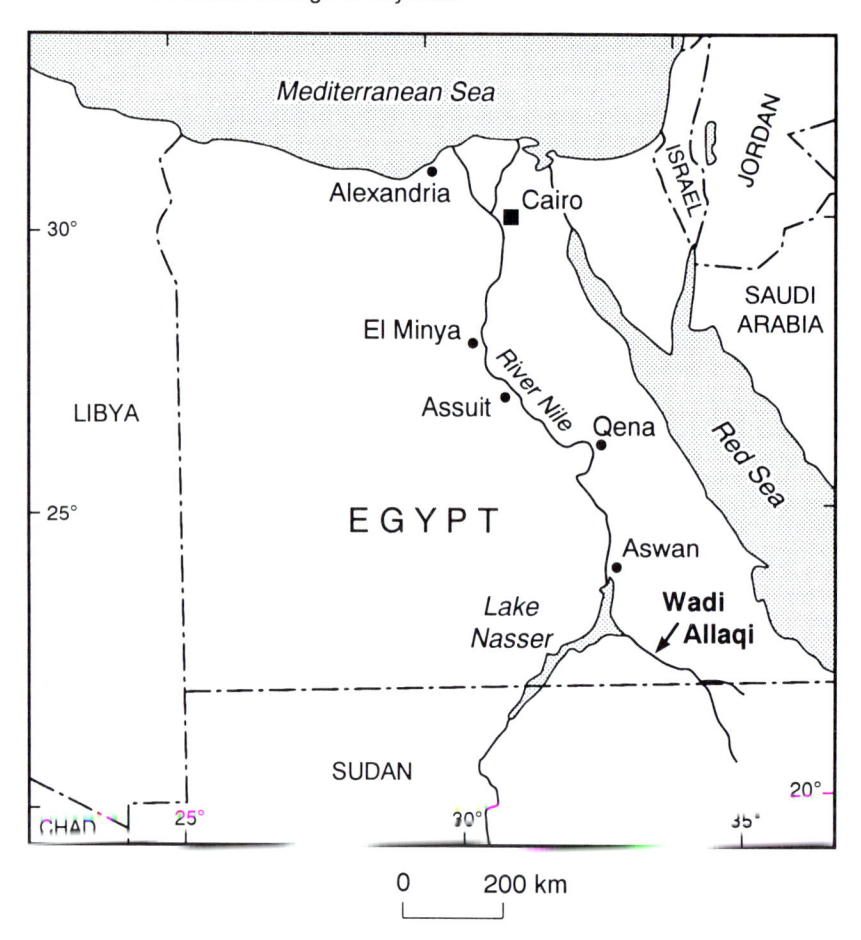

Figure 20.1 Location of the Wadi Allaqi study area in Egypt

between latitude 23°58′N and 20°27′N. That part of it that extends southwards into Sudan is known as Lake Nubia.

Lake Nasser stores water from the annual flood of the Nile with the object of increasing year-round supplies for downstream irrigation, and for hydroelectricity generation at the High Dam. The episodic and variable regime of the River Nile at Aswan is a product of the early-summer monsoonal rains which feed the upper catchment of the Blue Nile in the Ethiopian Highlands (Figure 20.2). The Blue Nile accounts for about 71% of the annual flow at Aswan (Briggs & Dickinson, 1988). As its upper catchment is located towards the northernmost range of the East African monsoon area a high degree of variability is normal. Over the past 100 years annual flow values have ranged from 151 billion cumecs in 1978 to 45 billion cumecs in 1913. When approaching its maximum planned level Lake Nasser holds about 162 billion cumecs. The actual amount of water that flows through the dam is controlled by the High Dam Lake Authority according to irrigation and power generation needs, but cannot exceed 50.5 billion cumecs, a value fixed by international treaty with Sudan and Ethiopia (the Nile Water Agreements of 1929 and 1959). This is 90% of Egypt's current water demand, which is about 55.5 billion cumecs (White, 1988). Pressure on all

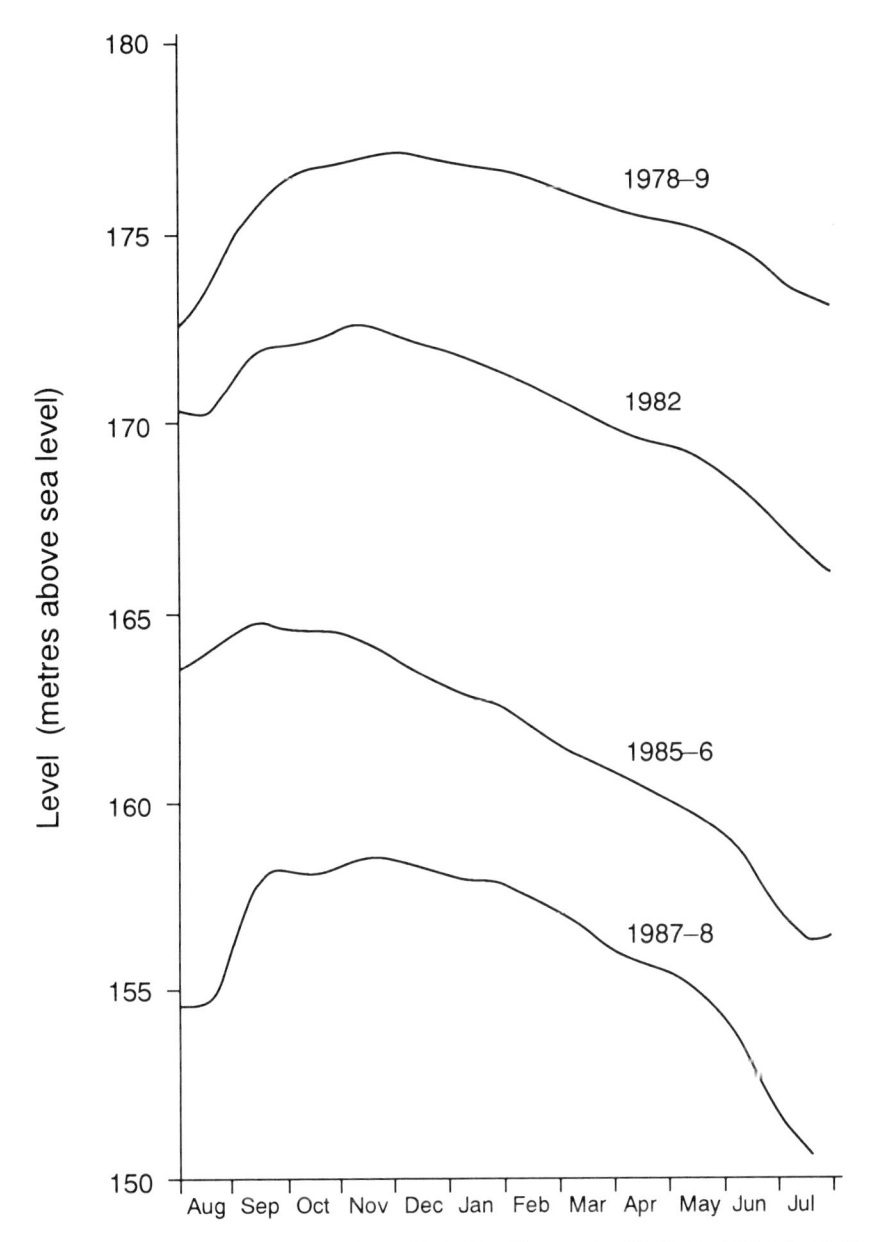

Figure 20.2 Fluctuations in water level in Lake Nasser in 1978–9, 1982–3, 1985–6 and 1987–8 (data from Aswan High Dam Lake Authority)

water resources in Egypt is acute, and new resource development coupled to better use of existing resources is a top national priority.

The formation of Lake Nasser has had substantial effects on the shoreline marginal zone lands, as a result of the periodic fluctuation in water levels and due to seepage from the lake into shallow groundwater immediately adjacent to the lake. Although the total shoreline is extensive (5300–7800 km, depending on water level; White, 1988) the greatest effects have been observed in the near-lake sections of shallow-sloping wadis entering the lake, such as Wadi Allaqi.

Upstream of the High Dam there is increasing pressure to develop these shorelines. Maintenance of water quality in the lake is of crucial importance, and any development of lake-margin lands has to be considered in the light of potential impacts on lake water quality, e.g. resulting from nutrient runoff in drainage water from cultivated land. Development which follows sustainable principles has, perhaps, the best chance of minimising potential impacts on the lake water quality.

The key to sound development lies in utilising the fluctuations in the level of Lake Nasser. These have given rise to a major freshwater penetration of the surrounding desert through tributary wadis entering the lake. In the case of Wadi Allaqi, although the lowest 50 km of the wadi are now permanently inundated, above this there is a zone some 40 km in length and 1–2 km in width which has experienced periodic submergence. Although the shifting shoreline presents some problems for development, there are also positive aspects. The problems relate to the need to move cultivated areas and infrastructure. This is less of a problem than it may appear at first sight. The maximum level of Lake Nasser cannot exceed 180 m a.s.l. If the lake should receive water when at this level it is carried out into the Western Desert by a concrete sluice and channel on the western bank of the lake. Excess water will then be lost by evaporation and percolation from the outfall of the overflow. It is worth noting, however, that since the dam was completed water deficit rather than surplus has been the main issue. Strict water rationing had to be introduced when the lake fell to its low point in the years 1987–8. The fact that the upper level of the lake is fixed means that any location within Wadi Allaqi above 180 m a.s.l. will not be subjected to inundation. Fixed infrastructure such as roads and buildings can thus be sited at the margins of the wadi, which are above this critical level.

The opportunities that are offered by utilising the fluctuating level of the Lake are considerable. Flooding will give an immediately accessible source of water which can be easily exploited using very basic technology. Shallow wells and simple pumps will allow irrigation of plots that have been shown to be capable of growing good yields of locally usable crops. Grazing potential is good if desirable species are promoted (Springuel, 1991). Careful management of these plant species will be necessary as their distribution is confined to the wetter locations and they are very susceptible to overgrazing. However, when used in conjunction with the traditional grazing resources of the area, the new potential offered by the environmental change in the lower Wadi is considerable.

The aim of this paper is to outline the main ecological impacts of the formation of Lake Nasser on an adjacent desert wadi ecosystem and to discuss the implications for sustainable development in the lower part of the Wadi Allaqi system.

METHODS

Vegetation and soil data were collected over a five-year period (1987–91) along a 17 km section of Wadi Allaqi closest to the lake shore. Vegetation data (species presence, cover, density, phenology) were collected from 100 m-spaced 5×5 m quadrats, located along cross-wadi transects 500 m apart. Soils were sampled for loss on ignition, pH, calcium and magnesium contents, texture and conductivity. The soil sampling regime was based on the same set of transects as used for plant sampling, except that samples were collected from every second quadrat along each cross-wadi transect, from depths of 0–10 and 20–30 cm. Data on sub-surface hydrological conditions are limited to observations in existing wells and to inferential observations from vegetation distribution.

PHYSICAL AND BIOLOGICAL RESOURCES OF WADI ALLAQI

Water

Prior to the creation of Lake Nasser, water availability in the Wadi Allaqi region was both temporally unpredictable and very scarce: the typical state for a hyper-arid environment (Safriel et al., 1989). Water was available from only two sources, which are still present. The first source is deep underground percolating water moving slowly down the Wadi from the Red Sea Hills in the east, over which there is an intermittent, low rainfall. Such flows are sparse, are at the base of the sediment in the Wadi (typically 30 m or more below the surface) and are of indifferent quality. In September 1988 data from a well near the junction of Wadi Allaqi and Wadi Haimur gave a water depth of 42 m, a salinity of 4400 ppm and a conductivity of 11 000 μS cm^{-1} at 28°C.

Rare and unpredictable (Safriel et al., 1989) rains constitute the second source. These take the form of intense short-lived storms resulting in localised flash floods. One such event occurred in lower Wadi Allaqi in August 1988, two years after the previous rains. Between 1988 and 1992 no further rainfall occurred in this area. The 1988 event caused flooding to a water depth of ca. 1 m in Wadi Quleib, a tributary of Wadi Allaqi (Figure 20.4), as indicated by silt deposits on the rock walls of the wadi. The approximate boundary between that part of the Wadi Allaqi system where rain and deep groundwater remain the only source of water and the section newly influenced by periodic lake water inundation is shown in Figure 20.4.

Control of the flood surge of the Nile means that the volume of water in Lake Nasser now varies over two timescales. Figure 20.2 shows these patterns between 1970 and 1992. The long-term annual flow rate of the Nile at Aswan is 84 billion cumecs (White, 1988). Thus the high and low values recorded this century, respectively in 1978 and 1913, represent 180 and 54% of the long-term average. On to this long-term scale of variation is superimposed the more regular annual pattern of lake level fluctuation, which is a function of the balance between the seasonal pattern of water input to the reservoir and the release of water through the dam, with a typical annual amplitude in the range 6–7 m (Figure 20.2). The implications of this dynamic system of inundation of the wadi are profound. Figure 20.3 shows the areas of the wadi remaining above water at different lake levels. The probability of submergence, on the basis of water-level data for 1970–92 for different elevations within the wadi, has been calculated (Table 20.1).

Variations in lake level result in shoreline movements of tens of kilometres because of the very gentle gradient of the wadi. Areas close to the current shoreline (in late 1991, approximately 170 m a.s.l.) have been intermittently flooded since the lake filled. At higher elevations, further up the wadi, the probability of inundation is lower and environmental effects of this are in consequence less marked. The area that has been affected by lake floodwater on at least one occasion, but has not been permanently under water during the period 1970–92 occupies a stretch of Wadi Allaqi approximately 40 km in length.

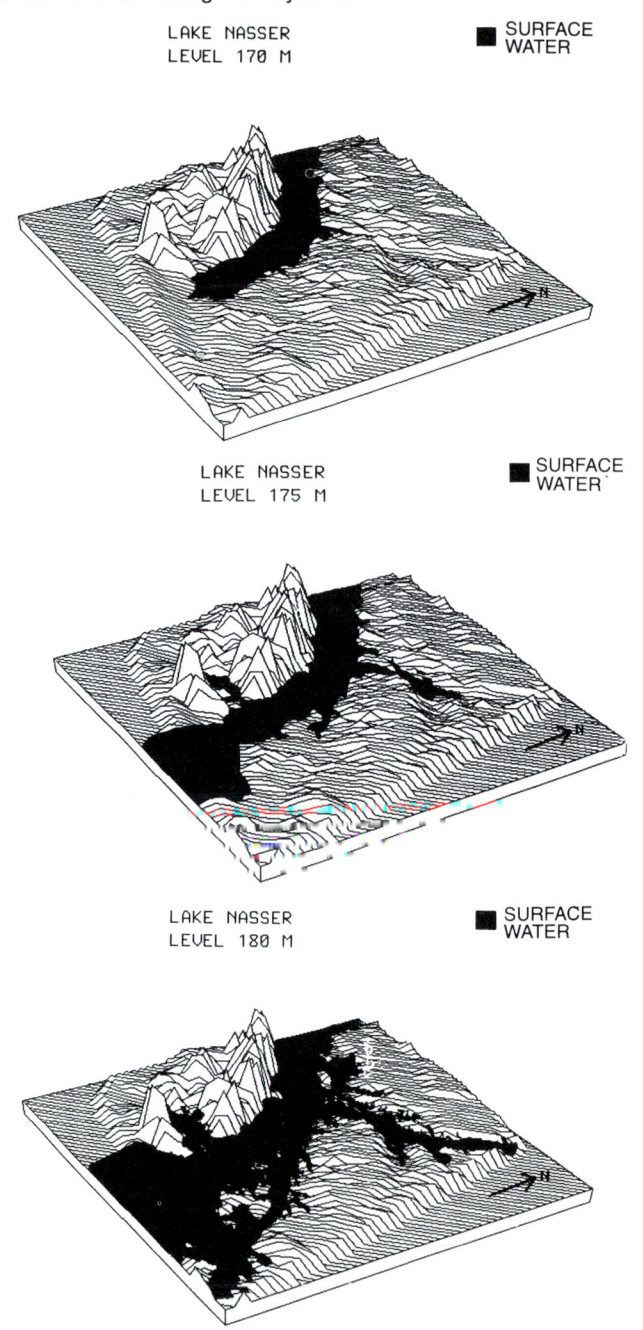

Figure 20.3 Digital terrain models of Wadi Allaqi land surface adjacent to the lakeshore (around the mouth of Wadi Um Ashira) at selected lake levels: (a) 170 m a.s.l., (b) 175 m a.s.l. and (c) 180 m a.s.l. (with acknowledgement to S.A. Selim and GRID, Geneva, Switzerland). Note: stippled areas in north-west and south-east corners of diagram are not digitised

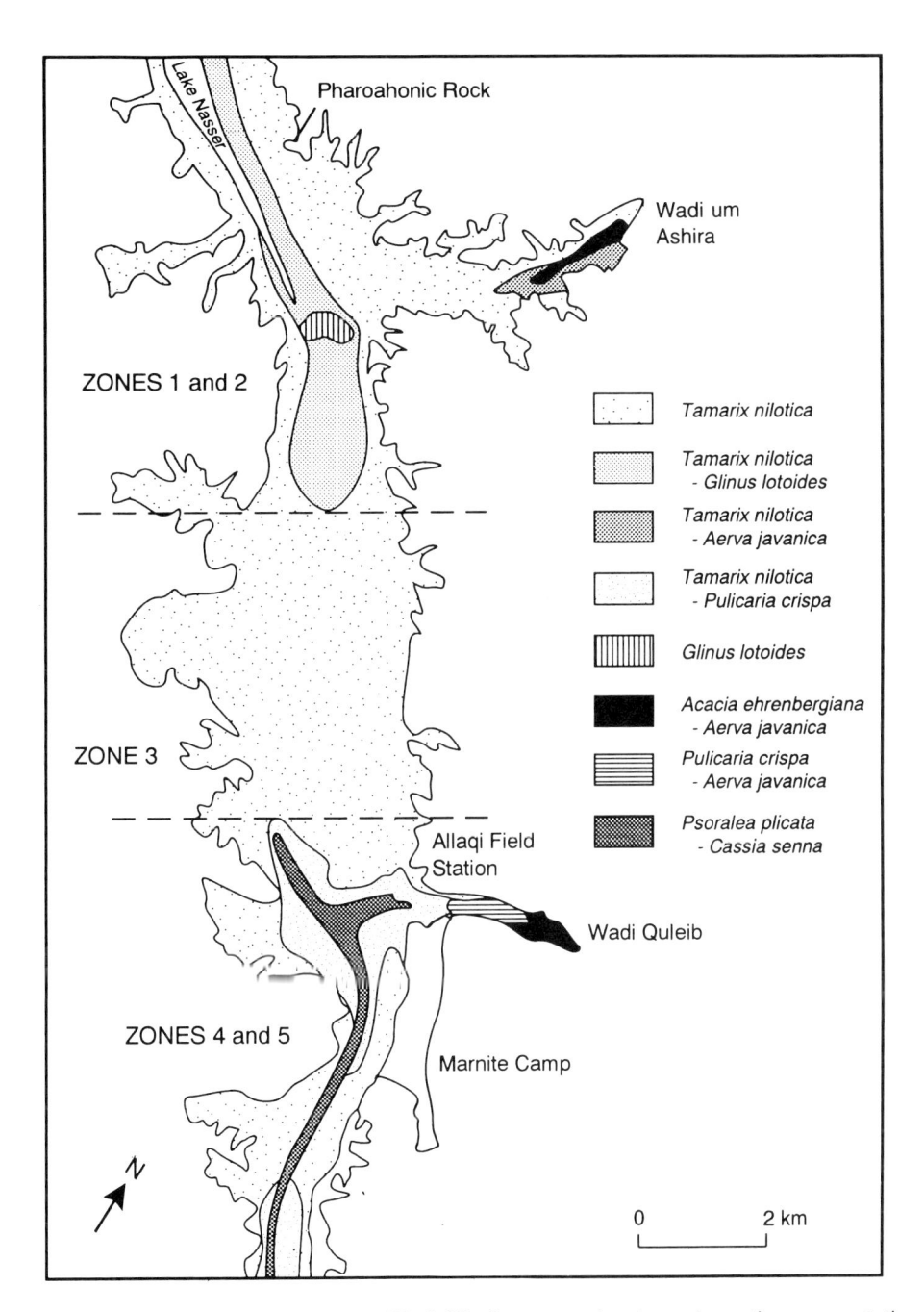

Figure 20.4 Outline map of the lower Wadi Allaqi system, showing schematic representation of water movement in the system (base map from Briggs, 1991)

Table 20.1 Probability of submergence in Wadi Allaqi

Contour level (m a.s.l.)	Probability of inundation (P_i)
150	1.00
155	0.95
160	0.86
165	0.67
170	0.21
175	0.12
180	0.00

Although few data are available, it is possible to suggest what the effects of the changing lake levels on the sub-surface hydrology of Wadi Allaqi would be. One possible consequence may be reverse sub-surface flow of water derived from the expanded water table which has accompanied the formation of Lake Nasser (Dickinson, 1989; Mekki & Dickinson, 1991). Any such flow of relatively low-salinity (i.e. less dense) water moving up the wadi would be expected to overlie the westward-flowing groundwater derived from sources higher in the wadi. Observations in a series of twelve shallow wells located up to 0.5 km from the shoreline of the lake in September 1990 indicated a significant positive correlation ($r = 0.695$; $p < 0.02$) between depth of water level below the soil surface (range: 1.70–5.90 m) and conductivity (range: 364–2500 μS cm^{-1}). There was no significant correlation between water pH (range: 7.7–8.4) and depth below the soil surface in the same set of samples. A single measurement taken from a well close to the lake edge in 1988, when the water level in the well was 2.5 m below ground level, gave a pH of 7.5 and a conductivity of 750 μS cm^{-1} (Mekki & Dickinson, 1991).

It is postulated that groundwater derived from the margins of Lake Nasser moves in two directions. Firstly, there may be a slow advance up-wadi through the unconsolidated sediments of the wadi floor. Virtually nothing is as yet known of the dynamics of any such water movement. Applications of general theories about water movements to this situation suggest that lateral water movement is unlikely to be much faster than a few tens of metres per annum (Hillel, 1982, pp. 107–32), with the physical properties of the substrate material potentially influencing the rate of movement. Secondly, there are vertical movements of the groundwater which is left behind as floodwaters retreat. These may allow lenses of lake water to descend quite rapidly to join the permanent water table at depth in the base of the wadi substrate. The long axis of Wadi Allaqi follows a fault-zone boundary between geosynclinal metasediments and volcanics, and Nubia sandstone (Mekki & Dickinson, 1991). The sandstone is an important aquifer and the fractured upper portions of the crystalline rocks may hold groundwater. Therefore the ultimate fate of this descending water may be to overlie the existing groundwater deep below the floor of the Wadi. Over an extended period of time it will be greatly augmented by the arrival of slow-moving lateral flows from the main body of groundwater expanding beneath Lake Nasser.

Soils

The flat floor of the lower 40 km of Wadi Allaqi to the junction with Wadi Haimur (Figure 20.4) comprises unconsolidated wadi sediments chiefly laid down during the early Holocene when the local climate was rather wetter. It has a very gentle average gradient of <1°, and is from 1 to 3 km in width between steep rocky margins.

Soil conditions may vary over short distances within the wadi. In general there is evidence to suggest that soil conditions are suitable to support a continuous yield of grazing and crop plants, given an appropriate resource-conserving management regime. In the surface layer, salt content, pH and conductivity tend to be higher than sub-surface conditions, probably because of the effects of floodwater evaporation from the soil surface, following periods of inundation. For example, surface pH values are generally in the range 7.5–8.0. A range of textures occurs in the sub-surface stratum of the Wadi. Coarse sands predominate, together with layers of finer silts (Table 20.2).

Vegetation

Prior to the filling of Lake Nasser, the only source of water supplementing the deep groundwater available to support plant growth in Wadi Allaqi was from very rare rainstorms and flash-floods. These sparse and intermittent supplies were sufficient to sustained a low cover of deep-rooted, perennial, xerophytic vegetation (e.g. *Acacia ehrenbergiana*, *Aerva javanica*; Springuel, *et al.*, 1989; Kassas & Girgis, 1965). This permanent cover was periodically supplemented by an ephemeral flora which developed immediately after rain events, characterised by species such as *Trianthema crystallina* and *Schouwia thebaica* (Springuel, 1991).

The plant communities of the lower Wadi Allaqi during the period 1987–91 were shown by Murphy *et al.* (1993) to comprise two primary vegetation types. Above the area affected by lake water inundation, the typical community remains a true desert-wadi vegetation characterised by *Acacia ehrenbergiana* and *Psoralea plicata*. This appears to have changed little during the period since completion of the lake (Kassas & Girgis, 1965).

Within the section of Wadi Allaqi closest to the lake there has occurred a striking example of vegetational change, in terms of both species composition and biomass. Within the area subject to periodic inundation the vegetation is currently dominated by *Tamarix nilotica*, more commonly found elsewhere in the Nile valley as a river-bank shrub. Here, *Tamarix*

Table 20.2 Summary of textural properties of topsoils and subsoils from Wadi Allaqi

Parameter	Minimum	Maximum	Median	Mean	σ
Topsoil samples					
Coarse sand	5.7	98.0	56.0	55.0	22.4
Fine sand	0.2	80.0	31.0	33.0	18.6
Silt	0.05	33.0	4.3	6.4	6.9
Clay	0.20	29.0	5.2	6.2	4.6
Subsoil samples					
Coarse sand	5.2	91.0	64.0	58.0	23.4
Fine sand	3.3	86.0	27.0	31.0	20.6
Silt	0.06	30.0	2.8	4.8	5.3
Clay	0.01	20.0	5.4	5.8	3.8

nilotica has greatly expanded its cover, to fill much of the wadi floor between about 156 and 177 m a.s.l. The *Tamarix* community may be subdivided into an area closest to the lake shore, where annuals such as *Glinus lotoides* and *Eragrostis aegyptiaca* are also present. Upstream of this is a zone where *Tamarix nilotica* forms near-monoculture stands, with luxuriant growth, and individual shrubs often >2 m in height. In the uppermost section of the *Tamarix* zone, species such as *Pulicaria crispa* become co-dominant, gradually excluding the *Tamarix* as the upper boundary of the inundation-influenced zone is reached. Overall, this vegetation change represents the ousting of the former hyper-arid desert wadi communities, by species more typical of the Nile Valley. Total plant biomass is now high in the lower Wadi, and indicates the potential for agricultural development there.

The current pattern of plant communities within the study area may provide useful clues to the availability and sources of water for supporting plant growth. *Acacia ehrenbergiana* and *A. raddiana* trees and shrubs are deeply rooted species, known for their ability to exploit deep sources of underground water (Vanderbelt, 1991; Loughenour *et al.*, 1990). From the occurrence of these plants in the upper part of the study area, the presence of such water supplies may be inferred. Closer to the lake the existence of vigorous stands of *Tamarix nilotica* suggests that, at least periodically, these plants have access to shallower supplies of underground water, since this species has wide-spreading but shallower roots than the *Acacia* spp. of the upper wadi. The rapid decline in the shallow-rooted annuals, such as *Glinus lotoides*, with distance from the shore indicates their low tolerance both to inundation (if the lake level rises) and to drought-stress as the lake level recedes. In effect these plants are confined to a shifting belt of land alongside the lakeshore, where there is an abundant supply of shallow soil water.

DEVELOPMENT POTENTIAL: SUSTAINABLE UTILIZATION OF NATURAL RESOURCES

The changes in environmental and biological conditions that have occurred in the Wadi Allaqi area as a result of the creation of Lake Nasser present an opportunity for the introduction of a programme of sustainable development based on the utilization of natural resources. Such opportunities are replicated at a number of other locations around the lake, and, potentially, in similar situations around man-made lakes in dryland regions elsewhere such as Lake Volta (Obeng, 1973). Already local Abbadi bedouins have recognised this new potential. Small crop plots of 0.1 ha or less are used for cultivation. Crops such as vegetables, lucerne and sorghum are grown using water applied by hand from shallow wells close to the prevailing shore of the lake. Plots are fenced by thorn hedges or old fish nets to protect them from grazing animals.

Grazing has been the traditional livelihood of local people. Camels from the region surrounding Allaqi are recognised as being of high quality in the Egyptian market. Furthermore, Allaqi is an important staging point in the *dabuka* (camel train route) between Atbara in north-east Sudan and the market of Daraw just north of Aswan. Now that the grazing potential of Allaqi has increased, the significance of the area to the regional grazing economy is considerably enhanced. However, the present new vegetation cover with dominance of Tamarisk does not provide good forage. *Tamarix nilotica* concentrates excess salts taken in by the extensive lateral root network, the salt being exuded to form crystals on the leaf surface. The concentration of salt in the leaves makes Tamarisk a poor forage species, only eaten by browsing animals such as camels and goats if other, more palatable, species are not available. The apparently low grazing pressure on Tamarisk may be a contributory factor to its current dominance in the wadi. Further modification of the vegetation of lower Allaqi, involving appropriate management to increase and sustain better pasturage species, is required to achieve the potential offered by the new environmental conditions.

Some form of agricultural development of the Allaqi area appears inevitable, since the pressure on resources in Egypt is such as to ensure that any new land open to utilisation is quickly, though not always optimally, exploited. Experimental irrigation schemes are under way, utilising water pumped from the lake to grow a range of horticultural and agricultural crops. Exploitation of high-quality marble and granite in upper Allaqi since the late 1970s has led to the construction of a metalled road almost to the lower wadi itself, thus greatly improving its accessibility. Though still a very sparsely populated area some new family groups of people have settled in the area. The area has a heritage appeal to those Nubian inhabitants of the area who were displaced by the Aswan High Dam scheme and resettled in Kom Ombo, north of Aswan.

Sustainable exploitation of the existing vegetation offers possibilities for development of the area. Springuel (1991) has identified seven categories of use for the natural vegetation of the area: medicinal use, timber, forage, food, fuel, commercial charcoal making and other uses such as production of oil and fibre. To these may be added a number of other potential activities for the region. Cropping is the most significant, but others such as fishing in Lake Nasser offshore of Allaqi and mineral exploitation will have implications for the renewable resources of the wadi.

If development is to follow the principles of sustainability, output must be at a level that can be maintained indefinitely by the natural resource base. Intensive agriculture dependent on sophisticated irrigation schemes, subsidised by artificial fertilisers and protected by pesticides, is unlikely to be compatible with this approach. Any possibility of contamination of lake water must be avoided to prevent worsening the pollution problems already encountered in the lower course of the Nile. It is important to protect the natural soil nutrient cycling systems, which are the main paths maintaining chemical fertility levels. This may be achieved through appropriate modification of soil microbiology as soil conditions change in response to environmental change.

The regular flooding of the lower part of the wadi has two important consequences for the maintenance of soil fertility which is crucial to sustainable agricultural use of the area. Firstly, the floods deposit silt on the soil surface, thereby enriching the surface layer with a new source of plant nutrients. In this way the floods of Lake Nasser are similar in effect, though not in scale, to the former floods in the lower Nile Valley, which have now been controlled by the Aswan High Dam. Secondly, the flushing of the soil surface layers during flood events will discourage salinisation. Salinisation is likely to be a problem if large-scale irrigated arable agriculture is practised in this area of extreme evapotranspiration. Any large-scale irrigation scheme would have to be accompanied by management systems to control salinisation, which are expensive and difficult to maintain properly.

Conservation of natural resources plays an important part in the proposed development of the wadi. The whole area is a protected area under the control of the Aswan Governorate branch of the Egyptian Environmental Affairs Agency. It has provisional status as a UN Man and Biosphere (MAB) programme reserve. Wadi Quleib is designated as the core protected area, the entire length of Wadi Allaqi as the buffer zone and the transition zone is a 2500 km^2 block around the buffer zone. A management plan for the protected area core zone has been proposed (Dickinson, 1991), which can provide guidelines for ecologically sound development for the whole region.

The challenge in Wadi Allaqi is to introduce viable sustainable development which is both productive and protective of the fragile resource base. The work reported here has gone some way towards establishing the necessary inventory of vegetation, soil and water resources, which is a prerequisite to establishing the carrying capacity and capability of the ecosystem for sustainable agricultural development.

ACKNOWLEDGEMENTS

The authors thank UNEP, The British Council and the Gilchrist Trust (Royal Geographical Society) for financial support of the work reported here. The work of S.A. Selim and GRID, Geneva, Switzerland in producing the terrain models shown in Figure 20.3 are also acknowledged. Thanks are also due to colleagues in the University of Glasgow and Faculty of Science at Aswan, of Assiut University, for their participation in the project.

REFERENCES

Ali, M. M., Hamad, A., Springuel, I. V. and Murphy, K. J. (1993). Environmental factors affecting submerged macrophyte communities in regulated water bodies in Egypt. Submitted to *Journal of Aquatic Plant Management.*

Briggs, J. (ed.) (1991). Provisional Atlas of Allaqi. Allaqi Project Working Paper Series 14, Department of Geography and Topographic Science, University of Glasgow.

Briggs, J. and Dickinson, G. (1988). Casting seed upon the wadi. *Geographical Magazine*, March 1988, pp. 30–1.

Dickinson, G. (1989). Water resources and eco-development. Allaqi Project Working Papers Series 1, Department of Geography and Topographic Science, University of Glasgow.

Dickinson, G. (1991). Conservation management strategies for the Wadi Allaqi Protected Area. Allaqi Project Working Paper Series 15, Department of Geography and Topographic Science, University of Glasgow.

Hillel, D. (1982). *Introduction to Soil Physics*, Academic Press, London.

Kassas, M. and Girgis, W. A. (1965). Habitat and plant communities in the Egyptian desert. VI. The units of desert ecosystem. *Journal of Ecology,* **53**, 715–28.

Loughenour, M. B., Ellis, J. E. and Popp, R. G. (1990). Morphometric relationships and development patterns of *Acacia tortilis* and *Acacia reficiens* in Southern Turkana, Kenya. *Bulletin of the Torrey Botanical Club,* **117**, 8–17.

Mekki, A. M. and Dickinson, G. (1991). Geology, mineral resources and water conditions in the Wadi Allaqi Area. Allaqi Project Working Paper Series 9, Department of Geography and Topographic Science, University of Glasgow.

Murphy, K. J., Springuel, I. and Sheded, M. (1993). Classification of desert vegetation using a multivariate approach: a case study of the Wadi Allaqi system (South-Eastern Desert of Egypt). Submitted to *Vegetatio.*

Obeng, L. E. (1973). Volta Lake: physical and biological aspects. In Ackerman, W. C., White, G. F. and Worthington, E. B. (eds), *Man Made Lakes: Their Problems and Environmental Effects*, Geophysical Monograph 17, American Geophysical Union.

Pulford, I. D., Murphy, K. J., Dickinson, G., Briggs, J. A. and Springuel, I. (1992). Ecological resources for conservation and development in Wadi Allaqi, Egypt. *Botanical Journal of the Linnaean Society,* **108**, 131–41.

Safriel, U. N., Ayal, Y., Kotler, B. P., Lubin, Y., Olsvig-Whittaker, L. and Pinshow, B. (1989). What's special about desert ecology? Introduction. *Journal of Arid Environments,* **17**, 125–30.

Springuel, I. (1991). Plant Ecology of Wadi Allaqi and Lake Nasser 4. Basis for the economic utilisation and conservation of vegetation in Wadi Allaqi conservation area, Egypt. Allaqi Project Working Paper Series 13, Department of Geography and Topographic Science, University of Glasgow.

Springuel, I., Ali, M. and Murphy, K. J. (1989). Plant ecology of Wadi Allaqi and Lake Nasser 2. Preliminary vegetation survey of the downstream part of Wadi Allaqi. Allaqi Project Working Paper Series 5, Department of Geography and Topographic Science, University of Glasgow.

Vanderbelt, R. J. (1991). Rooting systems of western and southern African *Faidherbia albida* (Del.) A. Chev. (syn. *Acacia aldiba* (Del.)) — a comparative analysis with biogeographic implications. *Agroforestry Systems*, **14**, 233–44

White, G. F. (1988). The environmental effects of the Aswan High Dam at Aswan. *Environment*, **30**, 4–11 and 34–40.

21 Soil-Forming Processes on Reclaimed Desertified Land in North-Central China

D. J. MITCHELL and M. A. FULLEN
School of Applied Sciences, University of Wolverhampton, UK

ABSTRACT

Desertification is a severe environmental problem in north-central China, where it is estimated that deserts are expanding at a rate of 1560 km^2 yr^{-1}, mainly due to human activities. The Institute of Desert Research of Academia Sinica (IDRAS) is co-ordinating efforts to combat desertification. Investigations into land reclamation techniques at two of the nine IDRAS field stations, namely Shapotou and Yanchi in Ningxia Autonomous Region, are discussed. The reclamation techniques considered are (i) irrigation with silt-laden Yellow River water; (ii) dune stabilisation using straw checkerboards and planted xerophytes; and (iii) redistribution of material from palaeosols. Detailed field investigations and laboratory analyses were made of soils and desert crusts that had been reclaimed at various times since 1956. Distinct and quite rapid improvements in the physical, chemical and biological properties of reclaimed desertified land were evident. However, reclamation involves considerable expenditure of time and resources.

INTRODUCTION

Deserts and desertified land cover 1.49 million km^2 of China, some 15.5% of the total land area (Zhu, 1989). During the Pleistocene, strong winds crossing the unvegetated northern areas winnowed out fine-grained sediment, leaving *gobi* or gravel deserts. Further downwind sands were deposited, forming *shamo* or sandy deserts. An estimated 41.8% of China's desert area is *gobi* and 58.2% *shamo*. In a very generalized way, the landscape of arid northern China can be described as a progressive north-west to south-east transition from *gobi*, through *shamo* to loess (Fullen & Mitchell, 1991).

In all, there are twelve deserts or sandy lands in northern China (Figure 21.1). Many of these are spreading at an alarming rate and 176 000 km^2 have been classified as 'desertified' over the past 25 years (Table 21.1) (Zhu et al., 1988; Zhu & Liu, 1988). These desertified lands have developed under common physical circumstances, namely frequent droughts, persistent winds and the exposure of sandy surfaces. Many interacting factors cause desert development and spread, with physical factors leading to the initial formation of deserts and land mismanagement being largely responsible for the extension of desertified land (Zhu et al., 1988) (Table 21.2).

Environmental Change in Drylands: Biogeographical and Geomorphological Perspectives.
Edited by A. C. Millington and K. Pye. © 1994 John Wiley & Sons Ltd

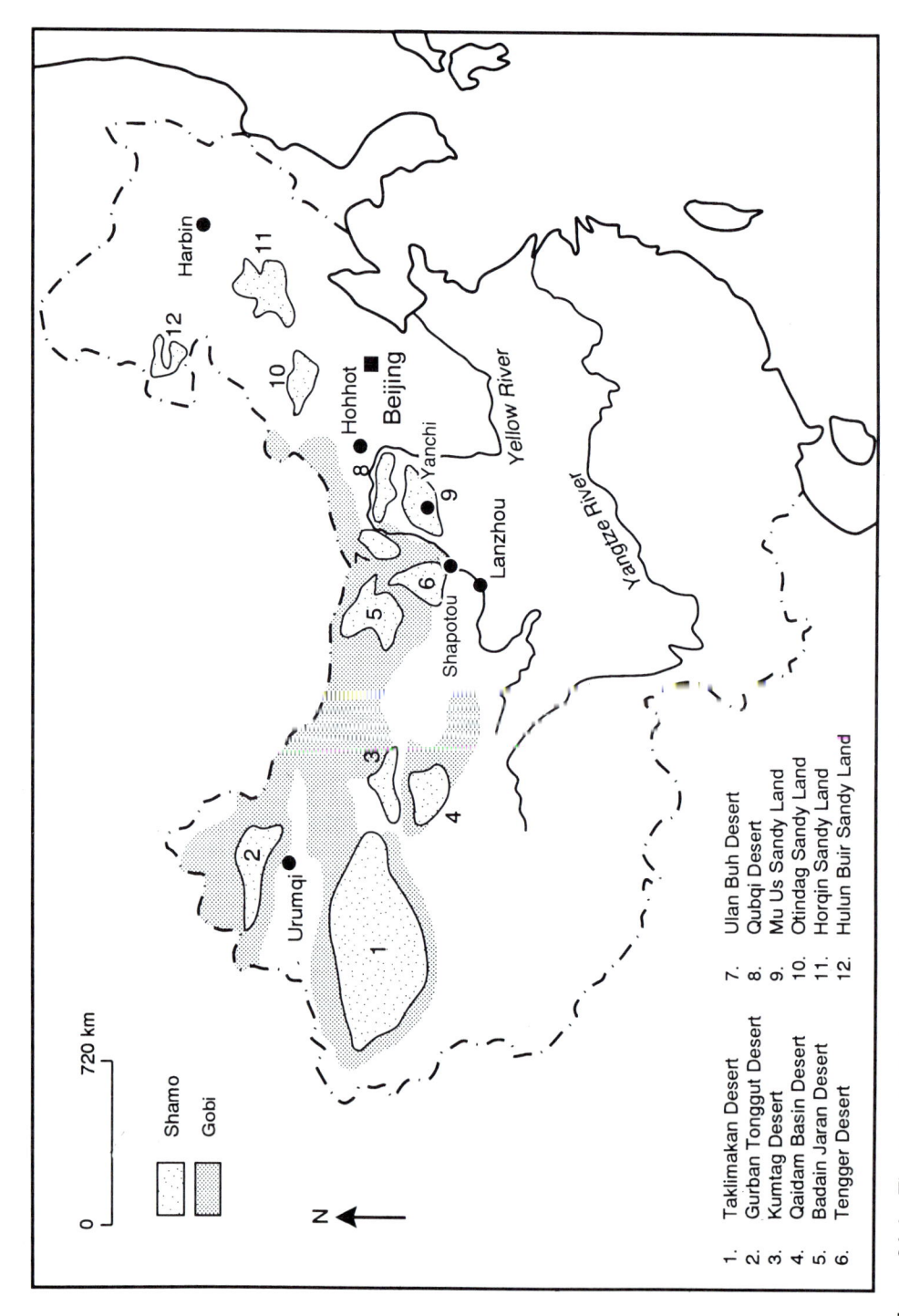

Figure 21.1 The deserts and desertified land of China. *Shamo* are sandy deserts and *gobi* are rock and gravel deserts. A distinction is drawn between desert and sandy lands; the latter are sub-humid area that are considered amenable to reclamation

Table 21.1 The nature and occurrence of desertified land. (Reproduced from Zhu et al., 1988, by permission of Institute of Desert Research)

Nature of desertification	Area (km²)	Proportion of all desertified lands (%)
Spread of desertification on sandy steppe	74 400	42.2
Reactivation of vegetated dunes (sandy lands)	92 200	52.3
Encroachment of mobile dunes	9 400	5.5

Table 21.2 Different desertified land types and their areas in northern China. (Reproduced from Zhu et al., 1988, by permission of Institute of Desert Research)

Cause of desertification	Area (km²)	Proportion of all desertified land (%)
Overcultivation on steppes	44 700	25.4
Overgrazing on steppes	49 900	28.3
Fuelwood collection	56 000	31.8
Technogenic factors	1 300	0.7
Misuse of water resources	14 700	8.3
Dune encroachment	9 400	5.5

Field observations by the authors on the Mu Us Sandy Lands accord with many studies which show desertification to be a 'blistering' process. Desert-like conditions develop in localized areas of rangeland, away from the desert margin. These 'blisters' enlarge, spread and merge, increasing the area of desertified land. In this region the fragile desert margin ecosystem was further disturbed by deforestation in historical times and the subsequent continual removal of shrubs and trees.

Efforts to combat desertification are coordinated by the Institute of Desert Research of Academia Sinica (IDRAS), which has its headquarters in Lanzhou and operates nine field stations in desertified areas. It researches into various techniques of desert reclamation and is thus able to recommend appropriate strategies for reclamation programmes. Some of the initial pioneering work was carried out at Shapotou Research Station and Yanchi Experimental Station for Co-operative Desertification Control Research, which are both in Ningxia Autonomous Region (Figure 21.1). At both stations, desertified land has been reclaimed using various techniques over different periods, thus providing an opportunity to investigate soil development over short time periods. The Shapotou Station was established in 1957 to find ways of protecting 40 km of the Lanzhou–Baotou railway line from sand encroachment (Figure 21.2). Yanchi Station is located in the Shabianzi region, part of the transitional area between the Ordos and the Loess Plateaus. In the past, this sub-humid (mean annual precipitation approximately 300 mm) area was covered by fertile steppe grasslands. Today, largely due to land mismanagement (Zhao, 1988), the area is characterised by a series of sandy ridges separated by vegetated depressions.

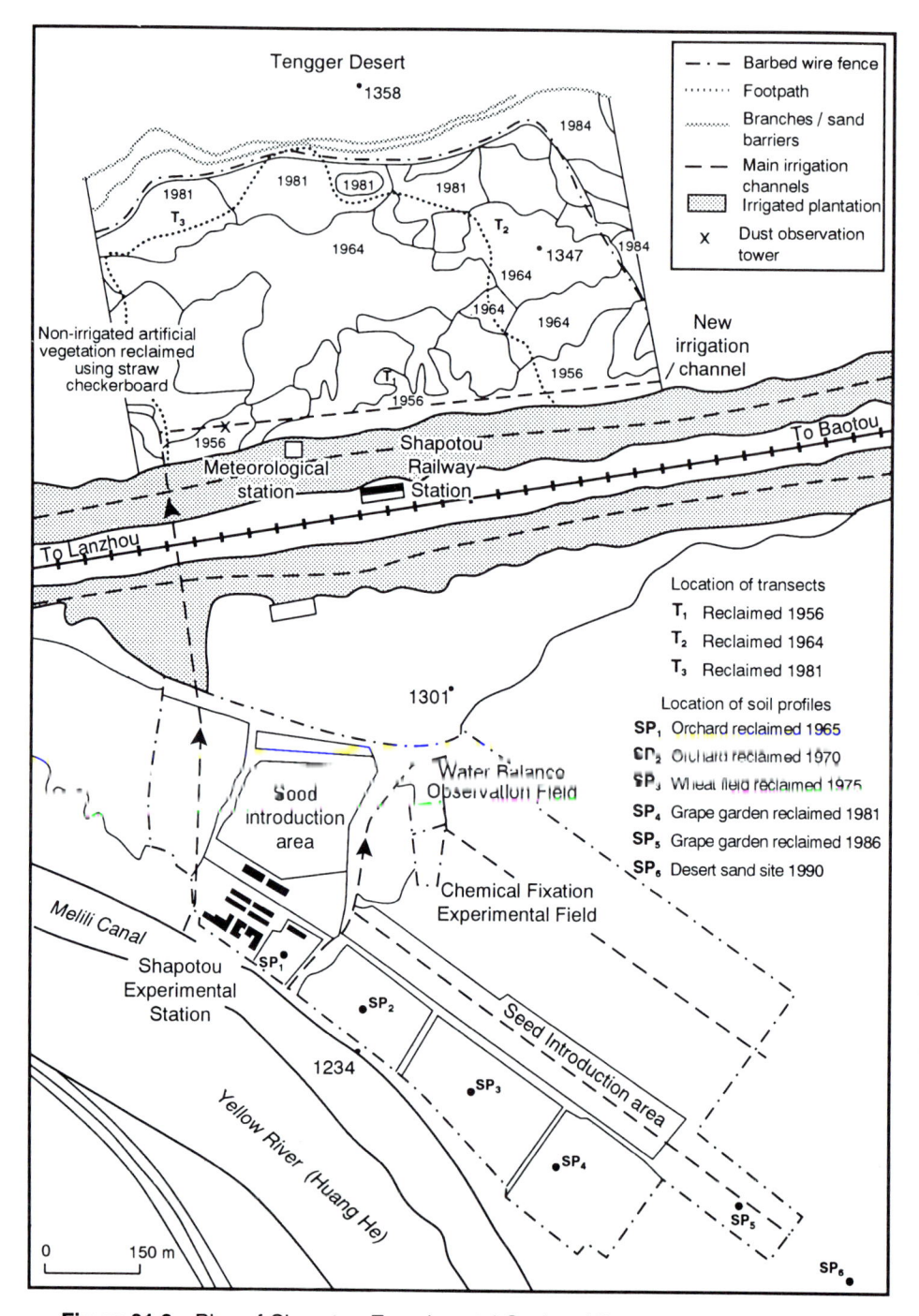

Figure 21.2 Plan of Shapotou Experimental Station, Ningxia Autonomous Region

At Shapotou, Yanchi and other IDRAS field stations, the following stabilisation and reclamation techniques are used:

1. *Windbreaks.* Aeolian deflation and deposition in north-central China have led to the extensive planting of trees as windbreaks. For example, at Shapotou pines, poplars and willows have been planted parallel to the railway line. Since 1985, biological controls and improved land management practices have led to a 22.7% increase in the woodland area at Yanchi, thus reducing the extent of severely desertified land by 10% (Zhu et al., 1986).
2. *Irrigation.* Irrigation not only adds water to reclaimed desert soils but, when silt-laden river water is used, it can improve the physical, chemical and biological properties of reclaimed soils.
3. *Straw and clay checkerboards.* To provide an environment for indigenous xerophytic plants to colonise and survive on sands, localised surface stabilisation is essential. Artificial checkerboards of either straw or clay are used to increase surface roughness and reduce wind erosion, thereby encouraging plant colonisation.
4. *Use of indigenous plants.* To hasten the process of colonisation, areas stabilised using straw or clay checkerboards are planted with indigenous xerophytic species.
5. *Land enclosure.* Overgrazing and the use of trees and shrubs for firewood have severely degraded semi-arid and steppe lands. Proper management of desert margins is essential for successful large-scale reclamation. At Yanchi, areas enclosed by fences and employing the recommended stocking density of 1 sheep per 0.87 ha have been stabilised and revegetated within a period of five years.
6. *Extracting palaeosols.* As a consequence of desertification in historical times, fertile steppe soils have often been buried by encroaching sand dunes. In the Shabianzi region, manual recovery of these palaeosols and irrigation with groundwater have resulted in highly successful land reclamation.
7. *Chemical treatment.* At Shapotou, various chemical agents are being tested on experimental plots to assess the feasibility of treatments for the stabilisation of dune sands. However, while results often indicate beneficial effects, material and application costs can make these treatments prohibitively expensive.

Soil development on areas reclaimed using three techniques were investigated at Shapotou and Yanchi Stations in 1990. The techniques investigated were (i) levelling of sand dunes and their subsequent irrigation with silt-laden water; (ii) dune stabilisation with straw checkerboards; and (iii) the excavation of palaeosols and their use with wind-blown sand and sheep manure as a soil.

RECLAMATION IN NINGXIA AUTONOMOUS AREA

Soils Developed on Irrigated Land

The Yellow River (Huang He) has one of the highest recorded sediment concentrations in the world (Long & Xiong, 1981) and is a major source of irrigation water in this region. The river deposits an estimated 1600 million tons of sediment into the Yellow Sea each year. In addition to the use of water in irrigation, other benefits accrue from the use of the silt-laden water, e.g. the addition of fine-grained sediment, nutrients and organic matter to sandy soils and a reduction in the sediment load of the rivers. Turbulent flow maintains these fine particles in suspension while they are transported in irrigation channels. In the period between 1971 and 1984, it was estimated that, on average, 63 million tons of sediment were diverted from the Yellow River each year by irrigation water, amounting to between 2.4 and 3.5% of the total sediment load (Chen, 1986).

Other organic inputs to reclaimed areas include sheep manure, added at rates up to 45 t ha^{-1} yr^{-1}, and crop residues. At Shapotou dunes are levelled (initially this work was done manually but bulldozers are now used) and a series of pumping stations transfers silt-laden water on to the levelled sands. This leads to the rapid development of minero-organic topsoils on the levelled sands. Pedological descriptions were performed on sites that had been reclaimed for 0, 4, 9, 15, 20 and 25 years, using Soil Survey of England and Wales field procedures (Hodgson, 1976). The mean accretion rate was 18 mm yr^{-1}, which broadly agrees with the 15 mm yr^{-1} measured by Chen & Quian (1984). There was a strong relationship between the number of years of irrigation and depth of cultivated topsoil (the Ap horizon) (Figure 21.3). Table 21.3 summarizes the salient pedological characteristics of the Ap and buried parent materials (bCu horizons).

The developing Ap horizons were considerably more agriculturally productive than bCu horizons. Ap horizons had finer textures, higher soil organic and nitrate contents and higher soil water conductivities and root densities, while bulk density and pH values were lower (Table 21.4). These differences between horizons were not evident in soil profiles at Yanchi that had only been irrigated with groundwater.

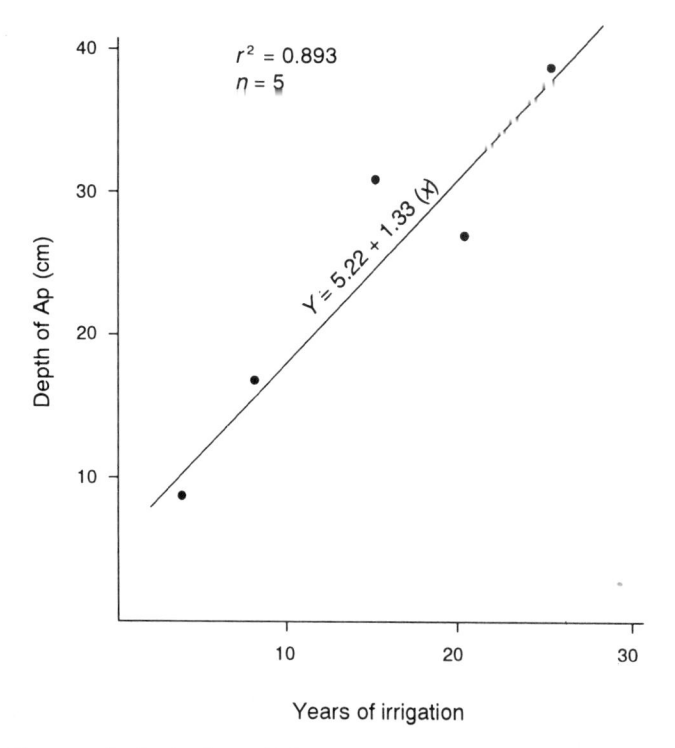

Figure 21.3 Changes in Ap horizon thickness with the number of years of irrigation treatment at Shapotou

Table 21.3 Typical profiles of desert sands reclaimed using Yellow River irrigation water at Shapotou. (Location: 37°27'N, 104°57'E; elevation: 1234 m)

Ap	Brown (air-dry 10YR 5/3); very slightly calcareous sandy loam; stoneless; slightly moist ($\bar{x}$ moisture content 10.1% by weight, $n = 25$ samples); weakly developed medium granular fragments; low packing density; medium to very fine simple tubular macropores associated with plant roots; very weak; deformable to semi-deformable; very weakly cemented to uncemented; slightly sticky to non-sticky; very to moderately plastic; many very fine and fine predominantly fibrous roots; sharp, usually smooth boundary
bCu	Yellowish brown (air-dry 10YR 5/4); very slightly calcareous medium sand; stoneless; slightly moist to dry ($\bar{x}$ moisture content 4.6%, $n = 25$ samples); apedal single grain; low packing density; extremely porous; fine simple tubular to no macropores; loose; semi-deformable; uncemented; non-sticky; non-plastic; many fibrous to no roots

Table 21.4 Mean soil properties of soil horizons at Shapotou

Horizon	Organic matter (%)	Nitrate[b] (mg l^{-1})	Conductivity (μs)	Bulk density (g cm^{-3})	Root density (g 100 cm^{-3})	pH
Irrigated Ap	1.36 (5)[a]	16 (4)	20 (5)	1.30 (25)	0.339 (5)	7.7 (5)
Irrigated bCu	0.23 (5)	L/0[c] (5)	5 (5)	1.40 (25)	0.079 (5)	7.9 (5)
Desert Cu	0.17 (1)	L (1)	5 (1)	1.62 (8)	0 (1)	8.1 (1)

[a]Numbers in brackets denote number of samples.
[b]Nitrate was determined by a Nitracheck meter on soil water extract.
[c]L denotes low ($< \sim 5$ mg l^{-1}).

At Shapotou, irrigated Ap horizons not only thickened with age but also developed, tending to become finer, more deformable, cemented, sticky and plastic. Soil organic content increased, especially in the first few years of reclamation, generally at a rate of between 0.1 and 0.2% yr^{-1} (Figure 21.4). Due to careful management and the addition of organic materials, soil aggregate structure became progressively more pronounced: micromorphological analyses of thin sections show larger and better interconnected macropores within older Ap horizons.

In order to assess the changes in the total concentration of specific elements, samples were oven-dried at Shapotou and later analysed using a Fisons Applied Research Laboratory 8410 X-ray fluorescence spectrometer at the University of Wolverhampton. Bulk soil samples from the Ap horizon of the five soil profiles were compared with desert sand (profile 6), Yellow River silt and sheep manure (Table 21.5). For older soils, all the macronutrients (Ca, Mg, K, P, S) increased, indicating a possible response to the application of Yellow River silt and sheep manure. Results showed that sheep manure had greater concentrations of P and S than the Yellow River silt. Both sources possessed relatively greater concentrations of micronutrients than desert sands, with the Yellow River silt having more Fe and Mn and less Cl, Cu and Zn than sheep manure. Al was found to increase and Si to decrease as the soils developed. In the topsoil, increases were evident in Cr, Zn and Pb. The geochemical composition of bulk samples from the bCu horizons of the five profiles remained fairly consistent.

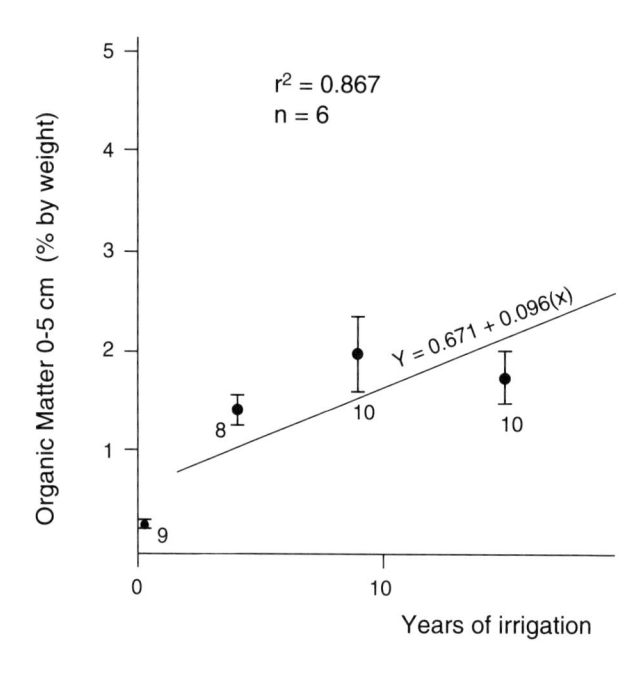

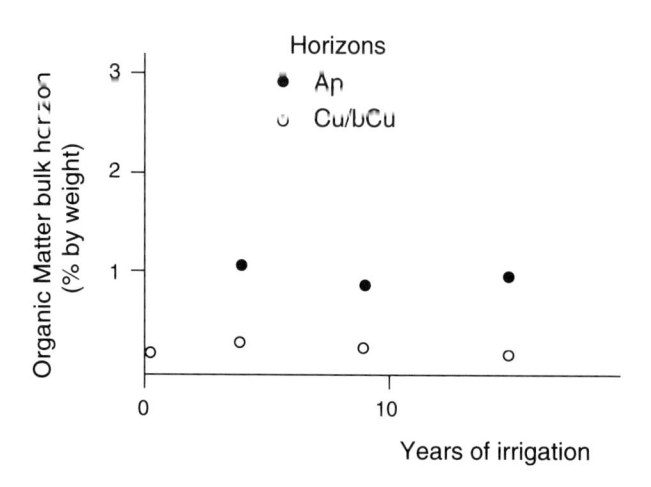

Figure 21.4 Changes in soil organic content with time (±SD), on irrigated sites at Shapotou

Soils Developed On Non-irrigated Land

Straw Checkerboards and Biocrusting

Using results from 30 years of physical and ecological experiments at Shapotou, IDRAS has devised a procedure for establishing an artificial ecosystem on mobile dunes (Figure

Table 21.5 Elemental geochemistry of Ap soil horizons, Yellow River silt and sheep manure from Shapotou. (All concentrations as dry weight of mineral matter SiO$_2$ to S as weight %. Cr to Zn as ppm. Loss-on-ignition of sheep manure (mean of two samples) 36.1%)

Sample source	SiO$_2$	Al$_2$O$_3$	Fe$_2$O$_3$	MnO	CaO	MgO	K$_2$O	Cl	Na	P$_2$O$_5$	S	Cr	Cu	Pb	Zn
Desert sand profile 6 (1990)	79.71	9.31	1.81	0.037	0.42	0.91	1.86	0.009	1.570	0.033	0.005	35	30	122	45
Grape garden profile 5 (1986)	70.72	9.84	3.02	0.074	4.35	1.88	2.26	0.023	1.225	0.095	0.015	50	35	105	65
Grape garden profile 4 (1981)	75.92	9.28	2.46	0.057	2.56	1.54	2.14	0.005	1.481	0.062	0.013	44	33	107	54
Wheat stubble profile 3 (1975)	62.73	9.51	2.88	0.075	5.56	1.90	2.19	0.021	1.214	0.078	0.018	45	32	100	60
Orchard profile 2 (1970)	66.79	9.68	2.92	0.075	5.38	2.07	2.19	0.035	1.285	0.091	0.020	52	33	99	61
Orchard profile 1 (1965)	69.91	11.51	2.89	0.073	5.42	2.05	2.20	0.029	1.302	0.135	0.033	51	54	100	71
Yellow River silt (1990)	57.65	13.61	4.14	0.105	8.36	2.90	2.57	0.080	1.467	0.106	0.041	59	34	94	81
Sheep manure	35.95	4.28	1.94	0.073	9.88	2.19	2.36	0.243	1.281	0.479	0.918	38	65	167	99

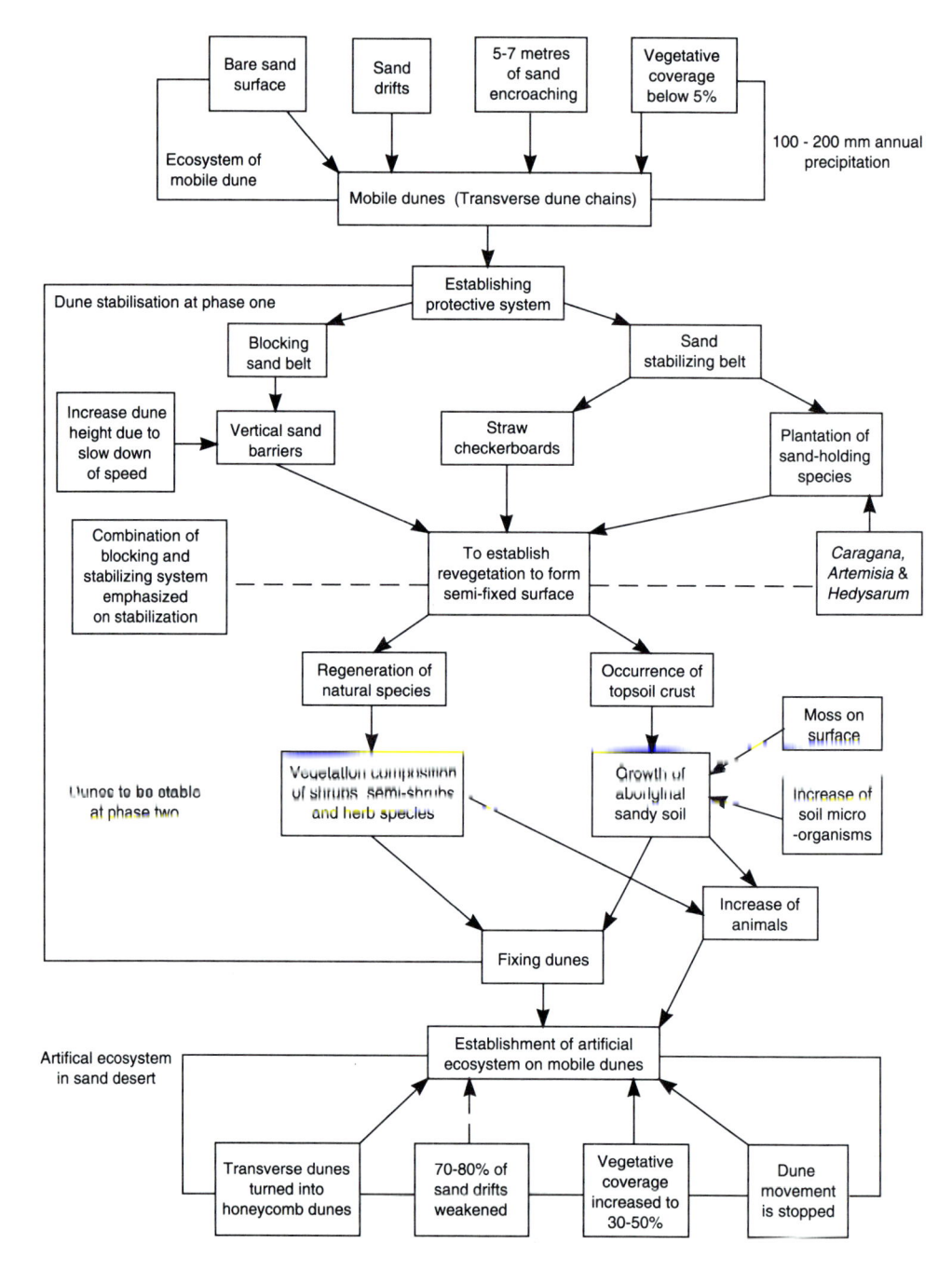

Figure 21.5 The artifical ecosystem on mobile dunes at Shapotou. (Reproduced from Zhu et al., 1988, by permission of Institute of Desert Research)

21.5) (Zhu et al., 1988). The process converts areas with shifting sands and less than 5% vegetative cover to areas of fixed dunes with 30–50% cover. The main aim of these techniques is to stabilise desert margins, thus preventing further desert encroachment and damage to neighbouring arable areas and rangelands.

A crucial initial stage in desert reclamation is the installation of a sand barrier, made from woven willow branches or bamboo, to act as a windbreak (Lin et al., 1984). Wind velocity and erosivity are thus reduced, causing aeolian deposition in the lee of the windbreak. Behind the sand barrier, straw checkerboards are constructed. These are made of wheat or rice straw which is embedded in the sand to a depth of 15–20 cm, in a grid of 1–3 metre squares, with the straw protruding 10–15 cm above the surface (Figure 21.6). The straw is applied at a rate of 4.5–6 t ha^{-1}. The checkerboards increase aerodynamic roughness, thereby reducing surface wind velocity and thus stabilising the sand. Where locally available, clay is used instead of straw to stabilise the surface, because of its greater durability.

In comparison to shifting sands, Liu (1987) showed that one metre squared straw checkerboards at Shapotou increased surface roughness length (Z_0) from 0.0025 to 0.89 cm. Measurements taken during a north-west wind showed that increased roughness resulted in a hundredfold decrease in sand transport (Table 21.6) (Shapotou Desert Research Station, 1986). Results of field experiments at Shapotou, using different sized checkerboards, emphasise the critical relationship between checkerboard size and wind erosion (Table 21.7). The checkerboards remain intact for 4–5 years, allowing time for cultivated xerophytic plants to establish themselves. In the Shapotou nursery there were 128 plant species used in desert reclamation. The main species used are *Artemisia ordosica*, *Caragana korshinskii*, *Ephedra przewalksii*, *Haloxylon ammodendron*, *Hedysarum scoparium* and *Tamarix* spp. In addition, *Agriophyllum squarrosum*, *Artemisia sphaerocephala*, *Hedysarum laeve* and *Salix psammophila* are used at Yanchi. *Agriophyllum squarrosum*, *Bassia dasyphylla* and *Corispermum hyssopifolium* are effective colonisers at desert margins. Within the checkerboard areas, reduced wind velocities, enhanced surface stability, organic matter inputs from biomass, improved moisture retention and aeolian dustfall encourage incipient soil development, which further stabilises the dunes.

The formation of surface crusts involves both physical and biological processes. Once the shifting sand has been stabilised, soil algae form the initial successional stage on a substrate low in plant nutrients. Typical accumulations at Shapotou consisted of a very dark grey (10YR 3/1) surface crust, often colonized by olive-grey bryophytes. These consisted

Table 21.6 Changes in aeolian characteristics due to straw checkerboards. (Reproduced from Shapotou Desert Research Station, 1986, and Liu, 1987, by permission of Institute of Desert Research)

Surface type	Wind speed (m s^{-1}) V_{2m}	Wind speed (m s^{-1}) $V_{0.5m}$	Roughness Z_0 (cm)	Sand movement (NW wind) (g cm^{-1} min^{-1})
Shifting sand	5.1	4.8	0.0025	1.374
Checkerboard (2m spacing)	5.2	4.4	0.088	n/a
Checkerboard (1m spacing)	5.0	3.7	0.89	0.013

Figure 21.6 Straw checkerboards at Shapotou planted with *Artemisia ordosica, Caragana korshinskii* and *Hedysarum scoparium* in 1981

Table 21.7 The relationship between checkerboard size and the depth of wind erosion. (Reproduced from Chapotou Desert Research Station, 1986, by permission of Institute of Desert Research)

Checkerboard size (m × m)	Depth of wind erosion (cm yr^{-1})
1 × 1	0
1 × 1.5	7.0
1 × 2	8.0
1 × 3	7.9
2 × 2	13.5
2 × 3	14.4
3 × 3	25.3

of *Barbula ditrichoides* Brth., typically 10–20 mm diameter, and *Bryum argenteum* Hedw., about 10 mm in diameter (Wang Shu Xiang, personal communication). A 1–3 mm thick crust overlies a brown to dark brown sandy area (10YR 4/3), typically 40 mm deep, mainly consisting of medium to fine sands. Due to aeolian inputs, the crusts are somewhat finer than dune sands and also contain more organic matter, Al, Mn, Ca, Mg, Cl, S and P (Tables 21.8, 21.9 and 21.10). Underneath lie yellowish-brown desert sands (10YR 5/4).

Table 21.8 Organic contents (% by weight) of Chinese desert samples (determined by loss-on-ignition at 375°C, 16 h)

Surface type	Maximum	Minimum	Mean	Standard deviation	Number of samples
Desert crust	3.59	0.85	1.85	1.08	5
Grey desert sand	2.73	0.34	1.02	0.95	5
Desert, beneath grey desert sand	0.70	0.14	0.34	0.13	44
Desert	0.30	0.13	0.20	0.05	25

Table 21.9 Depth of 'grey sand' (mm) on reclaimed desert margin sites at Shapotou

Number of years since reclamation	Maximum	Minimum	Mean	Standard deviation	Number of samples
34	77	0	40	13	71
26	180	0	39	29	63
9	110	0	43	25	63

In contrast to crusts formed by rain splash, algal crusts improve infiltration and decrease runoff (Booth, 1941; Bond & Harris, 1964; Shields & Durrell, 1964). These conditions increase surface humus, provide habitats for seeds to germinate, create solvent action on soil minerals, maintain a plant nutrient store and contribute to increased soil nitrogen levels. In Arizona, Fletcher & Martin (1948) found that flexible algal filaments and mould mycelia were important in increasing the tensile strength of biocrusts, thus binding soil particles together.

In South Australia, Bond (1964) found that sands with organic coatings tended to be hydrophobic, creating a mosaic of moist and dry patches. Similar conditions are found at Shapotou, but the moisture mosaic is dominated by the checkerboards. After rainfall the lines of the decomposed straw checkerboard remain moist, while the inter-spaces dry out rapidly. The Tengger Desert is near the northern limit of the summer monsoon, therefore rainfall occurs during July and August, but both the amount and duration are unreliable. During occasional, high-intensity summer storms, some sorting and rain splash crusting will occur, but there was limited evidence of surface runoff. An observed rain storm in July 1990 caused surface wetting, with no runoff and limited splash.

The fine silty particles, which contribute to crust formation at Shapotou, are associated with aeolian deposition. Desert dust storms usually occur in spring (March–May) (Ing, 1972), mostly from a north-westerly direction. A dust storm was observed at Shapotou on 29 July 1990. An exposed 500 cm^2 tray collected 0.44 g of fine sand and silt during the storm, equivalent to a deposition rate of 8.8 g m^{-2}. Once formed, biocrusts are durable and protect the underlying sand from wind erosion, as observed by Fletcher & Martin (1948) in Arizona and Bond & Harris (1964) in Australia. If the crust is damaged by trampling or burrowing animals, wind erosion may recur, forming blow-outs.

Algal growth is limited to the top 25 mm of soil (Shields et al., 1957); therefore survival rate is low in areas of rapid deposition. The crusts at Shapotou were found to form irregular

Table 21.10 Elemental geochemistry of desert crusts at Shapotou and Yanchi. (All concentrations as dry weight of mineral matter %. Soil depths in mm)

Sample source	SiO_2	Al_2O_3	Fe_2O	Mn_O	Ca_O	MgO	K_2O	Cl	Na	P_2O_5	S
Shapotou											
Desert crust reclaimed (1956)	70.03	11.33	3.10	0.076	3.80	2.04	2.10	0.013	1.488	0.085	0.033
Desert crust reclaimed (1964)	73.07	11.11	2.84	0.070	3.27	1.88	2.03	0.011	1.511	0.073	0.019
Desert crust reclaimed (1981)	72.10	11.29	2.80	0.067	2.99	1.88	2.05	0.004	1.550	0.071	0.014
Desert site reclaimed (1964)											
Top layer (0–1)	68.96	11.46	2.94	0.072	3.99	2.10	2.11	0.018	1.460	0.084	0.028
Crust (1–3)	73.07	11.11	2.84	0.070	3.27	1.88	2.03	0.011	1.511	0.073	0.019
Grey sand (3–15)	74.59	10.46	2.21	0.049	1.62	1.39	1.99	0.007	1.539	0.049	0.011
Top sand (15–25)	78.55	10.47	2.13	0.048	1.59	1.33	1.94	0.002	1.655	0.048	0.012
Lower sand (45–60)	77.69	9.81	1.85	0.038	0.43	0.94	1.93	0.002	1.547	0.036	0.007
Yanchi											
Desert crust reclaimed (1984–86)	73.08	10.66	2.39	0.063	4.47	1.67	1.97	0.000	1.501	0.075	0.014
Desert crust reclaimed (1978)	67.89	10.50	2.49	0.064	4.34	1.73	1.96	0.023	1.430	0.083	0.015

surfaces, with micromounds rising 15 mm, which were composed of dark desiccated moss, interspersed with microhollows and coated with fine aeolian dust. Algae are particularly well adapted to semi-desert conditions and have been known to survive long periods of drought (Shields & Durrell, 1964). Some species are very important as colonisers, because they are capable of fixing atmospheric nitrogen (Shields *et al.*, 1957; Shields & Durrell, 1964; MacGregor & Johnson, 1971).

At Shapotou, three south-east to north-west orientated transects were taken across checkerboard areas reclaimed in 1956 (T1 — 34 years), 1964 (T2 — 26 years) and 1981 (T3 — 9 years) (Figure 21.2). Every 1 m along the transects, the depth to the base of the humic-stained, accreting, aeolian 'grey sand' was measured. Sites appeared to require over six years of stabilisation before significant aeolian accumulation occurred. For example, a site reclaimed in 1984 only possessed a thin, discontinuous (approximately 10%) crust cover, while older reclaimed sites accreted fairly consistent depths of aeolian sand (Table 21.9). The mean accumulation depth of approximately 40 mm (Table 21.9) may be associated with the surface roughness in wind fields, induced by the straw checkerboards and planted xerophytes. Initial aeolian accretion was particularly evident on the 1981 site in the lee of the xerophytes, especially *Artemisia ordosica*. Incipient soils underwent development with time, becoming progressively darker and more distinct in older sites. Grey sands possessed particularly high organic contents (2.73%) on the 1956 site.

Geochemical analyses of the crusts from the sites reclaimed in 1956, 1964 and 1981 showed a progressive improvement in nutrient status, with Fe, Mn, Ca, Mg, Cl, P and S increasing and only the elements not associated with plant nutrition (Si and Na) decreasing (Table 21.10). Using the site reclaimed in 1964, geochemical analyses of the vertical zones showed that the nutrients decreased with depth, indicating the superficial nature of the nutrient store (Table 21.10). This concentration within a layer less than 40 mm thick would restrict natural plant colonisation. It was evident that many seeds germinated on the crust, but the majority of the young seedlings died. This may be attributed to the shallowness of the 'grey sand' and to insufficient moisture.

Two sites of desert crusting were investigated at Yanchi, for comparison with Shapotou. Elemental differences were apparent, the higher Ca and lower Al at Yanchi probably reflecting the variation in origin of aeolian fines (Table 21.10). In the area reclaimed by straw checkerboards and planted with *Artemisia ordosica* and *Salix psammophila* in 1978, the crust was found to be more irregular than at Shapotou, with as much as 25 mm between the micromounds and the microhollows. A field experiment showed that, following wetting, small algae and mosses rapidly became more apparent.

Within each transect at Shapotou, there were considerable variations in grey sand thickness. Generally, on dune crests the crust was thin and discontinuous (10–20% cover), probably periodically stripped by deflation, while further downslope, adjacent to crests, buried, laminated 'grey sand' layers were found. North-west facing slopes tended to possess thicker 'grey sand' accumulations than south-east facing slopes, probably due to greater aeolian accretion on slopes facing into prevailing dust-transporting winds. In addition, dead plants were much more evident on south facing slopes, especially *Artemisia ordosica*, most likely due to higher temperatures and greater evapotranspiration rates. Greater accumulations of 'grey sand' were evident at the base of steep slopes, probably representing localised colluvial accumulations.

Using 41 quadrats at Shapotou, Shen (1988) compared the changes in the dominance of *Artemisia ordosica, Hedysarum scoparium* and *Caragana korshinskii* with the progressive development of the soil on stabilised dunes. *Artemisia ordosica* and *Caragana korshinskii* achieved maximum dominance where 40 mm of soil had accumulated, while *Hedysarum scoparium* declined in dominance as soil depth increased. All three species declined in dominance when soil depth increased to 80 mm. The average soil depth at Shapotou was approximately 40 mm; therefore decreases in the dominance of the three main species appear to be associated with the degeneration of the planted species and the expected increase in natural colonisers.

In a comparison of drought-resistant characteristics of six psammophilous shrubs, Cai *et al.* (1988) found that *Caragana korshinskii* exhibited the strongest resistance in all the physiological indices, while *Hedysarum scoparium* depended on a long root system. At Shapotou, the root of a single *Hedysarum scoparium* shrub measured 6.2 m.

These reclamation methods have successfully changed an environment of mobile dunes to a stable ecosystem. At Shapotou, 25 years after the establishment of an artificial vegetation cover, more than ten additional species have naturally colonised the area and vertebrates have increased from two to more than 30 species (Chen, 1983).

Desert Reclamation Using Palaeosols

Excavation of buried topsoils can assist desert reclamation. In the Mu Us Sandy Lands, a fertile calcareous sandy silt loam frequently underlies a cover of desert sand (Figure 21.7). The Ordos grassland steppe was buried approximately 300 yr BP (Song Bingkui, personal communication). Sand mobility can be directly attributed to overgrazing on the semi-arid

Figure 21.7 Palaeosol on southern edge of Mu Us Sandy Land

steppes (Zhao, 1988). Table 21.11 shows the pedological characteristics of a representative soil profile. Reclamation efforts are in progress by 'mining' the palaeosols. Trenches are dug and the buried Ah horizon removed. The soil is then mixed with surface dune sands and its fertility improved by additions of sheep manure. On the southern Mu Us, areas reclaimed only a few years previously yielded productive crops of maize, wheat, soya beans and vegetables. Productivity was maintained by additions of sheep manure (approximately 110 t ha^{-1} yr^{-1}) and irrigation with groundwater. Soil samples were removed from areas reclaimed seven, five and three years previously. Soil organic matter contents increased with time, at a rate of approximately 0.1% yr^{-1}. For instance, the seven-year sample mean was 1.44%, compared with a mean of 0.97% for the three-year treatment. Although Ca and Mg in the 'mined' palaeosol are decreased when mixed with desert sand, all the nutrients in the cultivated soils increased with time (Table 21.12).

CONCLUSIONS

At Shapotou, Yanchi and other field stations of the Institute of Desert Research, it has been shown that desertified land can be successfully reclaimed using irrigation and natural stabilising methods. Using irrigation water with silt inputs, either from rivers or mined palaeosols, dune sands can become agriculturally productive. Careful management, using

Table 21.11 Profile description at Ma Hanzhang Farm, Yanchi (Location: 37°55′N, 107°33′E; elevation: 1326 m)

0–36 cm Cu1	Pale brown (air-dry 10YR 6/3); sandy loam; stoneless; slightly moist (1.78%); weakly developed medium granular peds; low packing density; extremely porous; fine simple tubular macropores associated with plant roots; loose; semi-deformable; uncemented; non-sticky; moderately plastic; common medium woody roots; common fine and very fine fibrous roots; root density 0.25 g 100 cm^{-3}; organic matter 0.80%; sharp smooth boundary
36–82 cm Cu2	Yellowish brown (10YR 5/4) loamy sand; stoneless; dry to slightly moist (5.31%); apedal single grain; low packing density; extremely porous; medium/fine simple tubular macropores associated with plant roots; loose; semi-deformable; uncemented; non-sticky; non-plastic; common fine and very fine fibrous roots; root density 0.003 g 100 cm^{-3}; organic matter 0.40%; sharp wavy boundary
82–148 cm bAh	Light brownish grey (10YR 6/2): sandy loam; stoneless; moist (14.08%); moderately developed coarse granular peds; medium packing density; moderately porous; medium, simple tubular macropores associated with plant roots; moderately firm; deformable; uncemented; slightly sticky; moderately plastic; few fine fibrous roots; root density 0.17 g 100 cm^{-3}; organic matter 1.01%; abrupt wavy boundary
148–206 cm bC(g)	Yellowish brown (10YR 5/4), with few fine yellowish red (5YR 5/8) distinct mottles in upper part and few medium very dark brown (10YR 2/2) distinct mottles in lower part of horizon; medium sand; stoneless; moist (6.48%); apedal single grain; low packing density; extremely porous; medium simple tubular macropores associated with plant roots; loose; semi-deformable; uncemented; non-sticky; non-plastic; few fine fibrous roots; root density 0.002 g 100 cm^{-3}, organic matter 0.31%

Table 21.12 Elemental geomchemistry of soils and desert crusts at Yanchi (All concentrations as dry weight of mineral matter, %. Analyses performed on bulk horizon samples. Reclaimed soils analyses based on 0–5 cm deep samples, n = number of replicate samples)

Sample source	SiO$_2$	Al$_2$O$_3$	Fe$_2$O$_3$	MnO	CaO	MgO	K$_2$O	Cl	Na	P$_2$O$_5$	S
Desert (1990)	85.10	9.41	1.53	0.039	1.43	0.87	1.71	0.004	1.547	0.041	0.020
Palaeosol (1990)											
Cu1 0–36 cm	72.21	9.98	2.19	0.057	4.78	1.94	1.88	0.018	1.422	0.066	0.012
Cu2 36–82 cm	78.26	9.09	1.74	0.042	2.75	1.29	1.77	0.005	1.363	0.048	0.012
bAh 82–148 cm	65.35	9.06	2.10	0.059	7.25	3.54	1.82	0.027	1.245	0.064	0.014
bCg 148–206 cm	77.78	8.24	1.41	0.050	4.74	1.10	1.72	0.010	1.328	0.042	0.021
Cultivated palaeosols											
Reclaimed (1987) 3 years ($n = 8$)	75.35	9.05	1.85	0.051	4.08	1.55	1.82	0.008	1.282	0.076	0.015
Reclaimed (1985) 5 years ($n = 10$)	75.46	9.24	1.93	0.053	5.00	1.60	1.84	0.008	1.315	0.076	0.018
Reclaimed (1983) 7 years ($n = 10$)	74.81	9.41	1.95	0.054	4.99	1.71	1.86	0.008	1.310	0.081	0.017

straw checkerboards and managed successions of xerophytes, can stabilise desert margins. Such stabilisation can halt further desert encroachment and reduce the environmental damage of aeolian inputs to rangeland and cropland downwind.

Reclamation efforts need to be carefully and continuously maintained for over six years before significant improvements in reclaimed lands are evident. If these efforts are continued, the physical, chemical and biological properties of these developing soils improve relatively quickly. While technically feasible, reclamation does involve considerable expenditure in time and resources.

ACKNOWLEDGEMENTS

This research was mainly funded by the Han Suyin Trust, the British Council and the University of Wolverhampton. The authors are grateful to Professors Zhu Zhenda, Song Bingkui, Cho Boaming Shi Qinhui, Wang Shu Xiang and Chen Hao of the Institute of Desert Research, Academia Sinica, Lanzhou. The XRF geochemical program was developed by Brian Bucknall and Dr Craig Williams at the University of Wolverhampton.

REFERENCES

Bond, R. D. (1964). The influence of the microflora on the physical properties of soils. II. Field studies on water repellent sands. *Australian Journal of Soil Research*, **2**, 123–31.

Bond, R. D. and Harris, J. R. (1964). The influence of the microflora on the physical properties of soil. I. Effects associated with filamentous algae and fungi. *Australian Journal of Soil Research*, **2**, 111–22.

Booth, W. E. (1941). Algae as pioneers in plant succession and their importance in erosion control. *Ecology*, **22**, 38–46.

Cai Yucheng, Wang Zhenqi, Liu Xu and Ja Fuqin (1988). Studies on drought-resistant physiology of several psammophilous shrubs (in Chinese, English Abstract). *Journal of Desert Research*, **8**(4), 39–45.

Chen Caifu & Qian Taitao (1984). Transformation and utilization of sand land and its beneficial results (in Chinese, English Abstract). *Journal of Desert Research*, **4**(3), 36–44.

Chen Shixiong (1983). The ecological effects of the artificial vegetation on shifting sand of the railway both sides in Shapotou area (in Chinese, English Abstract). *Journal of Desert Research*, **3**(4), 35–41.

Chen, Z. (1986). Analysis of changes in sediment delivery of the Huang He River (in Chinese). *Soil and Water Conservation in China*, **11**, 12–16.

Fletcher, J. E. and Martin, W. P. (1948). Some effects of algae and moulds in the rain-crust of desert soils. *Ecology*, **29**, 95–100.

Fullen, M. A. and Mitchell, D. J. (1991). Taming the Shamo 'Dragon'. *Geographical Magazine*, **63**(11), 26–9.

Hodgson, J. M. (1976). *Soil Survey Field Handbook*, Soil Survey Technical Monograph 5, Lawes Agricultural Trust, Harpenden.

Ing, G. K. T. (1972). A duststorm over central China, April 1969. *Weather*, **27**(4), 136–45.

Lin Yuquan, Jin Jun, Zou Benggong, Cong Zili, Wen Xiangle, Tu Xiong, Ye Jinrong, Zhang Baoshan and Wang Hanwu (1984). Effect of fence techniques in levelling sand accumulation around sandbreaks — case study in Shapotou District (in Chinese, English Abstract). *Journal of Desert Research*, **4**(3), 16–25.

Liu Yingxin (1987). The establishment and effect of protecting system along the Bautou–Lanzhou Railway in the Shapotou Study Area (in Chinese, English Abstract). *Journal of Desert Research*, **7**(4), 1–11.

Long Yuqian & Xiong Guishu (1981). Sediment measurement in the Yellow River. Erosion and sediment transport measurement. In *Proceedings of the Florence Symposium*, June 1981, International Association of Scientific Hydrology Publication 133, pp. 275–85.

MacGregor, A. N. and Johnson, D. E. (1971). Capacity of desert algal crusts to fix atmospheric nitrogen. *Soil Science Society of America Proceedings,* **35**, 843–4.

Shapotou Desert Research Station (1986). The principles and measures taken to stabilize shifting sands along the railway line in the south eastern edge of the Tengger Desert (in Chinese, English Abstract). *Journal of Desert Research,* **6**(3), 1–19.

Shen Weishou (1988). Community features of the artificial vegetation on sand land in the Shapotou area (in Chinese, English Abstract). *Journal of Desert Research,* **8**(3), 1–7.

Shields, L. M. and Durrell, L. W. (1964). Algae in relation to soil fertility. *Botanical Review,* **30**, 92–128.

Shields, L. M., Mitchell, C. and Drouet, F. (1957). Algae- and lichen-stabilized surface crusts as soil nitrogen sources. *American Journal of Botany,* **44**, 489–98.

Zhao Songqiao (1988). Human impact on northwest arid China: desertification or de-desertification? *Chinese Journal of Arid Land Research,* **1**(2), 105–16.

Zhu Zhenda (1989). Advances in desertification research in China (in Chinese, English Abstract). *Journal of Desert Research,* **9**(1), 1–13.

Zhu Zhenda and Liu Shu (1988). Desertification processes and their control in northern China. *Chinese Journal of Arid Land Research,* **1**(1), 27–36.

Zhu, Z., Liu, S., Wu, Z. and Di, X. (1986). *Deserts in China,* Institute of Desert Research, Academia Sinica, Lanzhou.

Zhu, Z., Liu, S. and Di, X. (1988). *Desertification and Rehabilitation in China,* The International Centre for Education and Research on Desertification Control, Lanzhou.

22 Facing Environmental Degradation in the Aravalli Hills, India

C. A. SCOTT

Catholic Relief Services, Tegucigalpa, Honduras

ABSTRACT

Environmental change in the Aravalli Hills of north-western India was investigated by examining the agroecology of the region. Because of their location on the eastern fringe of the Thar Desert, the Aravallis play a crucial role in impeding desertification. Although the climate in the Aravallis, particularly the onset of the monsoon, is variable, there has been no appreciable long-term decline in rainfall. The mix of vegetative species as well as hydrological processes, however, have changed as a result of land use.

The interrelations between resource use and environmental degradation are explored in the context of Dolpura, an *adivasi* (tribal) village in Udaipur District, Rajasthan. In this community, a promising environmental programme based on reforestation and the strengthening of traditional soil and water conservation techniques has been developed by Seva Mandir, a local non-governmental organisation.

INTRODUCTION

The Aravalli Hills (Figure 22.1) separate two very different agroecological zones: the Thar Desert to the west and the sub-humid to humid Gangetic and Malwa plains to the east. The importance of the Aravallis in impeding desertification can perhaps best be appreciated by considering the expanse of arid lands to the west. Southern Pakistan, Iran, the Persian Gulf region, the Arabian Peninsula and north Africa form an extensive dryland area, bisected only by the Indus, Tigris–Euphrates and Nile valleys. With elevations ranging from 300 m to just over 1100 m a.s.l, the Aravallis form not so much a physical, orographic barrier to the spread of the western desert, as an agroecological buffer between these two distinct climatic zones.

While an exhaustive analysis of desertification and environmental change in the region is neither the purpose, nor is within the scope, of this paper, it must be emphasised that the rapid biophysical (and social) transformations under way in the Aravalli Hills have implications for a geographical area beyond the hills themselves. The intent of presenting an agroecological profile of the Aravallis is rather to set the context for a more detailed investigation of specific responses to environmental change in one community in the southern Aravallis.

Environmental Change in Drylands: Biogeographical and Geomorphological Perspectives.
Edited by A. C. Millington and K. Pye. © 1994 John Wiley & Sons Ltd

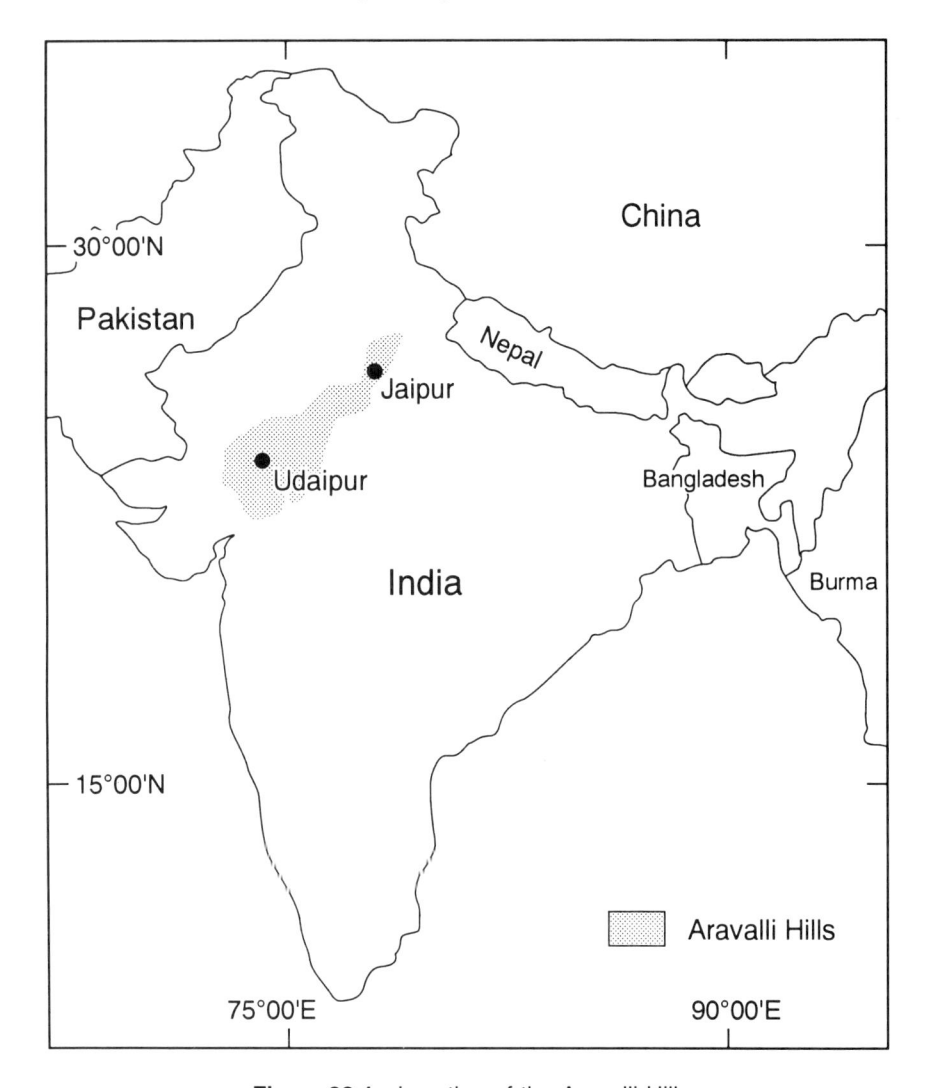

Figure 22.1 Location of the Aravalli Hills

BACKGROUND: A REGIONAL AGROECOLOGY

The Aravalli Hills extend over 1000 km in the north-west Indian states of Gujarat, Rajasthan and Haryana, and form the watershed divide between the Ganges and Indus basins. Surface discharge is intermittent in all but the most major rivers. The Luni River is the principal western drainage of the Aravallis, emptying into the Rann of Kutch, an estuary bordering the Arabian Sea. The Mahi and Sabarmati Rivers rise in the southern Aravallis and flow south to the Gulf of Khambat, while the Banas River drains the northern hills and eventually flows into the Ganges. The hills are severely eroded, and extensive quarrying for marble and limestone as well as ore extraction and refining operations have produced mining spoils, which, although limited in geographical extent, contribute to erosion and groundwater contamination.

In Rajasthan the average annual precipitation (P_{ann}) measured at recording stations to the east of the Aravallis is far greater than at stations to the west. For example, the 1009 mm recorded at Chittorgarh exceeds by 2.5 times that at Pali (411 mm), although the two are separated by only 160 km across the central Aravallis. Clearly the Aravalli Hills have an impact on the regional distribution of rainfall (IMD, 1988; Dhabriya, 1988b). Within the Aravallis, annual rainfall is relatively consistent, ranging from 520 mm at Ajmer to 780 mm at Kotra (IMD, 1988). However, intra-annual variability of rainfall at various stations in the region is pronounced (Dhabriya, 1988b) and will be discussed subsequently.

Prevailing climatic conditions in the region result in differences in agroecological capability. To the west where P_{ann} is less than 500 mm, land use is extensive with heavy reliance on livestock, particularly sheep, goats and camels. Here, cropping is restricted to sorghum and pearl millet in the monsoon *kharif* cropping season; unirrigated *kharif* yields are on the order of 200 kg ha^{-1} (Qureshi, 1989). In the *kharif* cropping season, seed is sown after the first monsoon rains in June. Crop growth is largely dependent on intermittent rainfall through October, with the harvest taking place in November. In the *rabi* cropping season, sowing takes place in November or December, with crop growth entirely dependent on residual soil moisture. Depending on the crop, the *rabi* harvest takes place in February or March. In the Aravalli Hills (P_{ann} of 500–800 mm), as further to the east, livestock are primarily buffaloes and cows. The soil moisture regime in the Aravallis allows cropping in two seasons: in *kharif*, maize, gram, dry paddy, sugar cane and fodder are cultivated; and in *rabi*, wheat, mustard, sesame and vegetables (the latter under irrigation).

It is estimated that only 20% of Rajasthan state's western (arid) region is suited to rainfed cropping; however, 30% was being cropped in 1951 and 60% in 1971 (Dhabriya, 1989). Pasture (usually common lands) and long fallows were the land uses most rapidly converted to agriculture. Jodha (1989) further points out that in western Rajasthan, the area of common lands including community forests has diminished significantly through land reform programmes. In other words, the land base on which the rural poor, particularly the landless, depended for subsistence has passed into the hands of those able to secure land allotments.

The intensification of land use brought about in part by human and livestock population growth is a matter of some concern here. For Rajasthan as a whole, the increase in human population in the 1901–81 period was 285% (Dhabriya, 1988a). Livestock population increased in a similar manner. Given the necessary heavy reliance on livestock in arid regions, it is not surprising to note that in 1981 the herd population was nearly double the human population. Livestock are increasingly pastured in forest areas, a process that has contributed to a dramatic change in the mix of vegetative species.

Natural vegetation in the Aravalli Hills may be characterised as sub-tropical, dry deciduous forest with an extensive range dominated by bunch grasses, typically *Heteropogon contortus*. Although somewhat arid for *sal* (*Shorea robusta*), the southern Aravallis borders on regions in Madhya Pradesh state with extensive *sal* forests. The climax forest species in the southern Aravallis are teak (*Tectona grandis*) and bamboo (*Dendrocalmus strictus*). Extensive barren lands are increasingly evident throughout the Aravalli Hills. In place of local species, numerous exotic species have been introduced on a widespread basis, particularly *Eucalyptus* spp. and mesquite (*Prosopis juliflora*). The latter grows as a hardy shrub and has colonised extensive tracts. According to local informants, the prevalence of the cactus-like *thur* (*Euphorbia* spp.) has increased significantly in the last decade.

Much pressure has been placed on the region's forest resources through increased commercial demand for timber, fuelwood and 'minor' forest produce, including leaves for the manufacture of such items as *bidis* (hand-rolled cigarettes) and disposable plates and bowls. The forest department's practice of exploiting resources through private contracting has resulted in the virtual elimination of teak from Aravalli forests.

The total forest area in the Aravalli Hills decreased by 42% from 1972 to 1984 (Dhabriya, 1988b). More seriously, according to the same researcher (Dhabriya, 1988a), who has interpreted recent satellite imagery, a series of twelve gaps, or corridors, has formed in the northeastern Aravallis. The gaps are dry valleys, 1–3 km in width, in which vegetative and soil moisture conditions increasingly resemble those in the arid west. There is evidence of 'desert encroachment' on to the sub-humid plains of Rajastan through the twelve corridors; sand dune formation and the increased incidence of dust storms have been detected to the east of the hills (Dhabriya, 1988a). The loss of forest cover and change in the vegetative mix in the Aravallis jeopardises their ability to impede the eastward advance of the desert. The loss of forest canopy on the high ridges may result in differential pre-monsoon heating and increased albedo. Variability in the onset of the monsoon is affected by the non-uniform solar radiation flux that results. It has been suggested that '. . .the effectiveness of the Aravalli Hills as a source of structural control for normal weather and climate can be revived by restoring its ecological status mainly by reforesting its barren peaks, slopes and foothills' (Dhabriya, 1988b).

THE CLIMATE DEBATE

The significance of climatic change in the Aravalli region is the subject of some controversy (for general discussions, see Gupta, 1989, and Meher-Homji, 1989). Dhabriya (1988b) quotes Irrigation Department sources which purport to show that P_{ann} in Jaipur (in north-central Rajasthan on the western fringe of the Aravalli Hills) has decreased 1% and monsoon rainfall 2% over the 1901–85 period. Furthermore, the duration of the monsoon rainy season is said to have decreased from 101 days to 55 days over the 1973–87 period (Figure 22.2). Additional data for the 1961–85 period are presented which indicate that twelve Rajasthan districts receive less than average annual rainfall and fourteen receive less than average monsoon rainfall. While the twelve districts that receive less than average annual rainfall are said to cover two-thirds of the land area of the state, eight are western (arid) districts with a low density of rainfall stations. As Rajasthan has a total of 27 districts, fifteen presumably had an increase (or no change) in average annual rainfall, while thirteen had an increase (or no change) in average monsoon rainfall.

From the data presented in Figure 22.2, it is evident that the 1973 monsoon withdrawal and onset were somewhat later than usual. Additionally, 1986 and 1987 were years of uncommonly short monsoon duration, and 1987 has been unequivocally recognised as a severe drought year in the region. Thus if the data from 1974–85 are assessed (see inset in dotted lines in Figure 22.2), the decline in monsoon duration is not so apparent.

Even within this shorter 12-year data set, however, it seems evident that variability in the onset of the monsoon persists. Thus, the data do not convincingly show that average annual rainfall in Rajasthan is declining. As Meher-Homji (1989) states, there is 'no declining tendency in precipitation of the arid zone of Rajasthan, bereft of forest growth'. While Meher-Homji's analysis does not cover Rajasthan's sub-humid zone (with more vegetative

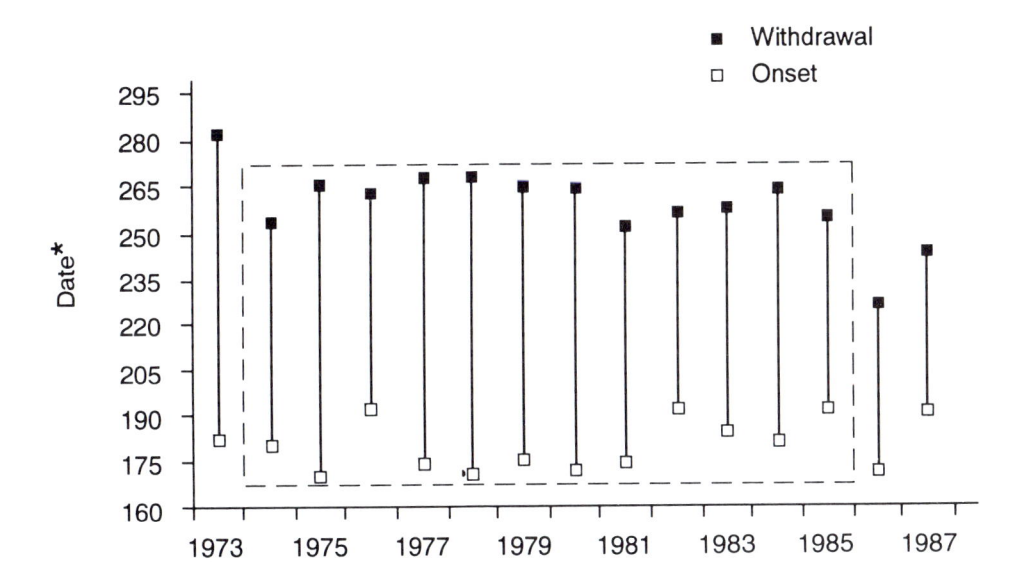

*Date is indicated in number of days past 1 January: 10 June = 160, 18 October = 290. Refer to text for comments regarding inset in dotted lines.

Figure 22.2 Onset and withdrawal of the Monsoon, Jaipur, 1973–87

cover), it does assess climatic data from other regions in India with high rates of forest loss. Meher-Homji suggests that forest clearance increases albedo (reflected solar radiation) and alters the atmospheric energy flux. In this manner, forest loss can be linked not to a secular decline in total rainfall, but to variability in its seasonal distribution.

THE COMPLEX HYDROLOGY OF THE ARAVALLI WATERSHEDS

While the data do not indicate a decline in annual rainfall, hydrological processes, particularly soil moisture available for biomass production, are changing in response to land use practices. The hydrology of the Aravalli Hills' principal western drainage, the Luni River, has been studied extensively in the aftermath of devastating floods in 1979 when 5-day rainfall values exceeded two times P_{ann} (CAZRI, 1982). Stream hydrographs with a very sharply rising discharge, an abrupt peak, a rapid initial decline, followed by a slowly tapering discharge, resulted from rainfall of unprecedented intensity occurring over a 5-day period. However, the ratios of runoff to total rainfall were as important to the nature of the flood event as was rainfall intensity; at numerous gauging stations measuring large drainage areas, runoff ratios exceeded 50%. Agricultural land with no crop cover and barren, sloping pastures were major contributors to the generation of runoff.

To convey some idea of the extreme variability of surface runoff in watersheds in the southern Aravallis, it may be noted that flow in the Wagwara Nala in Udaipur District, which was measured by the author, is minimal from September to May (on the order of 2–10 l s^{-1}, or 0.002–0.01 m^3 s^{-1}). Hydrological data for the 95 km^2 catchment area of the

418   Environmental Change in Drylands

Wagwara Nala, however, indicate that the 25-year maximum discharge may be as high as 400 m^3 s^{-1}. A major flood in June 1988 on the Wagwara Nala (not exceeded since 1973 according to local residents) had an estimated discharge of 200 m^3 s^{-1} based on Manning's equation. After three deficient monsoons, the flood of 1988 deposited 30 cm of sediment in a reservoir under construction.

Groundwater potential in the Aravallis is considerable. In the southern hills, shallow aquifers are 5–20 m below the ground surface, allowing relatively simple, low-cost irrigation techniques. Traditional lift technologies include the *rahat* or Persian wheel (Figure 22.3) and *chadas* (skin bags or metal buckets raised by oxen). Diesel, centrifugal pumps can draw down water levels quickly.

The fractured substrate throughout much of the Aravallis is composed of phyllite, schist and quartzite, and is highly permeable (Scott, 1988). The Aravallis' hydrogeology has important implications for surface water resources development. The small reservoir shown in Figure 22.4, for example, has such a fractured bed that the entire impounded volume infiltrates within four to six weeks after the withdrawal of the monsoon.

LAND USE AND DEGRADATION IN DOLPURA VILLAGE, UDAIPUR DISTRICT

The intensification of subsistence resource-use practices in the southern Aravalli Hills has contributed to environmental degradation. With the loss of vegetative cover and the increase in area dedicated to agriculture and grazing, sloping lands are increasingly subject to erosion.

Figure 22.3 A *rahat* or Persian wheel in operation

Figure 22.4 A reservoir catching runoff from a small watershed in the Aravalli Hills

Changing surface and sub-surface hydrological processes (increased runoff and decreased infiltration) result in soil moisture conditions adverse to traditional cropping strategies. Residents have adopted more risk averse behaviour, including crop diversification, increased reliance on livestock and rural–urban migration. To investigate the interrelations among resource use and environmental change in the Aravallis in more specific terms, land-use and climatic data available for Udaipur District, Rajasthan, are analysed, while observations on local practices in Dolpura village are presented.

For the period 1931–1980, the P_{ann} in Udaipur District was 638 mm (Sharma, 1987), while annual potential evapotranspiration at 1373 mm exceeds rainfall by a factor of 2.2. As seen in Figure 22.5, an intense seasonal drought occurs in April and May. The seasonal distribution of rainfall and potential evapotranspiration is critical for rainfed agriculture as only 28% of the district's cultivable land is irrigated (Seva Mandir, 1989). As the onset of the monsoon is variable, farmers in Dolpura do not sow all of their *kharif* crops at once, but rather stagger sowing to ensure that only part will be lost if the rains fail. All available fertile land is cropped, a practice that accelerates sheet erosion. The remaining land is used as pasture, with the prevalence of free grazing in the pre-monsoon season leading to soil compaction and subsequent increase in runoff as well as rill and gully erosion.

In 1981, 51% of Udaipur District's area (9740 km^2) was comprised of wasteland (1981 Census of India figures, quoted in Seva Mandir, 1989). Given that much wasteland is privately owned, government reforestation programmes have had little appreciable effect on limiting the expansion of wastelands or on converting them to productive uses, e.g. fodder production. Because resource use on government lands is not regulated, there is little conservation of resources on the part of local residents.

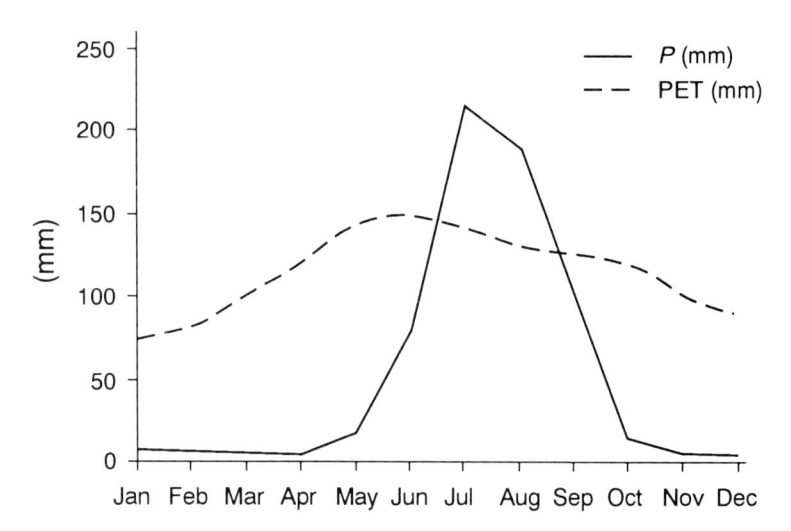

Figure 22.5 Mean monthly rainfall (P) and potential evapotranspiration (PET), Udaipur

At the 1981 population of 2 357 000, Udaipur District's population density averaged 123 persons km^{-2}, with a density on cropped land of 7 persons ha^{-1} (1981 Census figures, quoted in Seva Mandir, 1989). Udaipur heads the Rajasthan districts in absolute numbers of *adivasi* (tribal) inhabitants. The southern portions of the district, particularly Kherwara and Jhadol *tehsils* (administrative units), are predominantly *adivasi*. In Kherwara Tehsil (where Dolpura is located), over 85% of the population is made up of different *adivasi* groups, including Bhils, Garasias, Meenas and Saharias.

Bhils constitute the entire population of Dolpura village. Comprised predominantly of the Tamor clan, Dolpura is steeped in *adivasi* tradition. Vagadi dialect (a variation of Bhili intermixed with Gujarati) is spoken, although schooling and local media (primarily radio) are in Hindi, the official language of Rajasthan. Given its proximity, Gujarat state exerts a powerful cultural and economic influence on the local area. Because Bhils must marry outside the clans of both parents, the women of Dolpura are from distant villages.

Aspects of Bhil social organisation, particularly with respect to land use and ownership, are unique. The settlement pattern in Bhil communities is dispersed. Each household maintains a separate homestead, usually atop a hillock, surrounded by numerous plots of cultivated and pasture land. The equitability of landholding is particularly exceptional. While a nearby caste Hindu village, Kalyanpur, is characterised by skewed land concentration with pervasive tenancy, Dolpura has no landlessness. Each household maintains more than one separate parcel of land. In the traditional Bhil practice of land inheritance, every individual plot must be subdivided among a father's sons. For example, a household with five plots and five sons will divide each plot by five instead of giving each son one plot. Although this practice leads to fragmentation of holdings which are difficult to till, it is equitable in its distribution of risk, as each of the five plots possesses a different production potential.

Agriculture, livestock, minor forest produce use and migration are the principal survival strategies in Dolpura. Each has an effect on environmental change. Agriculture in Dolpura is largely subsistence-oriented. Due to the seasonality of male migration, women are the primary agricultural labourers in Dolpura. Labour mobilisation is most difficult during early monsoon land preparation, which coincides with the lean season of food shortage. What dung is not burned as fuel is used to fertilise fields. The use of nitrogen fertiliser is restricted to high-yielding varieties cultivated on irrigated land. Farmers rightly do not use fertiliser on rainfed land as they feel that such fertiliser 'burns' their crops in the absence of water to 'cool' the roots. Nutrient depletion of local soils is severe. Dolpura is not on a paved road, and is distant from markets; nonetheless, there is interest in growing onions, garlic and turmeric for market. However, irrigated landholdings are small (averaging less than 1 *bigha*, or 0.2 ha, per household). Rainfed farming is practised on both levelled land and sloping parcels better suited to other uses. Considerable erosion is generated by unsustainable farming and grazing on sloping lands.

Livestock consist primarily of goats, cows and buffaloes. Intensive pasture management is not prevalent, and bears significant potential for improvement, particularly fodder development. Common pasture (*charnot*), pooled private holdings (*chak*) and land owned by the forest department (though not necessarily forested) are the principal lands on which livestock are grazed. A number of local tree species which coppice well, particularly *neem* (*Azadirachta indica*) and *dhak* (*Anogeissus* spp.), are used as fodder sources in the pre-monsoon season. Because herders from Rajasthan's western districts move through Dolpura twice annually with their goats and sheep which graze unrestricted on all unfenced vegetation, local residents are compelled to graze their livestock in a similar manner or to lose it all.

In 1984, Udaipur District, centred in the southern Aravallis, accounted for 39% (or 2415 km^2) of Rajasthan's forest area. Prior to the nationalization of forests in 1947, forest cover in the Aravallis surrounding Udaipur was thick and supported a large wildlife population (Jagat S. Mehta, President of Seva Mandir, personal communication). In more recent times, forest cover and biodiversity have declined by all accounts. Historically, forests provided *adivasis* with a wide range of resources, including food, medicinal herbs, timber, fuel, fodder and fibre. With the nationalisation of forests and the post-independence emphasis on the institution of private contracting for resource extraction, local forest management practices have been neglected. At present, the decline in biodiversity resulting from deteriorating forest conditions has meant that *adivasis* are increasingly dependent on agriculture. Due to the virtual total lack of fuelwood in Dolpura, women are compelled to dry and burn *thur* (*Euphorbia* spp.), despite the harsh smoke it creates. According to local informants, this species was previously only used as a live fence.

With regard to rural–urban migration, it should be emphasised that men seasonally leave Dolpura in search of wage labour, usually in Gujarat. For four months during the intense pre-monsoon 1988 drought, few men in the age range of 16–45 remained in Dolpura. Men try to return for the planting season, unless they secure permanent, remunerative employment. In effect, migration mitigates pressure on resources in two ways: (i) migrants do not consume local resources while they are away and (ii) cash remittances allow for the purchase of some resources (notably kerosene and timber).

RESPONDING TO ENVIRONMENTAL DEGRADATION

While the subsistence resource-use practices in Dolpura contribute to declining forest cover and to soil loss, they can also serve as an effective means to rehabilitate resources. This section describes such local responses to changing environmental conditions in the Aravalli Hills.

In 1982, an Udaipur-based non-governmental organisation (NGO) (Seva Mandir) began working on watershed resource management in rural areas of the district. The primary vehicle was an adult literacy movement organised by Seva Mandir. Funding for community forestry, soil and water conservation, lift irrigation and the construction of a community centre in Dolpura was raised through domestic (Action for Food Production, or AFPRO) and international (Swiss Development Cooperation, SDC) sources and through community-wide *shramdan* (voluntary labour). While the programme objectives were straightforward, i.e. to enhance natural resource capability and local management, particularly of underused upslope land, the dynamics of implementation have been intricate. Programme outcomes have been mixed.

A reservoir was constructed on the Wagwara Nala which runs through the village. This served not only as the source for a lift irrigation scheme but also raised water levels in numerous wells scattered throughout the village. The village provided 25% of the labour costs of reservoir construction through *shramdan* and has a viable community organisation which directs soil and water conservation and reforestation activities.

Based on local practices for runoff control and soil moisture enhancement called *medbundi*, Seva Mandir promoted conservation measures that were initially implemented on individuals' agricultural holdings. As a support for agriculture, *medbundi* detains runoff and organic matter in the field. Ponded water increases the soil moisture necessary for seed germination. The water requirements of different crops are met through elaborate detention and drainage processes during the early monsoon (Scott, 1988). Paddy cultivation is possible in certain fields with good bunding and moisture retention (Figure 22.6).

Local ingenuity has devised a range of conservation techniques with significant innovation induced by the diversity of site conditions. For example, where soil depth is poor and stone is abundant, dry stone masonry retaining walls are constructed. Moisture retentive soil with high clay content is wetted and compacted against the upslope surface. Alternately, where soil depth is good, fields are usually bunded with wetted and compacted soil. If structural support is required, the earthen bund may be built with a dry stone masonry core wall (Scott, 1988). As soon as possible, bunch grasses are transplanted in an effort to reinforce the bund. Given that fallowed fields are opened to grazing, the establishment of effective vegetative reinforcement is found to be difficult. For this reason, farmers prefer structural techniques over vegetative conservation measures.

In order to assess Seva Mandir's district-wide programme supporting *medbundi* initiatives, a total of 1580 farmers were surveyed (Seva Mandir, 1987). Among the principal techniques, the highest preference (49%) was found to be for field bunds, both stone and earthen. Boundary walls around private pasture and forest plots were second at 30%. A good indication that conservation work actually benefited farmers and was performed not merely to receive the cash incentive is provided by the fact that 60% of households constructed more than twice the maximum payable volume of conservation structures stipulated by the programme. Considerably less interest was expressed in conventional soil conservation techniques; gully plugging and terracing were implemented by 8 and 4% of households respectively. The latter techniques are costly and have less direct impact on agricultural yields (Scott, 1987), but have been used successfully to enhance soil moisture for fodder production and reforestation on village commons (Figure 22.7).

Efforts at reforestation, particularly of upland areas, was undertaken in earnest by Seva Mandir in 1986. Because initial technical input came from the forest department, the species selected at the outset were primarily rapid-growing exotics, including *Eucalyptus* spp. In

Figure 22.6 Paddy cultivation through *medbundi*

Figure 22.7 Infiltration of impounded runoff behind a gully plug

the first year, one million saplings were planted (Seva Mandir, 1989). Sapling mortality rates at the nursery and once planted were high. Subsequently, popular local species including *neem* (*Azadirachta indica*), *bans* (*Dendrocalmus strictus*) and *vilayati babul* (*Acacia nilotica*) were introduced. Additionally, shallow trenches to harvest runoff from 3 m × 10 m microcatchments were dug upslope from the sapling to provide moisture to the root zone. In 1987, over two million saplings were planted and as many trenches were dug, while in 1988, 2.2 million saplings were planted. Survival rates once planted ranged considerably (from 0 to 80%) depending in part on local conditions and management intensity. Following the disastrous drought period of 1987–8, reforestation efforts were scaled back. In 1989, just 356 000 saplings were planted.

According to participants, the greatest benefit of reforestation activities has been the increased availability of fuel and fodder (Seva Mandir, 1989). In addition to meeting local resource needs, the programme has met with some success in re-establishing canopy cover on the Aravallis' ridges, although the survival of saplings in low-lying regions is clearly better than further upslope. Any farmer or conservationist knows, however, that land treatment, like water, proceeds downslope. Whatever transpires uphill will eventually manifest itself further down. Watershed treatment begins at the ridge, protecting meager soil resources. The effective establishment of saplings at the ridge level (Figure 22.8) must be supported by large pits (0.5 m × 0.5 m × 0.5 m) to allow moisture collection in the root zone. In this manner, techniques in moisture conservation aid in the reforestation of ridgelines which are critical in the conservation of watershed resources.

Figure 22.8 Reforestation of hilltops and ridges

CONCLUSIONS

Because the Aravalli Hills form the eastern limit to the Thar Desert, environmental degradation in the hills bears significance for the sub-humid and humid areas to the east. While the onset of the monsoon in the region is variable, average annual precipitation has not declined. Rather, changes in the mix of vegetative species and hydrological processes result from changing land-use practices. According to local informants, commercial forestry and mining have caused a significant loss in forest cover and groundwater contamination. Subsistence resource-use practices have exacerbated environmental degradation. The result of both types of resource-use practices has been an increase in arid-zone vegetation, indicating desertification in the Aravallis. With the loss of vegetative cover in Dolpura, an *adivasi* village in Udaipur District in Rajasthan, traditional agricultural and grazing practices on sloping lands have accelerated erosion. However, traditional soil and water conservation techniques in this community offer significant potential to respond to changing environmental conditions. These have been complemented by reforestation activities introduced by Seva Mandir. The programme is an effective means of meeting subsistence resource needs while protecting the agroecology of the Aravalli Hills.

REFERENCES

Central Arid Zone Research Institute (CAZRI) (1982). July 1979 Flash Flood in the Luni. CAZRI Technical Bulletin 6, Jodhpur.

Dhabriya, S. S. (1988a). *Desert Spread and Desertification: An Analysis of the Identified Aravalli Gaps on the Desert Fringe,* Environmentalist Publishers, Jaipur.

Dhabriya, S. S. (1988b). *Eco-Crisis in the Aravalli Hill Region,* Environmentalist Publishers, Jaipur.

Dhabriya, S. S. (1989). Problem of advancing rocky wastelands in the hill areas of Rajasthan. In Menaria, R. (ed.), *Environmental Conservation and Planning,* Ashish Publishing House, New Delhi.

Gupta, N. (1989). Increasing drought conditions in Southern Rajasthan: experience of Pratapgarh Tehsil. In Jayal, N. D. (ed.), *Deforestation, Drought and Desertification: Perceptions on a Growing Ecological Crisis,* Indian National Trust for Art and Cultural Heritage (INTACH), Studies in Ecology and Sustainable Development 2, INTACH, New Delhi.

Indian Meteorological Department (IMD) (1988). *Climate of Rajasthan State,* IMD Publications, New Delhi.

Jodha, N. (1989). A case study of the degradation of common property resources in India. In Blaikie, P. and Brookfield, H. C. (eds.), *Land Degradation and Society,* Methuen, London.

Meher-Homji, V. M. (1989). Trends of rainfall in relation to forest cover. In Jayal, N. D. (ed.), *Deforestation, Drought and Desertification: Perceptions on a Growing Ecological Crisis,* Indian National Trust for Art and Cultural Heritage (INTACH), Studies in Ecology and Sustainable Development 2, INTACH, New Delhi.

Qureshi, S. (1989). *Regional Perspective on Dry Farming,* Rawat Publishing House, Jaipur.

Scott, C. A. (1987). *Mitti Jal Ko Bachao* (in Hindi, *Save Soil and Water,* farmers' field manual), Seva Mandir Publications, Udaipur.

Scott, C. A. (1988). *Medbundl:* soil and water conservation in Rajasthan. *Indian Architect and Builder,* December.

Seva Mandir (1987). *A Report of Land Improvement Work,* Seva Mandir Publications, Udaipur.

Seva Mandir (1989). *Wasteland Development: A Review (1986–89),* Seva Mandir Publications, Udaipur.

Sharma, B. L. (1987). *Problems and Perspectives of Watering the Crops,* Concept Publishers, New Delhi.

23 Biogeographical and Geomorphological Perspectives on Environmental Change in Drylands

A. C. MILLINGTON[*]
Department of Geography, University of Reading, UK

and

K. PYE
Postgraduate Research Institute for Sedimentology, University of Reading, UK

INTRODUCTION

Although the papers in this volume only represent a small sample of a burgeoning biogeographical and geomorphological literature in this area (e.g. Mainguet, 1991; Thornes, 1990; Singhvi & Kar, 1992; Skujiņš, 1991; Pye, 1993; Middleton & Thomas, 1992; Warren & Khogali, 1992), they are indicative of the wide-ranging nature of research that is being undertaken in drylands under the banner of environmental change. Because this collection of papers, like those in any edited book, is *incomplete*, an apposite way of concluding this book is to provide a review of the main research trends highlighted by these papers and to comment on how these relate to other recent research.

The areal extent of the world's drylands has fluctuated greatly in the geological past (cf. the review by Glennie, 1987). Consequently both geological and contemporaneous geomorphological and biological evidence can be used to understand the processes leading to dryland expansion and contraction. The understanding we seek is important for the following reasons:

1. In a global warming scenario the effects of changing climate on geomorphological and ecological processes must be understood, predicted and managed; see, for example, Andrew Goudie's discussion which covers geomorphological processes in deserts (Chapter 1) and, for a corresponding analysis in terms of ecology and biogeography, Parry (1990).
2. The effect of human activities in drylands is well documented and appears to be a major influence on most types of dryland degradation (e.g. Hjort af Ornais & Shalih, 1989;

---

[*]*Present address*: Geography Department, University of Leicester, UK.

Environmental Change in Drylands: Biogeographical and Geomorphological Perspectives.
Edited by A. C. Millington and K. Pye. © 1994 John Wiley & Sons Ltd

Mainguet, 1991; Warren & Khogali, 1992); again an understanding of biogeographical and geomorphological functioning during change is relevant to better prediction and management.
3. Regardless of the cause, a major concern about environmental change is that it will affect people's livelihoods. The main problems appear to be (i) increased wind and water erosion, with their attendant on-site and off-site effects, (ii) loss of biodiversity, and (iii) the negative effects of vegetation change and soil erosion on the hydrological cycle (e.g. increased runoff and flooding, decreased groundwater recharge, increased soil salinisation). Therefore the processes leading to dryland degradation need to be understood so the effective remedial and preventative measures can be taken (Goudie, 1990). The relevance of such research is brought into focus in the final three chapters which report on the biogeographical, hydrological and pedological principles behind dryland reclamation and management in Eygpt (Chapter 20), China (Chapter 21) and India (Chapter 22).
4. At the geological timescale, an understanding of geomorphological processes in the arid zone has important repercussions for hydrocarbon exploration and other types of mineral expolitation.

Differentiating The Causes of Environmental Change

Some things are well understood by researchers concerned with dryland environmental change and are also very clear to those *outside* the subject. The first and most obvious point is that the triggers of environmental change can be both natural and/or human-induced. However, despite the wealth of research in this field our inability to differentiate between the natural and human signals in the palaeoenvironmental record (as well as in *some* contemporary environments) is still a major impediment to interpretation. This is particularly true of studies that focus on the middle to late Holocene, as illustrated by a number of contributions in this book. In particular, there appears to be difficulty in interpreting the causes of increased rates of soil erosion (or at least rapid sediment deposition), whether the study focuses on the last century (e.g. Pickard, Chapter 14) or the last few thousand years (e.g. Ballais, Chapter 10), and, possibly, regardless of the type of evidence used; compare, for example, the work of Sarah Metcalfe *et al.* (Chapter 7) which is based on lake sediments with that of Jean-Louis Ballais (Chapter 10) which based on aeolian landforms and sediments.

Spatial Scale and Environmental Change

Secondly, in the context of applied environmental research in drylands, with its important interface with social science, it is becoming ever more urgent that ecological and geomorphological responses to environmental change are understood at a variety of spatial and temporal scales. Furthermore, it is important that observations can be scaled-up or scaled-down so that different types of data can be integrated.

This is particularly important in the context of global change studies, e.g. predictions from global circulation models are made for cells of >0.5° latitude and longitude. In Chapter 11 Aaron Yair highlights the difficulties of applying the results of GCMs to drylands without detailed, small-scale knowledge of a wide range of surface variables. However, a major research effort (the HAPEX-Sahel Experiment in Niger) is underway and this includes scaling-up ground hydrological data, which are related to semi-arid and arid vegetation through the use of remotely sensed data. Nonetheless, the wider context of scaling-up and scaling-down experimental data sets in geomorphology and biogeography has largely been ignored. This leads us to question, in particular, the relevance of many small-scale process studies and palaeoenvironmental reconstructions if they cannot be placed in a larger framework.

In the preface to this book we stressed the division of papers based on the timescale covered by the studies. We return to this division to review research on the effects of environmental change in drylands.

LESSONS FROM THE PAST

Long-term palaeoenvironmental reconstructions in drylands have expanded in both number and scope in response to the need for evidence of past biogeographical and geomorphological responses to natural and human-induced change. Such evidence is required to validate, or at least set in context, climate change predictions and environmental responses to human activities. Palaeoecological investigations, once seemingly the preserve of scientists reconstructing high-latitude ecosystems, are now playing an increasing role in reconstructing ecosystem dynamics at desert margins, and in semi-arid and sub-humid areas.

However, palaeoecological reconstructions based on evidence from sites in arid and hyper-arid areas are still rare. This appears to be due to the lack of sites where *suitable* sediments have accumulated (for sediment accumulation *per se* is not uncommon in such environments) and the fact that the techniques at our disposal have not been able to take advantage of such sedimentary records.

Consequently, in the arid and hyper-arid cores of drylands geomorphological and sedimentological evidence remains the prime data source for reconstructing the effects of past environmental change (Table 23.1). However, the way in which much of this evidence has used in questionable (Thomas, 1989). For example, reconstructions based on alluvial cut-and-fill sequences and river channel morphology are hampered by (i) the influence of low-frequency–high-magnitude events on channel form, (ii) the seasonality of river regimes, (iii) the irregular movement of the coarse sediment, and (iv) a lack of equilibrium between channel form and process. A further dimension to this argument is provided in Chapter 2 where Andrew Nash *et al.* show that long-term changes in dryland valley form in the Kalahari can be explained in terms of groundwater activity rather than changes in discharge. Intriguingly, the most active phases of groundwater activity occur in drier periods whereas increased discharge would occur in wetter periods; however, the testing of this hypothesis in other areas may be hampered by the lack of groundwater data. Thomas (1989) and Livingstone and Thomas (1993) raise similar questions about the evidence provided by inactive and degraded vegetated palaeodunes from southern Africa and Australia. The third type of geomorphological evidence used — that from slopes — is the must equivocal of the three presented in Table 23.1. Arid zone mountains are dominated by erosional processes and evidence is generally lacking; moreover, slopes are poorly linked with channels in the dryland sediment cascade. However, the spatially intermediate suite of landforms between slopes and channels — alluvial fans and *glacis d'erosion* — provide depositional sequences that often include datable material and in which the geomorphological processes are reasonably well understood (Rachocki & Church, 1990), although these sequences are

Table 23.1 Geomorphological and sedimentological evidence used in the reconstruction of dryland palaeoclimates (modified after Thomas, 1989. Reproduced by permission of Belhaven Press. All rights reserved.)

Aeolian deposits
1. Deposits may vary considerably in thickness.
2. Laminae dips range from 0 to 34° (repose angle for sand) unless affected by post-depositional earth movements. Angles may be reduced by post-depositional compression.
3. Laminae bedding identifiable with structures in modern dunes.
4. Grain sizes range from coarse silt to coarse sand (*ca.* 60–2000 µm); majority in 125–300 µm range.
5. Low silt and clay content.
6. Large particles are often rounded, smaller are sub-rounded to sub-angular.
7. May be cemented (e.g. aeolianite) by haematite, calcium carbonate and gypsum.
8. May be reddened.
9. Distinctive surface micromorphology when viewed by SEM.

Water-lain deposits
1. Commonly calite cemented; locally may be cemented by gypsum or anhydrite.
2. Conglomerates may be common.
3. Sand fraction may be absent — removed by deflation.
4. Mudflow conglomerates present.
5. Sharp upward decrease in grain size indicating rapid fall in water level.
6. Clay pellets, pebbles and flakes common, may be due to the effects of salt efflorescence.
7. Mud cracks common.
8. Often inter-bedded with aeolian deposits.

complicated by erosional episodes and complex responses to external disturbances. Recent research on arid zone alluvial fans in the context of environmental change has, however, focused on dating surfaces using soil and rock varnish development: this is exemplified by the work of Kevin White and John Walden in Tunisia (Chapter 3) and Adrian Harvey and Stephen Wells in California (Chapter 4).

Within the context of palaeoenvironmental reconstruction two encouraging trends are evident; firstly, a holistic approach to the interpretation of lake sediments, taking it beyond palaeoecological reconstruction based on micro- and macrofossils to one in which the geomorphological and geochemical evidence locked in sediments is also used (e.g. O'Hara *et al.*, 1993), which in this book is exemplified by the work of Mark Macklin *et al.* (Chapter 6), Sarah Metcalfe *et al.* (Chapter 7), Henry Lamb *et al.* (Chapter 8), Neil Roberts *et al.* (Chapter 9) and Dave Gilbertson *et al.* (Chapter 12); secondly, the increased use of *new* techniques. In this book methodological innovations are exemplified by the widespread application of environmental magnetism to alluvial fan surfaces (White & Walden, Chapter 3; Harvey & Wells, Chapter 4) and sediments from sections and cores (Macklin *et al.*, Chapter 6; Metcalfe *et al.*, Chapter 7). However, such work also encompasses techniques such as thermoluminescence (TL) and optically stimulated luminescence (OSL) dating of quartz and feldspar in sediments (Nanson *et al.*, 1991, 1992; Page *et al.*, 1991; Chawla *et al.*, 1992; Stokes, 1992; Stokes & Breed, 1993; Wintle, 1993; Edwards, 1993; Rendell *et al.*, 1993), cosmogenic isotope dating (Cerling, 1990), uranium-series dating and ESR dating (Radtke *et al.*, 1988) and stable isotope analysis (Amundsen *et al.*, 1988; Clark & Fontes, 1990; Magaritz & Jahn, 1992). The application of such techniques is the key to unlocking the door to the information stored in inorganic sediments that previously were either ignored

is, perhaps, the greatest breakthrough in reconstructing dryland palaeoenvironments because they partially overcome the problems of:

(i) the paucity of suitable lake sediment records for palaeoecological research, by allowing sediments on palaeoenvironmentally significant geomorphological surfaces (e.g. alluvial fans and *glacis d'erosion*) to be dated and analysed; and
(ii) the poor organic matter preservation potential of many environments (e.g. highly alkaline salt-lakes, the lack of reducing environments, silica-poor environments due to the dominance of resistant calcareous bedrock in many mountainous areas), by allowing the inorganic component of lake and other sediments to be analysed in conjunction with the fossil and organic components.

The availability of absolute dating, used in conjunction with stratigraphic, pedological, sedimentological and archaeological evidence promises to revolutionise our understanding of the evolution of dryland landscapes (Page *et al.*, 1991; Goring-Morris & Goldberg, 1990; Kocurek *et al.*, 1991; Thomas & Shaw, 1991a, 1991b; Rendell *et al.*, 1993; Stokes & Breed, 1993). Despite these advances the research agenda for palaeoenvironmental reconstruction in drylands still includes some difficult tasks, most notably:

1. To increase the number of reconstructions based on lake sediment sequences from arid and hyper-arid areas. Moreover, palaeoenvironmental reconstructions based on lake sediments from the arid and hyper-arid zones must integrate the findings from recent research on contemporary geomorphological, geochemical and hydrological processes on playas and related environments (DeDeckker, 1983; Bowler & Teller, 1986; Last, 1989; Teller & Last, 1990; Rosen, 1991; Torgenson *et al.*, 1986). Studies of the geomorphology and geochemistry of both small salt pans and large playas have enabled facies models with diagnostic sedimentological characteristics to be established, and have also shown the importance of fluvial processes on large playas (Bryant, 1993; Millington *et al.*, 1989; Figure 23.1). It is time to apply this work to the palaeoenvironmental analysis of lacustrine sediments.
2. The identification of critical ecological margins in drylands and the analysis of the palaeogeography of these margins throughout the Quaternary. The following margins are particularly important in the context of climatic and human-induced changes:
    (i) the margins of dry woodlands with both drier (wooded grassland) and wetter (wet savanna and rainforest) ecosystems
    (ii) the montane treeline
    (iii) lake and swamp shorelines and
    (iv) floodplain communities.
3. Sedimentological and archaeological evidence from caves is beginning to be used in this context (Brook *et al.*, 1990; and Klien *et al.*, 1991). The potential for such work in many drylands is great because of the dominance of limestone rocks which, during wetter periods in the Quaternary, must have been subject to accelerated rates of solution weathering.
4. The integration of palaeoenvironmental reconstructions based on sediments from geomorphological surfaces and adjacent or related inorganic (and organic) lake sediments. In this context relevant cores may be obtained from playa margins where there will be fluctuations between playa and distal alluvial fan/*glacis d'erosion* facies.
5. The development of models and transfer functions to enable evidence from micro- and

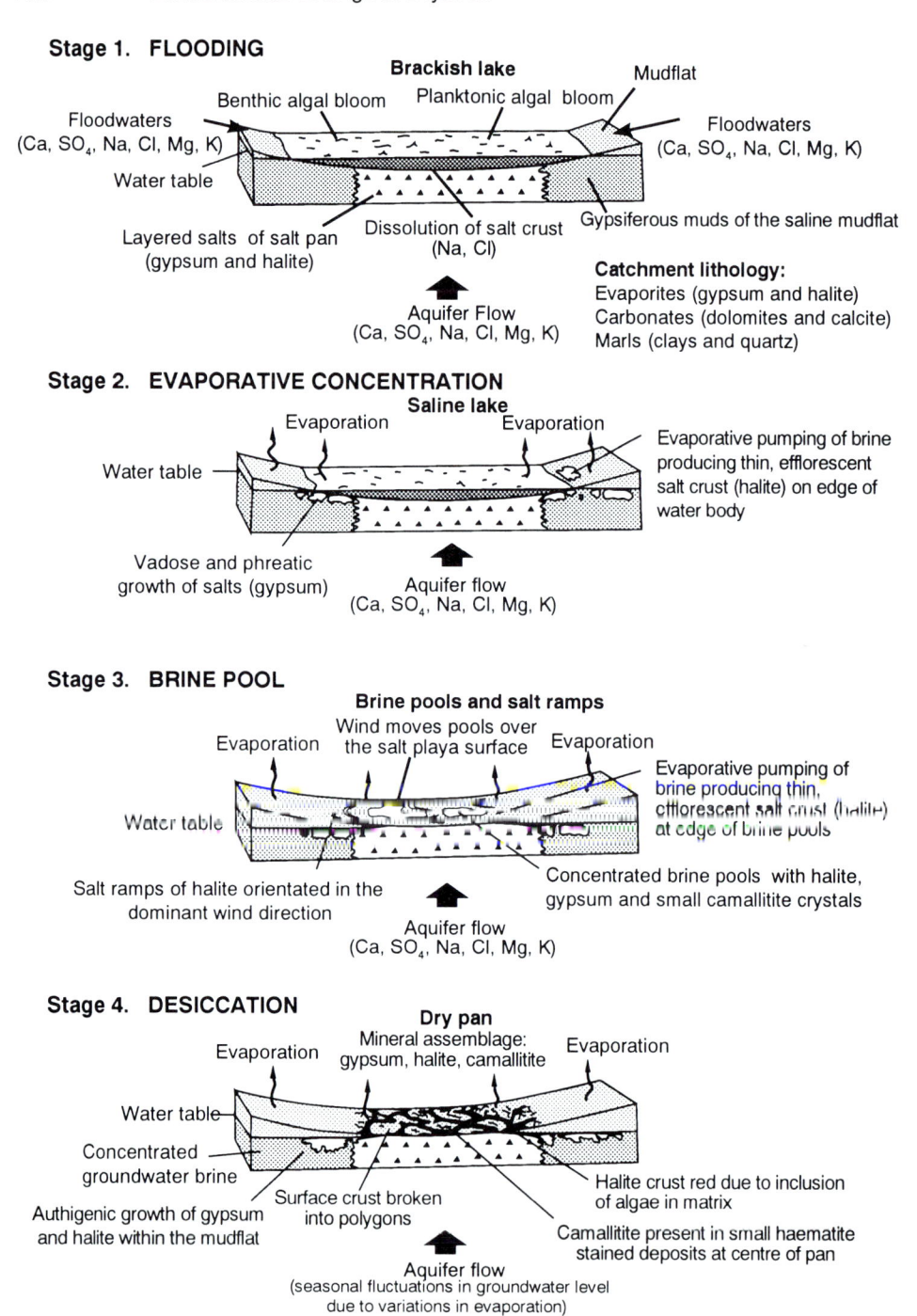

Figure 23.1 A chemical model for the Chott el Djerid. This model is based on that produced by Lowenstein & Hardie (1985) with added evidence from the flood on the Chott et Djerid in 1990 (from Bryant, 1993; Millington *et al.*, *in press*)

macrofossils found in dryland sites to be used in quantitative reconstructions of palaeoclimates or other types of disturbance peculiar to drylands.

LESSONS FROM THE PRESENT

Running in parallel with the body of work on palaeoenvironmental reconstruction is research into contemporary geomorphological and biogeographical processes. This research falls into two broad categories: (i) process observation and measurement; and (ii) process modelling.

The observation and measurement of geomorphological processes in drylands has always been a difficult task because of logistical considerations and the dominance of high-magnitude–low-frequency events that do not lend themselves readily to field monitoring. It is not surprising, therefore, that 'long' observational records from well-monitored dryland sites have been largely restricted to developed countries, or at least those with well-funded science programmes, e.g. the United States, Canada, Australia, Israel, Spain. The result has been that many geomorphological and ecological processes have not been adequately and extensively monitored (e.g. aeolian deflation and transportation, water and sediment discharge in channels, ephemeral lake filling-and-drying), or dryland landforms and ecosystems thoroughly observed, measured and quantified (e.g. spring mounds, aioun, dry channels). This situation has, however, begun to change in recent years both in the geomorphological (e.g. Millington et al., 1989; Stockton & Gillete, 1990; Littman, 1991; Leys, 1991; Goudie & Middleton, 1992; Enzel, 1992; Corbett, 1993) and ecological (Millington et al., submitted; Millington et al., 1993; Danaher et al., 1992) contexts. The papers in this book reflect this general observation.

The analysis of multitemporal remotely sensed data is therefore potentially important for monitoring both vegetation dynamics, geomorphological processes and the links between them (Millington, 1992). In this book, Rudi Goossens et al. (Chapter 19) and Nick Drake and Rob Bryant (Chapter 18) use multitemporal remotely sensed data to map the expansion of salinisation and waterlogging and to monitor flooding of playas respectively. In both cases the potential for calibration against ground data is clear, although they are typical of similar studies in that calibration is usually not undertaken. However, until remotely sensed data (whether it is from single datum or multi-temporal imagery) are routinely calibrated, their use in process measurement, as an input to models, or in verifying and extrapolating models, will be viewed with scepticism by many environmental scientists.

Monitoring is very much a key issue (whether it be from field or remotely sensed observations) because of the trend in environmental change research towards modelling the impacts of change on biogeographical and geomorphological processes. Such research has two important roles:

1. It provides a comprehensive statement of our knowledge about a particular process.
2. It enables the consequences of environmental changes on particular processes to be predicted.

The latter role is particularly important in the context of environmental management and planning. The potential for such modelling is great, but the applications (to both biogeography and geomorphology) are, like the use of remotely sensed data for process monitoring, limited. The main obstacles to process modelling in drylands appear to be:

(i) the perceived lack of historical records to provide the quantitative data inputs needed to run models (e.g. meteorological information, population census data, stocking rates), although it is apparent that the potential of archives to yield relevant data relates more to the insight of the scientist than to their actual content; and

(ii) a similar (perceived) lack of data to verify and validate model outputs.

The research agenda for the study of contemporary dryland biogeographical and geomorphological processes in the context of environmental change is wide ranging, but clear requirements include:

1. Increased observation and monitoring of biogeographical and geomorphological processes in a wide range of dryland environments to provide quantitative information which could be used (i) to develop and implement models; and (ii) to validate models.
2. Increased observation and monitoring requires the combined use of ground-based and remotely sensed observations over long time periods. Some headway is being made in this context with the IGBP Global Monitoring Sites. Nonetheless, other sites need to be established (either within or outside the IGBP framework) and sites with good ground observation networks need to have repeated remotely sensed data acquisition integrated into their data-collection schedules. New sensors, particularly those sensing the microwave part of the spectrum (e.g. the following sensors: ERS1, JERS1 and RADARSAT), are beginning to extend our ability to monitor parameters not available from sensors restricted to the visible and near infrared parts of the spectrum found on the sensors of the Landsat series of satellites, SPOT-HRV and NOAA-AVHRR. Microwave remote sensing has particular applications in soil and foliar moisture monitoring, terrain microtopography and vegetation canopy roughness; applications in dryland geomorphology and biogeography are beginning to appear. In the near future multisensor platforms (e.g. the US Polar Platform and the European COLUMBUS platform) will have multiple sensors covering the visible, near infrared and microwave parts of the spectrum, as well as high spectral resolution radiometers, which have proved, when mounted on aircraft, to have great value to the earth science and biological communities.
3. Model validation through the use of (i) contemporary observations and measurements; and (ii) recent historical material from archives.
4. Further research at the interface between vegetation dynamics and geomorphological processes must be conducted. It is becoming apparent that many geomorphological processes are critically dependent on the status of a biotic component. In addition, it is also known that vegetation patterns at certain scales in drylands have indirect links to the spatial distribution of specific geomorphological processes and surficial materials (see, for example, Yair, 1990, and Chapter 11; Thomas & Tsoar, 1990; Muhs & Matt, 1993; Alexander *et al.*, Chapter 5), although the relationships between spatial patterns of vegetation and soils/surficial materials in drylands are generally very poorly known (Wiesanga *et al.*, 1987).

RECENT ENVIRONMENTAL CHANGE

The effects of environmental change in the historical period are critical to model validation and therefore an understanding of the biogeographical and geomorphological responses to change is needed as it is a precursor to effective palaeoenvironmental reconstruction.

The value of increased use of data from archives in reconstructing recent palaeoenvironments was recognised by geomorphologists in the early 1980s, but its use in support of such reconstructions in drylands is a more recent phenomenon. Reports and maps from archives dating back to the last century were used by John Pickard (Chapter 14) to reconstruct the patterns of erosion related to the penetration of European settlers in New South Wales, and a related, though larger-scale, reconstruction based on 'travellers accounts' has been undertaken for a part of Algeria by Jean-Louis Ballais (Chapter 10). Other examples reinforce the potential of archive material in this area:

1. Reports from District Officers and Government Departments from the period between 1910 and 1950 have been used to validate predictions of flooding and low flows from a rainfall-runoff model for the Gambia River (Figure 23.2) (Amara, 1993).
2. A number of studies from eastern and southern Africa have used material from colonial and post-independence archives to develop qualitative models to explain patterns of environmental degradation (particularly woodland destruction, rangeland degradation and the resulting accelerated wind and water erosion) in drylands in the context of historical ecology (Anderson, 1984; Beinart, 1984; Kjekhus, 1977; Stocking; 1983; Webb, 1992).
3. Studies in the western United States have combined reports from a variety of sources; including travellers logs and aerial photography from the 1930s to assess the impact of human activity (population expansion, logging, forest grazing) on dry woodlands (e.g. Savage's, 1991, work on the Ponderosa pine forests of the Chuska Mountains, New Mexico).

Two important observations can be made from these studies of recent environmental change. Firstly, all of the studies, except those carried out by historians in Africa, used archive material in conjunction with other information: sediment analyses and ^{14}C dating (Pickard in Chapter 14 and Ballais in Chapter 10), rainfall and (limited) runoff records (Amara), and aerial photography, vegetation mapping and dendrochronology (Savage, 1991). Secondly, corroborating evidence is a strong point in all palaeoenvironmental reconstruction: archive material can corroborate field data that has already been collected and analysed, and field observations made in a ^{14}C-based chronology and the use of historical remotely sensed data (aerial and space-borne photography and imagery) can be used to validate archive data (e.g. dust storms: Middleton *et al.*, 1986; Goudie & Middleton, 1992; terrace degradation and increased soil erosion: Douglas *et al.*, 1994).

LINKS BETWEEN DRYLAND BIOGEOGRAPHY AND GEOMORPHOLOGY

The important overall control exerted by water erosion has been well known to gemorphologists since the seminal work of Langbein and Schumm (1958). Experimental work has supported the general relationship between climate, vegetation cover and sediment yield using sedimentation rates in reservoirs and plot-scale studies of hillslope erosion (Walling & Kleo, 1979). Nonetheless, there seem to be a number of issues related to this relationship which are, as yet, equivocal.

Firstly, there are the actual mechanisms by which different vegetation types (either at the scale of the individual plant or vegetation community) exert influence on different water erosion processes. For example, whilst experimental work has concentrated on the relation-

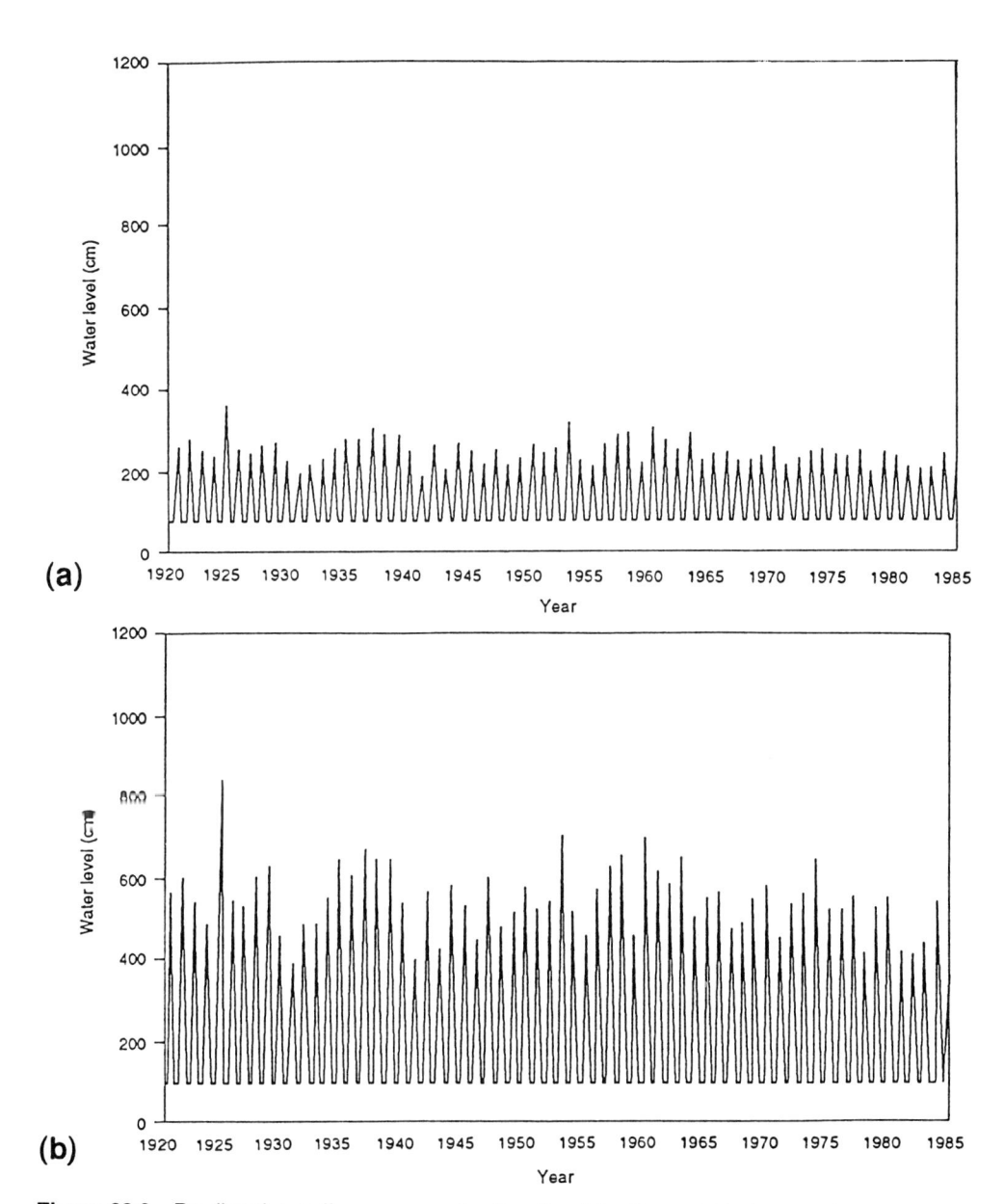

Figure 23.2 Predicted runoff sequences for the Gambia River at Kedougou, Senegal, for the period 1920–85: (a) maximum monthly water levels, (b) minimum monthly water levels. The predictions were based on a rainfall-runoff model derived from data from between 1970 and 1985, and the back predictions were checked against agricultural archive data (after Amara, 1993)

ship between rainfall and the canopy, the microenvironments created by plant stalks and protruding roots on overland flow, rill flow and gully flow are under-researched. Furthermore, the potentially important interactions between litter accumulation and decay in the topsoil and slope erosion are poorly understood in drylands. Roy Alexander *et al.* (Chapter 5) illustrated the important role of lichens in controlling runoff, and David Mitchell and Mike Fullen (Chapter 21) showed the importance of lichens in soil reclamation. Similar research needs to be undertaken in other environments and on other types of epiphytic crusts which are far more widespread on dryland soils than had previously been thought (Skujiņš, 1991).

Secondly, there is the acceptance that dryland woodlands afford a protective role in the context of water erosion and that woodland disturbance leads to increased erosion. Ecological studies show that many dryland woodlands have a sparse (and seasonally variable) canopy, thereby casting doubt on its overall efficiency in reducing raindrop impact. Moreover, dryland forests are often characterised by a low understorey and/or ground cover, there is always seasonally bare ground, the litter layer is patchy and the humic content of the topsoils is low. Another question lies in the response of dry woodlands to disturbance; recent studies on coniferous forests in the south-western United States show that some types of disturbance lead to increased tree density which, it must be assumed, reduces rather than accelerates erosion (Savage, 1991). However, such responses may not be confined to woody vegetation; Nobel (1992) has observed similar behaviour by a dryland perennial after disturbance.

In the context of aeolian erosion, recent work has also thrown doubt on traditional interpretations that areas of vegetated desert dunes are necessarily relict features which formed during a drier or more windy period in the past. Vegetated linear dunes, in particular, may actively form under conditions of partial vegetation cover (Thomas & Tsoar, 1990; Thomas & Shaw, 1991a, 1991b). Although the interaction between short-term climatic periodicity, episodic vegetation growth and sand movement on such dunes is still poorly understood, recent evidence from the Kalahari and elsewhere indicates its probable importance (Livingstone & Thomas, 1993). Soil microflora, particularly bacteria, algae, lichens and fungi found in epiphytic crusts may play a more important role in controlling dune stability than the abundance and distribution of higher plants. Earlier work by Ash & Wasson (1983) suggested that wind-blown sand transport falls close to zero if vegetation cover exceeds 30%, but more recent studies indicate that this is only the case if the intervening soil surface is crusted. If the surface crust is disturbed, sand transport may be significant. There are also many instances, including parts of the southern Negev, where dune and sand sheet surfaces are almost entirely stablised by surface crusts and have a vegetation cover significantly lower than 30% (Goring-Morris & Goldberg, 1990; Rendell *et al.*, 1993). Dune reactivation can occur if the crust is broken up due to trampling or off-road vehicle pressure and if present-day wind energy is sufficiently high, without any reduction in rainfall. Such an event is frequently accompanied by increased dust flux from the surface and enhanced deposition in areas downwind (Pye, 1989).

ENVIRONMENTAL CHANGE IN DRYLANDS — WAYS FORWARD

It is apparent from this chapter than the scientific research community now has many insights into a wide range of gemorphological and vegetation responses to different types of environmental disturbance. Furthermore, it has at its disposal many new techniques that will enable processes to be monitored and Quaternary and Holocene changes to be fixed in time.

However, such research is wide-ranging: many of the individual techniques are specialised and the vital linkage between vegetation and geomorphology all point to multidisciplinary research involving geomorphologists, ecologists, biogeographers, plant physiologists, hydrologists, climatologists and remote sensors. A further problem is that much of the research is carried out by scientists from outside the world's drylands. An important issue therefore may be the choice of a series of contrasting, fully equipped dryland sites supported by remotely sensed data where the international community can undertake collaborative research.

REFERENCES

Amara, S. S. (1993). Rainfall-runoff relationships for the Gambia River. Reading Geographical Papers 112, Department of Geography, University of Reading.

Amundsen, R. G., Chadwick, O. A., Sowers, J. M. and Doner, H. E. (1988). Relationship between climate and vegetation and the stable isotope chemistry of soils in the eastern Mojave Desert, Nevada. *Quaternary Research,* **29**, 245–54.

Anderson, D. M. (1984). Depression, dust bowl, demography and drought: the colonial state and soil conservation in East Africa during the 1930s. *African Affairs,* **83**, 321–43.

Ash, J. E. and Wasson, R. J. (1983). Vegetation and sand mobility in the Australian desert dunefield. *Zeitschrift fur Geomorphologie Supplementband,* **45**, 7–25.

Beinart, W. (1984). Soil erosion, conservationism, and ideas about development in Southern Africa. *Journal of Southern African Studies,* **11**, 52–38.

Bowler, J. M. and Teller, J. T. (1986). Quaternary evaporites and hydrological changes, Lake Tyrell, north-west Victoria, *Australian Journal of Earth Sciences,* **33**, 43–63.

Brook, G. A., Burney, D. A. and Cowart, J. B. (1990). Desert palaeoenvironmental data from cave speleotherms with examples from the Chihuahuan, Somali-Chalbi and Kalahari Deserts. *Palaeogeography, Palaeoclimatology and Palaeoecology,* **76**, 311–29.

Bryant, R. G. (1993). The sedimentology and geochemistry of non-marine evaporites on the Chott el Djerid, using both ground and remotely-sensed data. Unpublished PhD thesis, University of Reading.

Cerling, T. E. (1990). Dating geomorphological surfaces using cosmogenic $_3$He. *Quaternary Research,* **33**, 148–56.

Chawla, S., Dhir, R. P. and Singhvi, A. K. (1992). Thermoluminescence chronology of sand profiles in the Thar Desert and their implications. *Quaternary Science Reviews,* **11**, 25–32.

Corbett, I. (1993). The modern and ancient pattern of sandflow through the southern Namib deflation basin. In Pye, K. and Lancaster, I. (eds.), *Aeolian Sediments, Ancient and Modern,* IAS Special Publication 16, Blackwell Scientific Publications, Oxford, pp. 43–58.

Clark, I. D. and Fontes, J.-C. (1990). Palaeoclimate reconstruction in northern Oman based on carbonates from hyperalkaline groundwaters. *Quaternary Research,* **33**, 320–36.

Danaher, T. J., Carter, J. O., Brook, K. D., Peacock, A. and Dudgeon, G. S. (1992) Broad-scale Vegetation Mapping using AVHRR Imagery. *Proceedings of the 6th Australasian Remote Sensing Conference,* Wellington, New Zealand, Vol 3, 128–137.

DeDeckker, P. (1983). Australian salt lakes: their history, chemistry and biota — a review. *Hydrobiologica,* **105**, 231–44.

Douglas, T. D., Kirkby, S. J., Critchley, R. W. and Park, G. (1994). Agricultural terrace abandonment in the Alpujura, Andalucia. *Land Reclamation and Rehabilitation* (in press).

Edwards, S. R. (1993). Luminescence dating of sand from the Kelso Dunes, California. In Pye, K. (ed.), *The Dynamics and Environmental Context of Aeolian Sedimentary Systems,* Geological Society Special Publication 72, Geological Society Publishing House, Bath, pp. 59–68.

Enzel, Y. (1992). Flood frequency of the Mojave River and the formation of late Holocene playa lakes, southern California. *The Holocene*, **2**, 1–18.

Glennie, K. W. (1987). *Desert sedimentary environments*, present and past – a summary. *Sedimentary Geology*, **50**, 135–65.

Goring-Morris, A. N. and Goldberg, P. (1990). Late Quaternary dune migration in the southern Levant: archaeology, chronology and palaeoenvironments. *Quaternary International*, **5**, 115–37.

Goudie, A. S. (1990). *Techniques for Desert Reclamation*, Wiley, Chichester.

Goudie, A. S. and Middleton, N. J. (1992). The changing frequency of dust storms through time. *Climatic Change*, **20**, 197–225.

Hjort af Ornais, A. and Sahlih, M. A. M. (eds.) (1989). *Ecology and Politics: Environmental Stress and Security in Africa*. Scandinavian Institute of African Studies, Uppsala.

Kjekhus, H. (1977). *Ecology Control and Economic Development in East African History*, Heinemann, London.

Klien, R. G., Cruz-Uribck, K. and Beaumont, P. B. (1991). Environmental, ecological and palaeoanthropological importance of the late Pleistocene mammalian fauna from Equus Cave, Northern Cape Province, South Africa. *Quaternary Research*, **36**, 94–119.

Kocurek, G., Havholm, K. G., Deynoux, M. and Blakey, R. C. (1991). Amalgamated accumulations resulting from climatic and eustatic changes, Akchar Erg, Mauritania. *Sedimentology*, **38**, 751–22.

Langbein, W. B. and Schumm, S. A. (1958). Yield of sediment in relation to mean annual precipitation. *Transactions American Geophysical Union*, **39**, 1076–84.

Last, W. M. (1989). Continental brines and evaporites of the northern Great Plains of Canada. *Sedimentary Geology*, **64**, 207–21.

Leys, J. F. (1991). The threshold friction velocities and soil flux rates of selected soils in S. W. New South Wales, Australia. *Acta Mechanica Supplementum*, **2**, 103–12.

Littman, T. (1991). Dust storm frequency in Asia: climatic control and variability. *International Journal of Climatology*, **11**, 393–412.

Livingstone, I. and Thomas, D. S. G. (1993). Modes of linear dune activity and their palaeoenvironmental significance: an evaluation with reference to southern African examples. In Pye, K. (ed.), *The Dynamics and Environmental Context of Aeolian Sedimentary Systems*, Geological Society Special Publication 72, Geological Society Publishing House, Bath, pp. 91–101.

Lowenstein, T. K. and Hardie, L. A. (1985). Criteria for the recognition of salt pan evaporites. *Sedimentology*, **32**, 627–44.

Mainguet, M. (1991). *Desertification. Natural Background and Human Mismanagement*, Springer Verlag, Berlin.

Magaritz, M. and Jahn, R. (1992). Pleistocene and Holocene soil carbonates from Lanzarote, Canary Islands, Spain: palaeoclimatic influences. *Catena*, **19**, 511–19.

Middleton, N. J. and Thomas, D. S. G. (1992). *World Atlas of Desertification*, Edward Arnold and UNEP, London and Nairobi.

Middleton, N. J., Goudie, A. S. G. and Wells, S. G. (1986). The frequency and source areas of dust storms. In Nickling, W. G. (ed.), *Aeolian Geomorphology*, Allen and Unwin, New York, pp. 237–59.

Millington, A. C. (1992). Changing environments: some remote sensing perspectives on land degradation. *Proceedings of the 6th Australasian Remote Sensing Conference*, Wellington, New Zealand, Vol 1, 1–10.

Millington, A. C., Drake, N. A., Townshend, J. R. G., Quarmby, N. A., Settle, J. J. and Reading, A. J. (1989). Monitoring salt playa dynamics using Thematic Mapper data. *IEEE Transactions on Geoscience and Remote Sensing*, **27**, 754–61.

Millington, A. C., Wellens, E. J., Saull, R. J. and Settle, J. J. (1993). Explaining and monitoring land cover dynamics in drylands using multitemporal analysis of NOAA-AVHRR imagery. In Curran, P. and Foody, G. (eds.) *Environmental Remote Sensing from Regional to Global Scales*, (in press), Wiley, Chichester.

Millington, A. C., Crosetti, M., Saull, R. J. and Archer, G. (1994). Mapping land cover in changing environments using AVHRR-NDVI data: the development and implementation of a methodology in Pakistan. *Remote Sensing of Environment* (in press).

Muhs, D. R. and Maat, P. B. (1993). The potential response of eolian sands to Greenhouse Warming and precipitation reductions on the Great Plains of the United States. *Journal of Arid Environments* (in press).

Nanson, G. C., Price, D. M., Short, A., Young, R. W. and Jones, B. G. (1991). Comparative uranium–thorium and thermoluminescence dating of weathered Quaternary alluvium in the tropics of northern Australia. *Quaternary Research*, **35**, 347–66.

Nanson, G. C., Chen, X. Y. and Price, D. M. (1992). Lateral migration, thermoluminescence chronology and colour variation of longitudinal dunes near Birdsville in the Simpson Desert, central Australia. *Earth Surface Processes and Landforms*, **17**, 807–20.

Nobel, P. S. (1992). Annual variations in flowering percentage, seedling establishment and ramet production for a desert perennial. *International Journal of Plant Sciences*, **153**, 1.

O'Hara, S. L., Street-Perrott, F. A. and Burt, T. P. (1993). Accelerated soil erosion around a Mexican highland lake caused by prehistoric agriculture. *Nature*, **362**, 48–51.

Page, K. J., Nanson, G. C. and Price, D. M. (1991). Thermoluminescence chronology of late Quaternary deposits on the Riverine Plain of S. E. Australia. *Australian Geographer*, **22**, 14–23.

Parry, M. (1990). *Climate change and world agriculture*, Earthscan, London.

Pye, K. (1989). Processes of fine particle formation, dust source regions, and climatic change. In Leinen, M. and Sarnthein, M. (eds), *Palaeoclimatology and Palaeometeorology: Modern and Past Patterns of Global Atmospheric Transport*, NATO ASI Series C, Volume 282, Kluwer, Dordrecht, pp. 3–30.

Pye, K. (ed.) (1993). *The Dynamics and Environmental Context of Aeolian Sedimentary Systems*, Geological Society Special Publication 72, Geological Society Publishing House, Bath.

Rachocki, A. H. and Church, M. (1990). *Alluvial Fans: A Field Approach*, Wiley, Chichester.

Radtke, U., Bruckner, H., Mangini, A. and Hausmann, R. (1988). Problems encountered with absolute dating (U-series, ESR) of Spanish calcretes. *Quaternary Science Reviews*, **7**, 439–45.

Rendell, H. M., Yair, A. and Tsoar, H. (1993). Thermoluminescence dating of periods of sand movement and linear dune formation in the northern Negev, Israel. In Pye, K. (ed.), *The Dynamics and Environmental Context of Aeolian Sedimentary Systems*, Geological Society Special Publication 72, Geological Society Publishing House, Bath, pp. 69–74.

Rosen, M. R. (1991). Sedimentologic and geochemical constraints on the evolution of Bristol Dry Lake Basin, California, USA. *Palaeogeography, Palaeoclimatology, Palaeoecology*, **84**, 229–57.

Savage, M. (1991). Structural dynamics of a southwestern pine forest under chronic human influence. *Annals of the Association of American Geographers*, **81**, 271–89.

Singhvi, A. K. and Kar, A. (eds.) (1992). *Thar Desert in Rajastan: Land, Man and Environment*, Geological Survey of India, Delhi.

Skujiņš, J. (ed.) (1991). *Semi-arid Lands and Deserts. Soil Resource and Reclamation*, Marcel Dekker, New York.

Stocking, M. (1983). Farming and environmental degradation in Zambia: the human dimension. *Applied Geography*, **3**, 63–77.

Stockton, P. and Gillete, D. A. (1990). Field measurements of the sheltering effect of vegetation on erodible land surfaces. *Land Degradation and Rehabilitation*, **2**, 77–85.

Stokes, S. (1992). Optical dating of young (modern) sediments using quartz: results from a selection of depositional environments. *Quaternary Science Reviews*, **11**, 153–9.

Stokes, S. and Breed, C. S. (1993). A chronostratigraphic re-evaluation of the Tusayan Dunes, Moenkopi Plateau and southern Ward Terrace, northeastern Arizona. In Pye, K. (ed.), *The Dynamics and Environmental Context of Aeolian Sedimentary Systems*, Geological Society Special Publication 72, Geological Society Publishing House, Bath, pp. 75–90.

Teller, J. T. and Last, W. M. (1990). Palaeohydrological indicators in playas and salt lakes, with examples from Canada, Australia and Africa. *Palaeogeography, Palaeoclimatology, Palaeoecology*, **76**, 215–40.

Thomas, D. S. G. (1989). Reconstructing ancient arid environments. In Thomas, D. S. G. (ed.), *Arid Zone Geomorphology*, Belhaven, London, pp. 311–34.

Thomas, D. S. G. and Shaw, P. (1991a). *The Kalahari Environment*, Cambridge University Press, Cambridge.

Thomas, D. S. G. and Shaw, P. (1991b). Relict dune systems: interpretations and problems. *Journal of Arid Environments*, **20**, 1–14.

Thomas, D. S. G. and Tsoar, H. (1990). The geomorphological role of vegetation in desert dune systems. In Thornes, J. B. (ed.), *Vegetation and Erosion*, Wiley, Chichester, pp. 471–89.

Thornes, J. B. (1990) (ed.) *Vegetation and Erosion*, Wiley, Chichester.

Torgenson, T., DeDeckker, P., Chivas, A. R. and Bowler, J. M. (1986). Salt lakes: a discussion of processes influencing palaeoenvironmental interpretation and recommendations for future study. *Palaeogeography, Palaeoclimatology, Palaeoecology*, **54**, 7–19.

Walling, D. E. and Kleo, A. H. A. (1979). Sediment yields of rivers in areas of low precipitation. In *The Hydrology of Areas of Low Precipitation*, IAHS-AISH Publication 128, Wallingford, 479–93.

Warren, A. and Khoglai, M. (1992). *Assessment of Desertification and Drought in the Sudano-Sahelian Region 1985–1991*, United Nations Sudano-Sahelian Office, New York.

Webb, J. L. A. (1992). Ecological and economic change along the middle reaches of the Gambia River, 1945–1985. *African Affairs*, **91**, 543–65.

Wiesanga, P. J., Hendrickx, J. M. H, Nash, M. H., Ludwig, J. and Duagherty, L. A. (1987). Variation of soil and vegetation with distance along a transect in the Chihuahuan Desert. *Journal of Arid Environments*, **13**, 53–63.

Wintle, A. G. (1993). Luminescence dating of aeolian sands: an overview. In Pye, K. (ed.), *The Dynamics and Environmental Context of Aeolian Sedimentary Systems*, Geological Society Special Publication 72, Geological Society Publishing House, Bath, pp. 49–58.

Yair, A. (1990). The role of topography and surface cover upon soil formation along hillslopes in arid climates. *Geomorphology*, **3**, 287–300.

Index

Abri II de Puechmargues (France)   288, 289
Acacia ehrenbergiana   387, 388
Acacia raddiana   388
Achir (Algeria)   188
Adige Valley   301
adivasi (tribal)   413
Advat-Mitze Ramon (Israel)   220
aeolian
   activity   2–3, 179, 190, 197
   deposition   403
   erosion   437
   processes   239, 247
   saltation   184
   suspension   184, 186
aerodynamic roughness   403
Aerva javanica   387
age classes of trees   329
agricultural
   development   390
   frontier, in Mexico   132
   origins   254, 261
   self-sufficiency, promotion of   231
   systems, abandoned   220
   systems, ancient   220–222
Agriophyllum squarrosum   403
agro-ecology   413
Aïn Ben Noui (Algeria)   179
Alberta   329, 330, 331, 337
Alcañiz (Spain)   114–115
aleppo pine, *see Pinus halepensis*
algal micro-fossils   241
Algeria   177, 192
alluvial fan
   aggradation   68, 78, 81
   aggradation and dissection sequences   68, 70
   apex   78
   dissection   68, 78
   fan-head dissection   68
   fan-head trenching   1, 71, 77, 78, 82
   progradation   69, 71, 79, 82
   systems   67
   telescopic dissection   45–50
alluvial fans   1
   mountain front   67
   switch from a sediment-rich debris flow regime to a sediment-poor fluvial regime   75

alluviation   114
alluvium   184
Almeria (Spain)   85
almonds   248
Alnus (alder)   139
Alnus glutinosa   332
Altithermal   7
amorphous inorganic iron   100
AMS ^{14}C dating   114, 139
Anabasis articulata   97, 98
ancient agricultural systems, spatial distribution   203, 220–221
Angolan Plateau   333
animal husbandry   190
Animas River, Colorado   324, 339
Animas River floodplain   323, 334, 330
annual plants   259, 261, 262, 263, 264, 388
annuals, shallow-rooted   388
Anomoeoneis costata   136
Anomoeoneis sphaerophora   136
anthocyanin   266
anthropogenic disturbance, effects of   300
anthropogenic impacts   17–23
anthropogenic interpretation   190–191
anthropogenic interference in the landscape   288
Anthyllis spp.   309
Anthyllis terniflora   94, 97
Ap soil horizon   398, 399
Aqua-Viva 'Centenarium'   188
aragonite   153, 159
Aral Sea   19, 20
Aravalli Hills (India)   413–425
archaeological material in valley bottomlands   129
archives, use of   434, 435
Argaric (Early Bronze Age)   301
arid floor, in bioclimatological classification   184
arid Neolithic climate   191
aridification   177, 191, 246
aridisols   44
aridity indices   201, 202, 223
arroyo   1
   infilling   114
   problem, the   281
Artemesia   119, 125, 128, 155, 159, 243

Artemesia barrelieri 94, 97
Artemesia ordesica 403, 407–8
Artemesia sphaerocephaia 403
Artemisia steppe 116, 125
artificial ecosystem 402
Asparagus horridis 97, 98
Asphodelus microcarpus 94, 97
Aswan 379, 380, 383, 389
Aswan High Dam (Egypt) 21, 379, 380, 389
Augéry (France) 288, 289
Auob 26–28, 32, 34
Aurés (Algeria) 179, 184
Australia 12 16
auto-stabilisation processes 87, 91
AVHRR 347, 348, 352, 353, 359, 362, 363
Azadirachta indica 421

badland
  evolution 86
  stabilisation 86
badlands 86
  multiple-age 86
Baheira Governorate (Egypt) 366, 368, 369
Baheira Province (Egypt) 378
balsam poplar, *see Populus balsamifera*
bankful level 323
Baotou 395
Barbula ditrichoides Brth. 404
bare soil 370
barley 241
  cultivation 235
Basin of Mexico 141
Bassia dasyphylla 403
Baume de Montclus (France) 288, 289
bazinas 187
bCu soil horizons 399
Beatton River (British Columbia) 323, 336
bedouins 388
Beer Sheva (Israel) 201, 202, 207, 221
Beer Sheva depression 203, 215
Beer Sheva plain 224
Bell Beaker style 286
Belly River (Canada) 337
Beni Ulid (Libya) 230, 231, 233, 248
Bentîous (Algeria) 188
Benue River (Nigeria) 333
Bhils 415
biocrusting 210, 400–408
biodiversity, loss of 428
biological
  activity, limiting factor 200
  diversity, of playas 348, 361, 363
  production, of playas 348, 361, 363
  productivity, of playas 363
  transformations 232

biomass 306–20, 387
Bir Labrach (Algeria) 191
Biskra (Algeria) 179, 181, 184, 188, 190
Bled Daya (Algeria) 188
Bled el Mahder (Algeria) 187, 188, 191
Bled el Mazoucchia (Algeria) 187
Bled Sallaouine (Algeria) 188
blow-outs 405
Blue Nile 380
Boteti River (Botswana) 27, 36
Botswana 26–38
bottomland forests 326
boulder-built 'sluices or overflows' 233
Boussargues (France) 289
Bow River (Alberta) 334
Breckland 325
Bromus spp. 94, 97
bryophytes 403
Bryum argenteuum Hedw. 404
Buellia spp. 94, 96, 97, 98, 99
buried fences 273, 278–280
burrowing animals 218
BWh climate 366
Byzantine period 188

calcite 138, 153, 159
calcium 256, 264, 382
  carbonate 100, 368, 373
    accumulation 71
    leached 368, 373
calcrete 86
caliche rubble 72
California 7, 10, 19
California Juniper, disappearance of 79
camels 388
Camprafaud (France) 288, 289
capillary fringe 369
Caragana korshinskii 403, 407–408
Cardial level 288
catastrophic events 281
catastrophic floods 199
catchment hydrology, historical changes in 362
caves, archaeological and sedimentological evidence 431
Cedrus atlantica 160
centre pivot irrigation 367
centric diatoms 243
ceramic pottery assemblages 231
cereal crops 248
cereal cultivation 246, 247, 248
cereal-type pollen 246
Cerro del Gallo (Spain) 301
Chaetoceros muelleri 136
Chalcolithic 286, 288, 300, 301

change in the surface properties 199
channel
  entrenchment 114
  migration 274, 276, 278
  migration, rate of 275, 276
  processes 323
channels, changes in 273, 274–278, 323
Chara 150
Chara oogonia 153, 168
charcoal making 389
Chasséen 286, 288
check dams 235
chemical treatment, for stabilisation and reclamation 397
Chenopodiaceae 199, 122, 125, 128, 155, 159, 243, 246
Chenopodiaceae-Amaranthaceae (cheno-ams) 134
Chergui (Algeria) 192
Chichimecs 136
China 192, 393–411
Chinese Green Wall 192
Chobe River (Botswana) 326
Chott Djerid (Tunisia) 350, 351, 359, 360, 361, 362, 363
Chott el Fedjaj (Tunisia) 350, 351, 359, 361, 362, 363
Chott el Guettar (Tunisia) 350, 351, 359, 361, 362, 363
Chott el Hodna (Algeria) 177, 178, 188
Chott Melrhir (Algeria) 179
Chott Merouane (Algeria) 179
Chupícuaro 132
Chuska Mountains (New Mexico) 435
Chydorus sphaericus 154
Cima volcanic field 68
cisterns 222, 233, 235, 239
  for water-storage 233
citrus 365, 370 3
Classic Culture, of Mexico 132
clay horizon, impermeable 373
climate change 290
climatic change 199, 281, 290, 352, 415
  and human disturbance, affect on vegetation 134
  effect on ecological processes 427
  effect on geomorphological processes 427
  inability to differentiate in the palaeoenvironmental record 428
  recent 254
climatic changes, possible impacts of 305
climatic factors and erosion 301
climatic reconstruction 191
climbing dunes 239
closed
  basins 347

surface water hydrology 349
clover 368
coarse-textured soils 369
Cocconeis placentula 153
colluvial
  deposits 2, 289, 407
  loess soils 206
  slopes 73–79
colluvio-aeolian accumulations 184
colluvio-alluvial aprons 179
colluvium 179, 184
colonizing ability 263, 263
Colorado Plateau 29
Colorado River 19, 337
commission-omission matrix 370, 371
competing species 306
complex response 68, 79, 82
Complex Terminal (aquifer in Tunisia) 351
Compositae 243, 247
Condamine (France) 286
Continental Intercalaire (aquifer in Tunisia) 351
Corispermum hyssopifolium 403
cosmogenic isotope dating 430
cotton 188, 261
coupling channel and fan systems 67
creek channels 274–78
critical ecological margins 431
  dry woodlands 431
  floodplain communities 431
  lake shorelines 431
  montane treelines 431
  swamp shorelines 431
crop
  cultivation 188, 190–91
  growth 368
cropping patterns 370
crops, domesticated 255
cross-wadi walls 235, 237
Cruciferae 243, 246
crustose lichens 109
crystalline iron oxides 100
cultivation during Roman Era 190, 191
cut-and-fill 1
Cyclotella krammeri 153–154

dabuka (camel train route) 388
dam construction 16, 21, 322, 337
  long-term impacts 339
dambos 32
dams 188, 336–9
  contribution to the failure of poplar regeneration 337
  downstream impacts 322, 336–9
Daphnia 154, 160

date-palms 248
dating of geomorphological surfaces 44
Dead Sea (Israel) 166
Death Valley (USA) 68, 70, 79
debris cones 70
debris flow deposition 71
debris flows 70–71, 75, 76–77, 78, 81, 301, 302
 transitional and fluvial facies 71
Deception Valley (Botswana) 26
DECORANA 92–93
deep-rooting species 309, 387
deep-weathering 32, 36
deferration 51, 53, 62
deflation 348, 362
deforestation 289
deltas 17
denudation rates 3
desert
 margin ecosystem 395
 pavement 68, 71
 rock varnish 68
 scrub 79
desertification 19, 177, 191, 393, 413
 anthropogenic 191
 fight against 192
Desmids 154, 160
development
 ecologically-sound 390
 potential 388–90
diaspore 255, 256, 257, 259, 261, 263, 264
diatoms 136, 153, 168, 241
digital image processing 370–4
digital terrain model 290, 294
Diospyros mespiliformis 333
diploid 256, 257, 258, 261, 263, 266
Diploschistes dicapsis 96, 97, 98
Diplotaxis crassifolia 96, 97
divergence matrices 370–1
Djebel Tenia (Algeria) 179, 182, 184, 186, 187, 191
Djebel Tenia piedmont 191
Djedi Wadi (Algeria) 188, 189
Djerid–Fedjaj basin (Tunisia) 351
dolomite 153, 159
domestic stock 273, 275
domesticates 259, 262
Donezan (France) 288
downcutting, rate of 278
drainage
 canal systems 367
 channels, lack of maintenance of 374
 density 2, 3, 6
 network, transporting capacity of 213
Drotsky's Cave (Botswana) 28, 29
drought 273, 274

 centred on 4600 yr BP in Mexico 141
drought stress, on floodplain vegetation 334
dryland valley system development 26, 29–38
dum 26
dune
 fields 2–3
 stabilization 397, 400–408
dunes 179
 stabilization 397, 400–408
duricrust development 29
durum 264, 266
dust storms 2, 4

Early Bronze Age 286, 289, 290, 301
Early Holocene 247
Early Iron Age 301
Early/Middle Holocene 248
Early to Middle Neolithic 302
East African monsoon 380
eastern Algeria 177
Ebro Basin (Spain) 113–114
ECD (electrical conductivity of drainage water) 368
ECI (electrical conductivity of irrigation water) 368
ecological impacts, of lake formation 382
ECP (electrical conductivity of saturated soil paste) 368
edaphic 256–7, 266, 320
edaphics 257, 264
Egypt 365–376, 379–390
Egyptian Environmental Affairs Agency 389
Eleagnus angustifolia 192
elevation, within a floodplain 322
elimination of ligneous perennials 192
embodied energy 323
emmer 254, 264, 265, 266
energy-related approach, for the examination of vegetation and geomorphology 323
entrenchment, of Holocene river valleys 129
environmental change 28, 231, 235, 248, 254, 321, 379
 effect on people's livelihoods 428
 in floodplains 339
 long-term 339–40
 recent 434–435
 triggers of 428
environmental degradation 413
 anthropogenic and climate-induced 114
 models to explain patterns of in eastern and southern Africa 435
environmental fluctuations and stresses 249
environmental research, interface with social science 428
Ephedra 243, 246

Ephedra przalwalskii 403
ephemeral lakes 348, 353, 354
　areal and volumetric changes 348
　residence times of 360
ephemerals 387
Epicardial 288
epiphytic crusts 437
　role in dune stabilisation 437
episodic uplift 87
Eragrostis aegyptiaca 388
erosion 271, 272, 273, 274, 280, 281, 294, 295, 300, 305–321
　effects on productivity 308
　rates 213, 308
　simulation 286, 290
erosion/stabilisation
　cycles 87
　sequences 86
erosion-resistant species 290
erosional response, vegetation effects on 306
Ethiopia 380
Ethiopian Highlands 380
Eucalyptus camaldulensis 274, 276, 278, 326
Eucalyptus spp. 415
Euphorbia spp. 415
Euzomodendron bourgeanum 94, 97, 98, 99
evapotranspiration 407
excess stream power 78, 82
extreme rainfall events 294, 298, 300, 302

Fagonia cretica 96–97
farming systems 233
Faulkanhagan Creek (Australia) 277, 280
Faulkanhagan Tank (Australia) 277, 280
feldspar dating 430
fence posts 280
ferns 246
Fertile Crescent 253, 259, 263
fibre production 389
Ficus sycomorus 333
figs 248
Final Bronze Age 301
finite difference model 290
flash floods 383
flood
　irrigation 367
　magnitude and frequency 331
　regime 322
　surge, of the Nile 383
flood-water
　farmers 231, 232
　farming 231, 232–237, 239, 248, 249
　　systems 233
　　abandonment 248
flooding 382

flooding ratio 347–63
　geomorphological and biological inferences 348
flooding regime 322
floodplain
　disturbance 322
　　patterns 322
　　processes 322
　　regimes 322, 325, 339
　　thresholds 325, 336, 339
　elevation 322
　forest ecology 322, 336
　forest regeneration 332
　forests 322, 325, 326, 329, 331, 333, 334
　geomorphology 322
　landforms 322
　sediments 322, 323, 329
　species distribution 329
　trees 322
　turnover 323
　vegetation 322, 331, 332, 334, 336, 339
　　disturbance processes 323
　　mosaics 323
　water tables 337
floodplains 321–336
　disturbance models 322
fluid abstraction 17
flushing of salt from topsoil 389
fluvial activity, enhanced 79
Font Juvénal (France) 288, 289
Fontbouïsse culture 286
Fontbuxian 290
food production 389
forage production 388, 389
forest pioneer stage on new sediment 323
fossatum 190
fossil soil 179, 183, 184
Foun d'en Peyre II (France) 288
Frankenia thymifolia 97, 98
free iron oxides 100
freeze–thaw 81
freshwater penetration 382
frost action 79, 81
fuel production 389
Fulgensia fulgens 94, 96, 97, 98, 99
fungal remains 241

Gambia River 435
gamma distribution 313
garrigue 286, 291–293
Garrison Dam (North Dakota) 337
Gasr Mm10 (Libya) 233
Gazel (France) 288
Gemellae castrum (Algeria) 107, 108
General Circulation Models (GCMs) 7, 9, 199

geochemical
  analysis   116–117
  composition   399
geological timescale, importance in
    understanding geomorphological
    processes   428
geomorphological
  change   273
  evidence of environmental change   429
  surfaces   44
Gilf Kebir (Egypt)   29
glacis d'erosion   429
Glen Canyon Dam (Colorado River)   337
Glinus lotoides   388
global warming   4, 7–9, 427
Gobi desert   192
gobi deserts   393
goethite   53, 57, 62, 63
Gomphonema gracile   154
grain amaranth   134
Gramineae   153, 159, 243, 246
grape presses   232
grapes   248
grasses, the world's largest-seeded   260
grazing   281, 379, 382, 388
  banning   192
  of sheep   280
  potential   382, 388
  resources, traditional   382
Great Erg Oriental   177
Great Plains (USA)   7, 10–11
Great Sandy Desert (Australia)   16
Great Wall of China   192
Greece   302
Green Dam (Algeria)   192–197
greenhouse effect   199, 281
greenhouse gases   4
Grerat D'nar Salem (Libya)   237–248, 249
Grotte de Collier (France)   301
ground subsidence   17
groundwater   21, 28–29, 386, 429
  circulation   32
  erosion   29, 37
  extraction, increase in   362
  recharge   28–29
  sapping   29, 32, 34, 36–38
  seepage   349, 351, 360, 381
  shallow   381
  table   369, 372–3
    depth   368–9
    position   369
groundwater-fed lakes   352
growth models, of plants   311
gullying   184, 233, 235
gypsiferous
  crust   44, 184
  soil   186–187
gypsum   100

halophytes   122
Haloxyron ammodendron   403
Halutza sands   221
Hammada tamariscifolia   94, 97
HAPEX-Sahel Experiment   428
Harmattan   192
Hedysarum laeve   403
Hedysarum scoparium   403, 407–408
Helianthemum almeriense   94, 97
hematite   53, 57, 62, 63, 153, 159
Hemilepistus reamuri   219
herbivores
  domestic and feral   273–4
  introduced   280–281
heterocarpy   257, 258, 262, 263
High Atlas Mountains   147, 160
High Plains (USA)   2
hillslope
  debris flows   73, 75–78, 81
  dissection, by gullying   78
  erosion   288
  processes   68
  runoff collection (water harvesting)   222
  sediment supply   67
  stripping   79
hillwash   301
  sequences   289
historical (archive) data   348, 434, 435
Hodna Basin (Algeria)   177, 192
Hoh River valley (USA)   333
holm oak, see *Quercus ilex*
holm oak association   288, 290
Holocene   162–171, 186, 215, 331, 332
  alluvial sediments   333
  Atlantic Optimum   7
  environmental change   113, 271
  lakes   114
  superficial deposits   238
  vegetational history   237, 247
Hordeum spp.   94, 97
Hovav Plateau (Israel)   218–220
Huelva (Spain)   301
human
  impacts, on flora   254, 261, 262, 263, 267, 427
  influences, on dry woodlands   435, 437
    on plant evolution   253
    on the landscape   285
  mismanagement, leading to desertification   177
human-induced flooding   352
Hume River (Canada)   326

Hunirice 188
hunter-gatherers 238
hydraulic installations 188
hydro-electricity generation 380
hydrology 415
hydroperiod 321, 322, 331
hydrophobic organic coatings 405

Iberian (Iron Age) 301
Ile Corrège-Leucate (France) 288
Imilchil (Morocco) 149, 150
increased soil salinity 224
inertial k-type dependency 319
inertial persistent k-type species 309
infiltration 307
　capacity 85, 207
　rate, bare soil 306
　rates 85, 307–308
Institute of Desert Research, Academia Sinica (IDRAS) 393
intense erosion, during Purépecha period 142
inter-specific plant interaction 306
intermittent degradation 301
intermittent erosion 302
inundation 359, 387, 388
inundation, probability of 383
invasive opportunistic r-type strategy species 309
iron oxides 44
irrigated citrus 370, 371, 373
irrigation 367, 373, 380, 382, 389
　water, quality of 366
Islamic Era 188

jet-streams positions 81
Juniperus (juniper) 138, 155, 288
k-dependent species 315, 319
k-strategy 313
Kairouan (Tunisia) 354, 356, 357
Kalahari Desert 26, 28
Kalahari Group sediments 26, 34, 36
Kalahari–Zimbabwe Axis 36, 38
karstic lake system 168
Kherbet Djouala (Algeria) 188
Kullenberg piston corer 152
Kuruman River (Botswana) 37
Kuruman Valley (Botswana) 26–27, 37–38

La Hoya de San Nicolás de Parangueo (Guanajuato) (Mexico) 132–136
La Piscina de Yuriria (Guanajuato) (Mexico) 133, 136–138
laagte 26

Labiatae 243
lacustrine
　carbonates 153
　sedimentation 147
　sediments 68
Lago de Pátzcuaro (Michoacán) (Mexico) 133, 139–141
lake
　levels 81, 148, 163, 166, 168
　sediments, holistic approach to interpretation 430
　shorelines 68
Lake Abhé (Ethiopia–Djibouti) 166, 167, 170
Lake Azigza (Morocco) 168
Lake Bizerte (Tunisia) 349, 350, 361
Lake Bonneville (USA) 332
Lake Bosumtwi (Ghana) 166, 167, 170
Lake Chad 166, 167, 170
Lake Ichkeul (Tunisia) 349, 350, 361, 363
Lake Isli (Morocco) 148, 153, 160
Lake Mojave (USA), shorelines 71
Lake Naivasha (Kenya) 334, 335
Lake Nasser (Egypt) 379–390
Lake Nubia (Sudan) 380
Lake Sidi Ali (Morocco) 168
Lake Tigalmamine (Morocco) 163, 166, 167, 168–169, 171, 173
Lake Tukana (Kenya–Ethiopia) 166, 167, 170
Lake Victoria 334, 335
Lake Ziway–Shala (Ethiopia) 166, 167, 170
laminated turbidite facies 152
land
　clearance 300
　degradation 280
　enclosure 397
　mismanagement 393
　use 187–188, 190, 192
Landsat MSS 371, 373
landscape degradation history 288
Languedoc 288, 308, 301
Lanzhou (China) 395
Late Bronze Age 301
Late Classic (1400–900 yr BP) period intense drought 142
Late Holocene 248
Late Neolithic 289, 301
Late Pleistocene climate, of Moroccan Atlas Mountains 160
Late Pleistocene climate, of southwestern USA 79
Laval de la Bretonne (France) 286, 290, 294, 298, 301
Le Lébous (France) 286
leaching 373
Les Companelles (France) 286
Les Fraignants (France) 288, 289

Letlhakane Valley (Botswana)   25–27
Letlhakeng   27–28, 35–36
Libyan Pre-Desert   231, 237, 247
Libyan Valleys Project   231
lichen
   colonisation   85, 109
   cover   94, 106
   crust   106, 107
lichens   85, 106, 437
limes   187–188, 189
limestone   256, 264, 266
limiting runoff generation   206
Limonium insigne   97, 98
lineaments   35
linear dunes, vegetated   437
litter   306
livestock rearing   379
loess   203, 213–215, 218–220, 220, 222, 223, 225
   penetration   200, 213–215, 223
Loess Plateaux (China)   395
log skew Laplace distributions   239, 240
longitudinal dunes   203, 205, 211
Los Millares   301
loss-on-ignition (LOI)   241, 302, 405
lower Holocene dunes   179
lucerne   388
Lymnaea spp.   38

M'Doukal (Algeria)   179, 188
MAB Project (Man and the Biosphere)   389
Macedonia   302
Mackenzie Delta (Canada)   326
Mackenzie River (Canada)   321, 329
macronutrients   399
macroporosity   368
maghemite   53, 62, 63
Maghreb   164, 179, 190
Maghrebian margin of Sahara   163
magnesium   382
magnetic susceptibility   134
magnetite   53, 62, 63
maize
   cultivation   131
   domestication   131
Makgadikgadi Basin, *see* Makgadikgadi Depression
Makgadikgadi Depression (Botswana)   27, 28, 36
manuring   247
mapping and inventory, of soil salinization   366
mathematical simulation models   305
maximum
   likelihood classification   370–372

   vegetation capacity   306
mean
   reflectance vector   352–363
   storm infiltration rate   306
   storm rainfall rate   306
meander lobes   329
meander sinuosity   337
mechanical (salt) weathering   81
medbundi   422–423
medicinal use of vegetation   389
Medieval Era   188, 301
Mediterranean   87
Mediterranean chorotypes   218
mekgacha   26
Mesarfelta (Algeria)   188
meso-Mediterranean open woodland   116
Mesoamerica   131, 143
METEOSAT   347, 351
Mexica (Aztecs)   136
Mexico   131–132
micro-fossils, organic-walled   243, 246
microbial processes, in rock varnish formation   61
micronutrients   399
microwave remote sensing   434
midden deposits   232
middens   231
Middle Atlas   166, 168, 171
Middle Bronze Age   290, 301
Middle Pleistocene   184
milk thistle, *see Silybum marianum*
Milk River (Canada)   329, 331
mineral magnetic analysis   50–51, 52, 56–57, 101, 116–117
minero-organic topsoils   398
Miocene (Tortonian)   86
Miocene marls   100
miombo woodlands   333
Missouri River   337
Mitzpe Ramon   201
mixed olive-cereal-grass-animal husbandry   247
Mmone/Quoxo Valley system (Botswana)   26, 27, 34
mobile dunes   400
model validation   434
modelling   433
   microfossil and macrofossil distributions   431–3
modern
   analogues   235
   pollen rain   246
Mojave Desert   68, 79, 81
Mojave River, discharge at end of Pleistocene   81
Molopo Valley (Botswana)   26, 27, 32

Momba Station (Australia)   272–273
monitoring geomorphological and ecological processes   433
Montagne d'Alaric (France)   288, 289
Montpellier (France)   286
Montpellier garrigue   290
Morocco   147, 160, 163, 164, 166, 168
Moselebe Valley (Botswana)   26, 27
Msila (Algeria)   188
Mu Us Sandy Lands (China)   395
mudflats   359
multitemporal remotely sensed data   433
Munsell soil colours   51, 55–57
Murcia (Spain)   306, 311, 313, 315, 318, 319
Myriophyllum   155

NADW (North Atlantic Deep Water)   171, 173
Namibia   26
natural resources, conservation of   389
Navicula elkab   136
Navicula halophila   136
Ncamasere Valley (Botswana)   27, 32, 34
Near East   165
Nefzaoua (Tunisia)   351
Negev Desert (Isreal)   20
Negev Highlands (Israel)   199, 215, 220
Nemencha (Algeria)   179
Neogene   86
Neolithic Era   179, 184, 187, 254, 263, 267, 285, 288
Neovolcanic Axis (Mexico)   131, 133, 143
network orientation analysis   34
New South Wales   273
Newly Reclaimed Areas (Egypt)   366, 369, 376
Nile Delta   17, 366–367
Nile River   21, 380, 383
Nile Valley   17, 379, 389
Nile Water Agreements   380
Ningxia (China)   393
nitrate content   398
Nitzschia spp.   136
nomadic Arabs   188
non-governmental organisations   413
North American prairie rivers   325, 329
North Atlantic Deep Water formation   163
North-east Spain   113–116, 301
Nossop Valley (Botswana)   26, 27, 28, 32, 33, 34
nutrient
   runoff, in drainage   380
   stocks   308

Ogallala Aquifer (USA)   21

oil production   389
Okavango Delta (Botswana)   26, 27, 36
Okwa Valley (Botswana)   26, 27, 28, 34, 36
Olea spp.   119, 122, 128, 243, 246, 247, 248
olive
   cultivation   248, 248–9
   presses   232, 239
olives   248
omiramba   26
optically stimulated luminescence dating   114, 430
optimum humidity, of the Holocene in North Africa   186
Orange River (Southern Africa)   26
Ordos (China)   395
organic matter   308
   decay rate   308
organic-bound iron   100
ostracods   168
Oued Merguellil (Tunisia)   349
Oued Sed (Tunisia)   349
Oued Zeroud (Tunisia)   349
overexploitation, of the environment   300
overgrazing   191, 246, 382, 395
overland flow   233, 239, 307, 309, 315
oxbow   323
oxido-reduction zone depth   368–369

packrat middens   68
   fossils   81
palaeoclimatic interpretation, of desertification   184–187
palaeodunes   429
palaeoenvironmental reconstruction   68
palaeohydrology   322
palaeolake MegaChad   332
palaeosediments   332
palaeosols   210, 393, 408–409
palygorskite   184, 185
Panamint Mountains (USA)   79
pans   2
pastoral agriculturalists   238
pasture   191
pedogenesis   44, 81, 86
pedogenic calcrete   79
pedological descriptions   399, 409
Peiro Signado (France)   288, 289
perched water table   373
perennial
   irrigation, without adequate drainage   366
   species   218
   vegetation   210, 218
periodic
   disturbance of vegetation   323
   fluctuations, in water levels   380–381

lake inundation 383
submergence, of vegetation and soil 382
Persian Gulf 16–18
petrocalcic horizons 73
pH 382, 387
phenotypic plasticity 263
Phillyrea spp. 160, 288
Phragmites australis 124
phyllosilicate 153
Picea glauca 323
Pinus (pine) 122, 128, 134, 136, 138, 155, 243, 147
Pinus halepensis 288
Pistacia spp. 160, 243, 246
Plain of Kairouan (Tunisia) 349
plant
 biomass 388
 domestication 261, 262
 interactions 306
 response to climatic change 311
plant-erosion interactions 306
Plantago spp. 243
planting, of herbs to consider desertification 192
plastic wire mesh layout 192, 195
plateau-walls 233
plateaux-soils, loss of 247
playas 2, 347, 348, 349–51, 352–363
 contemporary geomorphological and hydrological processes 421
 fluctuations between margins and alluvial fans 431
Pleistocene Lake Mojav, see also pluvial Lake Mojave 70
Pleistocene–Holocene transition 68, 69, 70, 71, 75, 78, 79, 81
Pliocene 7, 86
Pluchea dioscoridis 331
pluvial Lake Mojave 68, 79, 81
point-bar
 regeneration 331
 sediments 323
point-bars 323, 327
polje 238, 247
pollen
 analysis 116, 117, 132–141, 152, 155, 241
 preservation, differential 241
Polpah Station (Australia) 271, 274, 275, 278, 281
Polygonaceae 243
polyploids 257, 258, 261, 264, 266, 267
Polypodiaceae 243, 246, 247
Polypodiaceae spores 247
Pongolo River (South Africa) 326
poplar 323, 325, 326, 329
 models of regeneration 329

mortality 334
regeneration 329, 337
replenishment mechanisms 329
poplars, decline in species downstream of dams 337
population decline 301
Populus angustifolia 334
Populus balsamifera 323, 326
Populus deltoides 329, 334
Populus deltoides x *Populus balsamifera* 330
Populus ilicifolia 326, 331
porcupines (*Hystrix indica*) 220
post-European
 geomorphological changes 272
 sedimentation 271
Postclassic Culture of Mesoamerica 132, 134, 142–143
Potamogeton 150, 155
Populus balsamifera 323, 326
Preclassic Culture of Mesoamerica 132, 134, 141
preservation potential of organic matter 431
primary domesticates 262
prismatic gypsum structure 184, 185
Prosopis juliflora 415
Protohistoric era 187
psammophilous shrubs 408
Psora decipiens 96, 97
Psoralea plicata 387
Puente del Regallo (Spain) 116, 117, 118
Pulicaria crispa 388
Purépecha (or Tarascans) 132
Pyrenees 302
pyrite 153, 159

Qal'a of the Bani Hammad 188
Qatar 17
quartz dating 430
Quaternary 86, 164, 166, 173
Quercus coccifera 116
Quercus ilex 288
Quercus spp. (oak) 122, 134, 155, 243
Quercus woodland 132, 136
quick response to rainfall 207

r-dependency 309, 315, 319
r-type strategy species 309, 319
radiocarbon dating 68, 160, 163, 288
rain, high intensity 186
rainfall 206–207
 interception 85
 properties 207
 simulation 106–107
 variability, inter-annual 306

rainfall–runoff modelling 435
rainwater harvesting 231
Rajasthan 413
Rambla de Tabernas (Spain) 86
Ramon Ridge (Israel) 201
rangeland degradation 280
reactivation of dune erosion 437
reclamation 366–367, 374
  of dune topography 366–367
Red Deer River (Canada) 331
Red Sea Hills 383
reduced
  runoff 224
  water availability 224
reduction zone depth 368–369
reforestation 413
regeneration
  levels 334
  niches 332, 326
  of trees 322, 325, 326, 336, 337
  patterns 332
  potential 326
  requirements 322, 326
  sites 322, 325
  strategies 326
regional pollen rain 246
regolith 73
remotely sensed data, calibration with ground data 433
removing ground cover, under olive trees 246
replenishment of floodplain water tables 337
research agenda for the study of dryland processes 434
reservoir construction, see dam construction
reverse sub-surface groundwater flow 386
Rhamno lycioidi–Querceto cocciferae sigmetum 116
Rhopalodia gibberula 136, 265, 266
rice 370
Rio Regallo (Spain) 114–116
riparian communities 322
River Red Gum, see *Eucalyptus camaldulensis*
rock/soil ratio 225
rock varnish 68, 71, 73
Roman Era 187–188
Roman wells 188
Romano-Libyan Era 231, 239
Romano-Libyan farming 247
Rooibrak Valley (Botswana) 26, 27
rooting depth, average 368
Ruisseau de Fournas (France) 288, 289
runoff 7, 9, 319
  changes in due to global warming 5
  changes on Russion Plain 21
  coefficient 222
  energy 213

  decrease in 213
  farms, ancient 222
  generation 203, 213, 225
  processes 203
  rates 213
  threshold of daily rain necessary to generate 206
runoff-contributing area 222
Ruppia 125, 128, 129

S'Gag (Aurés Range) (Algeria) 184
sabkhas 16–18
Sahara Desert 7, 163, 164, 167, 171, 173, 192, 197
Saharan Atlas 192
Saharan floor, in bioclimatological classification 187, 191
Saharo-Arabian species 218
Sahel 163, 164, 192, 197
Sahelian drought 163
Sahelian margin of Sahara 163
Salada Pequeña (Spain) 114
saladas (saline lakes systems) see also playas 114
saline lakes, endoric 114
salinity
  expansion of 373
  high levels of 369
salinization
  monitoring 373
  of soils 366, 368–369, 389
Salix psammophila 407
Salsola genistoides 94, 96, 97, 98
Salsola papillosa 97, 98
salt
  balance 199, 200
  content of sediment 241
  crust 349, 359
  exuded 389
  input 223
    by rainwater 214
  leaching 222, 224
  marshes 16–17
  oscillator model 163, 171
  pan 359
  scalding 280
salt-affected soils 365, 366
saltation 186
sand
  accumulations 179, 186
  barrier 408
  penetration 200
  seas 192
  storms 179
sand-strewn slopes 179

sandy wind-blown veils   179
scale
  spatial   428
  temporal   428
scaling-down   428–9
scaling-up   428–9
Schouwia thebaica   387
Schumm–Langbein curve   305, 348, 363
sea level rise   16–17
seasonality of climate   254, 259, 263, 264, 267
Sebkah el Hani (Tunisia)   349, 350, 354–358, 361, 363
Sebkah Kelbia (Tunisia)   349, 350, 354–358, 361–362, 363
Sebkah Moknine (Tunisia)   350, 351, 354–358, 361, 363
Sebkah Noual (Tunisia)   350, 351, 359, 360, 361, 363
Sebkha Mellala (Algeria)   166
Sede Boqer (Israel)   207, 215, 216, 217, 218, 219, 220
sediment
  accumulation rates   151
  conductivity   241
  cores   231
  production and transport mechanisms   79
  supply   78
    and transport relationships   68
  yield   6, 19, 151, 213, 271, 272, 348
sedimentary records of rivers   114
sedimentation   271, 273, 274, 277, 280, 423
  acceleration of   349
  rates   348, 349, 361, 362–363
sedimentological
  evidence for environmental change   429
  studies   239–241
seed
  bank   237
  dispersal   256, 257, 258, 259, 263
  size   257, 258, 259, 260, 261, 263, 264, 265, 266, 267
seedling establishment   322
seedlings, regeneration and establishment   322, 329
seismic survey   151–152
semi-arid/arid climatic boundary   351
semi-arid
  Australia   271, 273, 274, 280, 281
  Australian floodplains   326
  creeks   280
  ecosystems   273
  North American prairies   323
Senegal River   333
sensitivity to change   319
Serorome Valley (Botswana)   27, 35
Serratula flavescens   97, 98

Seva Mandir   413
Shabianzi region (China)   395
shamo deserts   393
Shapatou (China)   393
shoreline   68
  marginal zone   381
  movements   318–312
side-scan sonar   152, 154, 155
Sideritis spp.   94, 97
Sidi Abderhamane (Algeria)   188
Sigmopollis   243
silcrete   31, 35, 37
silt deposition   389
Silybum marianum   254, 263
simulated rainfall, hydrological response   206
Sinai sand field (Israel)   203
slope angle reduction   86, 87
slopes   429
slump facies   152
smectite   184
snails   218, 220
snowfall, changes in due to global warming   9
social stresses   301
Soda Mountains (USA)   68, 70, 79, 81
sodium, water-soluble   100
soil   70
  and water conservation techniques, traditional   413, 420–425
  conductivity   382, 387
  development, see also pedogenesis   68, 86
  erosion   150, 191, 225, 235, 306–320, 427
  fertility, maintenance of   389
  moisture   306, 309, 315
  profiles   68
  properties   206
  salinity   369–370
    measurements   370
  salinization   200, 214, 224
  texture   382, 387
  thickness   306
  truncation   306
sorghum   388
South-east Michigan (USA)   323
South-east Spain   86, 301
South-east USA   322, 326
South-west Aurés (Algeria)   187
South-west France   286
Southern Argolid (Greece)   302
sparker seismic reflection profiles   152
species
  biomass   306
  composition   388
  diversity, of snails   220
  extinction   306
  replacement   306
Sphincterochila zonata   220

sponge spicules   241
spores   241
SPOT XS imagery   367, 370–374
springs   222
Squamarina spp.   96, 97, 99
Sr/Ca ratio   136, 138, 141
St Mary's River (Canada)   337
St Pierre de la Fage (France)   288, 289
St-Guilhem-le-Désert (France)   289
stable isotope analysis   430
staggered erosional response   87
standing water, area on playas   353, 359
Station de Rosier (France)   286
Staurastrum pingue   154
Staurastrum planctonicum   154
Staurosirella leptostauron   154
Staurosirella pinnata   154
steppe   159, 191, 192, 247
  grassland   238
  grassy   247, 248
  vegetation   122, 191
Stipa tenacissima   94, 96, 97
stock numbers   273
stocking rates   273
stone
  accumulation   86
  cover development   85
  pavement   44
storm water management   231
storms, intense short-lived   383
straw checkerboards   393, 397, 400–8
structural control of valleys   34–36
sub-surface hydrology   382, 386
substantial arable and pastoral husbandry, development
successional
  patterns, observed on floodplains   322
  trends, through time   322
Sudan   380
surface
  properties   199, 206
    rapid alteration of   200
sustainable
  development   381, 388–40
  utilization of natural resources   388–340
synaptospermy   258

Tabernas Basin (Spain)   86
talus   73
talus slopes   75, 76, 78
Tamarindus indica   333
Tamarix africana   124
Tamarix nilotica   379, 387, 388
Tamarix spp.   403
Tana floodplain   323, 324, 331, 337

Tana floodplain forests   333–334
Tana River (Kenya)   323, 324, 327, 328, 329, 331, 334, 335, 337
Tectona grandis   415
Teloschistes lacunosa   97, 98
Tengger Desert (China)   405
tephra   68
Terrera Ventura (Spain)   301
Teruel (Spain)   114, 115
Teucrium polium   94, 97
Thar Desert (Pakistan)   7
thermoluminescence-dating   4, 239, 430
Thessaly (Greece)   302
Thubunae (= Tobna) (Algeria)   188
Thymus spp.   309
Thymus vulgaris   94, 97
timber production   389
Tobna (see Thubunae)
tolerance
  to flooding   326, 328
  to prolonged flooding   321
Toninia spp.   96, 97
total accumulated energy   323
Tournemire   289
trample burs   257
tree and cereal crop production, in wadi floors   235, 236, 238
Trianthema crystallina   387
Trichilia emetica   333
Triticeae   259
Triticum/Aegilops, see also wheat group   254
Triticum boeoticum, see also wild einkorn   254, 255, 257
Triticum dicoccoides, see also wild emmer   254, 255, 256
tropical cyclones   12–16
trout (*Salmo trutta*)   150
tumuli   188
Tunisia   164, 191, 347–363
Tunisian Southern Atlas   44–50
turbidite facies   150
turnover time   323
TWINSPAN   92, 94, 95, 97
Typha angustifolia   124
Tzintzuntzan   139

Udaipur (India)   413
United States of America, arid south-west   281
Upper Pleistocene   184, 187
Upper Río Lerma Basin (Mexico)   141–2
Uranium series dating   430
urban communities, decline of   188
USLE   150, 305

Valle de Santiago, Guanajuato (Mexico)   132

valley
   floor water management system   122
   terraces   32, 37
vegetation
   changes, anthropically-induced   288
   clearance   190
   colonisation   86
   cover   214–215
      affect of increased infiltration and surface protection   218
   dynamics, links with geomorphological processes   433, 434
Vérazian   290
vesicular arbuscular micorrhyza   241–242
VESPAN II   92
Vitis   119, 122

wadi
   breaching   184
   entrenchment   188
   sediments   386
   vegetation communities   387–388, 389
   walls   235
Wadi Allaqi (Egypt)   379–390
Wadi Biskra (Algeria)   179
Wadi Djedi (Algeria)   179
Wadi el Abiod (Algeria)   179
Wadi Gobbeen (Libya)   233, 248
Wadi Haimur (Egypt)   383, 386
Wadi Maïtar (Algeria)   192
Wadi Matrat Kebir (Algeria)   179, 184
Wadi Merdum (Libya)   239
Wadi Mimoun (Libya)   233
Wadi Naïma-el Ahmar (Algeria)   188
Wadi Quleib (Egypt)   383, 389
wadi-edge field systems   235
wall engineering   248
walls   231, 233–237, 248
Wannara Creek (Australia)   274, 275, 276, 278, 280
Wannara Dam (Australia)   275, 276
water
   and sediment supply to alluvial fans   71
   availability   215
      spatial distribution of   207
   quality   382
   spreading   235
   table   322, 329, 331
water-harvesting   220

waterlogged soils   370–376
   growth   374–376
waterlogging
   expansion of   373, 374–376
   loss of land due to   376
   monitoring of   372–374
Waterton River (Canada)   337
weathering   61
weeds   254, 262–262
western Algeria   192
wheat   367
   cultivation   235
   group   253–67
white spruce, *see Picea glauca*
widespread land clearance   288
wild
   einkorn   254, 255, 257, 259, 260, 261, 263
   emmer   254, 255, 256, 257, 258, 259, 260, 261, 262, 263, 264, 266
   fires   281
willow   323, 326, 329
wind flow   177
wind-blown
   deposits   179
   sand   179
windbreaks   397
wooded Sahelian steppe   184

X-ray
   diffraction   51, 152
   fluorescence spectrometer   399
Xaudum Valley (Botswana)   27, 33–34
xerophytes, planted   393
xerophytic vegetation   387

Yanchi (China)   393
Yellow River (China)   393
Yellow Sea (China)   397
Younger Dryas   171

Zabi Justiniana (Algeria)   188
Zacapu Basin (Mexico)   132
Zahrez Rharbi (Algeria)   192
Zapata (Spain)   301
Zea mays   131, 134, 138, 139, 141, 143
Ziban Range (Algeria)   177, 179, 188, 191, 192, 197
Zygnemataceae   243